New Strategies to Advance Pre/Diabetes Care: Integrative Approach by PPPM

Advances in Predictive, Preventive and Personalised Medicine

Volume 3

Series Editor:

Olga Golubnitschaja

Managing Editor:

Kristina Yeghiazaryan

For further volumes:
http://www.springer.com/series/10051

Mahmood S. Mozaffari
Editor

New Strategies to Advance Pre/Diabetes Care: Integrative Approach by PPPM

Springer

Editor
Mahmood S. Mozaffari
Department of Oral Biology
College of Dental Medicine
Georgia Regents University
Augusta, GA, USA

ISSN 2211-3495 ISSN 2211-3509 (electronic)
ISBN 978-94-007-5970-1 ISBN 978-94-007-5971-8 (eBook)
DOI 10.1007/978-94-007-5971-8
Springer Dordrecht Heidelberg New York London

Library of Congress Control Number: 201393341

Printed on acid-free paper

Springer is part of Springer Science+Business Media (www.springer.com)

What This Book Series Is About...

Current Healthcare: What Is Behind the Issue?
For many acute and chronic disorders, the current healthcare outcomes are considered as being inadequate: global figures cry for preventive measures and personalised treatments. In fact, severe chronic pathologies such as cardiovascular disorders, diabetes and cancer are treated after onset of the disease, frequently at near end-stages. Pessimistic prognosis considers pandemic scenario for type 2 diabetes mellitus, neurodegenerative disorders and some types of cancer over the next 10–20 years followed by the economic disaster of healthcare systems in a global scale.

Advanced Healthcare Tailored to the Person: What Is Beyond the Issue?
Advanced healthcare promotes the paradigm change from delayed interventional to predictive medicine tailored to the person, from reactive to preventive medicine and from disease to wellness. The innovative Predictive, Preventive and Personalised Medicine (PPPM) is emerging as the focal point of efforts in healthcare aimed at curbing the prevalence of both communicable and non-communicable diseases such as diabetes, cardiovascular diseases, chronic respiratory diseases, cancer and dental pathologies. The cost-effective management of diseases and the critical role of PPPM in modernisation of healthcare have been acknowledged as priorities by global and regional organisations and health-related institutions such as the Organisation of United Nations, the European Union and the National Institutes of Health.

Why Integrative Medical Approach by PPPM as the Medicine of the Future?
PPPM is the new integrative concept in healthcare sector that enables to predict individual predisposition before onset of the disease, to provide targeted preventive measures and create personalised treatment algorithms tailored to the person. The expected outcomes are conducive to more effective population screening, prevention early in childhood, identification of persons at-risk, stratification of patients for the optimal therapy planning, prediction and reduction of adverse drug-drug or drug-disease interactions relying on emerging technologies, such as pharmacogenetics, pathology-specific molecular patters, sub-cellular imaging, disease modelling, individual patient profiles, etc. Integrative approach by PPPM is considered as the medicine of the future. Being at the forefront of the global efforts, the European

Association for Predictive, Preventive and Personalised Medicine (EPMA, http://www.epmanet.eu/) promotes the integrative concept of PPPM among healthcare stakeholders, governmental institutions, educators, funding bodies, patient organisations and in the public domain.

Current Book Series, published by Springer in collaboration with EPMA, overview multidisciplinary aspects of advanced bio-medical approaches and innovative technologies. Integration of individual professional groups into the overall concept of PPPM is a particular advantage of this book series. Expert recommendations focus on the cost-effective management tailored to the person in health and disease. Innovative strategies are considered for standardisation of healthcare services. New guidelines are proposed for medical ethics, treatment of rare diseases, innovative approaches to early and predictive diagnostics, patient stratification and targeted prevention in healthy individuals, persons at-risk, individual patient groups, sub-populations, institutions, healthcare economy and marketing.

Prof. Dr. Olga Golubnitschaja

Book Series Editor

Dr. Golubnitschaja, Department of Radiology, Medical Faculty of the University in Bonn, Germany, has studied journalism, biotechnology and medicine and has been awarded fellowships for biomedical research in Paediatrics and Neurosciences (Medical Centres in Austria, Russia, UK, Germany, the Netherlands, and Switzerland). She is well-cited in the research fields of "gene hunting" and "subtractive hybridisation" applied to predictive prenatal and postnatal diagnostics published as *O. Labudova* in years 1990–2000. Dr. Golubnitschaja is an expert in molecular diagnostics actively publishing in the fields of perinatal diagnostics, Down syndrome, diabetes mellitus, hyperhomocysteinemia, cardiovascular disease, neurodegenerative pathologies, and cancer.

She is the *cofounder* of the theory of multi-pathway organ-related blood fingerprinting with specific molecular patterns at epi/genomic, transcriptional and post/translational levels, author of fundamental works in *integrative medicine*. Dr. Golubnitschaja holds appointments, at the rank of Professor, at several European Universities and in International Programmes for Personalised Medicine and is author of more than 300 international publications in the field. Awards: National and International Fellowship of the Alexander von Humboldt-Foundation; Highest Prize in Medicine and Eiselsberg-Prize in Austria; She is *Secretary-General* of the "European Association for Predictive, Preventive & Personalised Medicine" (EPMA in Brussels, www.epmanet.eu), Editor-in-Chief of *The EPMA-Journal* (BMC, in London); Book Editor of *Predictive Diagnostics & Personalised Treatment: Dream or Reality*, Nova Science Publishers, New York 2009; Book Co-editor *Personalisierte Medizin*, Health Academy, Dresden 2010; Book Series Editor *Advances in Predictive, Preventive and Personalised Medicine*, Springer 2012; *European Representative* in the EDR-Network at the NIH/NCI, http://edrn.nci.nih.gov/; and *Advisor and Evaluator* of projects dedicated to personalised medicine at the EU-Commission in Brussels, NIH/NCI, Washington, DC, USA, and at Foundations and National Ministries of Health in several countries worldwide.

Preface

The writings of earliest civilizations reveal that diabetes has afflicted man for a very long time. The first known clinical description of the disease is attributed to the Roman encylopaedist, Aulus Cornelius Celsus (ca 25 BC–ca 50), although the more accurate description and the introduction of the term, "*diabetes*", are attributed to Aretaeus, a celebrated ancient Greek physician (second century AD). Despite the early recognition of the disease, recommended treatment options (e.g., dietary modifications) of ancient times were not conducive to improved quality of life. Interestingly, however, the suggestion to those afflicted with the disease to exercise is discernible from the writings of Avicenna, a Persian polymath (960–1037 AD). Centuries later, Thomas Willis (1621–1675) of England noted the sweet taste of urine of those affected by the disease which prompted physicians of the time to rely on this feature as a diagnostic sign of the disease. This was followed by important observations of Mathew Dobson (1766) who identified sugar as the culprit for the sweet taste of urine and also noted the presence of excess sugar in the blood. These observations strengthened the rationale for dietary recommendations. Indeed, Apollinaire Bouchardat (1809–1866) recognized that fasting reduces urinary glucose excretion, an observation which led him to advocate dietary approach and restriction of calories for management of diabetes. Importantly, he speculated that the cause of diabetes related to the pancreas which proved to be true when Banting and Bess isolated insulin from the pancreas (1921) thereby revolutionizing the field of diabetes.

The earlier seminal observations regarding dietary modifications in management of diabetes are particularly of interest given the increasing recognition over the past two to three decades that overweight and obesity largely fuel the pandemic of type 2 diabetes. Although genetic factors are important, the prevalence of obesity is related to behavioral and environmental changes that have resulted from technological advances. Clearly the abundance of diets containing excess saturated fat, coupled with large servings and life style-related factors contribute to this epidemic. The worldwide increase in the incidence and prevalence of diabetes mellitus continues to exert undue burden on the healthcare systems. The cost-impact of the disease poses a major challenge for both developed and developing countries and their economies. These prevailing conditions have prompted a paradigm shift from reactive to perspective medicine which is crystallized in the concept of predictive, preventive and personalized medicine (PPPM) as an integrative medical approach. The rationale for PPPM relates to the recognition that prediction of populations at risk for diabetes mellitus should help devise strategies to prevent its target organ complications thereby reducing morbidity and mortality and associated costs. Indeed, aside from diabetes mellitus, PPPM is emerging as the focal point of efforts in healthcare aimed at curbing the prevalence of other non-communicable diseases such as cardiovascular diseases, chronic respiratory diseases and cancer. The expected outcomes of the integrative approach by PPPM include detection and diagnosis of disease at early stage which is conducive to its more effective control (i.e., emphasis from reactive to preventive medicine and from disease to wellness), identification and stratification of patients that would enable selection of optimal therapy and prediction and reduction of adverse drug-drug or drug-disease interactions relying on relevant information including pharmacogenetics, among others.

The effective management of non-communicable diseases and the critical role of PPPM have been acknowledged as priorities by global and regional organizations such as the World Health Organization, United Nations General Assembly, the Center for Disease Control, the European Union and the European Association for Predictive, Preventive and Personalized Medicine (EPMA), among others. Since 2009, EPMA (http://www.epmanet.eu/) has been at the forefront of efforts in promoting PPPM among healthcare professionals. Accordingly, in 2010, the *EPMA Journal* was launched to provide a scientific platform for rapid dissemination of emerging scientific information with PPPM as the focal point. This was followed by the first World Congress on PPPM in September of 2011 in Bonn, Germany, with participants from over 40 countries. The commitment of EPMA to PPPM is further exemplified by the publication of a series of books on advances in PPPM, published by *Springer*, including the present one entitled *New Strategies to Advance Pre-Diabetes Care*: *Integrative Approach by PPPM*. I do thank authors for their contributions and believe that the readership would find this book to be a timely and helpful addition to the growing lists of scientific contributions in the field.

Mahmood S. Mozaffari

Prof. Dr. Mahmood S. Mozaffari

Editor

Dr. Mahmood S. Mozaffari is a Professor of Oral Biology and Pharmacology/Toxicology at the Georgia Regents University (Augusta, Georgia, USA). He received his Ph.D. in Basic Medical Sciences, with specialization in Pharmacology and Therapeutics, from the University of South Alabama School of Medicine in 1988 followed by his DMD from the University of Alabama at Birmingham (UAB) in 1991 and his certificate in Hospital Dentistry from the same institution (1992).

Dr. Mozaffari has a long-standing interest in the pathogenesis of diabetes mellitus and its impact on the renal and the cardiovascular systems as well as the outcome of interventional modalities. In addition, he has carried out seminal studies focused on (a) the role of taurine in the kidney and the vasculature and (b) the impact of pressure overload on the outcome of myocardial ischemia-reperfusion injury. Current research interests include the contribution of immune and inflammatory mechanisms to the impact of pressure overload on the outcome of an ischemia-reperfusion insult to the heart. His research has been supported by the American Heart Association and the National Institutes of Health. He has served as an Associate Editor for *The EPMA Journal*, an ad-hoc reviewer for over a dozen scientific journals and as a member of the study section of the American Heart Association. He is a Fellow of the American Heart Association and the European Association for Predictive, Preventive and Personalised Medicine.

Contents

Contributors

Worku Abebe Department of Oral Biology, CL 2140, College of Dental Medicine, Georgia Regents University, Augusta, GA, USA

Mohamed Al-Shabrawey Oral Biology and Anatomy, College of Dental Medicine, Georgia Regents University (GRU), Augusta, GA, USA

Ophthalmology and Vision Discovery Institute, Medical College of Georgia, GRU, Augusta, GA, USA

Vascular Biology Center, Medical College of Georgia, GRU, Augusta, GA, USA

Babak Baban Department of Oral Biology, College of Dental Medicine, Georgia Regents University, Augusta, GA, USA

Manuela Bartoli Department of Ophthalmology, Georgia Regents University, Augusta, GA, USA

Department of Pharmacology and Toxicology, Medical College of Georgia, Augusta, GA, USA

Carla F. Bento IBILI, Faculty of Medicine, University of Coimbra, Coimbra, Portugal

Centre of Ophthalmology, Faculty of Medicine, University of Coimbra, Coimbra, Portugal

Sergio Li Calzi Department of Pharmacology and Therapeutics, College of Medicine, University of Florida, Gainesville, FL, USA

Melanie Cebioglu Department of Radiology, Rheinische Friedrich-Wilhelms-University of Bonn, Bonn, Germany

Jorge F.J. Coelho Department of Chemical Engineering, University of Coimbra, Coimbra, Portugal

Departamento de Engenharia Química, University of Coimbra, Polo II/Rua Silvio Lima, Coimbra 3030-790, Portugal

Rosemeyre A. Cordeiro Department of Chemical Engineering, University of Coimbra, Coimbra, Portugal

Christopher W. Cutler Department of Periodontics, Georgia Regents University, College of Dental Medicine, Augusta, GA, USA

Dimiter V. Dimitrov Diavita Ltd, Bulgaria

Robert Dinsmore Section of Plastic Surgery, Medical College of Georgia at Georgia Regents University, Augusta, GA, USA

Ahmed A. Elmarakby Department of Oral Biology, College of Dental Medicine, Georgia Regents University, Augusta, GA, USA

Paulo Ferreira Department of Chemical Engineering, University of Coimbra, Coimbra, Portugal

Department of Health Sciences, Portuguese Catholic University, Viseu, Portugal

Ana C. Fonseca Department of Chemical Engineering, University of Coimbra, Coimbra, Portugal

Maria H. Gil Department of Chemical Engineering, University of Coimbra, Coimbra, Portugal

Joana R. Góis Department of Chemical Engineering, University of Coimbra, Coimbra, Portugal

Olga Golubnitschaja Department of Radiology, Rheinische Friedrich-Wilhelms-University of Bonn, Bonn, Germany

Maria B. Grant Department of Pharmacology and Therapeutics, College of Medicine, University of Florida, Gainesville, FL, USA

Philip J. Hanes Dental Academic Administration, Georgia Regents University, College of Dental Medicine, Augusta, GA, USA

Paul Bernhard Henrich Eye Clinic, University Hospital, Basel, Switzerland

Tatjana Josifova Eye Clinic, University Hospital, Basel, Switzerland

Jadranka Koehn Department of Cranio-Maxillofacial and Oral Surgery, Medical University of Vienna, Vienna, Austria

Kurt Krapfenbauer Department of Otorhinolaryngology, Head and Neck Surgery, Medical University of Vienna, Vienna, Austria

Ranjitha Krishna Department of Periodontics, Georgia Regents University, College of Dental Medicine, Augusta, GA, USA

Folami Lamoke Department of Pharmacology and Toxicology, Medical College of Georgia, Augusta, GA, USA

Jun Yao Liu Department of Oral Biology, College of Dental Medicine, Georgia Regents University, Augusta, GA, USA

Filipa Marques IBILI, Faculty of Medicine, University of Coimbra, Coimbra, Portugal

Centre of Ophthalmology, Faculty of Medicine, University of Coimbra, Coimbra, Portugal

Mohamad Masoumy Department of General Surgery, Medical College of Georgia at Georgia Regents University, Augusta, GA, USA

Patrícia V. Mendonça Department of Chemical Engineering, University of Coimbra, Coimbra, Portugal

Mahmood S. Mozaffari Department of Oral Biology, College of Dental Medicine, Georgia Regents University, Augusta, GA, USA

Matthew Neu Department of Pharmacology and Therapeutics, College of Medicine, University of Florida, Gainesville, FL, USA

Paulo Pereira IBILI, Faculty of Medicine, University of Coimbra, Coimbra, Portugal

Centre of Ophthalmology, Faculty of Medicine, University of Coimbra, Coimbra, Portugal

Hans H. Schild Department of Radiology, Rheinische Friedrich-Wilhelms-University of Bonn, Bonn, Germany

Raquel Seiça Institute of Physiology, Faculty of Medicine, University of Coimbra, Coimbra, Portugal

IBILI, Faculty of Medicine, University of Coimbra, Coimbra, Portugal

Cristina M. Sena Institute of Physiology, Faculty of Medicine, University of Coimbra, Coimbra, Portugal

IBILI, Faculty of Medicine, University of Coimbra, Coimbra, Portugal

Jacqueline Sequoia Department of Family Medicine, Medical College of Georgia at Georgia Regents University, Augusta, GA, USA

Lynn C. Shaw Department of Pharmacology and Therapeutics, College of Medicine, University of Florida, Gainesville, FL, USA

Sylvia Smith Ophthalmology and Vision Discovery Institute, Medical College of Georgia, GRU, Augusta, GA, USA

Cellular Biology and Anatomy, Medical College of Georgia, GRU, Augusta, GA, USA

Kristina Yeghiazaryan Department of Radiology, Rheinische Friedrich-Wilhelms-University of Bonn, Bonn, Germany

Jack C. Yu Section of Plastic Surgery, Medical College of Georgia at Georgia Regents University, Augusta, GA, USA

Plastic and Reconstructive Surgery, Georgia Regents University, Augusta, GA, USA

Global Figures Argue in Favour of Preventive Measures and Personalised Treatment to Optimise Inadequate Diabetes Care

Kristina Yeghiazaryan, Melanie Cebioglu, and Olga Golubnitschaja

Abstract Diabetes mellitus (DM) is projected to be one of the leading health problems of the twenty-first century. New prognoses calculated for DM prevalence are becoming year-by-year more pessimistic. The pandemy contributors can be summarised as the upward trends of the population ageing, environmental risk factors, urbanisation, additive effects of diverse stress factors, badly chosen life-style including unfavourable nutritional habits, increasing prevalence of obesity, low physical activity, etc. Urgent measures are required to reduce the diabetes related burden. Thus advanced predictive diagnostic tools and personalised treatment strategies in (pre)diabetic care are critical and should exert beneficial impact on public health.

The experts recommend:

- Risk assessment for predisposed individuals;
- Early prevention of DM in early childhood, youth and young adults;
- Professional consultations to improved physical activity, individual nutrition and lifestyle in the groups of risk;
- Creation of personalised treatment algorithms in (pre)diabetes;
- Creation of specialised centres for integrative medical approaches in predictive, preventive and personalised medicine to assure individualised prevention and treatment at any pre/diabetes stages;
- Innovative educational programmes for professionals involved in advanced pre/diabetes care;
- Preventive educational programmes in the general population.

K. Yeghiazaryan • M. Cebioglu • O. Golubnitschaja (✉)
Department of Radiology, Rheinische Friedrich-Wilhelms-University of Bonn, Sigmund-Freud-Str. 25, Bonn D-53105, Germany
e-mail: olga.golubnitschaja@ukb.uni-bonn.de

M.S. Mozaffari (ed.), *New Strategies to Advance Pre/Diabetes Care: Integrative Approach by PPPM*, Advances in Predictive, Preventive and Personalised Medicine 3,
DOI 10.1007/978-94-007-5971-8_1, © Springer Science+Business Media Dordrecht 2013

Keywords Diabetes • Patient • Obesity • Childhood • Youth • Adults • Teenager • Nutrition • Epidemic • Pandemy • Life-style • Environmental • Urbanisation • Additive effect • Stress • Economy • Morbidity • Mortality • Risk factor • Healthcare industry • Policy makers • Complication • Pathology • Predictive, preventive and personalised medicine • Public health • Education

1 *Diabetes Mellitus* as the Global Pandemy of a Chronic Disorder

Diabetes mellitus (DM) is a lifelong progressive disease. It is classified as a group of heterogeneous metabolic disorders characterised by hyperglycaemia as a common feature. The vast majority of cases of DM falls mainly into two broad aetiopathogenetic categories, classified as type 1 and type 2 DM. Type 1 DM is induced by beta-cell destruction of autoimmune nature. In contrast, pancreatic beta cell dysfunction and insulin resistance are hallmark-features of type 2 DM.

The incidence of type 1 DM accounts for 5–10% of all new diagnosed DM-cases, while the dominating type 2 DM accounts for 90–95% of all cases registered worldwide [1]. The cumulative prevalence of type 1 and type 2 DM has already achieved the pandemic scale with currently registered more than 366 million patients comprising almost 8.3% of the adult population around the globe [2]. As published in the year 2004, the diabetes prevalence among all age groups was estimated to more than double within 30 years rising from 2.8% in 2000 up to 4.4% in 2030 [3]. Already now the prevalence among adults aged between 20 and 79 years rises above these estimations, currently demonstrating 6.4% rates prognosed to reach 7.7% by the year 2030 with the absolute number of DM-patients up to 439 million by the year 2030 [4, 5]. Furthermore, the overall prevalence for all age groups is estimated to overcome 550 million by the year 2030 [2]. New prognoses calculated for DM prevalence are becoming more and more pessimistic with year-by-year increasing acceleration coefficients (see Fig. 1).

2 A Number of Factors Have Been Identified as Contributing to the Global Diabetic Pandemy

Pandemy contributors can be summarised as the upward trends of the population ageing, environmental risk factors, urbanisation, additive effects of diverse stress factors, uncorrectly chosen life-style including unfavourable nutritional habits, increasing prevalence of obesity, low physical activity, etc. [6, 11–14]. As a consequence, the number of diabetic patients is continuously increasing worldwide.

The increased prevalence of DM is predicted to occur virtually for every nation, while the greatest increase is expected in developing countries [3, 15]. The prospective estimates are that in 2 decades from 2010 to 2030, there will be an increase of

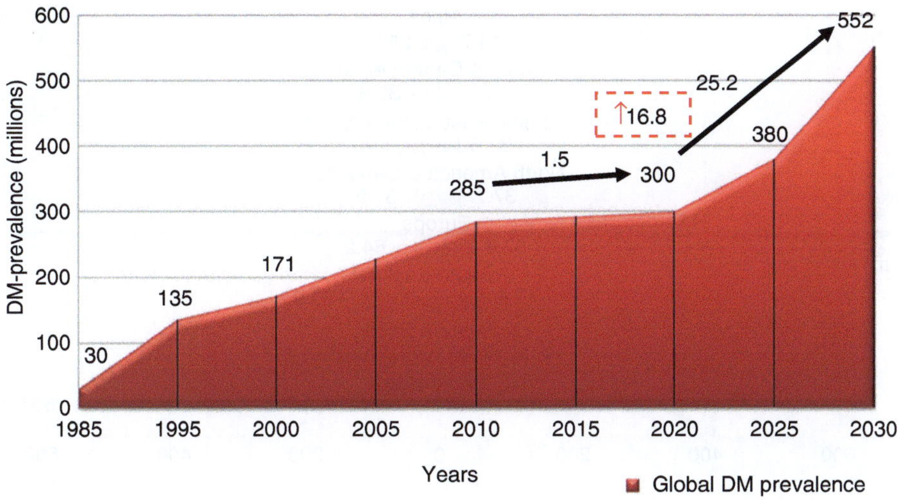

Fig. 1 Global prevalence of diabetes mellitus (DM) of collected statistical data in the years 1985–2010 and its prognose till the year 2030; under the current upsurge, DM-affected population is prognosed to rise from 285 million in 2010 to 439 million by the year 2030. Worldwide, a strong increase in diabetes-related healthcare and economical burden is predicted for the next two decades [3, 4, 6–10]. The acceleration coefficient calculated to characterise the increasing DM prevalence in the current decade is 1.5 that is, however, 16.8-times higher for the coming decade 2020–2030

69% in the number of adult diabetic patients for the developing countries, while for developed countries it is expected to increase by 20% [4]. Also the assumed density of patient is different for the developing *versus* developed countries when urban and rural areas are compared. Hence, whereas in developed countries the prevalence of DM has been shown to be similar in urban and rural areas, in contrast, in developing countries significantly higher DM prevalence occurs to the urban populations. The latter may relate to some area characteristic factors such as unbalanced diet, progressing obesity among all age groups, low physical activity and variety of stress factors more specific for the urban area in developing countries [3]. Versus global context, the regional particularities in the development of diabetic pandemy are summarised in corresponding numbers by Fig. 2.

Despite the above listed particularities, the list of the top ten countries includes both the developing and developed countries that have been selected as demonstrating the highest rate of pandemy in the age group between 20 and 79 years – see Fig. 3.

3 Acknowledged Factors Contributing to the Disease Development

Genetic predisposition is well-acknowledged for both type 1 and 2 of DM [17, 18]. According to the statistical data collected by the National Institute of Diabetes and Digestive and Kidney Diseases at the National Institute of Health in USA, individuals

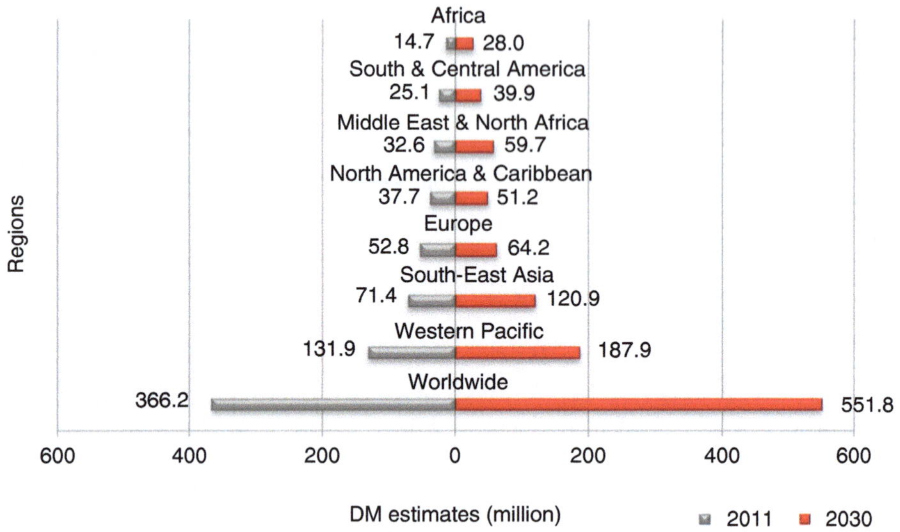

Fig. 2 Worldwide regional estimations for the real pandemic rates in the populations affected by DM in the year 2011 *versus* those prognosed for the year 2030 [16]

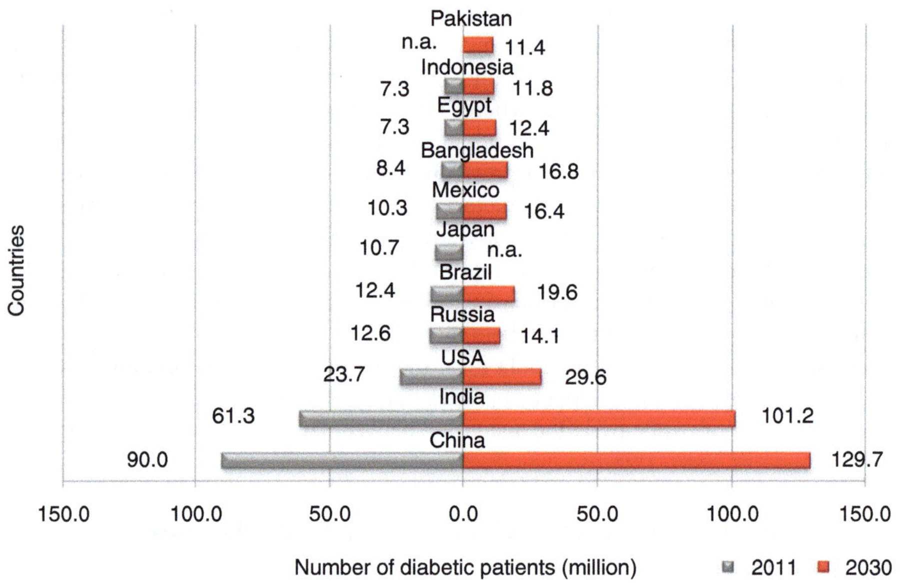

Fig. 3 Worldwide top ten countries selected with the highest rate of DM-affected populations in the age group between 20 and 79 years as it has been registered in the year 2011 and prognosed for the year 2030 [2]. Collectively these countries have been prognosed to reach the highest pandemic level of diabetes with about 400 million of patients by the 2030

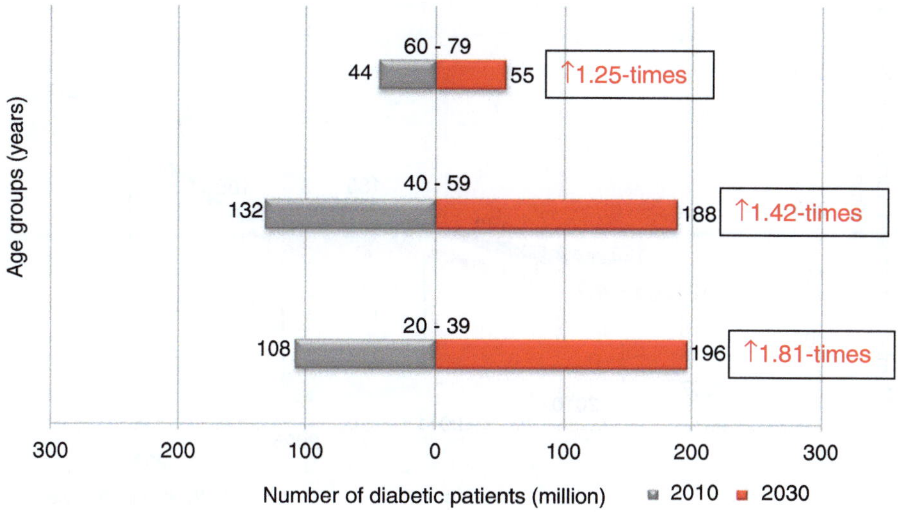

Fig. 4 Global DM prevalence in different age groups for the years 2010 and 2030 [23]. Current DM prevalence is likely to be largest among the age group of 40–59 years with 132 million people, followed by the age group 60–79 years with 108 million. In 2030, an increase is expected in all age groups, however, with the greatest contribution to the DM-pandemy by the age group 60–79 years of life. Estimates indicate a 1.25 and 1.42-times increase for the age groups of 60–79 years and 40–59 years, respectively. Noteworthy, the prognosed DM-rates progression is inversely increasing with the age (younger age – higher progression), and the youngest group of 20–39 years old people is prognosed to deliver the highest rates of diabetic progression that will double the DM-cohort of this age group by the year 2030 compared to 2010

of African-Caribbean, Asian, and Hispanic origin are more frequently affected by the disease [19]. Furthermore, current prognosis by the year 2030 assumes that countries with abundant populations such as China, India and USA, will have more significant increase in both absolute and relative DM-rates compared to other countries. Worldwide, India has the highest absolute number of patients registered in 2010: currently estimated population affected by DM is 50.8 million, while the expected one is rising up to 87 million of DM-affected people by the year 2030 [4]. As for Europe, Germany is listed among top ten countries having the highest estimated number of diabetic patients worldwide [20, 21].

Ageing populations is one of the major factors promoting the pandemic scale of DM [14, 22]. Currently diabetes is seen to be highly prevalent in the two age groups of 40–59 and 60–79 years with 131 and 108 million affected individuals, respectively (see Fig. 4). By 2030, it is predicted that there will be 188 million people with diabetes aged 40–59 years and considerably even more people in the age group over 60 years old, estimated to be about 196 million [24]. In addition, both industrialisation and globalisation affect the rising prevalence of diabetes in the younger age categories including teenagers [25].

Although ageing is the well-acknowledged factor contributing to the disease development, however, there are completely new epidemiologic factors characteristic

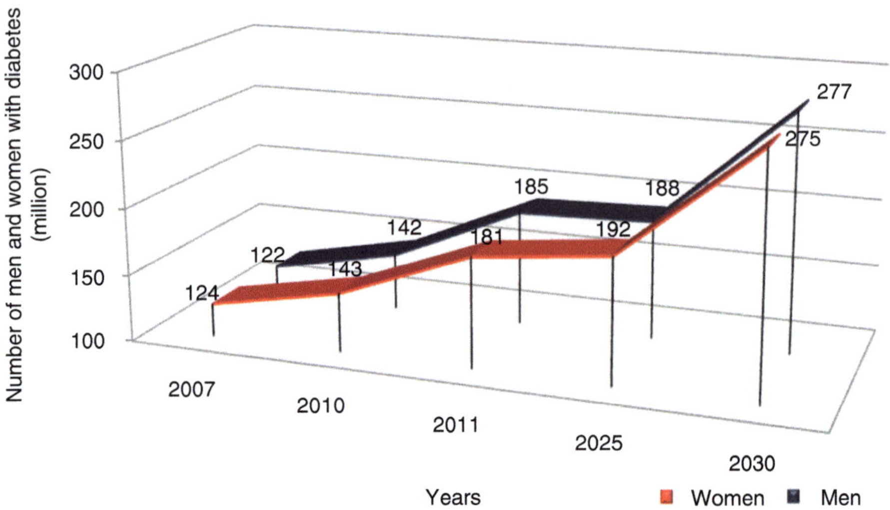

Fig. 5 Global prevalence of diabetes in men and women aged between 20 and 79 years, registered and prognosed in the time-frame of 2007–2030 [24, 31, 32]

for the twenty-first century that speed up the disease progression particularly in youth and young adults. Hence, Fig. 4 demonstrates that the prognosed DM-rates progression will be inversely increasing with the age (younger age – higher progression), and the youngest group of 20–39 years old people will be delivering the highest rates of diabetic progression that will double the DM-cohort of this age group by the year 2030 compared to 2010. This is a completely new situation and very big challenge for the societies around the globe requiring special competencies of several groups of professionals as well as innovative approaches in healthcare and health economy.

In most countries, a lower educational level was associated with higher prevalence of diabetes compared to better educated groups. In Western countries, the risk for developing diabetes is strongly linked to a low socioeconomic status. Most of the studies have shown that people with higher socioeconomic position have a lower prevalence of diabetes, while people with low socioeconomic status have a higher prevalence of the disease [26–30]. Further, the influence of socioeconomic factors on DM prevalence has been shown to be higher among women compared to men.

4 Gender-Related Particularities in Prevalence of Diabetes

Even though global prevalence of Diabetes mellitus is similar in men and women, some gender particularities is registered and have been prognosed for the next two decades as demonstrated in Fig. 5. In the age above 65 years, diabetes tends to be more frequent in women than men [3, 33]. By the year 2030 female diabetic patients will be prevalent in this patient cohort. Consequently, the targeted primary and secondary prevention will be of highest priority for the specialised measures in general and female healthcare, in particular.

4.1 Secondary Complications and Mortality in Diabetic Patients as the Multiplicator of Burden Related to Inadequate Diabetes Care

Major chronic complications, secondary to diabetes (see Fig. 6a, b) are affecting a number of organ systems, being responsible for a considerably shortened life-expectancy, decreased life-quality and sudden death. Specific complications of diabetes depend on many factors including an individual epi/genetic predisposition, duration and severity of the metabolic derangements [36].

Microvascular complications, due to the extensive damage to small blood vessels, are involved in the pathomechanisms of all chronic DM-related pathologies. Diabetic retinopathy, the earliest manifestation of the organ damage developed due to microvascular complications and proliferative new vascularisation, is the leading cause of the blindness in human beings. Recent studies in the United States implicate chronic nephropathy as the leading cause of renal failure. It is affecting also up to 40% of patients with diabetes worldwide [37]. Peripheral sensory neuropathy is the most common form of diabetic polyneuropathy in average affecting 40–50% of the overall patient cohort with diabetes [38]. The pathology affects the feet, leading to foot lesions, numbness and consequent foot/toe amputation.

Macrovascular complications, is the main risk factor for all cardiovascular mortalities in diabetic patients [39]. The pathology relates to the heart and large blood vessels combining peripheral vascular diseases, coronary artery diseases, arthrosclerosis and stroke [40].

Cancer provocation in diabetes: Recent research and population studies both demonstrate that diabetic patients are generally predisposed to the vast majority of the cancer types [41–43].

Mortality: Worldwide every 10 sec a person dies due to a diabetes-related pathology. Furthermore, DM is documented to be the fourth leading cause of death. In the year 2010, the global mortality attributed to diabetes was estimated to be 3.96 million, showing an increase of 5.5% when compared to the related mortality in the year 2007 [33]. The rates of all deaths in the cohort of diabetic patient aged between 20 and 79 years was lowest (6%) in the poorest African countries and highest in the North American region (15.7%) as demonstrated in Fig. 7. Doubling of the world populations is supposed to be one of the reasons for the global burden related to diabetes in the next 25 years [44]. The highest number of deaths due to diabetes is expected to occur in the countries with large populations, namely China, India, USA and the Russian Federation. More women than men are expected to die from diabetes and related complications, due to the highest prevalence of the pathology and more frequent co-morbidities in the ageing female patient cohorts [45, 46].

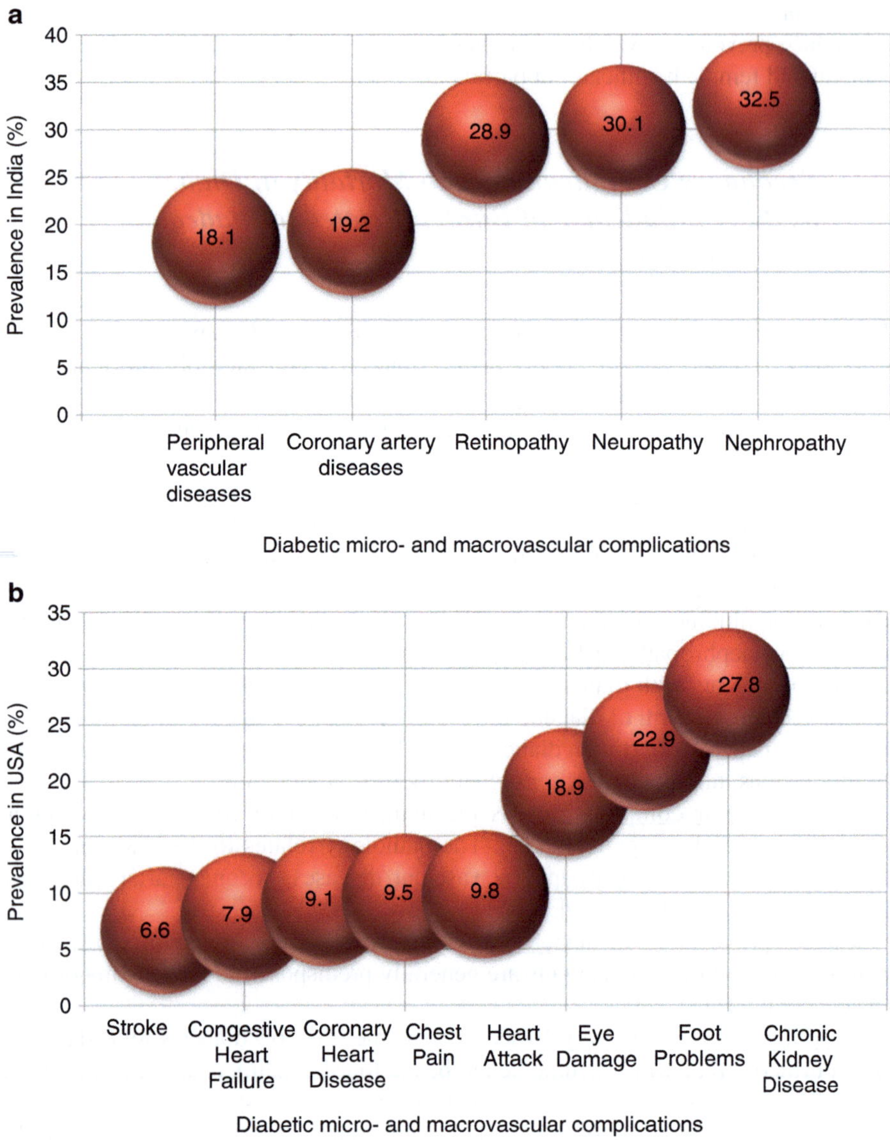

Fig. 6 (**a**) Prevalence of micro- and macrovascular complications secondary to diabetes type 2; corresponding rates have been registered for North-Western (Bikaner) population of India in the time frame January 1999 – December 2000 [34, 35]. (**b**) Prevalence of micro- and macrovascular complications secondary to diabetes type 2; corresponding rates have been registered by National Health and Nutrition Examination Survey, USA in the time frame 1999–2004 [33]

When comparing similar age groups of general *versus* DM-affected populations, the annual mortality rates have been documented to be more than two-times higher in latter ones. These rates vary depending on the regional particularities as demonstrated in the Fig. 8. Each year nearly three million people die because of diabetes,

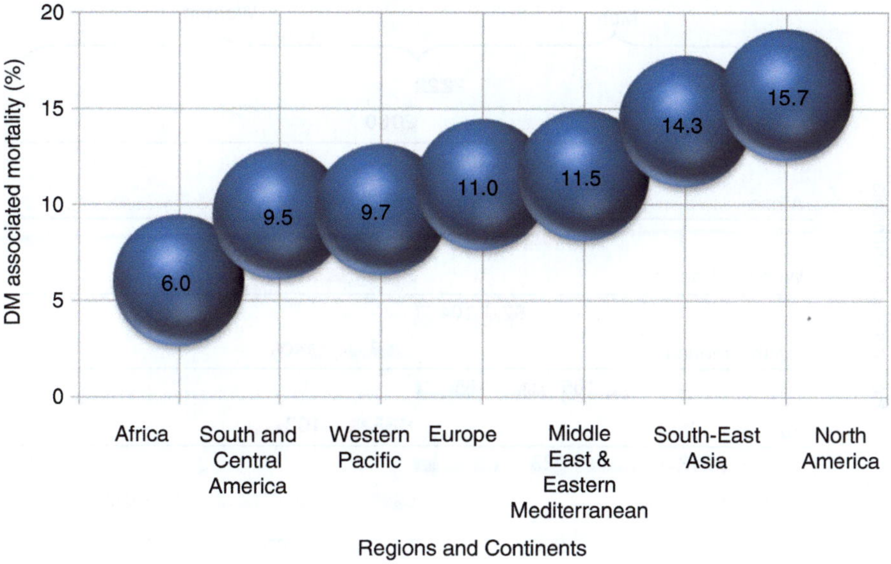

Fig. 7 Worldwide documented DM-associated mortality for the population aged between 20 and 79 years; the regional particularities as updated for the year 2010 [33]

and two thirds of them die in developing countries [45]. The diagrams in Fig. 8 demonstrate the statistical data collected for the death rates attributable to diabetes in groups of comparison considering geographic regions, gender and age. The lowest mortality rates are registered in the Eastern Mediterranean region; the highest mortality – in the South-Eastern Asia with the difference of ten-times between the lowest and highest mortality. When considering gender as the risk factor, a significantly higher prevance in the DM-related mortality is generally characteristic for men in the group of 35–64 years old people. In contrast, a DM-related mortality seems to be more typical for females in the age group over 64 year for all regions with the exception for Europe, where the DM-associated mortality rates are equal for both genders in advanced age. Particularly alarming are the rates collected for the young groups (below 34 years of age) demonstrating the highest mortality in the South-Eastern Asia followed by African regions [44].

5 Conclusions and Recommendations

According to the information provided above, the experts conclude:

- Diabetes mellitus and its secondary complications are one of the leading health problems of the twenty-first century;

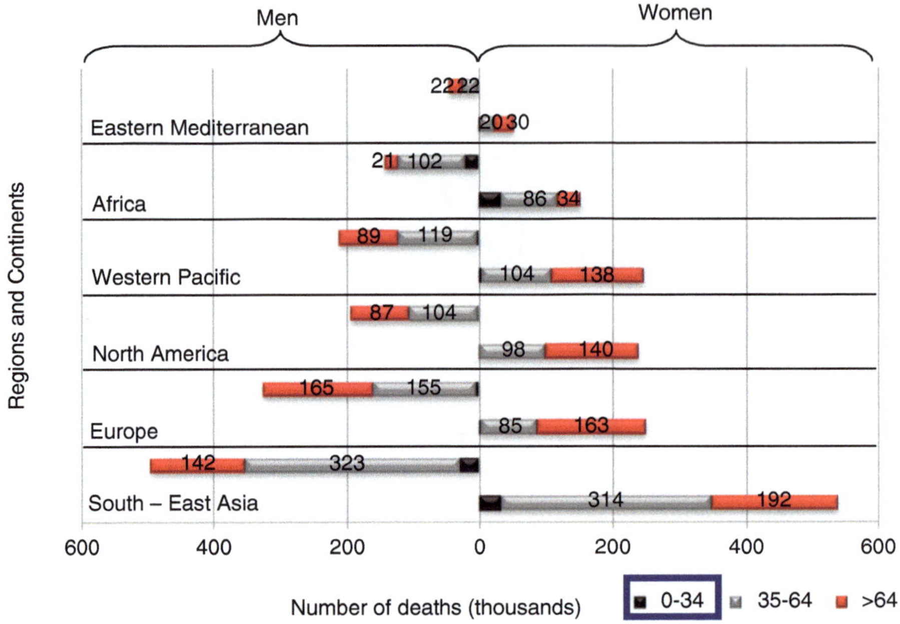

Fig. 8 Diagram of the death rates (in thousands) attributable to diabetes by corresponding age groups and gender; the mortality rates are the lowest ones in the Eastern Mediterranean followed by the African regions. In contrast, the highest mortality rates are registered in the South-Eastern Asia followed by the Northern America. The difference between the lowest and highest mortality rates is extremely big, namely more than ten-times for both men and women. Considering the gender particularities a significantly higher prevance in the DM-related mortality is generally characteristic for men in the group of 35–64 years old people. In contrast, a DM-related mortality appears to be more typical for females in the age group over 64 year for all regions with the exception for Europe, where the DM-associated mortality rates are equal for both genders in advanced age. Particularly alarming are the statistical data collected for the young groups (below 34 years of age) demonstrating the highest mortality rates in the South-Eastern Asia (29.1 thousand men and 33.1 thousand women) followed by African regions with 20 thousand men and 32.2 thousand women [44]

- New prognoses calculated for DM prevalence are becoming year-by-year more pessimistic;
- The pandemy contributors can be summarised as the upward trends of the population ageing, environmental risk factors, urbanisation, additive effects of diverse stress factors, uncorrectly chosen life-style including unfavourable nutritional habits, increasing prevalence of obesity, low physical activity, etc.;
- Over the next 20 years, DM-affected population is likely to be increase by 50%;
- Although the prevalence of diabetes continues to rise in every nation, the highest increase is expected to occur in the countries with large populations – China, India and USA;
- Urban population in developing countries is expected to double during the next two decades that is likely to contribute to a significant increase in diabetes prevalence;

- Both DM prevalence and DM-related mortality demonstrate significant geographic differences as well as age- and gender-dependent particularities;
- Although, the most important contributor to the DM prevalence still appears to be the changing demographic profile of societies with the prompt upgrowth in older populations and longer life span, however, the new phenomenon of the twenty-first century and big challenge for healthcare and health economy is the highest upgrowth of the patient rates observed in the young groups aged below 34 years of life;
- The prognosed DM progression will be inversely increasing with the age (younger age – higher progression): the youngest group of 20–39 years old people will be delivering the highest rates of diabetic progression that will double the DM-cohort of this age group by the year 2030 compared to 2010;
- Consequently, the spectrum and severity of DM-related pathologies will be increasingly progressing over the next few years;
- The progressive prevalence of DM as well as the increasing spectrum and severity of secondary complications will lead to dramatic economical burden in a global scale.

The experts recommend

- Urgent measures should be considered to advance diabetes care at any level of the pathology development;
- Risk assessment for predisposed individuals;
- Early prevention of DM in early childhood, youth and young adults;
- Professional consultations to improved physical activity, individual nutrition and lifestyle in the groups of risk;
- Creation of personalised treatment algorithms in (pre)diabetes;
- Creation of specialised centres for integrative medical approaches in predictive, preventive and personalised medicine to assure individualised prevention and treatment at any pre/diabetes stages;
- Innovative educational programmes for professionals involved in advanced pre/diabetes care;
- Preventive educational programmes in the general population.

References

1. Centers for Disease Control and Prevention (2007) National diabetes fact sheet: General information, national estimates on diabetes in the United States, Atlanta, GA: U.S. In: Department of Health and Human Services, Centers for Disease Control and Prevention. http://www.cdc.gov/diabetes/pubs/pdf/ndfs_2007.pdf. Accessed 10 Apr 2012
2. The diabetes atlas, 5th edn. The global burden. International Diabetes Federation, Brussels. http://www.idf.org/diabetesatlas/5e/the-global-burden. Accessed 10 Apr 2012
3. Wild S, Roglic G, Green A, Sicree R, King H (2004) Global prevalence of diabetes: estimates for the year 2000 and projections for 2030. Diabetes Care 27:1047–1053
4. Shaw JE, Sicree RA, Zimmet PZ (2010) Global estimates of the prevalence of diabetes for 2010 and 2030. Diabetes Res Clin Pract 87:4–14

5. George B, Cebioglu M, Yeghiazaryan K (2010) Inadequate diabetic care: global figures cry for preventive measures and personalized treatment. EPMA J 1:13–18
6. Seidell JC (2000) Obesity, insulin resistance and diabetes–a worldwide epidemic. Br J Nutr 83(Suppl 1):S5–S8
7. Zaini A (2000) Where is Malaysia in the midst of the Asian epidemic of diabetes mellitus? Diabetes Res Clin Pract 50(Suppl 2):S23–S28
8. WHO. Diabetes. http://www.who.int/mediacentre/factsheets/fs312/en/. Accessed 10 Apr 2012
9. International Diabetes Federation. The diabetes atlas, 2006 3rd edn. Prevalence and projections. International Diabetes Federation, Brussels. http://www.idf.org/sites/default/files/IDF%20Diabetes%20Atlas-2007%20%283rd%20edition%29_0.pdf. Accessed 10 Apr 2012
10. Muoio DM, Newgard CB (2008) Mechanisms of disease: molecular and metabolic mechanisms of insulin resistance and beta-cell failure in type 2 diabetes. Nat Rev Mol Cell Biol 9:193–205
11. Magliano DJ, Shaw JE, Shortreed SM, Nusselder WJ, Liew D, Barr ELM, Zimmet PZ, Peeters A (2008) Lifetime risk and projected population prevalence of diabetes. Diabetologia 51:2179–2186
12. Scheen AJ (2003) Pathophysiology of type 2 diabetes. Acta Clin Belg 58:335–341
13. Mokdad AH, Ford ES, Bowman BA, Dietz WH, Vinicor F, Bales VS, Marks JS (2003) Prevalence of obesity, diabetes, and obesity-related health risk factors, 2001. JAMA 289:76–79
14. Bogaert YE, Schrier RW (2011) Into the future: prevention of diabetes. Contrib Nephrol 170:256–263
15. Hossain P, Kawar B, El Nahas M (2007) Obesity and diabetes in the developing world–a growing challenge. N Engl J Med 356:213–215
16. International Diabetes Federation. Regional overviews. http://www.idf.org/diabetesatlas/5e/regional-overviews. Accessed 13 Apr 2012
17. Mehers KL, Gillespie KM (2008) The genetic basis for type 1 diabetes. Br Med Bull 88:115–129
18. Gonzalez-Bulnes A, Ovilo C (2012) Genetic basis, nutritional challenges and adaptive responses in the prenatal origin of obesity and type-2 diabetes. Curr Diabetes Rev 8:144–154
19. Chan JCN, Malik V, Jia W, Kadowaki T, Yajnik CS, Yoon K-H, Hu FB (2009) Diabetes in Asia: epidemiology, risk factors, and pathophysiology. JAMA 301:2129–2140
20. Rathmann W, Giani G (2004) Global prevalence of diabetes: estimates for the year 2000 and projections for 2030. Diabetes Care 27:2568–2569, author reply 2569
21. Frese T, Sandholzer H, Voigt S, Voigt R (2008) Epidemiology of diabetes mellitus in German general practitioners' consultation–results of the SESAM 2-study. Exp Clin Endocrinol Diabetes 116:326–328
22. Kolb H, Mandrup-Poulsen T (2010) The global diabetes epidemic as a consequence of lifestyle-induced low-grade inflammation. Diabetologia 53:10–20
23. International Diabetes Federation. The diabetes atlas, 4th edn. Diabetes estimates excel tables. DM 2010_7 regions.xls. DM 2030_7 regions.xls. International Diabetes Federation, Brussels. http://www.idf.org/node/23640. Accessed 13 Apr 2012
24. International Diabetes Federation. The diabetes atlas, 4th edn. Diabetes and impaired glucose tolerance. The global burden. International Diabetes Federation, Brussels. http://www.idf.org/sites/default/files/Diabetes%20and%20Impaired%20Glucose%20Tolerance_1.pdf. Accessed 10 Apr 2012
25. Zimmet P, Alberti KG, Shaw J (2001) Global and societal implications of the diabetes epidemic. Nature 414:782–787
26. Zaman R (2009) High prevalence of diabetes mellitus and promoting factors among human urban population of Bahawalpur-district, Pakistan: cross-sectional study. Res J Med Sci 3:62–69
27. Espelt A, Borrell C, Roskam AJ, Rodríguez-Sanz M, Stirbu I, Dalmau-Bueno A, Regidor E, Bopp M, Martikainen P, Leinsalu M, Artnik B, Rychtarikova J, Kalediene R, Dzurova D, Mackenbach J, Kunst AE (2008) Socioeconomic inequalities in diabetes mellitus across Europe at the beginning of the 21st century. Diabetologia 51:1971–1979

28. Dalstra JAA, Kunst AE, Borrell C, Breeze E, Cambois E, Costa G, Geurts JJ, Lahelma E, Van Oyen H, Rasmussen NK, Regidor E, Spadea T, Mackenbach JP (2005) Socioeconomic differences in the prevalence of common chronic diseases: an overview of eight European countries. Int J Epidemiol 34:316–326
29. Connolly V, Unwin N, Sherriff P, Bilous R, Kelly W (2000) Diabetes prevalence and socioeconomic status: a population based study showing increased prevalence of type 2 diabetes mellitus in deprived areas. J Epidemiol Community Health 54:173–177
30. Rabi DM, Edwards AL, Southern DA, Svenson LW, Sargious PM, Norton P, Larsen ET, Ghali WA (2006) Association of socio-economic status with diabetes prevalence and utilization of diabetes care services. BMC Health Serv Res 6:124
31. International Diabetes Federation. The diabetes atlas, 5th edn. Gender distribution. http://www.idf.org/diabetesatlas/5e/diabetes
32. International Diabetes Federation. The diabetes atlas, 4th edn. Gender distribution. http://da3.diabetesatlas.org/index74cc.html
33. Roglic G, Unwin N (2010) Mortality attributable to diabetes: estimates for the year 2010. Diabetes Res Clin Pract 87:15–19
34. Agrawal R, Ranka M, Beniwal R, Sharma S, Purohit V, Purohit D, Kothari R (2004) Prevalence of micro and macro vascular complications in type 2 diabetes and their riskfactors. Int J Diabets Dev Countries 24:11–16
35. Brill MT (2011) Diabetes. Twenty First Century Books, Tunbridge Wells
36. Yamagishi S (2011) Cardiovascular disease in recent onset diabetes mellitus. J Cardiol 57:257–262
37. Shinada M, Akdeniz A, Panagiotopoulos S, Jerums G, Bach LA (2000) Proteolysis of insulin-like growth factor-binding protein-3 is increased in urine from patients with diabetic nephropathy. J Clin Endocrinol Metab 85:1163–1169
38. Richardson LC, Pollack LA (2005) Therapy insight: influence of type 2 diabetes on the development, treatment and outcomes of cancer. Nat Clin Pract Oncol 2:48–53
39. Noh H, King GL (2007) The role of protein kinase C activation in diabetic nephropathy. Kidney Int Suppl 72:S49–S53
40. Ban CR, Twigg SM (2008) Fibrosis in diabetes complications: pathogenic mechanisms and circulating and urinary markers. Vasc Health Risk Manag 4:575–596
41. Golubnitschaja O (2010) Advanced diabetes care: three levels of prediction, prevention & personalized treatment. Curr Diabetes Rev 6:42–51
42. Cebioglu M, Schild HH, Golubnitschaja O (2008) Diabetes mellitus as a risk factor for cancer: stress or viral etiology? Infect Disord Drug Targets 8:76–87
43. Cebioglu M, Schild HH, Golubnitschaja O (2010) Cancer predisposition in diabetics: risk factors considered for predictive diagnostics and targeted preventive measures. EPMA J 1:130–137
44. Roglic G, Unwin N, Bennett PH, Mathers C, Tuomilehto J, Nag S, Connolly V, King H (2005) The burden of mortality attributable to diabetes: realistic estimates for the year 2000. Diabetes Care 28:2130–2135
45. Roglic G (2009) Diabetes in women: the global perspective. Int J Gynaecol Obstet 104(Suppl 1):S11–S13
46. Golubnitschaja O (2012) Changing long-held beliefs is never easy: a proposal for multimodal approaches in female healthcare – an integrative view. In: Costigliola V (ed) Healthcare overview: new perspectives. In: Golubnitschaja O (book series ed) Advances in Predictive, Preventive and Personalised Medicine. Springer, Dordrecht, Heidelberg, New York, London

Three Levels of Prediction, Prevention and Individualised Treatment Algorithms to Advance Diabetes Care: Integrative Approach

Olga Golubnitschaja

Abstract Currently every 10 sec one patient dies of diabetes-related pathologies – the epidemic scale of *Diabetes mellitus* (DM) is still underestimated in the healthcare sector. Generally, there are three levels desirable for an optimal (pre)diabetes healthcare:

– First level: prediction of the familial predisposition, targeted prevention of DM early in childhood;
– Second level: prediction of early/premature ageing and targeted prevention of DM pre-stages;
– Third level: prediction of secondary complications in the cohort of DM-patients and creation of treatments tailored to the patient to prevent cardiovascular, neurodegenerative and cancer diseases frequently developed in diabetic patients.

Communication among the related professional groups such as researchers, educators, healthcare providers, policy-makers, politicians, etc. is essential in the overall process of reorganisation of (pre)diabetes healthcare.

Keywords Diabetes mellitus • Morbidity • Mortality • Cardiovascular • Cancer • Neurodegenerative • Complications • Biomarker • Omics • Patient profile • Disease modelling • Childhood • Teenager • Youth • Adult • Socio-economical burden • Patient care • Predictive, preventive and personalised medicine • Healthcare industry • Policy-makers • Education

O. Golubnitschaja (✉)
Department of Radiology, Rheinische Friedrich-Wilhelms-University of Bonn, Sigmund-Freud-Str. 25, Bonn D-53105, Germany
e-mail: Olga.Golubnitschaja@ukb.uni-bonn.de

M.S. Mozaffari (ed.), *New Strategies to Advance Pre/Diabetes Care: Integrative Approach by PPPM*, Advances in Predictive, Preventive and Personalised Medicine 3, DOI 10.1007/978-94-007-5971-8_2,

1 Unsatisfactory Results of Current Diabetes Care

Following facts argue for themselves:

- Worldwide, every 10 sec one patient dies from DM and DM-related complications [1];
- The worldwide estimated impact of undiagnosed DM and DM-related complications ranges between 30 and 50% for DM type 2 [2];
- A dynamic increase in DM-affected subpopulations is well-documented [3–6];
- An increase of about one million DM-diseased individuals each year has been registered in the USA since 2002. As a reflection of the permanently increasing prevalence, the cost burden of 174 billion dollars in 2007 is greater than 132 billion dollars in 2002 [5, 7]; these statistics do not include intangible costs (e.g., care provided by unpaid caregivers, pain and suffering) or individuals with undiagnosed diabetes or impaired fasting glucose;
- As for the current prognosis of the epidemic scale, year-by-year it is getting more pessimistic as demonstrated in Fig. 1, and the overall number of diabetic patients is expected to double over the next two decades [1, 4, 5, 7, 8];
- DM type 2, historically characterised as an adult-onset disorder, now comprises approximately 30% of new DM-cases already in the second decade of life, i.e. among teenagers [9–12];
- Early onset of DM type 2 leads to a dramatic shift in the oveall appearance of complications characteristic for DM-patients, namely early and highly increased risk as well as poor prognosis for cardiovascular, neurodegenerative and cancer diseases developed secondarily to DM [13–16]. Direct consequences include: significantly shorter life-span expectation, markedly lower life-quality and

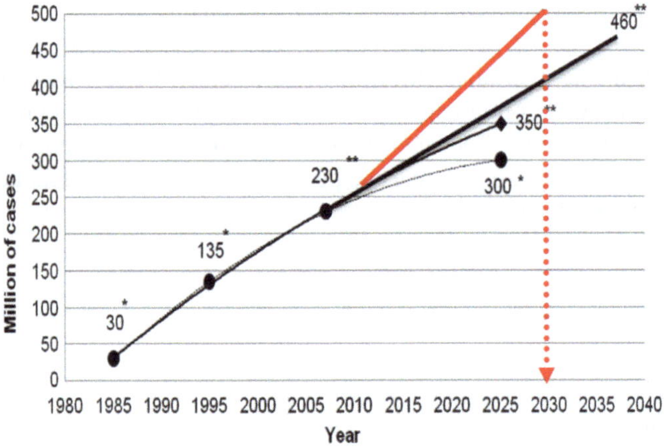

Fig. 1 Worldwide prognosis of the epidemic scale in diabetes; * estimations as published around the year 2000; ** worsening prognosis as published in 2003–2008 [2, 4, 5, 7]; current prognosis is marked in red colour [8]

a dramatic increase in the cost burden per DM-patient multiplied by the currently high and epidemically increasing part of the DM-affected subpopulation in the society.

Taken together, diabetes care faces a whole spectrum of problems including the necessity for population screening, targeted preventive measures, ethics, economics, and broad dissemination of the related information, with each aspect awaiting a dramatic improvement. Due to the high prevalence of the pathology and severe chronic complications, the application of predictive diagnostics, targeted preventive measures and treatment approaches tailored to the patient are emerging to advance pre/diabetes care.

2 Three General Levels for Predictive Diagnostics and Targeted Preventive Measures in Desirable Pre- and Diabetes Care

Generally, there are three levels desirable for an optimal pre- and diabetes healthcare:

- First level: prediction of the familial predisposition, targeted prevention of DM early in childhood;
- Second level: prediction of early/premature ageing and targeted prevention of DM pre-stages;
- Third level: prediction of secondary complications in the cohort of DM-patients and creation of treatments tailored to the patient to prevent cardiovascular, neurodegenerative and cancer diseases frequently developed in diabetic patients.

The principle difference between the current and highly-desirable pre/diabetes care is the progress from the unsatisfactory curative to the advanced predictive and preventive approaches as summarised in Fig. 2.

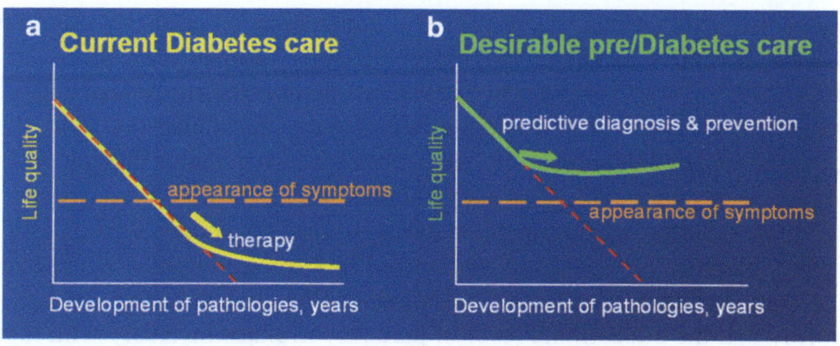

Fig. 2 Schematic presentation of the principle difference between the currently applied curative approach (**a**) and the highly-desirable predictive diagnostics followed by targeted preventive measures (**b**) in pre-stages and diabetes care systems (data adapted from Costigliola et al. [17])

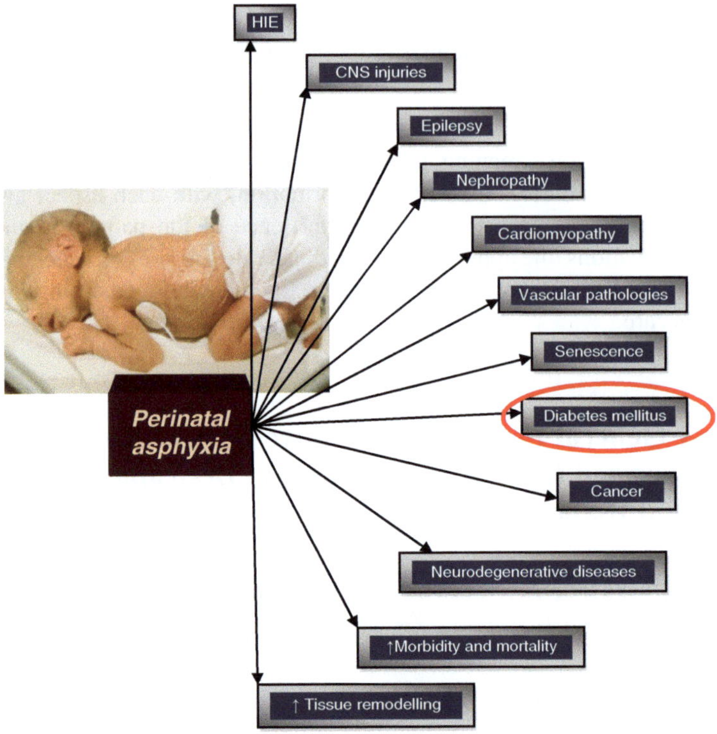

Fig. 3 Birth asphyxia increases the risk of severe pathologies including diabetes (Data adapted from Peeva and Golubnitschaja [18])

3 First Level – Familial Predisposition to Diabetes: Postnatal Diagnostics, Educational and Preventive Measures Early in Life

The newborn is the primary level for the application of predictive diagnostics, well-timed preventive measures and individualised treatment (Fig. 3).

Further, there is a growing body of evidence demonstrating certain endogenous and exogenous factors to be strong contributors to the prevalence of DM worldwide. Thus, the influence of economic aspects with respect to a higher DM-prevalence in developing countries is evident as demonstrated in Fig. 4 [6, 14].

Current population studies in perinatal and postnatal diagnostics clearly show the highest statistics for the newborn affected by birth asphyxia, and predispositin to short- and long-term complications including DM, in developing countries. It is noteworthy that the educational level of parents correlates inversely with high risk and incidence of asphyctic babies [18].

How can this knowledge be used in favour of targeted preventive measures? Parents should be educated about the risk factors and potential long-term outcomes. In the case

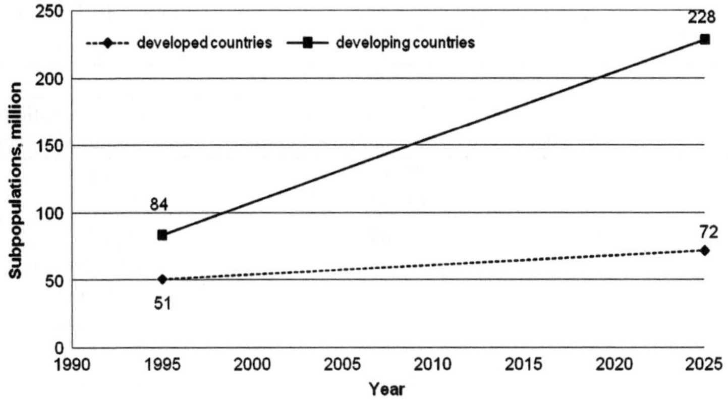

Fig. 4 The role of economical aspects in worldwide DM-prevalence [6, 14]

Table 1 Common exogenous/endogenous factors and their role in DM-risk

Exogenous/endogenous factors	Recommendations[a]
1. Common dietary rules	
Fish intake	> twice weekly
Vegetables intake	> 4-times weekly
Fruits intake	> once daily
Vitamin supplement	Regularly
2. Drug abuse	
Alcohol consumption	1–2-times dose weekly is reasonable (individually consulted)
Tobacco consumption	None
3. Body activity	
Sport	Preferably regular exercises (individually consulted) avoiding occasional ones
4. Overall Stress	
Stressful job	Weak till moderate
Stress at home	Decrease as much as possible

[a]The recommendations are based on the published data [23]

of the asphyxiated newborn, complementary postnatal diagnostic approaches should be applied to estimate the degree of the organ-damage and potential complications. Promising non-invasive blood tests are currently under development which would allow a more precise diagnosis and reliable prognosis for severe pathologies (diabetes, cancer, etc.), which can potentially be developed later in life [19].

Further educational measures should consider the known pathology-relevant exogenous and endogenous factors such as genetic background, environmental aspects, nutrition, life style, stress, infections, job-related and body-activities [6, 20–22]. Some of the central factors and consequent recommendations are summarised in Table 1.

Table 2 SOD-induction in leukocytes by in vitro incubation with 2.45×10^{-2} mmol/L paraquat during 18 h [25]

	SOD induction		
Life decade	Median	Values deviation within the subgroup	Characteristic for subgroup's majority
20–30	82	75–85	80–85
31–40	84	65–85	83–85
41–50	64	45–80	60–65
51–60	54	25–75	Two characteristic parts: 1. Around 65; 2. Around 50
61–70	24	10–50	20–25
71–80	5	1–15	Around 6
81–86	8	5–15	Around 8
over 86	13	10–17	Around 12

4 Second Level: Ageing Processes and DM Pre-stages

Ageing is one of the strongest contributors to the development of the majority of human pathologies including DM. In particular, early ageing supports or even triggers the (pre)lesions, pre-stages and the consequent pathologies early in life. Early ageing can have diverse origins and usually results from the interaction of genetic, epigenetic and various environmental factors. The well-known causes of the early/premature ageing are inborn genetic disorders, such as Down's Syndrome, suppression of longevity-driving genes, uncoupled functional interaction between nuclear and mitochondrial genomes and unfavourable environmental factors. The common resulting effect is an imbalance in the increased production of aggressive reactive species and decreased antioxidants and detoxification pathways activity [24].

The activity of detoxification pathways consistently decreases over the life. The decrease correlates well with ageing that is individual and can be estimated non-invasively by in vitro tests measuring the kinetics of SOD-activity under artificially produced stress conditions (Table 2).

The most important messages from the test are the following [26]:

1. The biggest differences in the induction of the stress-reaction demonstrate individuals within the fifth, sixth and seventh life-decades;
2. Some individuals in the sixth life-decade exhibit the parameters similar to those in the fourth and even third ones; others, however, tend to show parameters more characteristic for the ninth and tenth decades of life;
3. Noteworthy, the individuals in the tenth life-decade demonstrate better values compared to the majority in the eight decade that is obviously relevant for the biology of longevity.

Recent innovative diabetes-related studies offer a more precise definition of the diabetic phenotype at the molecular level that may prove useful in delineating diabetogenic genes or gene products in diabetes pre-stages. Thus, an identification of secreted and pancreas specific proteins/peptides responsible for abnormalities of insulin secretion in type 2 DM which might serve as predictive markers for β-cell dysfunction in diabetes pre-stages has been reported [27]. Further strategies for the detection and identification of additional pancreas specific biomarkers and characteristics of an ideal biomarker for deficiency in insulin secretion are necessary. Since an ideal biomarker is difficult to find or even does not exist at all, one potential solution seems to be plausible, namely to use the combinatorial power of a set of biomarkers, each of which alone may not offer satisfactory specificity and sensitivity. The creation of individualised care algorithms is closely linked to the identification of a set of molecular-targets that would serve as the basis for the early diagnosis of diabetes pre-stages, therapy selection and disease monitoring that together could lead to personalised patients treatment.

Currently, for the individuals highly predisposed to the early ageing/extensive ageing-processes generally and, in particular to diabetes pre-stages, the following preventive and therapeutic measures are recommended [26]:

- the life-style should be personally discussed with the specialists to develop individualised nutrition and optimise body-activity taking into account all related stress contributors including relevant environmental factors;
- Targeted metabolic corrections might be performed;
- Gene-therapy and alternative forms of individualised therapy approaches can be potentially applied.

5 Third Level: Development of Severe Complications Secondary to DM – Prediction, Prevention and Personalised Treatments Algorithms

5.1 Chronic Diseases Developed Secondary to Diabetes

According to the current worldwide statistics, DM is currently the fourth leading cause of death, in particular, due to fatal DM-related complications. Since our knowledge concerning all DM-triggered secondary pathologies is currently incomplete, the real figures for DM-related mortality are assumed to be significantly higher than considered till now [13].

The common onset of DM type 2 in early adulthood leads to early onset of diverse severe complications such as angiopathy, retinopathy, nephropathy, polyneuropathy, silent ischemia, cancer and “diabetic foot”. As reviewed earlier [13, 28] chronic degenerative processes appear as a “domino-effect”. Thus, the clinical manifestation of an actively proliferating diabetic retinopathy (DRP) is the early indicator of the down-stream developing cascade of severe chronic complications, see Fig. 5.

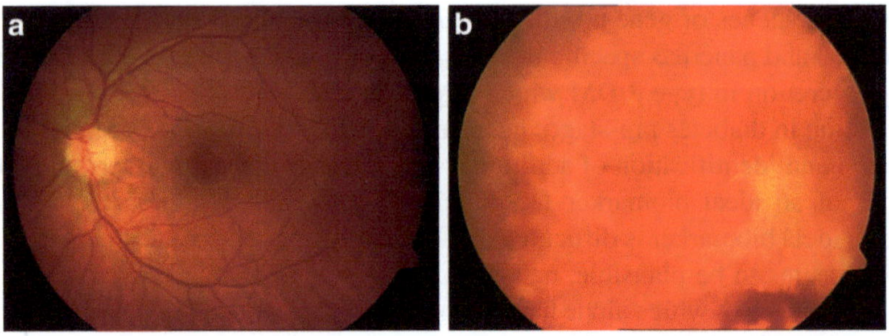

Fig. 5 Optical imaging of the human eye with (**a**) healthy retina (*left*) and (**b**) proliferative diabetic retinopathy (*right*); diffuse vascular structure, extensive tissue remodelling and neovascularisation of the retina are the attributes of highly proliferative diabetic retinopathy

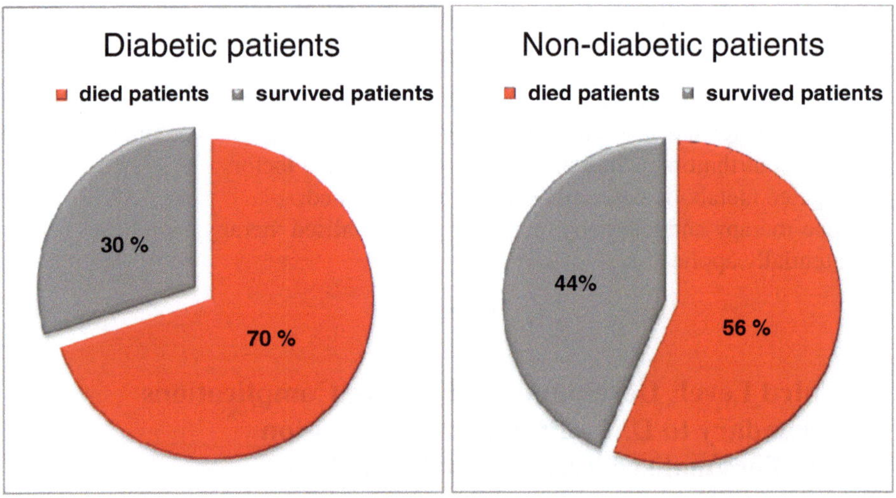

Fig. 6 Current statistics to cancer patients with and without DM: the treatment success and prognosis are significantly worse in DM-obused cancer patients

Therefore, early onset of DRP is the indication for immediate secondary prevention in DM and application of individualised treatments such as anti-VEGF therapeutic approaches; the treatment algorithms may be personally created depending on the individual DRP parameters [29].

5.2 Cancer Predisposition in Diabetes: Contributing Factors

Diabetic patients are more frequently predisposed to almost all cancer types compared to the general population [29]. Also the prognosis is worse compared to cancer patients without DM-background (see Fig. 6) [30].

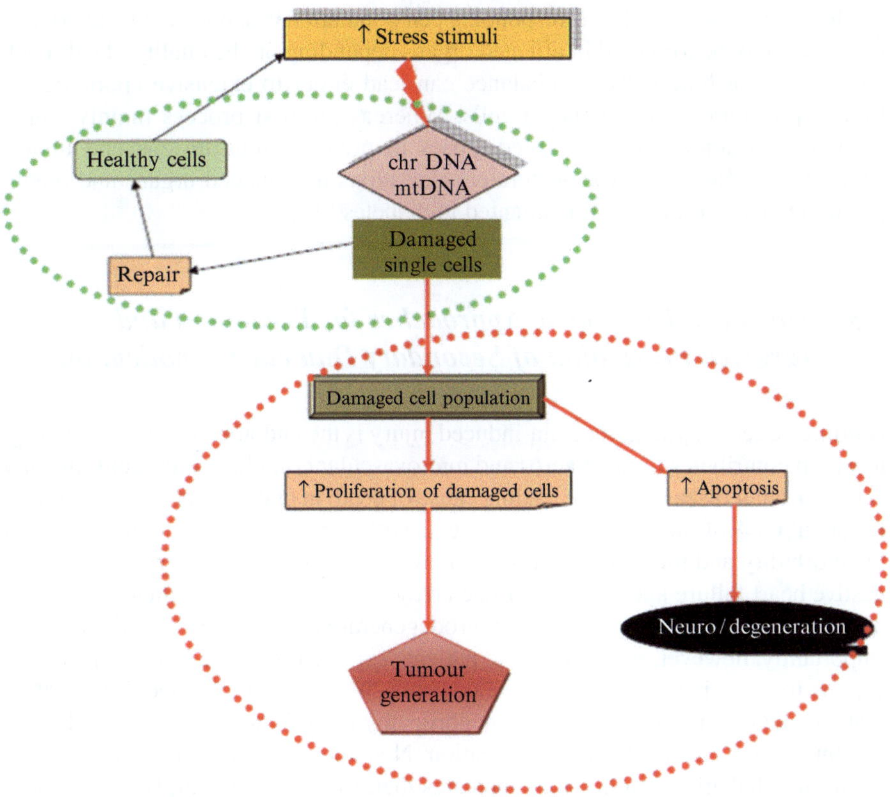

Fig. 7 Diagram demonstrates potential mechanisms for the increased risk of both degenerative processes and malignancies under stress conditioned by the impaired glucose/insulin metabolism in diabetes mellitus. The first part circled in *green colour* shows physiologic repair of damaged single cells. The second part circled in *red colour* demonstrates pathophysiologic processes leading either to (neuro)degeneration or malignant transformations [33]

The aetiology of cancer in diabetes is not yet completely understood. It is evident that cancer in diabetes can have diverse origins. The overall contributing factors have been reviewed by us elswhere [15, 30, 31].

Although the molecular pathomechanisms of secondary diabetic complications require further extensive investigations, recently collected research data have made a good contribution to our understanding of the pathology in its whole complexity. Oxidative damage to DNA is well documented for cells isolated from DM-patients [32] and can be readily explained by impaired glucose/insulin metabolism. These findings indicate an imbalance between the increased production of reactive oxygen species (ROS) and decreased DNA-repair capacity in diabetics. As long as single damaged cells can be repaired, the integrity of a corresponding cell population remains intact (see the green circle in Fig. 7).

However, as soon as the corresponding cell population is damaged, two alternative processes may be triggered in affected organs depending on the quality of cell-cycle controlling machinery. This imbalance can lead either to extensive apoptotic cell loss or proliferation of damaged cells. Whereas the first process mainly causes tissue-degeneration, the latter predisposes to pre-cancerous lesions and tumour development. Both (neuro)degenerative alterations in damaged organs and predisposition to cancer are well documented in diabetes [33].

5.3 Advanced Diagnostic Approaches for Prediction and Targeted Prevention of Secondary Diabetic Complications

A prime target of hyperglycaemia-induced injury is the endothelial cells, the damage to which contributes to the macro- and microvascular complications. Consequently, cardiovascular diseases are the leading cause of death in diabetic patients, accounting for about 65% of the deaths among patients with type 2 diabetes mellitus; much of the morbidity and mortality relates to atherosclerotic coronary artery disease, congestive heart failure and sudden cardiac arrest [34]. A further significant portion of diabetic patients develop cancer or neurodegenerative diseases as described above. Importantly, however, not all diabetic patients experience severe long-term complications irrespective of blood glucose control [29]. For example, about 20–40% of diabetic patients ultimately develop nephropathy but it is not clear why all diabetic patients do not manifest this complication. Nonetheless, differential susceptibility to chronic diabetic complications indicates that, aside from hyperglycaemia, other factors (e.g., genetics) modulate an individual's risk for vascular complications. These aspects emphasise the need for an individualised approach to detection and management of diabetes mellitus.

Although urinary proteomics is considered as a potentially powerful diagnostic tool for the early detection of diabetic renal damage, there are some major limitations of the technology such as the analysis of low level proteins and correctness of data interpretation. Further, the sample collection is rather problematic: peptide patterns for first-void and mid-stream female urine are different as well as the contamination from non-urinary passage in female subgroups is an obvious obstacle [35]. Since disease specific changes in blood-plasma and vascular proteomes occur up-stream towards the majority of organ systems in type 2 diabetes [36], it should be clarified as to which changes monitored in the renal proteome originate from the kidney and which from the blood-plasma [28]. A detection of pathology-specific circulating nucleic acids in blood-plasma is considered as one of the most potent diagnostic approaches in the individualised treatment of diabetics [37, 38]. A non-invasive disease-specific gene expression profiling in circulating leukocytes has been proposed for the predictive diagnosis of secondary diabetic complications [39]. This approach is based on disease- and severity-specific alterations in the gene expression patterns of circulating leukocytes and protease activity in serum that can be *ex vivo* evaluated in individual blood samples. The gene products, proposed to play a role in the pathology of type 2 diabetes, belong to the stress-proteome and have a function in the following pathways: cell-cell communication, cell-adhesion, apoptosis, tumourigenesis, transcription regulation, tissue remodelling and neurodegeneration. A clinical application of the test is currently under consideration, and a nanotechnology which should provide a possible easy and cheap routine application of the test is under development in our laboratory. The predictive biomarkers may be further used as molecular targets for personalised treatment of affected individuals.

5.4 Creation of Individualised Treatment Algorithms

In light of the above, an early and predictive diagnosis followed by targeted preventive measures and individualised therapy approaches would dramatically improve the overall outcome of secondary complications in diabetic patients. An excellent example can be given by advanced ophthalmo-diabetological care for individualised patient treatment, e.g. when therapeutic approaches such vitrectomy are applied [40]. The technique of vitrectomy is rapidly improving and clinical guidelines should take its progress into consideration. Novel techniques of intravitreal drug application show some positive effects; however, there is a lack of randomised, well-controlled trials that can potentially result in improved guidelines for clinical practice. Severe and moderate vision loss in diabetes is essentially preventable by advanced early and predictive diagnostics and personalised patient treatment. Advanced treatment approaches are possible only by understanding the molecular and pathophysiological mechanisms of the disease and precise prediction of individual complications. The interplay between three vasoactive systems, VEGF, prostoglandins and nitric oxide, have predictable role in the progression of the ischaemic retinal changes [41].

A further example can be given for the prediction and individualised treatment of aortic valve degeneration and replacement in diabetics. On average, 20–30% of implanted bioprostheses show dysfunction after about 10 years post implantation. Recent reports predict that a greater than 50% incidence of failure will be seen in bioprostheses at 12–15 years [42]. These figures are much worse for diabetic pateints, who are at high risk of degeneration of both native valvular tissue and bioprostheses [43]. To predict the progressive valvular degeneration and necessity of accelerated reimplantation, comparative expression analyses have been undertaken demonstrating the high predictive power of some marker-candidates and novel therapeutic targets [44, 45]. Among them – the ECM-degrading enzymatic-core, the expression profile of which changes dramatically at the remodeling of valvular geometry and complications such as accelerated degeneration of bioprostheses [45, 46]. Moreover, the activity of the tissue-remodelling key-enzymes – metalloproproteinases MMP-2 and MMP-9 – can be non-invasively measured in blood by *ex vivo* zymography [28]. This opens good perspectives for predictive diagnostics and new target therapy approaches, in order to avoid these common complications well known for diabetic patients [16].

6 Outlook

Application of high-tech in the drug delivery systems has significantly improved metabolic control over glucose/insulin/glucagon homeostasis in prestages and diabetes. Further advanced systems are currently under development [47–50]:

- glucose biosensors: intelligent stimuli-responsive delivery systems using glucose sensitive hydrogels, in order to release insulin in a pulsatile profile in response to blood glucose levels;
- transplantation of insulin producing tissue;
- creation of artificial pancreas;
- individualised gene-therapy approaches.

An understanding and possibly a complete description of the factors underlying the burden of diabetes mellitus will provide policy-makers, healthcare providers and educators with an opportunity to guide primary and secondary preventive initiatives at both individual and community levels. Communication among the professionals involved in the overall process of restructuring the (pre)diabetes care is of paramount importance.

References

1. Kowluru RA, Chan P-S (2007) Oxidative stress and diabetic retinopathy. Exp Diabetes Res 2007:43603
2. Borch-Johnsen K, Beck-Nielsen H, Christiansen JS, Heickendorff L, Brandslund I, Faber JO, de Fine Olivarius N (2003) Guidelines on diagnosis of type 2 diabetes. Investigation, diagnosis and glucose measurement. Ugeskr Laeger 165:1558–1561
3. Mohan V, Sandeep S, Deepa R, Shah B, Varghese C (2007) Epidemiology of type 2 diabetes: Indian scenario. Indian J Med Res 125:217–230
4. Chaturvedi N (2007) The burden of diabetes and its complications: trends and implications for intervention. Diabetes Res Clin Pract 76(Suppl 1):S3–S12
5. American Diabetes Association (2008) Economic costs of diabetes in the U.S. in 2007. Diabetes Care 31:596–615
6. King H, Aubert RE, Herman WH (1998) Global burden of diabetes, 1995–2025: prevalence, numerical estimates, and projections. Diabetes Care 21:1414–1431
7. Hogan P, Dall T, Nikolov P (2003) Economic costs of diabetes in the US in 2002. Diabetes Care 26:917–932
8. Golubnitschaja O, Yeghiazaryan K (2012) Opinion controversy to chromium picolinate therapy's safety and efficacy: ignoring 'anecdotes' of case reports or recognising individual risks and new guidelines urgency to introduce innovation by predictive diagnostics? EPMA J 3:11
9. Glaser NS (1997) Non-insulin-dependent diabetes mellitus in childhood and adolescence. Pediatr Clin North Am 44:307–337
10. Pinhas-Hamiel O, Dolan LM, Daniels SR, Standiford D, Khoury PR, Zeitler P (1996) Increased incidence of non-insulin-dependent diabetes mellitus among adolescents. J Pediatr 128:608–615
11. Huang TT, Goran MI (2003) Prevention of type 2 diabetes in young people: a theoretical perspective. Pediatr Diabetes 4:38–56

12. Rosenbloom AL, Joe JR, Young RS, Winter WE (1999) Emerging epidemic of type 2 diabetes in youth. Diabetes Care 22:345–354
13. Golubnitschaja O (2009) Advanced technologies for prediction of secondary complications in diabetes mellitus. In: Golubnitschaja O (ed) Predictive diagnostics and personalized treatment: dream or reality? Nova Science Publishers Inc, New York
14. Cebioglu M, Golubnitschaja O (2009) New tendencies in the epidemiology of diabetes mellitus. In: Golubnitschaja O (ed) Predictive diagnostics and personalized treatment: dream or reality. Nova Science Publishers Inc, New York
15. Cebioglu M, Schild HH, Golubnitschaja O (2009) Diabetes mellitus as a risk factor for cancer: is predictive diagnosis possible? In: Golubnitschaja O (ed) Predictive diagnostics and personalized treatment: dream or reality? Nova Science Publishers Inc, New York
16. Yeghiazaryan K, Skowasch D, Bauriedel G, Schild HH, Golubnitschaja O (2009) Prediction of degeneration of native and bioprosthetic aortic valves. In: Golubnitschaja O (ed) Predictive diagnostics and personalized treatment: dream or reality. Nova Science Publishers Inc, New York
17. Costigliola V, Gahan P, Golubnitschaja O (2009) Predictive medicine as the new philosophy in health care. In: Golubnitschaja O (ed) Predictive diagnostics and personalized treatment: dream or reality. Nova Science Publishers Inc, New York
18. Peeva V, Golubnitschaja O (2009) Birth asphyxia as the most frequent perinatal complication. In: Golubnitschaja O (ed) Predictive diagnostics and personalized treatment: dream or reality. Nova Science Publishers Inc, New York
19. Yeghiazaryan K, Peeva V, Morelli M, Herrera-Marshitz M, Golubnitschaja O (2009) Potential targets for early diagnosis and neuroprotection in asphyxiated newborns. In: Golubnitschaja O (ed) Predictive diagnostics and personalized treatment: dream or reality. Nova Science Publishers Inc, New York
20. Nakano T, Ito H (2007) Epidemiology of diabetes mellitus in old age in Japan. Diabetes Res Clin Pract 77(Suppl 1):S76–S81
21. du Prel J-B, Icks A, Grabert M, Holl RW, Giani G, Rosenbauer J (2007) Socioeconomic conditions and type 1 diabetes in childhood in North Rhine-Westphalia, Germany. Diabetologia 50:720–728
22. Płaczkiewicz-Jankowska E, Szybi ski Z, Huszno B (2007) Environmental factors in the development of type 1 diabetes – a new insight. Przegl Lek 64:175–179
23. Lesgards J-F, Durand P, Lassarre M, Stocker P, Lesgards G, Lanteaume A, Prost M, Lehucher-Michel M-P (2002) Assessment of lifestyle effects on the overall antioxidant capacity of healthy subjects. Environ Health Perspect 110:479–486
24. Golubnitschaja O, Fountoulakis M (2007) Gene hunting by substractive hybridization in Down syndrome correlation with proteomics analysis. In: Lajtha A, Youdim MBH, Riederer P, Mandel SA, Battistin L (eds) Handbook of neurochemistry and molecular neurobiology. Springer US, Boston
25. Niwa Y, Ishimoto K, Kanoh T (1990) Induction of superoxide dismutase in leukocytes by paraquat: correlation with age and possible predictor of longevity. Blood 76:835–841
26. Golubnitschaja O (2010) Advanced diabetes care: three levels of prediction, prevention & personalized treatment. Curr Diabetes Rev 6:42–51
27. Koehn J, Turhani D, Krapfenbauer K (2009) Potential predictive biomarkers for the dysfunction of ß-cells in type 2 diabetes. Predictive diagnostics and personalized treatment: dream or reality. Nova Science Publishers Inc, New York
28. Golubnitschaja O (2008) Clinical proteomics in application to predictive diagnostics and personalized treatment of diabetic patients. Curr Proteomics 5:35–44
29. Golubnitschaja O, Jaksche A, Moenkemann H, Yeghiazaryan K, Karl SE, Trog D, Schild HH, Löffler KU (2005) Molecular imaging system for possible prediction of active retinopathy in patients with diabetes mellitus. Amino Acids 28:229–237
30. Cebioglu M, Schild H, Golubnitschaja O (2010) Cancer predisposition in diabetics: risk factors considered for predictive diagnostics and targeted preventive measures. EPMA J 1:130–137
31. Cebioglu M, Schild HH, Golubnitschaja O (2008) Diabetes mellitus as a risk factor for cancer: stress or viral etiology? Infect Disord Drug Targets 8:76–87

32. Frustaci A, Kajstura J, Chimenti C, Jakoniuk I, Leri A, Maseri A, Nadal-Ginard B, Anversa P (2000) Myocardial cell death in human diabetes. Circ Res 87:1123–1132
33. Golubnitschaja O (2009) Diabetes mellitus. In: Golubnitschaja O (ed) Predictive diagnostics and personalized treatment: dream or reality? Nova Science Publishers Inc, New York
34. Mozaffari M, Abdelsayed R, Schaffer S (2009) Diabetic complications: pathogenic mechanisms and prognostic indicators. In: Golubnitschaja O (ed) Predictive diagnostics and personalized treatment: dream or reality? Nova Science Publishers Inc, New York
35. Schaub S, Wilkins J, Weiler T, Sangster K, Rush D, Nickerson P (2004) Urine protein profiling with surface-enhanced laser-desorption/ionization time-of-flight mass spectrometry. Kidney Int 65:323–332
36. Mayr M, Mayr U, Chung Y-L, Yin X, Griffiths JR, Xu Q (2004) Vascular proteomics: linking proteomic and metabolomic changes. Proteomics 4:3751–3761
37. Gahan PB (2008) Circulating nucleic acids in plasma and serum: roles in diagnosis and prognosis in diabetes and cancer. Infect Disord Drug Targets 8:100–108
38. Gahan PB (2009) Circulating nucleic acids in plasma and serum: roles in diagnosis and prognosis. In: Golubnitschaja O (ed) Predictive diagnostics and personalized treatment: dream or reality. Nova Science Publishers Inc, New York
39. Golubnitschaja O (2006) Method for early diagnosis of proliferative diabetic retinopathy. Patent: 06115230.2
40. Josifova T, Schneider U, Henrich PB, Schrader W (2008) Eye disorders in diabetes: potential drug targets. Infect Disord Drug Targets 8:70–75
41. Josifova T, Henrich PB, Schrader W (2009) Prevalence, targeted treatment and prevention of eye disorders in diabetes. In: Golubnitschaja O (ed) Predictive diagnostics and personalized treatment: dream or reality. Nova Science Publishers Inc, New York
42. Schoen FJ, Levy RJ (2005) Calcification of tissue heart valve substitutes: progress toward understanding and prevention. Ann Thorac Surg 79:1072–1080
43. Golubnitschaja O (2007) Cell cycle checkpoints: the role and evaluation for early diagnosis of senescence, cardiovascular, cancer, and neurodegenerative diseases. Amino Acids 32:359–371
44. Golubnitschaja O, Yeghiazaryan K, Skowasch D, Schild H, Bauriedel G (2006) p21WAF1/CIP1 and 14-3-3 sigma gene expression in degenerated aortic valves: a link between cell cycle checkpoints and calcification. Amino Acids 31:309–316
45. Yeghiazaryan K, Skowasch D, Bauriedel G, Schild H, Golubnitschaja O (2007) Could activated tissue remodeling be considered as early marker for progressive valve degeneration? Comparative analysis of checkpoint and ECM remodeling gene expression in native degenerating aortic valves and after bioprosthetic replacement. Amino Acids 32:109–114
46. Yeghiazaryan K, Bauriedel G, Schild HH, Golubnitschaja O (2008) Prediction of degeneration of native and bioprosthetic aortic valves: issue-related particularities of diabetes mellitus. Infect Disord Drug Targets 8:88–99
47. Coelho JFJ, Ferreira P, Gil MH (2008) New approaches in drug delivery systems: application for diabetes treatment. Infect Disord Drug Targets 8:119–128
48. Coelho JFJ, Ferreira P, Fonseca AC, Fernandes AF, Seiça R, Gil MH (2009) Advanced drug delivery systems in diabetes treatment. In: Golubnitschaja O (ed) Predictive diagnostics and personalized treatment: dream or reality. Nova Science Publishers Inc, New York
49. Harrison PT (2008) Application of gene therapy in diabetes care. Infect Disord Drug Targets 8:129–133
50. Harrison PT, Flynn R, Lee CM (2009) Gene therapy and individualized medicine. In: Golubnitschaja O (ed) Predictive diagnostics and personalized treatment: dream or reality. Nova Science Publishers Inc, New York

Diabetes Mellitus: New Challenges and Innovative Therapies

Cristina M. Sena, Carla F. Bento, Paulo Pereira, Filipa Marques, and Raquel Seiça

Abstract Diabetes is a common chronic disease affecting an estimated 285 million adults worldwide. The rising incidence of diabetes, metabolic syndrome, and subsequent vascular diseases is a major public health problem in industrialized countries. This chapter summarizes current pharmacological approaches to treat diabetes mellitus and focuses on novel therapies for diabetes mellitus that are under development.

There is great potential for developing a new generation of therapeutics that offer better control of diabetes, its co-morbidities and its complications. Preclinical results are discussed for new approaches including AMPK activation, the FGF21 target, cell therapy approaches, adiponectin mimetics and novel insulin formulations. Gene-based therapies are among the most promising emerging alternatives to conventional treatments. Therapies based on gene silencing using vector systems to deliver interference RNA to cells (i.e. against VEGF in diabetic retinopathy) are also a promising therapeutic option for the treatment of several diabetic complications.

In conclusion, treatment of diabetes faces now a new era that is characterized by a variety of innovative therapeutic approaches that will improve quality of life in the near future.

Keywords Endothelial dysfunction • Endothelium • Nitric oxide • Oxidative stress • Therapeutic approaches

---

C.M. Sena (✉) • R. Seiça
Institute of Physiology, Faculty of Medicine, University of Coimbra, Subunit 1, polo 3, Azinhaga de Santa Comba, Celas, Coimbra 3000-354, Portugal

IBILI, Faculty of Medicine, University of Coimbra, Coimbra, Portugal
e-mail: csena@ci.uc.pt

C.F. Bento • P. Pereira • F. Marques
IBILI, Faculty of Medicine, University of Coimbra, Coimbra, Portugal

Centre of Ophthalmology, Faculty of Medicine, University of Coimbra, Coimbra, Portugal

M.S. Mozaffari (ed.), *New Strategies to Advance Pre/Diabetes Care: Integrative Approach by PPPM*, Advances in Predictive, Preventive and Personalised Medicine 3,
DOI 10.1007/978-94-007-5971-8_3,

1 Introduction

Diabetes mellitus is increasing globally affecting more than 285 million people worldwide [1]. Of these, over 90% is type 2 diabetes and, because of the increase in the aging population and massive rise in prevalence of obesity, the incidence is likely to be more than doubled by 2030 [1, 2]. Diabetes is a global problem with devastating human, social and economic impact. A major concern with the diabetes epidemic is the anticipated increase in mortality and morbidity related to the complications of the disease [3, 4]. Insufficient drug therapies, poor patient compliance to therapy and inadequate management regimes may be contribute to the increasing incidence of complications linked to the disease.

1.1 Glycemic Control

Glycemic control is fundamental to the management of diabetes. It is now established that the risk of diabetic complications is dependent on the degree of glycemic control in diabetic patients. Clinical trials such as the Diabetes Control and Complications Trial [5] and the Stockholm Diabetes Study in type 1 diabetes [6], and the UK Prospective Diabetes study [7, 8] and Kumamoto study [8] in type 2 diabetes, have demonstrated that tight glycemic control achieved with intensive insulin regimes can reduce the risk of developing or progressing retinopathy, nephropathy and neuropathy in patients with all types of diabetes. However, the Third National Health and Nutrition Examination Survey (NHANES III) showed that only 50% of diabetics have been able to achieve a glycosylated hemoglobin (HbA1C) level, as an index of chronic glycemia, less than 7%; therefore, the only way to ensure the long-term health of diabetic patients is to maintain constant normoglycemia [2].

The goal of antidiabetes therapy is to reduce hyperglycemia to either prevent or minimize the complications associated with this disease. HbA1C levels are a measure of glycemic control, and antidiabetes treatment aims to reduce HbA1C to as close to normal (<6.0%) as possible, without inducing hypoglycemia. Improved glycemic control is associated with significant reductions in microvascular complications, such as retinopathy, nephropathy, and neuropathy [7, 8]. In a meta-analysis, tight glycemic control has also been shown to reduce cardiovascular disease [9]. As a result, a clinical trial is currently investigating whether reducing HbA1C to less than 6.0% decreases the rate of cardiovascular disease and death [10, 11], which would be clinically beneficial as macrovascular disease accounts for approximately 50% of deaths in individuals with type 2 diabetes.

In type 2 diabetes, despite the clinical benefits of tight glycemic control and the availability of many distinct antidiabetic agents, treatments tend to become less effective over time as insulin resistance increases and pancreatic β-cell function deteriorates. Patients with type 1 diabetes are unable to maintain normoglycemia at all times, even with an intensive insulin therapy.

Table 1 American Diabetes Association – recommended therapeutic goals for patients with type 2 diabetes [13]

Glycosylated hemoglobin	<7% (as close to 6% as possible for each individual patient)
Preprandial plasma glucose	90–130 mg/dl (5.0–7.2 mmol/L)
Postprandial plasma glucose	<180 mg/dl (<10 mmol/L)
Blood pressure	<130/80 mmHg
Low-density lipoprotein cholesterol	<100 mg/dl (<2.6 mmol/L)
High-density lipoprotein cholesterol	>40 mg/dl (>1.1 mmol/L)
Triglycerides	<150 mg/dl (<1.7 mmol/L)
Body mass index	<25.0 kg/m^2

Insight into the pathophysiology of diabetes and its complications has led to the development of improved therapeutic strategies. Proper glycemic control and attainment of other nonglycemic management targets (e.g., blood pressure, lipids, body weight) are essential to the prevention of long-term complications of diabetes and to the reduction of overall disease management costs. Two prime examples of such therapy are treatment of hypertension and early stages of diabetic renal dysfunction with antihypertensive agents, and specifically with angiotensin-converting enzyme inhibitors, and laser photocoagulation of proliferative retinopathy and clinically significant macular edema. The latest American Diabetes Association (ADA) guidelines recommend a range of interventions to better improve type 2 diabetes outcomes. The ADA algorithm advocates a stepwise therapeutic approach that is based upon reduction in the plasma glucose concentration to achieve an HbA1C <7.0% [12]. Several therapeutic goals are recommended as summarized in Table 1 [13].

2 Type 1 Diabetes

Type 1 diabetes is a metabolic disorder that results from the progressive destruction of insulin-secreting β cells in the islets of Langerhans of the pancreas, leading to insulin insufficiency and hyperglycemia [14]. People with type 1 diabetes are dependent on insulin for the rest of their life. But insulin is not a cure, and people with diabetes are at significant risk for a wide range of serious complications, including heart and kidney diseases and blindness [15].

To avoid the variations in blood glucose levels in those with type 1 diabetes, and to reduce the chance of long-term complications, it would be helpful for new β-cells to be produced. Ideally, new β-cells should be derived from cell sources already existing within the person with diabetes so avoiding immunosuppression [16]. Alternatively exogenous sources of surrogate β-cells have also been described, including adult human pancreases donated after death, fetal pancreas, pluripotent and multipotent stem cells, and cells that reside in the liver. Whilst most of these exogenous sources are heterologous to the recipient; some are autologous such as cord blood stem cells and pluripotent stem cells induced from fibroblasts derived from skin [17].

Transplantation of human pancreatic tissue derived after death is the alternative therapy currently available that allows the possibility of even transient cessation of exogenous insulin.

2.1 Pancreas Transplantation

There has been a great deal of excitement and investigation in islet transplantation, but at the current time only whole-organ pancreas transplantation can be counted on for long-term exogenous insulin independence and normalization of HbA1C. The goal of pancreas transplantation is to safely restore normoglycemia by the provision of sufficient β cell mass [18]. Transplantation of a pancreas is not a life-saving operation but it improves quality of life because patients do not need to inject insulin on a daily basis or regularly monitor glucose concentrations with finger sticks, and hypoglycemic unawareness is no longer a problem. The majority of pancreas transplants are simultaneous pancreas-kidney transplantation, the preferred treatment option for diabetics with approaching or coexisting end-stage renal disease [19]. The benefits of pancreas transplantation are clear: improved quality of life, prevention of recurrent diabetic nephropathy, freedom from exogenous insulin with euglycemia and normalization of HbA1C, less stringent dietary restrictions, less frequent blood glucose monitoring and stabilization of or improvement in secondary complications. The major disadvantages to the patient are the operative risk, the need for chronic immunosuppression, and the inherent side effects of chronic immunosuppression [18, 20]. The long-term advantages of this surgical procedure have to be balanced against the potential morbidity and mortality associated with diabetes, and the side effects from the long-term immunosuppression that is needed to prevent alloimmunity and autoimmune recurrence [18].

2.2 Islet Transplantation

A promising way of treatment of diabetes is using β-cell replacement therapy. Significant advances in islet isolation techniques have allowed progress in research using transplanted human islets [21].

Enthusiasm for the promise of human islet transplantation started in 2000 with the publication of the Edmonton protocol (Fig. 1). Because many traditional immunosuppression regimens were known to be toxic to islets, Shapiro and coworkers [23] employed a modified immunosuppression protocol with a monoclonal antibody against interleukin 2, daclizumab, and eliminated glucocorticoids. Notably, both host-versus-graft and autoimmune reactions were avoided. However, despite

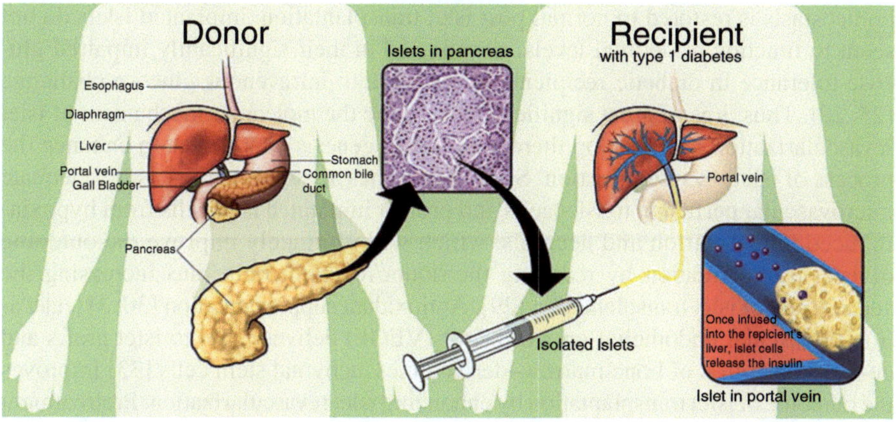

Fig. 1 The process of islet transplantation (Illustration by Giovanni Maki) [22]

short term success, long term insulin independence is usually unsustainable. Several follow-up studies indicate that even with the rigorous application of steroid-free immunosuppressive regimens, only a small percentage of the patients maintained insulin independence. Noteworthy, 80% of patients evaluated had fewer hypoglycemic events and decreased their insulin requirements 2 years after transplant [24, 25].

The Edmonton protocol is limited by several major factors: the lack of a sufficiently large source of islets due to the scarcity of cadaveric pancreas donors, and the presence of persistent immune rejection as well as the potential for recurrence of autoimmunity. There is still a slow and progressive loss of insulin production from transplanted islets in diabetic recipients over time, as evidenced by reports that 30–40% of islet recipients may experience recurrence of autoimmune diabetes with re-acquisition of insulin dependence 1–2 years post transplantation [23, 25]. A third limiting factor is islet revascularization, which appears to play an important role in determining the long-term survival and optimal performance of functional islet mass post transplantation. Rapid re-establishment of an appropriate microvascular system in newly transplanted islets is crucial for survival and function of islet grafts. Unfortunately, islets implanted at ectopic sites, such as under the renal capsule or in the liver and spleen, are invariably associated with markedly reduced vascularization, in comparison with native islets in the pancreas [26, 27]. This impairment in islet revascularization accounts at least in part for the demand of sufficiently large quantities of islet mass for restoration of normoglycemia in type 1 diabetic subjects. In addition, delayed and inadequate islet graft vascularization can deprive islets of oxygen and nutrients, causing islet cells to undergo cellular apoptosis and subsequent cell death, particularly in the core of large islets or in the center of aggregated islet clusters post transplantation. Moreover, a lack of sufficient islet revascularization may also compromise the optimal performance of transplanted islets. Indeed, there are clinical data indicating that even after postabsorptive blood glucose homeostasis is restored to normal post islet transplantation, implanted islets do not seem to function at optimal levels, as reflected in their significantly impaired glucose tolerance in diabetic recipients in response to intravenous glucose challenge [25, 28]. Thus, it is of great significance to define the molecular mechanism of islet revascularization and develop therapeutic angiogenesis approaches to enhance the process of islet revascularization. Such approaches are expected to ensure adequate microvascular perfusion to islet cells and protect implanted islet cells from hypoxia-induced inflammation and necrosis, which will ultimately improve the outcome of islet transplantation by reducing the donor/recipient ratio thus increasing the success rate of islet transplantation [29]. Antioxidant supplementation [30, 31], elevation of vascular endothelial growth factor (VEGF) delivery [32] to islet grafts and co-transplantation of bone marrow–derived mesenchymal stem cells [33] improves the outcome of islet transplantation by enhancing islet revascularization. Pretreatment of islet donors with exenatide improved graft function compared to post-transplant treatment of recipients only [34]. In addition, a combination of thymoglobulin and CTLA4-Ig prevented autoimmunity recurrence after islet transplantation, reverting ongoing autoimmune diabetes in animal models [35]. On the other hand, some immunosuppressive therapies, notably ciclosporin and tacrolimus, may induce insulin resistance and hyperglycemia and/or impair β-cell function and proliferation [36, 37]. However, metformin was effectively used for reversing immunosuppressant-induced hyperglycemia in experimental animal models [38].

Many other interventions are being tested to improve islet survival, such as the use of growth factors [39], glucagon-like peptides (GLP) and GLP-1 analogs [40], oxygen carriers [41], substances that reduce inflammatory response [42], anticoagulants [43] and antiapoptotic drugs [44].

The limitations and long term failure of the early transplantation therapies indicate that the procedure in its current format is not suitable for all patients with type 1 diabetes [45]. Intensive research is being conducted to look for alternative sources of β-cells.

An alternative to avoiding immunosuppression is to deliver encapsulated islets; immobilized in selectively permeable casings, islets would be capable of sensing glucose and secreting insulin but would be protected from immune attack. In theory, these membranes would permit islet transplantation without immunosuppression and would also make xenotransplantation a viable option, potentially solving the problem of donor scarcity by making use of more abundant sources such as porcine islets [46, 47]. Microencapsulation strategies have been used successfully in animal studies; the delivery of bovine islets packaged in alginate gel spheres to the peritoneum of rats has been shown to normalize STZ-induced hyperglycemia [48]. Although complications with vascularization and fibrosis around encapsulated islets have not been totally resolved, clinical trials for encapsulated islet delivery to type 1 diabetics are moving forward [49].

Very recently, Living Cell Technologies received approval to progress its New Zealand Phase II human clinical trial with its groundbreaking DIABECELL®. It comprises neonatal porcine islets encapsulated in alginate microcapsules [50]. The approval follows a positive assessment from the New Zealand Data Safety and

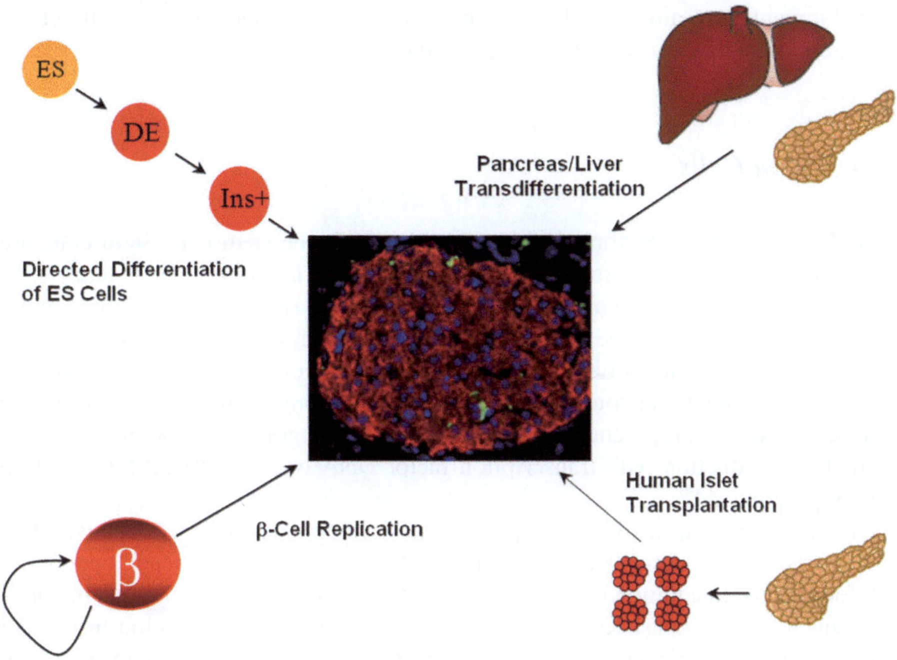

Fig. 2 Sources of new β-cells for the treatment of type 1 diabetes. Type 1 diabetes results from the autoimmune destruction of pancreatic β-cells. A cure for the disease may lie in the development of techniques to replenish islet mass. Some researchers have already directed the differentiation of ES cells toward an endocrine (*Ins*+) cell. The potential exists for liver and nonendocrine pancreatic cells to be reprogrammed by gene transfer or growth factor treatments to produce and release insulin in response to glucose. Recent evidence has suggested that adult β cells are mitotically active and can replicate in response to increasing insulin demands. A complete understanding of the factors and signals that regulate this process could allow this innate ability to be controlled in patients with diabetes. Finally, cadaveric human islets represent a promising source of therapeutic β cells through transplantation. Immunofluorescence staining of a primary adult mouse islet is shown in the center (insulin appears *red*). **Abbreviations**: *DE* definitive endoderm, *ES* embryonic stem, *Ins*+ insulin-positive

Monitoring Board of the first four patients to receive its DIABACELL® implants. Apparently, intraperitoneal islet transplantation has the potential to ameliorate type 1 diabetes mellitus and avert the long-term consequences of chronic diabetes which cannot be achieved by conventional insulin treatment [50].

As an alternative to islet transplantation, implantation of alginate-encapsulated insulin-producing cells differentiated from embryonic stem cells was demonstrated feasible and viable in diabetic animal models [51], although bone marrow transplantation per se also improved islet function through increases in nerve growth factor [52].

Thus, the development of alternative sources of islets, such as the promotion of transdifferentiation of nonbeta cells into an insulin-secreting phenotype, may be a viable option (Fig. 2). In addition, advances in the understanding of pancreas development have already informed the development of techniques to direct the differentiation of stem cells toward endocrine cells.

2.3 Stem Cells

An alternative resource for transplantable β-cells is the stem cell. Stem cells are cells that are able to proliferate while maintaining an undifferentiated status (self-renewal) and retaining a capacity to differentiate into specialized cell types under appropriate conditions. The sources of such stem cells for the purpose of generating β-cells were firstly identified through several studies on adult and fetal pancreas. There are two potential sources of stem cells: embryonic stem cells (ES) [53] and the new induced pluripotential stem cells that are reprogrammed from mature adult cells by transduction with transcription factor DNA or recombinant transcription factor proteins [54].

Recently, Kroon et al. [55] reported the development of functional islet-like structures in mice transplanted with human ES cell-derived pancreatic endoderm. Unfortunately, in addition to the generation of functional β cells in the vast majority of transplant recipients, Kroon et al. also observed teratoma formation in a small number of cases. This outcome highlights the delicate balance between the risks and benefits that will need to be managed in future stem cell replacement therapies.

New technologies for reprogramming of adult cells to an induced pluripotent state (iPS), similar to ES cells, may change the landscape of future β-cell therapy [56–60].

Theoretically, it should become straightforward to make autologous iPS cells starting with readily accessible skin or blood cells from any individual, and to use these as a source of transplantable cells, obviating the need for immunosuppression. At a practical level it is not clear that this degree of individualized medicine can be made cost effective for patient populations. However, assessments of cell banking strategies have indicated that as few as ten pluripotent cell lines carefully chosen for homozygosity of the most common alleles at the major histocompatibility genetic loci would provide considerable practical benefit for transplantation matching [61].

It remains to be shown that iPS cells have the same capacity as human ES cells for differentiation toward β-cells. However, a very recent report documents that human iPS cells, indeed, give rise to insulin-producing cells [62]. The iPS cells were induced to yield pancreatic islet-like clusters using one of the culture protocols previously described for human ES cells.

The recent demonstration that human somatic cells can be reprogrammed to a pluripotent stem cell fate suggests that ethical and legal issues will need to be circumvented [58, 59]. Two groups of investigators used genes known to be enriched in ES cells and involved in the establishment of pluripotency to reprogram human somatic cells. Using slightly different protocols, both groups established pluripotent lines that were capable of differentiating into all three germ layers and forming teratomas when injected into mice. The limitations of this finding are considerable, given that the reprogramming involved the use of viruses and the expression of the proto-oncogene c-Myc. However, alternative methods of cell transduction can be developed and somatic cell–derived insulin-producing cells for diabetes therapy may be possible in the not too distant future.

It is noteworthy that Nagy and co-workers discovered a new method to create pluripotent stem cells without disrupting healthy genes [63]. This method uses a novel wrapping procedure to deliver specific genes to reprogram cells into stem cells. Previous approaches required the use of viruses to deliver the required genes, a method that carries the risk of damaging the DNA. This new method does not require viruses, and so overcomes a major hurdle for the future of safe, personalized stem cell therapies in humans. This study accelerates stem cell technology and provides a road map for new clinical approaches to devastating diseases such as diabetes.

In a 2007 Brazilian study [64], a small number of people newly diagnosed with type 1 diabetes were able to stop using insulin after being treated with stem cells made from their own blood. Although stem cell transplants, which involve shutting down the immune system and then building it up again, can be risky though the technique may 1 day provide an additional treatment option for type 1 diabetes [64, 65].

More recently a pilot study has shown that transfusion of umbilical cord blood in children with type 1 diabetes may decelerate the loss of endogenous insulin production [66]. The experimental challenge in the future will be to unravel the optimal cell source; but there is growing reason for optimism in the replacement of islet transplantation from deceased organ donors by stem cell-derived highly functional β-cells as an achievable goal.

2.4 Transdifferentiation

Another method of β-cell replacement for diabetes therapy might involve gene transfer or growth factor treatments to redirect nonislet cells to manufacture insulin and respond to changes in extracellular glucose. Given the close embryonic origins of the pancreas and liver, some investigators have focused efforts on transdifferentiation of the liver to a pancreatic endocrine cell fate. Ferber et al. [67] demonstrated that adenoviral-mediated expression of PDX1 induced insulin expression in the liver and rescued STZ-induced hyperglycemia in mice. Although the reprogramming of liver or exocrine tissue to pancreatic endocrine cell fates rescues mouse models of hyperglycemia, the application to human disease may be limited by ongoing autoimmune attack on the β-cells that patients with type 1 diabetes face [68]. Insulin-producing cells in the liver might also be susceptible to autoimmune lymphocyte infiltration and apoptosis. Offering some optimism in this regard, Shternhall-Ron and coworkers [69] demonstrated that adenovirally delivered PDX1 rescued glucose tolerance in cyclophosphamide-accelerated nonobese diabetic mice.

The potential for gene therapy is enormous. For instance, gut K cells of the mouse were induced to produce human insulin by transfecting the human insulin gene linked to the 5′-regulatory region of the gene encoding glucose-dependent insulinotropic polypeptide (GIP) [70]. Also, research demonstrating that hepatocytes transfected with the PDX*1* gene under the control of the rat insulin 1 promoter were able to produce insulin [67, 71] attracted significant attention and has inspired new hope. In these studies, sufficient levels of insulin were secreted to satisfy the needs of a diabetic mouse, which, when treated, became and remained steadily euglycemic. These studies, however, have not yet been successfully repeated by other groups. This finding suggests that gene therapy–mediated transdifferentiation may prove to be a viable strategy for producing new β-cells, despite ongoing autoimmune attack (see Sect. 5).

It has recently been shown that infecting liver with a virus containing the gene for neurogenin, a transcription factor that is expressed as cells begin differentiating into insulin-producing β-cells, cells in the liver can take on the function of pancreatic cells and go on to reverse symptoms of diabetes in a mouse model of the disease [72]. Other researchers showed that adding genes for transcription factors can change cell differentiation [73].

2.5 *Prevention and Early Intervention*

In humans, the accumulation of islet antibodies with differential specificities for β-cell proteins, in combination with genotyping for susceptibility alleles, can predict the risk to develop clinical diabetes. However, researchers are still unable to arrest β-cell destruction in pre-diabetic patients, even though a lot of evidence collected from preclinical studies using various therapeutic regimens in different animal models for type 1 diabetes has been successful in preventing type 1 diabetes [74]. Some compounds (anti-CD3 antibodies, glutamic acid decarboxylase (GAD) of 65 kDa [GAD65], Diapep277, and anti-thymocyte globulin) that reestablished long-term tolerance in animal models after new-onset type 1 diabetes show promising effects in reducing β-cell decline in phase I and II clinical trials in humans with recently diagnosed type 1 diabetes, but none of them was able to cure the disease [75, 76]. Other antigen non-specific agents, such as the anti-thymocyte globulin, monoclonal anti-CD20 antibody (Rituximab) and monoclonal anti- interleukin-1 beta (Canakinumab), are currently under trial in recent-onset type 1 diabetes [74, 76].

The GAD-based diabetes vaccine for type 1 diabetes has been demonstrated to slow or arrest the destruction of insulin producing β-cells, which is characteristic of autoimmune diabetes. Studies have demonstrated that the Diamyd® vaccine is most efficacious early in the disease process, in recent-onset type 1 diabetes patients. Diamyd Medical is currently conducting two clinical Phase III studies on type 1 diabetes: one in Europe and one in the United States. These studies apply to people diagnosed with type 1 diabetes within the past 3 months. Prevention studies have already been initiated, where researchers intend to evaluate the Diamyd® vaccine in children and adults at risk of developing type 1 diabetes [77]. When people are first

Table 2 Summary of the novel approaches for the prevention and treatment of type 1 diabetes

Approach/drug	Studies/development stage
Antigen-specific	
Insulin (parental, oral, mucosal)	Efficacy studies, phase I/II
Insulin β chain	Animal studies, phase I
Proinsulin (peptide, vaccine)	Animal studies, phase I
Glutamic acid decarboxylase 65 (Diamyd)	Phase II
Antigen-non-specific	
Anti-CD20 monoclonal antibody (Rituximab)	Animal studies, phase II
Anti-CD3 monoclonal antibody (MoAB)	Phase I/II
Anti-thymocyte globulin (ATG)	Animal studies, phase I
DiaPep277 (HsP60 peptide)	Phase II
Non-antigen-specific	
Cyclosporin	Various
Nicotinamide (± combinations)	Pilot, efficacy studies
Vitamin D3/Fish oil	Pilot ongoing
Hydrolyzed cow's milk	Phase I, ongoing
Dietary gluten elimination	Pilot
Potassium channel openers (Diazoxide)	No effect
Rapamycin	Ex vivo
Combinations	
Anti-CD3 MoAB + insulin peptide	Animal studies, phase II planned
Anti-CD3 MoAB + exenatide	Animal studies, phase II planned
Mycophenolate mofetil + anti-CD25 MoAB (Daclizumab)	Open-label randomized trial, phase II ongoing
Rapamycin (Sirolimus) + interleukin-10	Animal studies
Rapamycin (Sirolimus) + interleukin-2	Animal studies, phase II planned
Stem cell therapy	
Autologous umbilical cord	Pilot
Autologous bone marrow-derived stem cells	Pilot
Transplantation	
Pancreatic islet transplantation + glucocorticoid free immunosuppressive regimen	Various

Adapted from [74]

diagnosed with type 1 diabetes, they still have cells that produce insulin. Diamyd® treatment is intended to halt or slow the autoimmune destruction process and save the remaining β–cells. Diamyd® can also be combined with other drugs that stimulate the generation of new β–cells, or with β–cell transplants.

Recent advances in research for immunotherapeutics resulted in the proinsulin-based plasmidic vaccine BHT-3021, which proved well tolerated while preserving β-cell function in a placebo-controlled trial [78].

The novel approaches for the prevention and treatment of type 1 diabetes are summarized in Table 2 and extensively reviewed elsewhere [74].

3 Type 2 Diabetes

Type 2 diabetes mellitus is a multifactorial metabolic disorder characterized by chronic hyperglycemia due to relative or absolute lack of endogenous insulin. Patients have elevated fasting and postprandial plasma glucose levels. Type 2 diabetes development appears to involve defects in insulin action and secretion [79, 80]. Type 2 diabetes is the result of a combined defect in insulin resistance, β-cell dysfunction, increased hepatic glucose production [79–82].

Glucose levels remain normal or only mildly impaired when islet β-cells remain able to compensate with sufficient insulin output to overcome insulin resistance (IR). A progressive decline in β-cell compensation in a subset of insulin-resistant subjects eventually leads to overt hyperglycemia. IR affects the main insulin target tissues, namely, skeletal muscle, liver, and adipose tissue. The progressive loss of β-cell function and increased IR may be due to a number of factors such as genetic abnormalities and acquired defects. Although a degree of IR may be inherited, it may progress with additional factors such as nutrition and a sedentary lifestyle [83].

IR is likely due to the interplay of genetic factors, modifications to the insulin signaling pathway, and inflammation. Insulin secretion is affected by multiple factors, including the baseline secretion capacity, incretin effect, nutrient excess of glucose and free fatty acids (FFAs), and amyloid deposition [83, 84].

Treatment of type 2 diabetes has centered on (1) increasing insulin levels, either by direct insulin administration or oral agents that promote insulin secretion (insulin secretagogues, such as oral sulfonylureas), (2) improving tissue sensitivity, such as with insulin sensitizer biguanide metformin or thiazolidinediones (TZDs), or (3) reducing the rate of carbohydrate absorption from the gastrointestinal tract by the use of α-glucosidase inhibitors or agents that decrease gastric motility (amylin analogues). Progressive β-cell dysfunction and β-cell failure are fundamental pathogenic features of type 2 diabetes, and, ultimately, the development and continued progression of diabetes is a consequence of the failure of the β-cell to overcome insulin resistance. Current therapies become less effective over time as a result of progressive loss of β-cell function and number, with the result that a majority of type 2 diabetic patients do not achieve current glycemic goals as reflected by HbA1c greater than 7% in more than 60% of treated patients [7, 8, 85–88].

3.1 Conventional Therapies

Diet and exercise are the first step of therapy for type 2 diabetes; if these do not keep blood sugar at goal levels, then antihyperglycemic agents are added. Drug therapy for type 2 diabetes aims to control blood sugar levels both in the basal (fasting) state and postprandially; rational combinations of agents with different mechanisms of action can be used (Fig. 3).

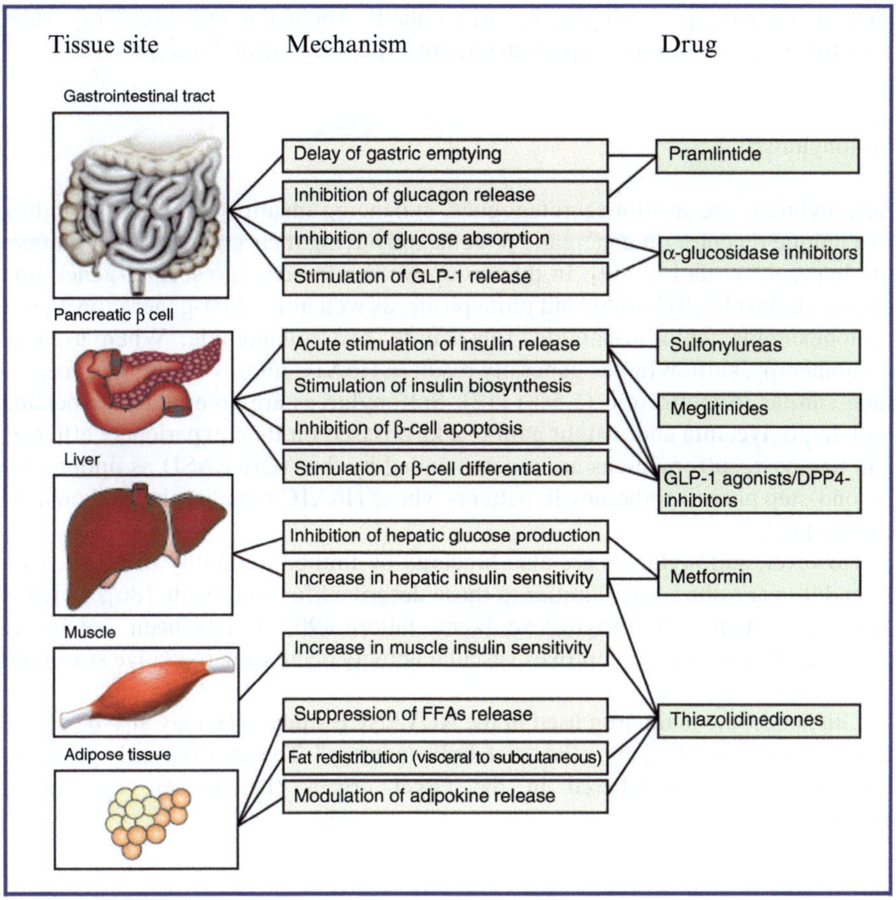

Fig. 3 Pharmacological treatment of hyperglycemia according to site of action. *GLP-1* glucagon-like peptide 1, *DPP 4* dipeptidyl peptidase 4, *FFAs* free fatty acids (adapted from Stumvoll et al. [83])

Four major classes of antihyperglycemic agents can be used, either as monotherapy or, more appropriately, in combination with one another:

- Insulin secretogogues
- Insulin sensitizers
- α-glucosidase inhibitors
- Insulin

3.1.1 Insulin Secretogogues

Insulin secretogogues correct hyperglycemia by stimulating insulin secretion—but only if the patient still has enough functioning β–cells. They close ATP-sensitive potassium channels in the β–cells of the pancreas, increasing insulin production;

slow-acting and rapid-acting agents are available. The major side effects of insulin secretogogues (and insulin replacement) are hypoglycemia and weight gain.

Sulfonylureas

Sulfonylureas are insulin secretagogues, enhancing insulin secretion by binding to a unique receptor on pancreatic β–cells and having their greatest effect on fasting hyperglycemia [12, 89]. In this group we can include the second-generation agents gliclazide, glyburide, and glimepiride, as well as the first-generation agents acetohexamide, chlorpropamide, tolazamide, and tolbutamide. When used as monotherapy, sulfonylureas generally result in HbA1C improvements of a magnitude similar to metformin (1.5%) [12]. Sulfonylureas are commonly associated with hypoglycemia and weight gain (~2 kg). Based on long experience, efficacy, and low cost, sulfonylureas are recommended by the ADA-EASD as options for second-step pharmacotherapy in patients whose HbA1C remains elevated on metformin [12].

However, sulfonylureas are also hindered by limited durability of effect, with 3- and 9-year failure rates similar to those described for metformin [86], which is again consistent with progressive β-cell failure [90]. It has been previously described that gliclazide improves vascular activity decreasing oxidative stress and inflammation.

Gliclazide, the main drug used in the ADVANCE study [91], may also be potentially beneficial in abolishing the "metabolic memory". The idea that early glycemic environment is remembered in the target organs (i.e. kidney, eye, heart, extremities).

Meglitinides

The recently introduced class of meglitinides consists of nateglinide, which binds to the same site of sulphonylurea receptor 1 as do the sulfonylurea derivatives, and repaglinide, which binds to a nearby site of the receptor, both leading to insulin release. They stimulate rapid, short-lived, insulin secretion [92]. They lower postprandial glucose levels, although fasting hyperglycemia is also improved. Meglitinides are more specific that sulfonylureas and are associated with lower risk of hypoglycemia but clinical experience remains limited. These agents cannot further stimulate insulin release in patients on maximal doses of sulfonylurea derivatives. These drugs can be used in patients with decreased renal function [93] and for individuals with varying daily meal patterns. Experimental evidence suggests a better preservation of β-cell function compared to sulfonylureas [94] and improved vascular effects that remain to be clinically proven [95]. The NAVIGATOR (Nateglinide and Valsartan in Impaired Glucose Tolerance Outcomes Research) study aims to assess the utility of these drugs in prevention of diabetes and their impact on cardiovascular morbidity and mortality.

3.1.2 Insulin Sensitizers

Metformin

Metformin, is an insulin sensitizer well accepted as a first-line agent for treatment of type 2 diabetes [12]. It is a biguanide derivative that exerts an antihyperglycemic effect with minimal risk of hypoglycemia. Metformin lowers blood glucose concentration and improves insulin sensitivity by reducing hepatic gluconeogenesis and enhancing insulin-stimulated peripheral glucose uptake [12, 89]. In addition, metformin reduces insulin resistance in muscle tissue and the liver, decreasing postprandial hyperglycemia and inhibits adipose tissue lipolysis thereby reducing circulating levels of FFAs [89, 96]. Metformin may also suppress inflammation independently of action on glucose, insulin and FFAs [97]. When used as monotherapy, metformin typically reduces HbA1C by about 1.5% [12]. Metformin also improves the lipid profile and lowers blood pressure and plasminogen activator inhibitor-1 levels in both patients and animals with impaired glucose tolerance and type 2 diabetes [8, 96, 98]. In overweight type 2 diabetic patients, metformin use is associated with decreases in macrovascular morbidity and mortality, effects that appear to be independent of the improvement in glycemic control [8, 96, 99, 100].

Caution must be exercised when using metformin in a number of other patient populations, including those with impaired hepatic function or requiring drug therapy for heart failure, and metformin should be temporarily discontinued prior to surgical procedures or intravascular radiocontrast studies [101]. In addition, metformin has been associated with rare cases of fatal lactic acidosis and is contraindicated in patients with renal dysfunction [101]. Furthermore, a study by Ting et al. noted that metformin use was associated with a risk of vitamin B12 deficiency [102]. The frequency of gastrointestinal adverse events may limit the ability to reach maximally effective doses, with rates of diarrhea and nausea/vomiting of up to 53 and 26% of patients, respectively [101]. Because of its efficacy, infrequency of weight gain or hypoglycemia, and low cost, metformin has been recommended by the ADA-EASD as first-line pharmacotherapy for type 2 diabetes [12]. However, the durability of metformin's effectiveness as monotherapy is limited consistent with the progressive loss of β-cell function seen in type 2 diabetes [90].

Thiazolidinediones

Thiazolidinediones, another class of insulin sensitizers, currently include pioglitazone and rosiglitazone. They enhance insulin sensitivity reducing hyperglycemia and ameliorating the dyslipidemia and inflammatory milieu of type 2 diabetes [103]. TZDs improve insulin sensitivity in muscle and adipose tissue and in the liver by activating peroxisome proliferators activated receptor-γ (PPAR-γ) [12, 89, 104]. TDZs alter adipose metabolism and distribution and redistribution of tissue triglyceride from visceral stores, reducing levels of circulating FFAs apparently by sequestration in a less lipolytic subcutaneous compartment [105]. TDZs also reduce circulating concentrations of pro-inflammatory cytokines that promote insulin resistance (e.g., tumor necrosis factor-α and interleukin 6) and at the same time increase concentrations of adiponectin, which has insulin-sensitizing and anti-inflammatory properties. TZDs have also been shown to improve β-cell function, probably as a result of PPAR-γ–mediated decreases in insulin resistance and β-cell fatty acid concentrations [104]. TZDs have demonstrated the ability to conserve β-cell function, delaying or preventing the development of type 2 diabetes in a high-risk population of women with gestational diabetes in the TRIPOD (TRoglitazone In the Prevention Of Diabetes) and PIPOD (Pioglitazone In the Prevention Of Diabetes) studies [106, 107]. A large number of studies show that the TZDs are effective as monotherapy or in combination with metformin, sulfonylureas, and insulin in the treatment of type 2 diabetes. When used as a single agent, TZDs decrease HbA1C in the range of 1.0–1.6% and have a lower incidence of treatment failure than with metformin or the sulfonylurea, glyburide [88, 108, 109]. However, the TZDs are more effective when used in combination with metformin, sulphonylureas, or insulin [110–112]. The TZDs should not be given to patients with heart failure because they typically cause fluid retention and peripheral edema, worsening heart disease [113, 114].

The multiple effects of thiazolidinediones on adipose tissue metabolism and cross-talk of these signals with liver and skeletal muscle, as well as pancreatic β-cells and the vascular endothelium, might account for the enhancement of insulin action and improvement in insulin secretion with these agents, as well as several beneficial effects on vascular function [115]. Renal and vascular benefits of thiazolidinediones have been demonstrated in controlled studies, for example, showing significant improvement in albumin excretion above that observed with a similar degree of glycemic lowering with sulfonylureas [116].

Unlike metformin, the thiazolidinediones can be used in patients with reduced renal function, and they are better tolerated without significant gastrointestinal side effects. A major adverse effect associated with clinical use of the thiazolidinediones is weight gain, which seems to be coupled to the effects of the drugs on adipose cell differentiation and triglyceride storage. Fluid retention is also linked to the PPARγ-agonist activity of the thiazolidinediones, leading to peripheral edema and a mild hemodilution in some patients [113, 117]. Fortunately, congestive heart failure is quite rare with use of thiazolidinediones, but remains a serious concern [118]. TZDs may cause hepatotoxicity and should not be used in patients with impaired liver function [119]; the U.S. Food and Drug Administration (FDA) recommends regular liver-function testing in all patients on TZDs. The use of TZDs has also been reported to (1) increase the risk of myocardial infarction and other CV diseases [120, 121], (2) decrease bone formation, accelerate bone loss, and increase the risk of fractures [122, 123], and (3) induce macular edema [124].

TZDs are recommended by the ADA-EASD as an alternative agent in second-step therapy after metformin failure [12]. Although more expensive than generic versions of metformin and sulfonylureas.

3.1.3 α-Glucosidase Inhibitors

The α-glucosidase inhibitors, acarbose, miglitol and voglibose, slow digestion of oligosaccharides, thereby providing an alternative to reduce postprandial glucose levels [125]. The α-glucosidase inhibitors do not cause weight gain, and have been shown to lower plasma triglyceride levels in some studies [126]. They must be dosed multiple times per day and are associated with frequent gastrointestinal side effects [12].

They generally have less potent glucose-lowering effects than other oral antidiabetics [12]. In the STOP-NIDDM (Study to Prevent Non-Insulin Dependent Diabetes Mellitus) trial, acarbose reduced the incidence of new cases of type 2 diabetes in high-risk subjects with impaired glucose tolerance [127]. Acarbose therapy is also associated with a reduction in incidence of cardiovascular events [128] although some controversy is present.

3.1.4 Insulin

Several different insulin analogs are available for type-1 and advanced type-2 diabetic patients. The injected insulin types differ in their onset and duration. Insulin therapy often has two components, an intermediate acting or long-acting insulin given at bedtime, and a rapid-acting insulin given before meals [129].

Insulin administration is the most effective means of restoring glycemic control; because there is no maximum dose, any HbA1C level can be reduced to the target range if insulin is dosed adequately [12]. However, insulin has a number of limitations. It is typically administered by subcutaneous injection, often requiring multiple injections per day. Insulin carries a tangible risk of hypoglycemia, and regular self-monitoring of blood glucose is usually required [12]. It is typically associated with weight gain that significantly increases in patients on intensive insulin therapy [130] with adverse cardiovascular consequences [107]. The ADA-EASD guidelines recommend addition of insulin as a second-step option for type 2 diabetes patients who are not adequately controlled on metformin monotherapy or as a third-step option for patients who still do not reach the HbA1C target goal on oral combination therapy [12]. Insulin is also the treatment of choice for patients with severely uncontrolled or symptomatic type 2 diabetes [12].

3.2 Novel Antidiabetic Agents

The currently available therapies used for type 2 diabetes do not significantly improve β-cell function. In addition, the current approach does not address defects in hormonal secretion thought to play key roles in the pathophysiology of type 2 diabetes.

New emerging therapies for type 2 diabetes have become available in recent years (Table 3). As a result of their recent availability, long-term studies are lacking

Table 3 Comparison of glucose-lowering agents

Class	Examples	Primary mode of action	Route of glycemic control	Adverse effects	Effect on weight	Effect on β cell function
α-glucosidase inhibitors	Acarbose, miglitol [45, 121]	Inhibit enzyme central to digestion of carbohydrates [45, 79, 121]	Postprandial glucose [45, 79, 121]	Diarrhea, abdominal pain, flatulence, ↑transaminases [45]	[131]	?
Biguanides	Metformin [45]	↓Hepatic glucose production, ↑muscle sensitivity to insulin [45, 79, 121]	Fasting glucose, insulin sensitivity [121]	Diarrhea, nausea, lactic acidosis [45, 79]	[12, 89, 132]	?
DPP-4 inhibitors	Sitagliptin [45]	Inhibition of DPP-4 results in ↑GLP-1 [45, 95, 121]	Postprandial glucose [45, 121]	Upper respiratory infection, nasopharyngitis, headache [129]	[134, 134]	[131, 133]
Meglitinides	Nateglinide, repaglinide [45, 121]	β-cell secretagogue [45, 121]	Postprandial glucose [45, 121]	Hypoglycemia [45, 79]	[12, 89, 131]	?
Sulfonylureas	Glimepiride, glipizide, glyburide [45, 121]	β-cell secretagogue [45, 79, 121]	Fasting and postprandial glucose [95, 121]	Hypoglycemia, weight gain [45, 79, 128]	[12, 131]	?

Thiazolidinediones	Pioglitazone, rosiglitazone [45, 121]	Enhanced peripheral insulin sensitivity, improved hepatic insulin sensitivity [45, 79, 121]	Insulin sensitivity, postprandial and fasting glucose [95, 121]	Fluid retention, weight gain, heart failure [45, 79, 95, 128]	[12, 89, 104, 131]	[12, 89]
Amylin analogues	Pramlintide [45]	↓Glucagon secretion, gastric emptying, and food intake [45, 128]	Postprandial glucose [45, 128]	Nausea, hypoglycemia [128]	[89, 131]	[131]
Incretin mimetics	Exenatide [45]	↓Glucagon secretion, gastric emptying, and food intake; ↑insulin secretion [45, 128]	Postprandial glucose [45]	Nausea, diarrhea, hypoglycemia, pancreatitis [128, 130]	[131]	[89]
Insulin	—	—	Fasting and postprandial glucose [128]	Hypoglycemia, weight gain [79, 128]	[47, 131]	?

↓ decreased, ↑ increased, *DPP-4* dipeptidyl peptidase 4, *GLP-1* glucagon-like peptide-1 and full safety profiles of these compounds are largely unknown, even though they are being used in large numbers of patients.

These drugs offer a range of different mechanisms of action that complement established therapies (Fig. 3). Several of the novel drugs are based on the incretin hormone, GLP-1. GLP-1 controls glucose levels through various mechanisms including glucose-mediated insulin secretion, suppression of inappropriate glucagon release, slowing gastric emptying and increasing satiety [135–137]. The natural hormone is rapidly degraded by the enzyme dipeptidyl peptidase 4 (DPP-4), and the two approaches to new agents are inhibitors of DPP-4 activity and development of GLP-1 analogues resistant to degradation. An important property of these agents is their neutral or beneficial effect on body weight [138].

3.2.1 Incretin System

Gastrointestinal polypeptide hormones that are secreted in response to ingestion of a meal augment postprandial insulin secretion; this is known as the incretin effect and can account for up to 70% of postprandial insulin secretion [139]. The two most important incretin hormones are GLP-1 and GIP; these hormones are secreted by the L cells of the distal ileum and colon, and the K cells of the duodenum and upper jejunum, respectively. Circulating concentrations of incretins rise within minutes of eating, implying likely stimulation via activation of neuro-endocrine pathways. Incretins act via specific G-protein-coupled receptors on β-cells to enhance glucose-stimulated insulin secretion [133].

The incretin effect is almost completely deficient in patients with type 2 diabetes, mainly because of reduced postprandial GLP-1 secretion that is accompanied by a markedly reduced insulinotropic action of GIP [131]. Investigators have observed that infusing GIP or GLP-1 into patients with type 2 diabetes recovers the first phase insulin response, and this indicates an improvement in β-cell function. In animal models, it has been suggested that GLP-1 stimulates neogenesis of β-cells and inhibits β-cell apoptosis. Thus, agents that increase the GLP-1 signal raise the intriguing possibility of replenishing endogenous insulin by increasing the β-cell population, potentially preventing or even reversing disease progression.

Novel therapies that exploit the incretin effect of GLP-1 include the injectable incretin mimetics and the orally active DPP-4 inhibitors. Their development arose from the understanding and extensive research into incretin physiology and metabolism.

3.2.2 GLP-1 Receptor Agonists

The only GLP-1 receptor agonist currently available is exenatide, although a second agent, liraglutide, is in late-phase clinical development [140]. These agents are not effective orally and require administration by subcutaneous injection.

In clinical trials, they lowered HbA1C concentration by 0.5–1.0% and had the additional advantage in producing a significant weight loss. The current indication for these drugs is for use after patients have experienced failure of glycemic control with metformin, sulfonylureas and/or TZDs [134, 136, 141]. Exenatide has been evaluated as an alternative to insulin in patients with type 2 diabetes who have inadequate glycemic control. It leads to weight loss, at least over a 3-year period. Long-acting forms of once-weekly injection, such as exenatide LAR, are under development. When used in combination with metformin, sulfonylureas, TZDs, or metformin plus a sulfonylurea, exenatide has been shown to result in weight loss [142]. Daily liraglutide therapy had similar effects with respect to blood glucose control, β-cell function, and weight loss as twice-daily exenatide, and its effects were sustained for at least 1 year.

These agents are not generally associated with hypoglycemia but a large proportion of treated patients experience nausea and vomiting, although these gastrointestinal adverse effects tend to become less over time. There have been some recent reports of hemorrhagic or necrotising pancreatitis and patients should be monitored for signs and symptoms. Exenatide is not recommended for patients with severe renal impairment [143].

3.2.3 Dipeptidyl Peptidase 4 Inhibitors

Pharmacological inhibitors of DPP-4 are orally active agents that acutely increase postprandial plasma levels of endogenous GLP-1 typically twofold. In contrast with incretin mimetics, and perhaps reflecting the relatively modest increases in incretin levels, DPP-4 inhibitors show little or no evidence of a deceleration in gastric emptying [144], and mild gastrointestinal adverse effects.

There are two DPP-4 inhibitors currently available, which are sitagliptin, available in Europe and the USA, and vildagliptin, currently only available in Europe; both are orally active small molecules that result in increased incretin hormone activity. In clinical trials of these agents, HbA1C concentration was reduced by 0.5–0.9%, and body weight was not significantly altered. They may be used either as monotherapy or in combination with metformin or sulfonylureas. When used as monotherapy, they are generally not associated with hypoglycemia. DPP-4 is found in many tissues throughout the body and there are a large number of potential substrates [145]. As a result, DPP-4 inhibitors can affect many processes that may lead to side effects; for example, they may interfere with the immune system and side effects of upper respiratory tract infections have been reported [145]. While these drugs are being used increasingly and frequently, data on long-term use are lacking and the continued safety profile remains to be established. We do not yet know whether these new agents will have significant, long-term disease-modifying effects.

DPP-4 inhibitors would almost certainly be more useful in type 2 diabetes management if used in addition to agents that would increase incretin secretion (incretinotropics).

Metformin actually increases incretin secretion [146], and this factor probably explains why the addition of DPP-4 inhibitors to metformin therapy has a synergistic effect on reductions in HbA1C more pronounced than simply adding together the effects of each as a monotherapy: metformin prevents the negative feedback of DPP-4 inhibition on incretin secretion.

3.2.4 Pramlintide

Amylin (islet amyloid polypeptide) is a 37-aminoacid peptide hormone that is co-secreted with insulin from islet β cells in response to meals [147]. The hormone acts centrally via the area postrema and other regions to activate neural pathways that decrease glucagon release from pancreatic α cells [148] and slow gastric emptying [149], contributing to a glucose-lowering effect. Amylin also appears to act centrally to promote satiety, thereby reducing caloric intake [150]. Human amylin precipitates and aggregates to form amyloid fibres, which play a putative role in the progressive β-cell destruction [147].

Pramlintide, a human amylin analogue, is currently only approved for use in the USA and as an adjunct to insulin therapy [151, 152]. It is administered by subcutaneous injection before meals and decreases post-meal glucose excursions by slowing gastric emptying and reducing glucagon secretion in a glucose-dependent manner. Clinical trials showed a relatively small reduction in HbA1C of 0.5–0.7%, although there was also a decrease in body weight of approximately 1–1.5 kg over 6 months, despite the concomitant treatment with insulin. The weight reduction was most likely caused by the satiety effect.

The α-glucosidase inhibitors, pramlintide, and DPP4 inhibitors generally result in smaller reductions in HbA1C when compared with other treatment options recommended by the ADA-EASD [131, 133].

Incretin-based therapies give new options for lowering blood glucose and should be placed alongside, and added to, older options.

3.2.5 Bromocriptine

Bromocriptine is a sympatholytic D2-dopamine agonist that has been recently approved for the treatment of type 2 diabetes. Based on animal and human studies, daily bromocriptine administration within 2 h of awakening is believed to augment low hypothalamic dopamine levels and inhibit excessive sympathetic tone within the central nervous system, resulting in a reduction in postmeal plasma glucose levels due to a decrement in hepatic glucose production [153]. Addition of bromocriptine to poorly controlled type 2 diabetic patients treated with diet alone, metformin, sulfonylureas, or thiazolidinediones produces a 0.5–0.7 decrement in HbA_{1c}. Bromocriptine also reduces fasting and postmeal plasma free fatty acid and triglyceride levels by approximately 30% [154].

3.2.6 Colesevelam

Colesevelam, a bile acid sequestrant used for the treatment of hyperlipidemia, was recently approved by the FDA for the treatment of type 2 diabetes. The Glucose-Lowering Effect of WelChol Study (Welchol, GelTex Pharmaceuticals Inc) showed overall hemoglobin A1c level reduction of up to 1.0% in patients with baseline hemoglobin A1c level greater than 8% over 12 weeks of treatment. This combined improvement in glycemic control and lipid profile gives colesevelam an advantage in the treatment of type 2 diabetes. It is administered orally, and its main side effects are gastrointestinal, particularly constipation. The mechanism of action is thought to be delayed or altered absorption of glucose in the intestines [155].

Although several classes of anti-diabetic drugs are available, achieving and maintaining long-term glycemic control is often challenging, and many current agents have treatment-limiting side effects. So, a significant need for novel therapeutical approaches remains.

3.2.7 Bariatric Surgery

The use of bariatric surgery to achieve and maintain weight loss and ameliorate or resolve diabetes is generating great interest in light of the current epidemic of obesity and diabetes worldwide [156, 157]. Bariatric surgery is recognized as the most effective therapy for treating severe obesity and its associated metabolic abnormalities in patients who have been unsuccessful with dietary, behavioral, and medical interventions [158]. Obesity-related diabetes is controlled by bariatric surgery [159], especially restrictive and malabsorptive surgeries [160, 161]. However, its use as an intervention for type 2 diabetes is not yet recognized in patients with a BMI below 35 kg/m^2 [156, 157].

The mechanisms by which bariatric surgery achieve improvement of diabetes are not well understood. It is suggested that caloric restriction, weight loss, diminished nutrient absorption, hormonal changes in the entero insular axis [159, 162], and rearrangement of the gastrointestinal anatomy are possible features of such mechanisms [156].

Rubino et al. [163, 164] advocated that the presence of an intestinal feature derived from excessive stimulus of the upper digestive tract would cause diabetogenic effects demonstrated in rats that bypassing a short segment of proximal intestine directly ameliorates type 2 diabetes, independently of effects on food intake, body weight or malabsorption, i.e., the foregut theory [165]. Strader et al. [166], also in a rat model, suggested that lower intestinal stimulation through ileal interposition, the hindgut theory, increased synthesis and release of peptide YY and GLP-1.

Bariatric surgery presently offers the highest chances of revealing weight independent mechanisms of diabetes resolution [158, 161]. Hypotheses have been postulated to explain the improved insulin sensitivity witnessed postsurgery and include the altered secretion of gut hormones [167, 168], modifications in intestinal gluconeogenesis [169], alterations in gut microbiota composition [170] and changes in intramyocellular lipid content [171].

Blood sugar levels return to normal in 55–95% of people with diabetes depending on the procedure performed. However, the surgery is expensive and there are risks involved, including a slight risk of death. Additionally, drastic lifestyle changes are required and long-term complications may include nutritional deficiencies especially vitamin D and iron deficiency anemia and osteoporosis [159].

4 Experimental Drugs in Development for the Treatment of Diabetes

Many new drugs have entered the market in the past 10 years, and it is controversial that these new drugs are superior to more established therapies [172, 173] probably because many of these drugs are active on the same molecular targets as previously approved drugs. Importantly, with the escalating global diabetes pandemic, developing new medicines that reduce the cost of long-term treatment for this chronic disease will become a necessity.

The preclinical and clinical data recently presented at the ADA and EASD meetings gives a preview of the new drugs and technologies to come. Many of these drugs are several years away from commercialization. However, because of the new technology, new target or new delivery method, the outlook remains positive for these drugs and companies in the expanding market of diabetes control.

4.1 PPAR and GLP-1 Receptor Agonists

New technology and approaches for treating type 2 diabetic patients is changing and the competition to secure the growing market is enormous. Eli Lilly and Co. and Amylin Pharmaceuticals Inc. (AMLN), makers of Byetta, recently announced the FDA's rejection of their new Byetta formulation. The proposed drug, Bydureon (Byetta LAR), would allow users to take one dose a week to regulate their diabetes.

Injectable peptide agonists of the GLP-1 receptor have shown significant promise as anti-diabetic agents due to their ability to amplify glucose-dependent insulin release and preserve pancreatic β-cell mass (as previously discussed in Sect. 3.2). Byetta was the first FDA-approved agent in a class of diabetes therapies called incretin mimetics [143]. Lixisenatide, under development by sanofi-aventis, is a novel human GLP-1 receptor agonist for the treatment of type 2 diabetes mellitus. In clinical trials, lixisenatide improved glucose tolerance, resulted in weight loss and lowered HbA1C, thereby causing significantly more patients to achieve target HbA1C levels compared with the placebo. Lixisenatide exhibited well-established GLP-1-related gastrointestinal side effects, with mild nausea occurring most frequently;

a low frequency of hypoglycemia was also reported. The results of phase III clinical trial showed that in comparison with patients receiving placebo, a significant number of those given lixisenatide demonstrated HbA1c levels less than or equal to 6.5%. Lixisenatide therapy also significantly reduced the mean change in glucose levels from baseline 2 h after a meal. Treatment with the active drug was separately shown to reduce glucose excursion. All treatment groups demonstrated weight loss as a result of therapy [174].

The new GLP-1 analog (VRS-859), created by Versartis, is a recombinant fusion protein. That is, it is made up of exenatide attached to XTEN. XTEN is a tail of an hydrophilic amino acid sequence that covers exenatide, hiding it from DPP-4 and good work for a longer time, permitting less frequent dosing and fewer side effects. A clinical study of VRS-859 in patients with type 2 diabetes is now in Phase Ia.

Peroxisome proliferator-activated receptors (PPARs) are ligand-activated transcription factors, which belong to the nuclear receptor superfamily and which regulate adipogenesis, lipid metabolism, insulin sensitivity, inflammatory processes and blood pressure [175]. Fibrates, long used as triglyceride-lowering agents, are now known to function as activators of PPARα, while the already available thiazolidinediones, rosiglitazone and pioglitazone act via PPARγ to influence free fatty acid flux and reduce insulin resistance and blood glucose levels. Future innovations are expected whereby newer agents will combine PPARα, PPARγ and possibly other PPAR-mediated properties.

Roche recently advance aleglitazar (R1439) into phase III clinical testing for the management of cardiovascular disease in high-risk patients with type 2 diabetes. This decision is supported by data from the Phase II SYNCHRONY study [176]. Aleglitazar is a rationally designed molecule providing balanced dual PPAR α/γ activation.

Specifically it combines the improvements in peripheral insulin sensitivity (and therefore glycemic control) associated with PPAR γ activation, with improved management of dyslipidemia, which is commonly associated with PPAR α activation.

4.2 GPR119 Agonists

G protein-coupled receptor 119 (GPR119) is expressed in insulin-producing β-cells of pancreatic islets together with L- and K-cells of the gastrointestinal tract, and is involved in insulin and incretin hormone release.

Several studies report that stimulation of GPR119 under hyperglycemic conditions leads to a dual nutrient dependent insulinotropic and incretinotropic effect to maintain glucose homeostasis [177–179]. Unlike receptors for GLP-1 and other peptides that mediate enhanced glucose-dependent insulin release, GPR119 has proven amenable to the development of potent, orally active, small-molecule agonists [180]. GPR119-specific agonists appear to provide a completely novel and previously unexplored approach to incretin therapy in patients with type 2 diabetes, increasing glucose-dependent insulin secretion through two complementary

Table 4 GPR119 agonists

Pipeline of G-protein coupled receptor GPR119/glucose-dependent insulinotropic receptor (GDIR) agonist		
Development stage	Drug	Company
Phase II	MBX-2982	Metabolex/Sanofi-Aventis
	GSK1292263	GlaxoSmithKline
Phase I	APD597	Arena Pharmaceuticals/Johnson & Johnson
	PSN821	OSI Pharmaceuticals
Pre-clinical	GPR119 Agonists	Bristol-Myers Squibb
	GPR 119 Project	Swedish Orphan Biovitrum
	GPR119 Follow-on	Metabolex/Sanofi-Aventis
Research project	GPR119 Agonists	Neurocrine Biosciences/Boehringer Ingelheim
	GPR119	Exelixis
	GPR119 Agonist	LG Life Sciences
	PSN821 Backup (AS1535907)	OSI Pharmaceuticals
	GPR119 Research Program	Kalypsys

Data: *EvaluatePharma*

mechanisms: directly, through actions on the β-cell, and indirectly, through enhancement of GLP-1 and GIP release.

The first GPR119 agonist clinical candidates (Table 4) are now in Phase I/II trials (Metabolex MBX-2982; GlaxoSmithKline GSK1292263; Arena/Ortho McNeil APD597; Prosidion/OSI PSN821) and with confirmatory reports of clinical proof of concept with respect to glycemic control and incretin release, the spotlight has been set for this new class of therapeutic [180, 181].

Either alone or in tandem with approved DPP-IV inhibitors, they could herald a brand new treatment paradigm for type 2 diabetes mellitus.

4.3 *Interleukin-1β Inhibitors*

More recently, a new understanding of type 2 diabetes as a disease caused by inflammation as been explored as a source for new therapies [182, 183]. In type 2 diabetic patients, high glucose concentrations induce the production of interleukin-1 (IL-1) β in β-cells, leading to impaired insulin secretion, decreased β-cell proliferation, and increased apoptosis. Whereas most type 2 diabetes medicines focus on maximizing insulin production by the pancreas or increasing the body's sensitivity to insulin, IL-1 blockers target the inflammatory response that can devastate insulin-producing cells.

It has recently been reported that administration of anakinra (Kineret, a recombinant IL–1-receptor antagonist and a competitive inhibitor of IL-1 binding to type I IL-1 receptors) for 13 weeks to type 2 diabetic patients decreased HbA1c by 0.46%, increased C-peptide secretion, and decreased the proinsulin/insulin ratio and IL-6 levels, without an effect on body weight or hypoglycemia [182]. Blockade of the master pro-inflammatory signaling protein, IL-1, resulted in disease improvement for type 2 diabetes patients [183].

XOMA-52, is a clinical stage IL-1β antibody for patients with type-2 diabetes. XOMA-52 is unique to other IL-1 β inhibitors because of its high affinity and long half-life (22 days). As a result, patients need to be injected with XOMA-52 once monthly and should increase patient compliance. In addition to safety and pharmacokinetic data, the Phase I trial demonstrated decreased HbA1c levels and improved β–cell function [184]. However, XOMA recently announced that their XOMA-52 phase II trial for type-2 diabetes had missed its primary end point (which was better blood glucose control). XOMA 052 is currently in Phase II development for type 2 diabetes and type 1 diabetes and cardiovascular disease.

Therapies addressing inflammation as an underlying cause of this disease, instead of focusing almost exclusively on improving the body's ability to produce and process insulin can be a valuable tool for diabetes.

4.4 Halozyme

Co-administration of recombinant human hyaluronidase accelerated the pharmacokinetics and glucodynamics of insulin formulations [185]. For patients with type-1 diabetes, Halozyme (HALO) recently presented Phase II data that showed improved pharmacokinetics and glucodynamics with patients receiving Lilly's Humalog and co-administration of Halozyme's PH20 enzyme. PH20 is a recombinant hyaluronidase enzyme that catalyzes the hydrolysis of hyaluronic acid, a major constituent of the interstitial barrier, and increases tissue permeability. This study demonstrated an improved blood glucose metabolism profile in patients taking the combination therapy, reflecting a more physiologic glucose profile thereby likely reducing the amount of exogenous insulin needed along with some potential complications such as hypoglycemia [185, 186].

4.5 Appetite Control

Anti-obesity drugs facilitate weight loss and contribute to further amelioration of obesity-related health risks. Several pharmaceutical companies are interested in developing new targets and technologies [187]. VIVUS, Inc. (VVUS) presented data from a year-long Phase II trial with Qnexa demonstrating reduced HbA1c levels and helped patients achieve and maintain significant weight loss through appetite suppression. Qnexa is a combination therapy, of low dose phentermine and topiramate, which in combination have synergistic effects through unique molecular targets resulting in reduced appetite and increased satiety – the two main mechanisms that impact eating behaviour. Qnexa, an investigational drug, is being developed to

Table 5 Selective SGLT2 inhibitors currently in development

Development stage	Drug	Company
Phase III	Dapagliflozin	Bristol-Myers Squibb/AstraZeneca
	Canagliflozin	Johnson & Johnson/Mitsubishi Tanabe Pharma
	ASP-1941	Astellas/Kotobuki
Phase II	LX-4211	Lexicon Pharmaceuticals
	CSG-452	Roche
	BI-10773	Boehringer Ingelheim
	BI- 44847	Boehringer Ingelheim
Phase I	ISIS-SGLT2Rx	Isis Pharmaceuticals

Pipeline of selective SGLT2 inhibitors. SGLT-sodium-glucose co-transporter

SGLT Sodium-glucose co-transporter

address weight loss. In phase II and III clinical data to date, Qnexa has demonstrated significant weight loss, glycemic control, and improvement in cardiovascular risk factors.

Arena Pharmaceuticals, Inc. (ARNA) recently described Phase III data from its BLOOM (Behavioral modification and Lorcaserin for Overweight and Obesity Management), demonstrating significant weight loss and reduced secondary endpoints associated with cardiovascular disease after 1 year of treatment compared to the placebo. Lorcaserin is a novel serotonin 2C receptor agonist; selective activation of this Gq-coupled GPCR in the hypothalamus leads to appetite control and increased metabolism [188].

4.6 Sodium-Dependent Glucose Cotransporter-2 Inhibitors

The sodium-dependent glucose cotransporters (SGLTs) are a large family of membrane proteins involved in the transport of glucose, amino acids, vitamins, osmolytes and selected ions across the brush border membrane of the intestinal epithelium and the proximal renal tubules. There are two types of cotransporters: SGLT-1 and SGLT-2. SGLT2 is the primary transporter in the kidneys, and selective inhibition of SGLT2 has helped elucidate the functions of these glucose transport proteins. Inhibitors of SGLT-2 block the reabsorption of glucose (in the proximal tubule) from the renal filtrate, whereas the SGLT-1 inhibitors suppress absorption of glucose from the gut; these inhibitors help limit the amount of glucose that is reabsorbed and retained in the body [189–191]. The SGLT inhibitors do not interfere with glucose metabolism, and thus, they are being used complementarily to mainstream approaches to glucose regulation, such as with the PPAR-γ agonists, DPP-4 antagonists, and GLP-1 analogues. The majority of inhibitors being developed as antidiabetic agents are selective for SGLT- 2 (Table 5), but there are some inhibitors of SGLT-1 and a mixed type of inhibitors. Most of the SGLT-2 inhibitors, such as dapagliflozin (BMS-512148), remogliflozin, sergliflozin, AVE2268, 189075, 869682, and T-1095, are derived from the prototype phlorizin and structurally are glycosides (Table 5). The potential ability of SGLT2 inhibitors to reduce plasma glucose levels without inducing increased insulin secretion or sensitivity, hypoglycemia or weight gain could constitute a new approach in diabetes therapy [191]. They are also promising for other therapeutic uses such as in obesity because they cause net loss of calories from the body.

Inhibition of SGLT also has effects beyond glycemic control: exposure of tissues to chemical or physical stress induces SGLT1 expression, which might be mediated by the transforming growth factor (TGF)-β1 signaling pathway [192]. Chronic administration of the SGLT inhibitor T-1095 prevented the progression of diabetic nephropathy and suppressed the development of albuminuria and the expansion of the glomerular mesangial area [193]. Whether this occurs through a modulation of the TGF pathway is yet to be determined. It is also possible that such selective inhibition provides renoprotection through beneficial effects on glomerular haemodynamics, glycosuria, euglycemia and glucose toxicity of tubular cells [193]. The clinical importance of these findings is currently unclear. A recent study showed a reduction in infarct size and edema during induced ischemia in the middle cerebral artery in mice treated with an SGLT inhibitor [194], which suggests that SGLT inhibition can improve stroke outcome. SGLT receptors have also been found in human cardiac tissue [195], but the effects of SGLT2 inhibition on CVD are yet to be determined.

Dapagliflozin represents the first selective SGLT2 inhibitor that functions by regulating renal glucose reabsorption. In preclinical studies, dapagliflozin administration resulted in a decline in fasting and postprandial glucose levels, accompanied by glycosuria [196]. Such declines in fasting glucose levels were sustained in Zucker diabetic fatty rats after 15 days of repeated dosing of dapagliflozin. Similarly, fasting glucose was reduced in streptozocin induced diabetic rats following administration of a single dose of dapagliflozin [196]. Clinical trial data are limited, but available evidence supports clinically significant reductions in fasting and postprandial plasma glucose, hemoglobin A1c, and body weight with this agent [189]. In addition, dapagliflozin has demonstrated excellent tolerability with safety data demonstrated in both Phase I and Phase II studies. Results of ongoing phase III clinical trials are necessary to demonstrate efficacy and safety of this agent across various patient populations and clinical scenarios [197]. Dapagliflozin, and Canagliflozin are currently undergoing Phase III clinical trials (ADA 2010). Several double-blind clinical trials are under development examining canagliflozin. An example is the CANTATA-Su (canagliflozin treatment and trial analysis-sulfonylurea) trial, which is examining the efficacy and safety of canagliflozin compared with glimepiride in patients with type 2 diabetes, who are not optimally controlled on stable-dose metformin monotherapy [198]. Studies are also being conducted on canagliflozin as monotherapy in patients with type 2 diabetes who are inadequately controlled with diet and exercise [199], not controlled with metformin and sulphonylurea [200], and canagliflozin compared with placebo and an active-control (sitagliptin) in patients with type 2 diabetes not controlled with metformin [201].

Isis Pharmaceuticals, Inc. (ISIS) recently presented preclinical data on ISIS-SGLT2Rx [202], an antisense drug that inhibits the production of SGLT2 that resulted in a significant reduction in blood glucose levels in multiple animal species as well as amelioration of diabetic complications such as cataract formation. ISIS-SGLT2Rx is the first kidney-targeted antisense drug for type 2 diabetes and Phase I study has been initiated. Isis is experienced in RNA targeting drugs and has already commercialized the world's first antisense drug. This novel technology allows Isis to target dysfunctional proteins through decreased transcriptional levels thus modifying signaling pathways not attainable through small molecule inhibition. Antisense technologies may play a key role in finding drugs with unique targets previously unreachable ultimately leading to improved disease management and quality of life.

Competitive inhibition of SGLT2 therefore represents an innovative therapeutic strategy for the treatment of hyperglycemia and/or obesity in patients with type 1 or type 2 diabetes by enhancing glucose and energy loss through the urine [191]. As SGLT2 inhibitors do not target the major pathophysiological defects in type 2 diabetes — namely insulin resistance and impaired insulin secretion — they represent a potentially promising new therapy for improving glycemic control in type 2 diabetes.

4.7 Regulation of Endogenous Gene Expression Using Small Molecule-Controlled Engineered Zinc-Finger Protein Transcription Factors

Peripheral neuropathy is a common, irreversible complication of diabetes. Diabetic neuropathy (DN) is a severe physiological consequence of chronic elevated blood glucose levels and is seen in many patients with advanced diabetes. Currently, the question being addressed is whether or not gene transfer of an engineered zinc finger protein transcription factor (ZFP) designed to upregulate expression of the endogenous VEGF-A gene can protect against experimental DN [203–205]. VEGF has protective and regenerative effects on neuronal cells in culture and in animal studies. No regenerative therapies are currently available. Data from an earlier Phase I, placebo-controlled study using VEGF Zinc Finger Protein Activator (SB-509) in patients with diabetic peripheral neuropathy showed statistically significant improvements in neurological examination and quantitative sensory testing as well as clinically relevant changes in nerve conduction velocity [204].

Thus, Sangamo BioSciences, Inc. (SGMO) recently described the results of phase II trials for SB-509 as a treatment for DN resulting in statistically significant and clinically relevant improvements in subjects [205]. SB-509 is an injectable plasmid encoding a DNA- ZFP transcription factor designed to upregulate the endogenous expression of the gene encoding VEGF, a peptide responsible for angiogenesis.

4.8 11β-Hydroxysteroid Dehydrogenase Type 1 Inhibitor

11β-hydroxysteroid dehydrogenase type 1 (11β-HSD1) is an enzyme that converts inactive cortisone into the potent biologically active hormone cortisol. This conversion occurs within cells of key metabolic tissues including liver, adipose, muscle and pancreas. Chronically elevated glucocorticoid levels cause obesity, diabetes, heart disease, mood disorders, and memory impairments. Excess 11β-HSD1 or cortisol leads to insulin resistance and metabolic syndrome in animal models and in humans. Inhibiting 11β-HSD1 activity signifies a promising therapeutic strategy in the treatment of type 2 diabetes, insulin resistance, obesity, lipid disorders, metabolic syndrome, and other diseases and conditions that are mediated by excessive glucocorticoid action [206]. Cortisol elevates blood glucose levels by driving glucose production in the liver, and inhibiting the uptake and disposal of glucose in muscle and adipose. Thus, cortisol acts as an antagonist of insulin action, and 11β-HSD1 mediated production of cortisol has been hypothesized to contribute to human insulin resistance, type 2 diabetes, and the often-associated cardiovascular comorbidities.

Results from this double-blind, placebo-controlled Phase IIb trial involving over 300 patients with type 2 diabetes showed that treatment with once-daily doses of INCB13739 significantly improved glycemic control, as measured by HbA1c, insulin sensitivity and total-cholesterol levels in patients with type 2 diabetes inadequately controlled with metformin [207].

Selectively inhibiting 11β-HSD1 in humans might become a new and promising approach for lowering blood glucose concentrations and have a favorable effect on insulin resistance, dyslipidemia, and obesity in type 2 diabetes. The modulation of 11β-HSD type 1 activity with selective inhibitors has beneficial effects on various conditions including insulin resistance, dyslipidemia, and obesity. These compounds improve insulin sensitivity in the liver and adipose tissue and possibly may cause weight reduction, as shown in animal studies. A large variety of 11β-HSD1 inhibitor classes have been synthesized [65, 208], and some are under investigation by the pharmaceutical industry to treat type 2 diabetes and obesity [145]. A number of 11β-HSD1 inhibitors (carbenoxolone, AZD4017, BVT116429, PF-00915275, PF-915275, PF-877423) are being investigated by many major pharmaceutical companies for treatment of type 2 diabetes and other abnormalities associated with the metabolic syndrome.

4.9 FGF-21 for the Treatment of Type 2 Diabetes

Fibroblast growth factor 21 (FGF-21) is a recently discovered metabolic regulator. FGF-21 activity depends on β-klotho, a single-pass transmembrane protein. β-klotho physically interacts with FGF receptors 1c and 4. FGF-21 is produced predominantly by the liver and, to a lesser extent, also by the adipose tissue [146–150]. FGF-21 exerts a unique role in the regulation of carbohydrate and lipid metabolism in the liver, adipose tissue and pancreas [209]. FGF-21 stimulates glucose uptake in adipocytes in an insulin-independent fashion [147]. FGF21 injection in ob/ob mice results in an increase in Glut1 in adipose tissue. Another beneficial effect of FGF-21 is its ability to reduce the secretion of glucagon from the pancreas, and to lower blood glucose and triglyceride levels when administrated to diabetic rodents and monkeys. Its administration in diabetic primates led to a sustainable improvement in glucose control without occurrence of hypoglycemia, a significant improvement of diabetic dyslipidemia and a mild weight loss without any significant side effects [148]. FGF-21 preserves pancreatic insulin content and β-cell mass when administered to diabetic mice. Moreover, FGF21 protects animals from diet-induced obesity when overexpressed in transgenic mice The function of FGF-21 in lipid metabolism and energy balance was recently emphasized by Potthoff and coworkers. They described the effect of FGF-21 on fatty acid oxidation and on PPARγ coactivator protein-1α, a key regulator of energy homeostasis [210].

FGF21 concentrations are higher in both rodent and human obesity, and recent data suggest that obesity may be an FGF21-resistant state. Free fatty acids (FFAs) were described as physiologic stimulators of FGF-21 secretion in humans, an effect which was also mediated via PPARα, whereas rosiglitazone, a PPARγ agonist, did not modify FGF-21 in humans [211]. In humans, FGF-21 levels are associated with markers of lipid metabolism and abdominal adiposity. However, FGF-21 is not directly regulated by moderate change in fat mass and body weight; and baseline levels of FGF-21 do not predict subsequent weight loss.

Pre-clinical data support a role of FGF-21 in energy balance [210], although the precise mechanism and a potential role in humans are yet unclear. This suggests that FGF-21 might offer a promising new therapeutic approach to treat type 2 diabetes and clearly more studies need to be done in order to evaluate the role of FGF-21 in humans.

New compounds designed to improve therapeutic properties over the native FGF-21 protein are under development by Ambrx and Merck and by Eli Lilly and Company, Inc. (LP10152 is in Phase I/II).

4.10 AMPK Activation

Understanding the molecules involved in energy sensing and energy balance may lead to specific therapies for obesity and diabetes and for their complications.

AMP-activated protein kinase (AMPK) is a serine/threonine protein kinase that plays a critical role in the regulation of metabolism given its fundamental role as a sensor of cellular energy levels that acts on diverse signaling pathways. The inhibitory effect of AMPK activation on hepatic gluconeogenesis and its stimulatory effect on peripheral glucose uptake make it a prime therapeutic target in the treatment of metabolic disorders such as type 2 diabetes and obesity. This is further bolstered by its ability to bypass the defects in insulin signaling that are observed in these diseases. Most of the current AMPK activators act via an indirect mechanism, inhibiting mitochondrial function and altering cellular

AMP levels. Several antihyperglycemic agents including the biguanide metformin, and thiazolidenediones such as rosiglitazone and pioglitazone, exert beneficial effects on glucose homeostasis in type 2 diabetes via a mechanism that involves activation of AMPK [212].

AMPK is also the target of insulin-sensitizing adipokines such as adiponectin. Stimulation of AMPK by adiponectin reduces insulin resistance and data from the Hoorn Study show that high adiponectin levels are associated with a reduced risk of impaired glucose tolerance and type 2 diabetes [213, 214]. Injection of adiponectin increased blood glucose uptake in normal mice and prevented insulin-resistance in obese mice. Expression of globular adiponectin in transgenic ob/ob mice showed decreased insulin resistance and diabetes compared with transgenic ob/ob mice not expressing globular adiponectin [215]. Regulation of adiponectin plays a pivotal role in obesity and diabetes type 2 by affects on glucose uptake and insulin sensitivity. Thus, pharmacologic agents that could directly increase adiponectin's circulating concentrations (e.g. PPARc agonists [216–218]) and/or compounds that could directly act on specific tissue targets, such as adiponectin receptor agonists that induce signaling pathways downstream of adiponectin receptors, would be an important and valuable tool in diabetes [219]. Thus, adipokines also offer new targets for therapy [for review see 220].

Moreover, direct and specific AMPK activators such as A-769662 and PT1 indicate that it is possible to develop new candidates, and suggest that AMPK is a potential drug target useful in diabetes.

4.10.1 Direct Activators of AMPK: A-769662 and PT1

In 2006, Abbott Laboratories identified the first compound described as direct activator of AMPK, thienopyridone A-769662 [221]. Similar to AMP, A-769662 has two effects in cell free assays: it activates AMPK via an allosteric mechanism and by inhibiting dephosphorylation at Thr172 [222, 223]. Although the exact binding site remains unknown, it is likely to involve the β subunit. Because it is not an AMP mimetic, A-769662 lacks the off-target effects of 5-aminoimidazole-4-carboxamide-1-β-riboside (AICAR) on fructose- 1,6-bisphosphatase and glycogen phosphorylase [221]. In cultured cells, A-769662 activates AMPK in the absence of the upstream kinase LKB1 [222, 223], which may indicate that it is acting primarily via the allosteric mechanism. This is supported by observations that increased phosphorylation of the AMPK substrate, ACC, is not always associated with a substantial increase in Thr172 phosphorylation [222]. Importantly, treatment of ob/ob mice with A-769662 decreased plasma glucose and triglyceride levels, and reduced hepatic gluconeogenic and lipogenic gene expression [221]. Although it has beneficial effects on metabolic parameters in vivo, the promise of A-769662 as a useful compound for drug development is counteracted by its poor oral absorption [221] and some recently reported AMPK-independent effects [224].

Pang and colleagues have recently identified PT1 as a novel direct activator of AMPK [225]. PT1 appears to increase AMPK activity by interacting with the auto-inhibitory domain found in the α subunits, relieving inhibition of kinase activity by this domain. Treatment of cultured cells with PT1 activates AMPK via increased phosphorylation of Thr172 without altering the intracellular AMP:ATP ratio, and can occur in LKB1-deficient cells, when it requires CaMKKβ. In liver cell lines, the PT1 dependent activation of AMPK is associated with decreased lipid content. These results suggest that antagonizing inhibition of AMPK by the auto-inhibitory domain may be a valid strategy for treatment of metabolic disorders, although the effects of PT1 in vivo remain to be investigated.

SIRT1 Activators

SIRT1, a member of the sirtuin family of NAD+-dependent protein deacetylases, has been considered an important mediator of longevity that is induced by caloric restriction [226]. Resveratrol, a natural compound found in the skin of red grapes, was recently shown to activate SIRT1. Subsequently, several small molecule activators of SIRT1 have been characterized [227].

In animals, SIRT1 activators improve whole-body glucose homeostasis and insulin sensitivity in adipose tissue, skeletal muscle and liver [228].

SIRT1 has also been linked to aging in the vascular system [229]. Overexpression of SIRT1 prevented the apoptosis of cultured human oxidized LDL, and this effect was mediated by a marked increase in the expression of endothelial nitric oxide synthase (eNOS) [230]. In addition, resveratrol treatment of spontaneously hypertensive mice has been shown to improve endothelium-dependent vascular relaxation and increase the expression of eNOS [231, 232]. Altogether, these data suggest that SIRT1 activators have the dual effect of lowering blood glucose and decreasing cardiovascular risk, and further investigation is essential.

Thus, several pharmaceutical companies include in their development pipelines sirtuin modulators. Namely, Sirtris Pharmaceuticals (Cambridge, MA, a wholly-owned subsidiary of GlaxoSmithKline) is developing the SIRT1 activators SRT501 (a proprietary formulation of the natural product resveratrol) and SRT2104 (a novel synthetic small-molecule SIRT1 activator that is structurally unrelated to resveratrol and is up to 1,000-fold more potent). SRT501 is in Phase II clinical trials in type 2 diabetes. SRT2104 has been tested in Phase I trials in healthy volunteers, and was found to be safe and well tolerated. Elixir Pharmaceuticals (Cambridge, MA) is developing a preclinical-stage SIRT1 activator for treatment of type 2 diabetes and obesity. Elixir also has a research-stage SIRT2 inhibitor under development for treatment of type 2 diabetes and obesity.

4.11 Antidiabetic Effects of Glucokinase Activators

Glucokinase (GK) plays a fundamental role in glucose homeostasis by controlling the rate of glucose metabolism in many tissues. GK is believed to function as the primary glucose sensor in a range of neuro/endocrine sentinel cells (including β-cells) that are involved in glucose homeostasis. In pancreatic β-cells, GK acts as a sensor for glucose stimulated insulin release and in hepatocytes it catalyzes the first step of glucose metabolism. The relationship between changes in GK activity and fasting plasma glucose in humans with GK diseases has been well established for both loss and gain of function mutations [233].

Such strong biological rationale for targeting GK as a potential anti diabetic therapy led to the discovery of small molecule allosteric GK activators (GKAs). GKAs bind to an allosteric site 2 nm away from the bound glucose site and act as non-essential mixed-type enzyme activators [234].

Animal studies have demonstrated that when the glucokinase gene is disrupted, mice die within days of birth from severe diabetes. Heterozygous loss-of-function mutations in the glucokinase gene can result in maturity onset diabetes of the young – type 2, which is linked to defective insulin secretion. The over-expression of glucokinase results in decreased basal blood glucose levels, as well as resistance to developing type 2 diabetes. Evidence suggests that the glucose lowering effects in rodents are mediated by dual effects on increasing plasma insulin levels and suppression of hepatic glucose production as assessed by a pancreatic clamp. Early data in humans suggests that GKAs offer a promising new therapeutic approach to treat type 2 diabetes [235].

Ro-28-1675 was the first GKA noted to reduce blood glucose after an oral glucose tolerance test in several models of type 2 diabetes in rodents. This compound also lowered basal glucose in diet-induced obese mice accompanying with a rise in insulin levels. Ro-28-1675 attenuated endogenous glucose production and elevated hepatic glucose disposal [236].

Several GKAs are presently at various stages of clinical development [235]. Roche was the first pharmaceutical company to bring a GKAs into clinical trials. However, Roche's GKAs were discontinued from further development; the reasons have not been disclosed.

AstraZeneca, OSI/Prosidion in conjunction with Eli Lilly, and Merck have been in and are continuing phase II investigations for PSN010. AZD6370 is presently undergoing phase II trials and is being developed for the potential treatment of obesity and type 2 diabetes [237]. Other compounds are presently in phase I and II clinical trials (Table 6), based on their strong *in vitro* and *in vivo* preclinical data [238]. For instance, ID1101 enhanced glucose-stimulated insulin release from pancreatic islet cells and glucose disposition by the liver, thereby reducing blood glucose and body weight.

Preclinical results of compound NN9101 showed good glucose regulation, no weight gain and increased insulin content in pancreas. Its glucose-lowering effect was also observed in first human data, including reduction in both fasting and postprandial glucose. TTP399, which is also in preclinical development, belongs to a chemical class different from that of NN9101. A potent GK activator [239], ARRY403 showed significant blunting of the postprandial glucose excursion in acute OGTT dose–response study in C57BL/6J normal mice. It also reduced both nonfasted and postprandial glucose in *ob*/*ob* mice, in ZDF female rats and NONcNZO10/LtJ mice, without causing hypoglycemia or abnormal weight gain.

Table 6 Agonists of GKAs under development

Development stage	Drug	Company
Phase II	AZD1656	AstraZeneca
Phase II	AZD6370	AstraZeneca
Phase I	TTP355 (NN9101)	Novo Nordisk AS, Transtech Pharma
Phase II	LY2599506 (PSN010)	Eli Lilly and Company, Inc., OSI Pharmaceuticals, Inc.
Discontinued in phase II	R1440	Roche
Phase I	R1511	Roche
Discontinued	RO-28-1675	Roche
Phase I	TTP355	Transtech Pharma
Preclinical	TTP399	Transtech Pharma
Phase I	ARRY-403	Array BioPharma Inc.

It is now ready to enter human clinical trials. Another potent activator of human GK *in vitro*, ARRY588 presented glucose-lowering activity in a dose–response OGTT study in C57Bl/6J mice [239] and in ob/ob mice [240].

Although the preclinical and early phase 1 studies suggest promise for these agents, further studies will be needed to elucidate efficacy, safety and tolerability in large numbers of patients in clinical trials.

4.12 Gluconeogenesis Inhibitors

Fasting hyperglycemia in diabetes is well correlated with accelerated hepatic glucose production in type 2 diabetes patients; free fatty acids also induce gluconeogenesis [241]. In animal studies, it has been shown that glucose production in the liver can be inhibited both by substances that act directly on gluconeogenesis and by substances that inhibit fatty-acid oxidation. Therefore, inhibition of both free fatty acid release (such as by SDZ WAG 994) and fatty-acid oxidation (such as by β-aminobetaine [emeriamine]) in the liver may be efficient modalities of treatment to prevent fasting hyperglycemia in diabetics [242]. Benfluorex, a derivative of fenfluramine, lowers plasma glucose by inhibition of gluconeogenesis [242]. Inhibition of fructose 1, 6 bisphosphatase, the key hepatic enzyme in gluconeogenesis is also a therapeutic target for treatment of type 2 diabetes [243]. Specific inhibitors of this enzyme (such as [MB06322 (CS-017) discontinued in phase II] and MB07803 in Phase IIa of development by Metabasis Therapeutics) may be useful in type 2 diabetes [244].

4.13 Oral Insulin

The prospect of a non-injectable means of insulin delivery has always been very attractive. There have been a number of pharmaceutical approaches to developing a non-invasive delivery route for insulin, most of them topical (e.g. patches) or inhalation [245]. These share the same weakness as injectable insulin: the delivered protein passes first into the peripheral circulation and only thereafter, in much diluted form, into the liver. Thus, topically and subcutaneously administered insulin result in fat and muscle being exposed to higher insulin levels than the liver. This is in contrast to endogeneous insulin which is delivered from the pancreas into the portal system and the liver. Oral delivery would share this advantage.

The potential market for an oral form of insulin is assumed to be enormous, thus many laboratories have attempted to devise ways of moving enough intact insulin from the gut to the portal vein to have a measurable effect on blood sugar.

The various challenges associated with the production of a reliable insulin formulation for oral delivery includes (1) maintaining storage stability of the active protein, (2) avoiding enzymatic degradation in the gastrointestinal tract and (3) overcoming poor spontaneous insulin permeability through the gastrointestinal system. Despite the multiple technical challenges many biopharmaceutical companies (Apollo Life Sciences, Bicon, Biodel Inc, Oramed Pharmaceuticals Inc) are currently developing their oral insulin products.

Biodel, Inc. is developing an oral formulation of insulin (VIAtab) designed to be administered sublingually. In a Phase I study, VIAtab delivered insulin to the blood stream quickly and resembled the first-phase insulin release spike found in healthy individuals. The company claims that an oral insulin therapy would be more convenient than currently available injectable or inhalable therapies, and they expect that convenience to result in increased insulin usage among the currently underserved early-stage patients with type 2 diabetes, thus helping to create better long-term outcomes for that patient population.

Oramed Pharmaceuticals, Inc. is currently conducting Phase IIb clinical trials of its oral insulin capsule, ORMD-0801 on 30 patients diagnosed with type 2 diabetes [246]. Oramed's platform technology has two components: (1) A chemical make-up that protects insulin during passage through the gastrointestinal tract, and (2) Absorption enhancers so that insulin could be absorbed by the intestine. Oramed Pharmaceuticals, Inc. in phase I clinical trials, has demonstrated that its oral insulin is safe, well tolerated, and has consistently reduced glucose and c-peptide levels in patients [247].

An alkylated, PEGylated, amphiphilic insulin conjugate (HIM-2, Biocon Corp., India) increased oral formulation of insulin in dogs and has reached Phase II clinical studies [248]. One of the major challenges in biopharmaceutical development, therefore, continues to be the need for effective oral delivery systems and a number of different approaches have already been used to promote oral delivery of insulin [249].

4.14 Insulin Mimetics

The concept of developing insulin mimetics is a long established one, but recent years have seen important developments. For example, protein tyrosine phosphatases (PTPs) are enzymes that control signal transduction pathways, and inhibition of individual PTPs can result in activation of therapeutically relevant kinase cascades. Tyrosine phosphatase 1B (PTP1B) is implicated as a negative regulator of insulin receptor signaling, so small-molecule inhibitors of this enzyme are being investigated as potential insulin mimetics or insulin augmenting agents. Inhibition of PTP1B in muscle and liver, down-regulates insulin signaling, thereby improving insulin sensitivity [250]. These compounds have a low potential for hypoglycemia. A number of compounds synthesized [251] as well as isolated from various plants have been found to be potent inhibitors of protein tyrosine phosphatase 1B.

During the last decades, a large body of evidence has accumulated to suggest that organo-vanadium compounds exert various insulin-like effects in both in vitro and in vivo systems. These include their ability to improve glucose homeostasis and insulin resistance in animal models of type 1 and type 2 diabetes mellitus. In addition to animal studies, several reports have documented improvements in liver and muscle insulin sensitivity in a limited number of patients with type 2 diabetes. [252, 253]. Among these compounds, vanadium (IV) oxo bis(maltolato) was the first to be investigated for its higher potency over inorganic vanadium salts in eliciting insulin-like properties in both in vitro and in vivo systems [254]. There are some concerns about the potential toxicity of available inorganic vanadium salts at higher doses and during long-term therapy. Therefore, new organo-vanadium compounds with higher potency and less toxicity need to be evaluated for their efficacy as potential treatment of human diabetes.

5 Gene-Based Therapies for the Treatment of Diabetes

Gene-based therapy is an attractive therapeutic strategy that has emerged as an alternative to the traditional treatments of diabetes (e.g. injections of insulin and islet transplantation). This therapeutic approach has the potential to treat, cure and prevent diseases and, despite the limited boundaries to its application, is frequently highlighted as one of the most promising technologies of this century. The aim of gene therapy is the transfer of genetic material into diseased cells or tissues, allowing for the healthy phenotype to be recovered [255]. To this end, three basic elements are required: a profound knowledge of the disease pathogenesis; the genetic sequence of interest that might function as a target; and stable delivery systems providing the targeted tissues with the level of duration of the gene expression necessary to reproduce the desirable therapeutic effect [256].

Type 1 and type 2 diabetes are two examples of very common polygenic and multifactorial diseases that, along with cancer, neurologic and other metabolic diseases, are ideal targets for gene therapy application, since they provide targets that include suppression of autoimmunity, restoration of insulin responsiveness, functional replacement of pancreatic islets and correction of vascular and nerve damage associated with prolonged hyperglycemia. In fact, the risk-benefit balance of gene therapy in diabetes seems to be superior when compared to all currently used treatments.

The first steps towards the application of this innovative therapy in type 1 diabetes were made soon after the cloning of the insulin gene, in the late 1970s. Since then, a plethora of non-insulin producing cells that are not targeted by the immune system for autoimmune destruction (e.g. gut K cells, hepatocytes, myocytes, muscle cells, fibroblasts, keratinocytes, pituitary and exocrine cells) have been used to transform into artificial insulin-producing substitute cells. As true β-cell surrogates, they should be able to synthesize, package and store insulin, according to the metabolic needs, as well as to precisely respond to the blood glucose levels and avoid lag time in the secretory response. To achieve this, the pro-insulin gene should be engineered in order to incorporate a specific mutated amino acid sequence recognized by the furin protease to produce mature insulin, since the β-specific endoprotease involved in the cleavage and maturation of insulin in β-cells is absent in non-β-cells [257–259]. A construct encoding this engineered cDNA can be expressed after gene-delivery using cell-specific promoters, viral vectors or nanoparticles [258, 259]. The use of synthetic promoters to regulate glucose-responsive insulin production may confer advantages to these therapies since they can partially override the unresponsiveness of these cells to glucose with physiological secretion of insulin. Indeed, a synthetic promoter library containing hepatocyte nuclear factor-1 (HNF-1), a CAAT/enhancer-binding protein (C/EBP) response element and a glucose-response element was screened and it was found that the promoter SP23137 is highly effective on the production and secretion of insulin in response to glucose [260]. Still, this therapeutic approach faces other limitations. For instance, the fold-induction of insulin release achieved by these constructs is smaller than that of the islets and insufficient to normalize glucose after tolerance tests. Thus, one can question the capability of gene-therapy-constructed nonpancreatic tissues to keep a stringent control over blood glucose concentration similar to the one provided by functional β-cells.

In opposition to ectopic gene expression of insulin, transplantation of pancreatic endocrine lineages reprogrammed from other cell types for β-cell replacement is also a possible and promising option. For example, it was shown that ectopic expression of human neurogenin 3 (NGN3) and pancreatic duodenal homeobox-1 (PDX1) in bone marrow-derived mesenchymal stromal cells triggers pancreatic endocrine gene expression, notably insulin, glucose transporter 2 (Glut-2) and somatostatin, as well as other β-cell lineage-associated genes [261]. The systemic delivery of a PDX1 construct to the livers of streptozotocin (STZ)-treated mice, using a first generation adenovirus therapy vector, was also shown to induce the activation of the endogenous pro-insulin 2 gene in the liver. This resulted in increased levels of serum insulin, glucagon and somatostatin and decreased levels of blood glucose that, despite the transience of the adenoviral constructs, lasted for a period of 4–6 months [72]. The *ex vivo* transduction of adult human hepatocytes with PDX1 also resulted in the induction of several β-cell gene products, and, upon transplantation under the renal capsule of diabetic mice, these cells did ameliorate hyperglycemia for prolonged periods of time [262]. Of noteworthy is the fact that transplantation of STZ-treated mice pancreas with PDX1-transduced adipose tissue-derived stem cells (ASCs) significantly decreases blood glucose levels and increases cell survival. Moreover, PDX1-ASCs were shown to be stably engrafted in the pancreas, to acquire a functional β-cell phenotype and partially restore pancreatic function *in vivo* [263]. More recently, the transcription factor NKX6 was also found to activate immature pancreatic markers, such as NGN3 and insulin gene enhancer protein 1 (ISL1), in human liver cells, and NKX6 in complementation with PDX1 was found to promote insulin expression and glucose-regulated processed hormone secretion in hepatic cells [264]. Another relevant study showed that NGN3/Betacellulin (BTC) gene delivery and overexpression in the pancreas of STZ-diabetic mice is also able to restore glucose-stimulated insulin secretion and reverse hyperglycemia in these animals, rendering the neo-islets with a transcription profile similar to that of pancreatic islets [265]. The treatment further normalized the hepatic glucose secretion, reversed ketonemia and restored the hepatic glycogen content similar to levels observed in nondiabetics.

Type 1 diabetes is characterized by infiltration of autoreactive T-cells that mediate the destruction of pancreatic β-cells and, as such, several therapeutic approaches aim at controlling autoimmunity in this disease. One of these strategies consists of inducing and/or expanding regulatory T cells (Treg), more specifically of the Foxp3-$CD4^+$-expressing Treg ($Foxp3^+$Treg) population, which are potent immunoregulatory effectors that modulate T cell and antigen-presenting cell (APC) responses [266]. Deregulation of $Foxp3^+$Treg has indeed been reported in diabetic patients [267] and described to result in systemic T cell-mediated autoimmunity in mice [268]. Administration of cytokines, such as IL-2 and TGF-β1 has been used to manipulate T cell-mediated autoimmunity since they play critical roles in the maintenance of and/or differentiation of $Foxp3^+$Treg. This approach offers several limitations, including the half-life of these proteins in the serum and, as such, the need for repeated administrations. Thus, an alternative approach to this cytokine-based immunotherapy is an inducible gene therapy strategy, which allows for the expression of the transgene *in vivo* and obviates the need for production, purification, and storage of recombinant proteins. Indeed, it has been shown that the intramuscular injection of adeno-associated virus (AAV) encoding murine IL-2 and TGF-β, followed by a short pulse of IL-2, is sufficient to prevent β-cell autoimmunity and diabetes in non-obese diabetic mice by increasing the frequency of $Foxp3^+$Treg without nonspecifically activating and/or expanding T cells and other effectors cells [269]. Another study demonstrated that the transfer of the gene encoding the anti-inflammatory cytokine IL-4, using AAV vectors, directly to the pancreatic islets of non-obese diabetic mice prevents islet destruction by blocking autoimmunity, mainly through the maintenance of $Foxp3^+$Treg [270]. Despite the benefits of the systemic expression of cytokines and other molecules by gene transfer to different tissues, the concern on the possible adverse effects resulting from this systemic expression on normal immune function remains unclear. These findings, however, suggest that gene therapy employing vectors that encode cytokines and other immunomodulatory proteins may also prove to be an effective and safe strategy to manipulate type 1 diabetes.

Other promising targets of gene-based therapy of diabetes are glucagon-like peptide-1 (GLP1) and angiotensin I-converting enzyme type 2 (ACE2). GLP1, an incretin hormone that regulates postprandial blood glucose levels, has been suggested for the treatment of type 2 diabetes given its insulinotropic action. However, this protein exhibits an extremely short half-life due to its rapid degradation by the dipeptidyl peptidase IV protease. Gene delivery of GLP1 is a way to overcome this problem and effective experimental therapeutics were indeed achieved. For instance, delivery of GLP1 cDNA, using chitosan nanocomplexes, driven by a cytomegalovirus promoter/enhancer, in Zucker diabetic fatty animals resulted in increased levels of plasma insulin and efficient clearance of the blood glucose in glucose tolerance tests, when compared to untreated animals [271]. Moreover, *in vivo* expression of GLP1 mediated by dsAAV vectors induces significant pancreatic β-cell proliferation, enhances β-cell mass and insulin secretion, and significantly delays the onset of hyperglycemia in db/db mice [272].

Gene therapy using ACE2, which is thought to act as a negative regulator for the renin-angiotensin system that also exists in pancreatic islets, was also shown to be effective on glycemic control in diabetic mice. In 8-week-old db/db mice, overexpression of ACE2 in pancreas, through the direct injection of adenoviral vectors encoding ACE2 gene, showed improved fasting glycemia, enhanced intraperitoneal glucose tolerance, increased islet insulin content and β-cell proliferation, and reduced β-cell apoptosis, as compared to the control. These evidences point to ACE2 as a novel target for the prevention of β-cell dysfunction and apoptosis occurring in type 2 diabetes [273].

Improvement of health condition and quality life of diabetic patients may also pass by the direct treatment of diabetic complications, such as neuropathy, renal damage, retinopathy, impaired wound healing and cardiovascular disease. One of the best examples for a possible gene therapy approach is proliferative diabetic retinopathy (DR). This pathology is the leading cause of blindness worldwide and is characterized by increased vascular permeability and retinal neovascularization [274–276]. The eye constitutes a privileged organ to assess the effectiveness of anti-angiogenic therapies, since it is non-invasively available for imaging techniques that allow the study of progression of neovascularization. Vascular endothelial growth factor (VEGF) is considered a central player in the regulation of angiogenis, and, therefore, has been proposed as the therapeutic target for treatment of late stages of DR characterized by neovascularization. However, currently used therapeutic approaches targeting VEGF in DR carry several drawbacks. Some of the approaches, developed and clinically tested to reduce neovascularization in DR, include intravitreal injections of antibodies that bind to and block VEGF action and its receptors, such as Ranibizumab (*Lucentis*) and Bevacizumab (*Avastin*). However, intravitreal administration is associated with increased intraocular hypertension and inflammation, whereas intravenous injection was observed to have significant risks of thromboembolic events. Another disadvantage of these therapeutic approaches is the requirement of frequent and costly injections. Therefore, alternative therapies that silence VEGF, for example by the use of vectors carrying specific interference RNAs against VEGF, are considered as a new generation of more effective strategies [256, 277].

Another good example for the promising use of gene therapy is diabetic critical limb ischemia, which is one of the leading causes of disability among diabetic patients and is the most common cause of non-traumatic amputation in diabetes.

It is a disease manifested by diminished blood flow to the legs and appears to be associated, at least in part, with loss of cell and tissue response to ischemia, leading to impairment of the wound healing and angiogenic responses. A recent study using a diabetic model of critical limb ischemia have shown that intramuscular injection of an adenovirus encoding a constitutively active form of HIF-1α (the key regulator of cell response to low oxygen) into the ischemic limb of diabetic mice increases the recovery of limb perfusion and function, reduces tissue necrosis, rescues the diabetes associated impairment of circulating angiogenic cells, enhances endothelial nitric oxide synthase activation, and increases vessel density and luminal area in the ischemic limb [278].

In agreement with the previous study, several reports have been suggesting that HIF-1 deregulation, induced by high glucose, is directly associated with the loss of cell adaptation to low oxygen in diabetes [279–282], having serious consequences in various organs and tissues, including myocardial collateralization, wound healing, renal, neural and retinal functions. Of significance, some groups have been showing that impairment of the HIF-1 pathway under high glucose conditions may result from the increased availability of methylglyoxal [283, 284], which is a highly reactive by-product of glycolysis. Thus, a therapeutic approach based on the gene delivery and overexpression of glyoxalase 1 (GLO1), the rate-limiting enzyme in the detoxification of methylglyoxal, to diabetic injured tissues, would ameliorate the cell response to hypoxia in diabetic tissues. Indeed, GLO1 overexpression is able to stabilize HIF-1α [284], as well as to increase HIF-1 transactivation in response to low oxygen and high glucose conditions, contributing to increase the levels of VEGF, CXCR4 and stromal cell-derived factor-1 (SDF-1) [283], which are known to increase mobilization of endothelial progenitor cells, improve neovascularization and tissue repair.

Other recent studies also showed that local sustained release of VEGF, using adenovirus vector-mediated gene transfer, accelerates experimental diabetic wound healing, by stimulating angiogenesis, epithelialization and collagen deposition [285, 286]. Administration of adenoviral VEGF together with Angiopoietin-1 (ANG-1) has also been shown to be an effective and a promising therapeutic approach in the treatment of diabetic myocardial injury after an acute hypoxic challenge, such as a heart attack. Preclinical data demonstrates the efficacy of co-administration of adenoviral VEGF and ANG-1 in increasing angiogenesis, increasing heart collateral development and reducing ventricular remodeling in the infarcted diabetic myocardium [287].

Since obesity is associated with an increased risk of developing insulin resistance and type 2 diabetes, there has been great interest in the development of gene therapies for the control of obesity. Leptin gene therapy is one of the most exciting and promising approaches in this field. This has been achieved by two different systems based on the transfer of the leptin gene to muscles using viral vectors or enhanced electroporation [288]. Intracerebroventricular injection of recombinant AAV encoding leptin in rats consuming a high-fat diet (HFD) have shown that leptin treatment reduces food intake and blocks the HFD-induced increase in weight, adiposity and metabolic variables. Blood glucose was shown to be slightly reduced but within the normal range. Treatment with leptin significantly augmented uncoupling protein-1 (UCP1) mRNA expression in brown adipose tissue, indicating increased thermogenic energy expenditure [289]. Recently, it was also found that intracerebroventricular and hypothalamic injection of recombinant AAV vectors encoding the leptin gene abolishes chronic hyperglycemia and hyperinsulinemia in animal models of diabetes [290]. Gene therapy approaches to enhance leptin supply locally in the hypothalamus, although unconventional, may offer an efficient anti-diabetic treatment able to prevent and mitigate both type 1 and 2 diabetes [291]. Another example of a molecular therapy for obesity-induced diabetes relies on the hepatic gene transfer of CPT1A, the key enzyme in fatty-acid β-oxidation, or its constitutive active mutant form CPT1AM, to high-fat diet-treated and genetically obese mice using AAV vectors. CPT1A and, to a greater extent, CPT1AM, enhance hepatic fatty-acid oxidation, reducing the liver triacylglyceride content in these animals. Obesity-induced weight gain, increase in fasting blood glucose and insulin levels, and augmented expression of gluconeogenic genes were also restored to normal only 3 months after AAV treatment, suggesting a protective action of both CPT1 forms against diabetes and obesity-induced insulin resistance [292].

The goal of the most current therapies for diabetes is to provide a tight glycemic control in all patients in order to minimize the significant complications associated with hyperglycemia. However, glucose management remains a difficult problem. Thus, the interest on gene therapy approaches that might replace the ongoing therapies and avoid the development of diabetic complications is increasing. The preclinical gene therapy studies mainly aim at developing strategies for insulin replacement and suppression of molecular pathways underlying the loss of β-cell mass and hyperglycemic complications. Given the efficacy of these approaches in animal models of diabetes, the possibility of bringing them into the clinic is becoming more real. Despite this, there are still several obstacles to overcome. One of them lays on the identification of gene delivery systems that combine a high efficiency in cell targeting and gene expression with the least immunogenicity. To date, both viral and nonviral vectors failed to meet all these requirements. Viral vectors do not completely satisfy biosafety concerns due to their immunogenecity and to the random integration of their genetic material into the host genome, whereas non-viral vectors present a low transduction efficiency and lack of cell specificity. However, many efforts are being made towards the development of a good gene transfer technology, which, in parallel with the increasing knowledge on the pathogenesis of type 1 and type 2 diabetes, makes gene therapy move closer to the clinic.

6 Conclusions

The prevalence of diabetes has been increasing worldwide and is expected to continue to rise over the next 50 years [2, 293]. The accompanying increase in the prevalence of diabetes-related complications and the occurrence of diabetes among younger adults, children, and adolescents is likely to have a substantial impact on healthcare costs, with cardiovascular disease and other chronic complications of diabetes accounting for more than one quarter of all direct expenses [3]. Patients with diabetes should be followed closely to ensure that they achieve and maintain both glycemic and nonglycemic treatment objectives. Given the loss of 80% of β-cell function at diagnosis of diabetes, as well as the 50% prevalence of pre-existing coronary heart disease at diagnosis, a more aggressive approach from the earliest stages of disease is clearly indicated [294]. A number of new therapeutic strategies have been introduced [85], and a large number of new compounds are in various stages of development for the treatment of type 2 diabetes [295] and this reflects the enormous impact of diabetes on public health.

Thus, new approaches for the treatment of diabetes and availability of a variety of innovative therapeutic strategies will improve quality of life in the near future. Prevention and major changes in lifestyle factors such as diet and physical exercise will also be needed to contain the growing burden of diabetes.

References

1. World Health Organisation (2012) Diabetes fact sheet no 312
2. Centers for Disease Control and Prevention (2011) Diabetes data and trends. Available via; URL: http://apps.nccd.cdc.gov/DDTSTRS/default.aspx. Accessed 10 Jan 2013
3. Cusick M, Meleth AD, Agrón E, Fisher MR, Reed GF, Knatterud GL, Barton FB, Davis MD, Ferris FL 3rd, Chew EY, Early Treatment Diabetc Retinopathy Study Research Group (2005) Associations of mortality and diabetes complications in patients with type 1 and type 2 diabetes. Diabetes Care 28:617–625
4. Mohan V, Pradeepa R (2009) Mortality in diabetes mellitus: revisiting the data from a developing region of the world. Postgrad Med J 85:225–226
5. Diabetes Control and Complications Trial Research Group (1993) The effect of intensive diabetes treatment on the development and progression of long-term complications in insulin-dependent diabetes mellitus: the diabetes control and complications trial. N Engl J Med 329:978–986
6. Reichard P, Nilsson B-Y, Rosenqvist U (1993) The effect of long-term intensified insulin treatment on the development of microvascular complications of diabetes mellitus. N Engl J Med 329:304–309
7. U.K. Prospective Diabetes Study (UKPDS) Group (1998) Intensive blood glucose control with sulphonylureas or insulin compared with conventional treatment and risk of complication in patients with type 2 diabetes (UKPDS 33). Lancet 352:837–853
8. U.K. Prospective Diabetes Study (UKPDS) Group (1998) Effect of intensive blood glucose control with metformin on complication in overweight patients with type 2 diabetes (UKPDS 34). Lancet 352:854–865
9. Selvin E, Marinopoulos S, Berkenblit G, Rami T, Brancati FL, Powe NR, Golden SH (2004) Meta-analysis: glycosylated hemoglobin and cardiovascular disease in diabetes mellitus. Ann Intern Med 141:421–431
10. Rosenstock J, Kim SW, Baron MA, Camisasca RP, Cressier F, Couturier A, Dejager S (2007) Efficacy and tolerability of initial combination therapy with vildagliptin and pioglitazone compared with component monotherapy in patients with type 2 diabetes. Diabetes Obes Metab 9:175–185
11. Rosenstock J, Baron MA, Dejager S, Mills D, Schweizer A (2007) Comparison of vildagliptin and rosiglitazone monotherapy in patients with type 2 diabetes: a 24-week, double-blind, randomized trial. Diabetes Care 30:217–223

12. Nathan DM, Buse JB, Davidson MB, Ferrannini E, Holman RR, Sherwin R, Zinman B, American Diabetes Association; European Association for the Study of Diabetes (2009) Medical management of hyperglycaemia in type 2 diabetes mellitus: a consensus algorithm for the initiation and adjustment of therapy. A consensus statement from the American Diabetes Association and the European Association for the Study of Diabetes. Diabetologia 52:17–30
13. American Diabetes Association (2011) Standard of medical care in diabetes. Diabetes Care 34:S11–S61
14. Atkinson MA (2005) Thirty years of investigating the autoimmune basis for type 1 diabetes. Diabetes 54:1253–1263
15. Morales A (2009) A better future for children with type 1 diabetes: review of the conclusions from the diabetes control and complications trial and the epidemiology of diabetes interventions and complications study. J Ark Med Soc 106:90–93
16. Furth ME, Atala A (2009) Stem cell sources to treat diabetes. J Cell Biochem 106: 507–511
17. Couri CE, Voltarelli JC (2008) Autologous stem cell transplantation for early type 1 diabetes mellitus. Autoimmunity 41:666–672
18. White SA, Shaw JA, Sutherland DE (2009) Pancreas transplantation. Lancet 373:1808–1817
19. Wiseman AC (2009) Simultaneous pancreas kidney transplantation: a critical appraisal of the risks and benefits compared with other treatment alternatives. Adv Chronic Kidney Dis 16:278–287
20. Tufveson G (2009) An experience of pancreas and islet transplantation in patients with end stage renal failure due to diabetes type I. Curr Opin Organ Transplant 14:95–102
21. Ricordi C, Lacy PE, Finke EH, Olack BJ, Scharp DW (1988) Automated method for isolation of human pancreatic islets. Diabetes 37:413–420
22. Naftanel MA, Harlan DM (2004) Pancreatic islet transplantation. PLoS Med 1:e58
23. Shapiro AM, Lakey JR, Ryan EA, Korbutt GS, Toth E, Warnock GL, Kneteman NM, Rajotte RV (2000) Islet transplantation in seven patients with type 1 diabetes mellitus using a glucocorticoid-free immunosuppressive regimen. N Engl J Med 343:230–238
24. Badet L, Benhamou PY, Wojtusciszyn A, Baertschiger R, Milliat-Guittard L, Kessler L, Penfornis A, Thivolet C, Renard E, Bosco D, Morel P, Morelon E, Bayle F, Colin C, Berney T, GRAGIL Group (2007) Expectations and strategies regarding islet transplantation: metabolic data from the GRAGIL 2 trial. Transplantation 84:89–96
25. Berney T, Toso C (2006) Monitoring of the islet graft. Diabetes Metab 32:503–512
26. Beger C, Cirulli V, Vajkoczy P, Halban PA, Menger MD (1998) Vascularization of purified pancreatic islet-like cell aggregates (pseudoislets) after syngerneic transplantation. Diabetes 47:559–565
27. Mattsson G, Jansson L, Carlsson PO (2002) Decreased vascular density in mouse pancreatic islets after transplantation. Diabetes 51:1362–1366
28. Ryan EA, Paty BW, Senior PA, Bigam D, Alfadhli E, Kneteman NM, Lakey JR, Shapiro AM (2005) Five-year follow up after clinical islet transplantation. Diabetes 54:2060–2069
29. Mineo D, Pileggi A, Alejandro R, Ricordi C (2009) Point: steady progress and current challenges in clinical islet transplantation. Diabetes Care 32:1563–1569
30. Mohseni Salehi Monfared SS, Larijani B, Abdollahi M (2009) Islets transplantation and antioxidant management: a comprehensive review. World J Gastroenterol 15:1153–1161
31. Caumo A, Maffi P, Nano R, Bertuzzi F, Luzi L, Secchi A, Bonifacio E, Piemonti L (2008) Transplant estimated function: a simple index to evaluate beta-cell secretion after islet transplantation. Diabetes Care 31:301–305
32. Cheng Y, Liu YF, Zhang JL, Li TM, Zhao N (2007) Elevation of vascular endothelial growth factor production and its effect on revascularization and function of graft islets in diabetic rats. World J Gastroenterol 13:2862–2866
33. Figliuzzi M, Cornolti R, Perico N, Rota C, Morigi M, Remuzzi G, Remuzzi A, Benigni A (2009) Bone marrow–derived mesenchymal stem cells improve islet graft function in diabetic rats. Transplant Proc 41:1797–1800

34. Buss JL, Rajab A, Diakoff E, Wang J, Miller K, Osei K (2010) Exenatide pre-treatment improved islet graft function compared to treatment post-transplant only. Proc 70th scientific sessions meeting of the American diabetes association, Orlando, Abstract 141-OR
35. Vergani A, D'Addio F, Petrelli A, Williams J, Sayegh MH, Fiorina P (2010) A combined strategy of thymoglobulin and CTLA4-iG cure diabetes and prevent recurrence of autoimmunity in NOD mice. Proc 70th scientific sessions meeting of the American diabetes association, Orlando, Abstract 336-OR
36. Delgado TC, Barosa C, Nunes PM, Jones JG (2010) Effects of cyclosporine A on glucose metabolism: implications for post-transplant diabetes mellitus. Proc 70th scientific sessions meeting of the American diabetes association, Orlando, Abstract 2042-P
37. Velazquez-Garcia SA, Rodriguez A, Salido E, Torres A (2010) Temporal administration of FK506 results in impaired glucose homeostasis and decreased beta cell proliferation in a rat model of insulin resistance. Proc 70th scientific sessions meeting of the American diabetes association, Orlando, Abstract 2052-P
38. Shivaswamy V, Griffin B, Clure C, Reyelts C, Calleroz A, Bennett RG, Hamel FG, Borgeson C, Larsen JL (2010) Metformin improves immunosuppressant-induced hyperglycaemia. Proc 70th scientific sessions meeting of the American diabetes association, Orlando, Abstract 2038-P
39. Olsson R, Maxhuni A, Carlsson P (2006) Revascularization of transplanted pancreatic islets following culture with stimulators of angiogenesis. Transplantation 82:340–347
40. King A, Lock J, Xu G, Bonner-Weir S, Weir GC (2005) Islet transplantation outcomes in mice are better with fresh islets and exendin-4 treatment. Diabetologia 48:2074–2079
41. Avila JG, Wang Y, Barbaro B, Gangemi A, Qi M, Kuechle J, Doubleday N, Doubleday M, Churchill T, Salehi P, Shapiro J, Philipson LH, Benedetti E, Lakey JR, Oberholzer J (2006) Improved outcomes in islet isolation and transplantation by the use of a novel hemoglobin-based O_2 carrier. Am J Transplant 6:2861–2870
42. Contreras JL, Eckstein C, Smyth CA, Bilbao G, Vilatoba M, Ringland SE, Young C, Thompson JA, Fernández JA, Griffin JH, Eckhoff DE (2004) Activated protein C preserves functional islet mass after intraportal transplantation: a novel link between endothelial cell activation, thrombosis, inflammation, and islet cell death. Diabetes 53:2804–2814
43. Cabric S, Sanchez J, Lundgren T, Foss A, Felldin M, Källen R, Salmela K, Tibell A, Tufveson G, Larsson R, Korsgren O, Nilsson B (2007) Islet surface heparinization prevents the instant blood-mediated inflammatory reaction in islet transplantation. Diabetes 56:2008–2015
44. Emamaullee J, Stanton L, Schur C, Shapiro AM (2007) Caspase inhibitor therapy enhances marginal mass islet graft survival and preserves long-term function in islet transplantation. Diabetes 56:1289–1298
45. Witkowski P, Herold K (2007) Islet transplantation for type 1 diabetes–where should we go? Nat Clin Pract Endocrinol Metab 3:2–3
46. Yang YG, Sykes M (2007) Xenotransplantation: current status and a perspective on the future. Nat Rev Immunol 07:519–531
47. O'Connell PJ (2009) The international xenotransplantation association consensus statement on conditions for undertaking clinical trials of porcine islet products in type 1 diabetes–chapter 6: patient selection for pilot clinical trials of islet xenotransplantation. Xenotransplantation 16:249–254
48. Lanza RP, Ecker DM, Kühtreiber WM, Marsh JP, Ringeling J, Chick WL (1999) Transplantation of islets using microencapsulation: studies in diabetic rodents and dogs. J Mol Med 77:206–210
49. Calafiore R, Basta G, Luca G, Lemmi A, Racanicchi L, Mancuso F, Montanucci MP, Brunetti P (2006) Standard technical procedures for microencapsulation of human islets for graft into nonimmunosuppressed patients with type 1 diabetes mellitus. Transplant Proc 38:1156–1157
50. Living Cell Technologies (2009) Open-label investigation of the safety and effectiveness of DIABECELL(R) in patients with type I diabetes mellitus. ClinicalTrials.gov identifier: NCT00940173. http://clinicaltrials.gov/ct2/show/NCT00940173?term=porcine+islets&rank=1. Accessed 10 Jan 2013

51. Gao Y, Shao S, Xie B, Lim SK, Li G (2010) Transplantation of alginate-encapsulated insulin-producing cells differentiated from embryonic stem/progenitor cells. Proc 70th scientific sessions meeting of the American diabetes association, Orlando, (Abstract 2033-P)
52. Hathout E, Sakata N, Chan NK, Chrisler J, Obenaus A (2010) Co-transplantation with bone marrow improves islet function. Relation to nerve growth factor. 70th scientific sessions meeting of the American diabetes association, Orlando, (Abstract 2041-P)
53. Best M, Carroll M, Hanley NA, Piper Hanley K (2008) Embryonic stem cells to beta-cells by understanding pancreas development. Mol Cell Endocrinol 288:86–94
54. Müller R, Lengerke C (2009) Patient-specific pluripotent stem cells: promises and challenges. Nat Rev Endocrinol 5:195–203
55. Kroon E, Martinson LA, Kadoya K, Bang AG, Kelly OG, Eliazer S, Young H, Richardson M, Smart NG, Cunningham J, Agulnick AD, D'Amour KA, Carpenter MK, Baetge EE (2008) Pancreatic endoderm derived from human embryonic stem cells generates glucose-responsive insulin-secreting cells in vivo. Nat Biotechnol 26:443–452
56. Okita K, Ichisaka T, Yamanaka S (2007) Generation of germline-competent induced pluripotent stem cells. Nature 448:313–317
57. Park IH, Zhao R, West JA, Yabuuchi A, Huo H, Ince TA, Lerou PH, Lensch MW, Daley GQ (2007) Reprogramming of human somatic cells to pluripotency with defined factors. Nature 451:141–146
58. Takahashi K, Tanabe K, Ohnuki M, Narita M, Ichisaka T, Tomoda K, Yamanaka S (2007) Induction of pluripotent stem cells from adult human fibroblasts by defined factors. Cell 131:861–872
59. Yu J, Vodyanik MA, Smuga-Otto K, Antosiewicz-Bourget J, Frane JL, Tian S, Nie J, Jonsdottir GA, Ruotti V, Stewart R, Slukvin II, Thomson JA (2007) Induced pluripotent stem cell lines derived from human somatic cells. Science 318:1917–1920
60. Wernig M, Meissner A, Foreman R, Brambrink T, Ku M, Hochedlinger K, Bernstein BE, Jaenisch R (2007) In vitro reprogramming of fibroblasts into a pluripotent ES-cell-like state. Nature 448:318–324
61. Taylor CJ, Bolton EM, Pocock S, Sharples LD, Pedersen RA, Bradley JA (2005) Banking on human embryonic stem cells: estimating the number of donor cell lines needed for HLA matching. Lancet 366:2019–2025
62. Tateishi K, He J, Taranova O, Liang G, D'Alessio AC, Zhang Y (2008) Generation of insulin-secreting islet-like clusters from human skin fibroblasts. J Biol Chem 283:31601–31607
63. Woltjen K, Michael IP, Mohseni P, Desai R, Mileikovsky M, Hämäläinen R, Cowling R, Wang W, Liu P, Gertsenstein M, Kaji K, Sung HK, Nagy A (2009) PiggyBac transposition reprograms fibroblasts to induced pluripotent stem cells. Nature 458:766–777
64. Voltarelli JC, Couri CE, Stracieri AB, Oliveira MC, Moraes DA, Pieroni F, Coutinho M, Malmegrim KC, Foss-Freitas MC, Simões BP, Foss MC, Squiers E, Burt RK (2007) Autologous nonmyeloablative hematopoietic stem cell transplantation in newly diagnosed type 1 diabetes mellitus. JAMA 297:1568–1576
65. Vicker N, Su X, Ganeshapillai D (2007) Novel non-steroidal inhibitors of human 11 β-hydroxysteroid dehydrogenase type 1. J Steroid Biochem Mol Biol 104:123–129
66. Haller MJ, Viener HL, Wasserfall C, Brusko T, Atkinson MA, Schatz DA (2008) Autologous umbilical cord blood infusion for type 1 diabetes. Exp Hematol 36:710–715
67. Ferber S, Halkin A, Cohen H, Ber I, Einav Y, Goldberg I, Barshack I, Seijffers R, Kopolovic J, Kaiser N, Karasik A (2000) Pancreatic and duodenal homeobox gene 1 induces expression of insulin genes in liver and ameliorates streptozotocin induced hyperglycaemia. Nat Med 6:568–572
68. Meier JJ, Bhushan A, Butler AE, Rizza RA, Butler PC (2005) Sustained beta cell apoptosis in patients with long-standing type 1 diabetes: indirect evidence for islet regeneration? Diabetologia 48:2221–2228
69. Shternhall-Ron K, Quintana FJ, Perl S, Meivar-Levy I, Barshack I, Cohen IR, Ferber S (2007) Ectopic PDX-1 expression in liver ameliorates type 1 diabetes. J Autoimmun 28:134–142

70. Cheung AT, Dayanandan B, Lewis JT, Korbutt GS, Rajotte RV, Bryer-Ash M, Boylan MO, Wolfe MM, Kieffer TJ (2000) Glucose dependent insulin release from genetically engineered K cells. Science 290:1959–1962
71. Ber I, Shternhall K, Perl S, Ohanuna Z, Goldberg I, Barshack I, Benvenisti-Zarum L, Meivar-Levy I, Ferber S (2003) Functional, persistent, and extended liver to pancreas transdifferentiation. J Biol Chem 278:31950–31957
72. Yechoor V, Liu V, Espiritu C, Paul A, Oka K, Kojima H, Chan L (2009) Neurogenin3 is sufficient for in vivo transdetermination of hepatic progenitor cells into islet-like cells but not transdifferentiation of hepatocytes. Dev Cell 16:358–373
73. Zhou Q, Brown J, Kanarek A, Rajagopal J, Melton DA (2008) In vivo reprogramming of adult pancreatic exocrine cells to beta-cells. Nature 455:627–632
74. Reimann M, Bonifacio E, Solimena M, Schwarz PE, Ludwig B, Hanefeld M, Bornstein SR (2009) An update on preventive and regenerative therapies in diabetes mellitus. Pharmacol Ther 121:317–331
75. Agardh CD, Cilio CM, Lethagen A, Lynch K, Leslie RD, Palmér M, Harris RA, Robertson JA, Lernmark A (2005) Clinical evidence for the safety of GAD65 immunomodulation in adult-onset autoimmune diabetes. J Diabetes Complications 19:238–246
76. Bresson D, von Herrath M (2009) Immunotherapy for the prevention and treatment of type 1 diabetes: optimizing the path from bench to bedside. Diabetes Care 32:1753–1768
77. Agardh CD, Lynch KF, Palmér M, Link K, Lernmark A (2009) GAD65 vaccination: 5 years of follow-up in a randomised dose-escalating study in adult-onset autoimmune diabetes. Diabetologia 52:1363–1368
78. Gottlieb P, Colman PG, Solvason N, Otto K, Roep B, Harrison LC, Isenbarth G, Robinson W, Utz PJ, Steinman L, Garren H, Quan J (2010) One-year results from a phase 1/2 clinical trial of BHT-3021, a DNA plasmid vaccine for type 1 diabetes (T1D). 70th scientific sessions meeting of the American diabetes association, Orlando, (Abstract 66-OR)
79. DeFronzo RA (1988) The triumvirate: β-cell, muscle, liver. A collusion responsible for NIDDM. Diabetes 37:667–687
80. DeFronzo RA (2009) From the triumvirate to the ominous octet: a new paradigm for the treatment of type 2 diabetes mellitus. Diabetes 58:773–795
81. Ramlo-Halsted BA, Edelman SV (2000) The natural history of type 2 diabetes: practical points to consider in developing prevention and treatment strategies. Clin Diabetes 18:80–84
82. Bays H, Mandarino L, DeFronzo RA (2004) Role of the adipocyte, free fatty acids, and ectopic fat in pathogenesis of type 2 diabetes mellitus: peroxisomal proliferator-activated receptor agonists provide a rational therapeutic approach. J Clin Endocrinol Metab 89:463–478
83. Stumvoll M, Goldstein BJ, van Haeften TW (2005) Type 2 diabetes: principles of pathogenesis and therapy. Lancet 365:1333–1346
84. Petersen KF, Shulman GI (2006) Etiology of insulin resistance. Am J Med 119:S10–S16
85. Horton ES (2008) Can newer therapies delay the progression of type 2 diabetes mellitus? Endocrine Pract 14:625–638
86. Turner RC, Cull CA, Frighi V, Holman RR (1999) Glycemic control with diet, sulfonylurea, metformin, or insulin in patients with type 2 diabetes mellitus: progressive requirement for multiple therapies (UKPDS 49). UK prospective diabetes study (UKPDS) group. JAMA 281:2005–2012
87. Saydah SH, Fradkin J, Cowie CC (2004) Poor control of risk factors for vascular disease among adults with previously diagnosed diabetes. JAMA 291:335–342
88. Kahn SE, Haffner SM, Heise MA, Herman WH, Holman RR, Jones NP, Kravitz BG, Lachin JM, O'Neill MC, Zinman B, Viberti G, ADOPT Study Group (2006) Glycemic durability of rosiglitazone, metformin, or glyburide monotherapy. N Engl J Med 355:2427–2443
89. Uwaifo GI, Ratner RE (2007) Differential effects of oral hypoglycaemic agents on glucose control and cardiovascular risk. Am J Cardiol 99:51B–67B
90. United Kingdom Prospective Diabetes Study Group (1995) Overview of 6 years' therapy of type II diabetes: a progressive disease. Diabetes 44:1249–1258

91. The ADVANCE Collaborative Group (2008) Intensive blood glucose control and vascular outcomes in patients with type 2 diabetes. N Engl J Med 358:2560–2572
92. Black C, Donnelly P, McIntyre L, Royle PL, Shepherd JP, Thomas S (2007) Meglitinide analogues for type 2 diabetes mellitus. Cochrane Database Syst Rev 2:CD004654. doi:10.1002/14651858.CD004654.pub2
93. Hasslacher C (2003) Safety and efficacy of repaglinide in type 2 diabetic patients with and without impaired renal function. Diabetes Care 26:886–891
94. Blicklé JF (2006) Meglitinide analogues: a review of clinical data focused on recent trials. Diabetes Metab 32:113–120
95. Ceriello A (2000) The post-prandial state and cardiovascular disease: relevance to diabetes mellitus. Diabetes Metab Res Rev 16:125–132
96. Kirpichnikov D, McFarlane SI, Sowers JR (2002) Metformin: an update. Ann Intern Med 137:25–33
97. Dandona P, Aljada A, Chaudhuri A, Bandyopadhyay A (2003) The potential influence of inflammation and insulin resistance on the pathogenesis and treatment of atherosclerosis-related complications in type 2 diabetes. J Clin Endocrinol Metab 88:2422–2429
98. Verma S, Bhanot S, McNeill JH (1994) Antihypertensive effects of metformin in fructose-fed hyperinsulinemic, hypertensive rats. J Pharmacol Exp Ther 271:1334–1337
99. Mamputu JC, Wiernsperger NF, Renier G (2003) Antiatherogenic properties of metformin: the experimental evidence. Diabetes Metab 29:6S71–6S76
100. Grant PJ (2003) Beneficial effects of metformin on haemostasis and vascular function in man. Diabetes Metab 29:644–652
101. Bristol-Myers Squibb Company. Glucophage (metformin hydrochloride tablets); Glucophage XR (metformin hydrochloride extended-release tablets). Available at: http://packageinserts.bms.com/pi/pi_glucophage.pdf. Accessed 10 Jan 2013
102. Ting RZ, Szeto CC, Chan MH (2006) Risk factors of vitamin B(12) deficiency in patients receiving metformin. Arch Intern Med 166:1975–1979
103. Yki-Jarvinen H (2004) Thiazolidinediones. N Engl J Med 351:1106–1118
104. Ahmed I, Furlong K, Flood J, Treat VP, Goldstein BJ (2007) Dual PPAR α/γ agonists: promises and pitfalls in type 2 diabetes. Am J Ther 14:49–62
105. Ye JM, Dzamko N, Cleasby ME, Hegarty BD, Furler SM, Cooney GJ, Kraegen EW (2004) Direct demonstration of lipid sequestration as a mechanism by which rosiglitazone prevents fatty-acid-induced insulin resistance in the rat: comparison with metformin. Diabetologia 47:1306–1313
106. Buchanan TA, Xiang AH, Peters RK, Kjos SL, Marroquin A, Goico J, Ochoa C, Tan S, Berkowitz K, Hodis HN, Azen SP (2002) Preservation of pancreatic β-cell function and prevention of type 2 diabetes by pharmacological treatment of insulin resistance in high-risk Hispanic women. Diabetes 51:2796–2803
107. Xiang AH, Peters RK, Kjos SL, Marroquin A, Goico J, Ochoa C, Kawakubo M, Buchanan TA (2006) Effect of pioglitazone on pancreatic β-cell function and diabetes risk in Hispanic women with prior gestational diabetes. Diabetes 55:517–522
108. Aronoff S, Rosenblatt S, Braithwaite S, Egan JW, Mathisen AL, Schneider RL (2000) Pioglitazone hydrochloride monotherapy improves glycaemic control in the treatment of patients with type 2 diabetes: a 6- month randomized placebo-controlled dose-response study. The Pioglitazone 001 Study Group. Diabetes Care 23:1605–1611
109. Phillips LS, Grunberger G, Miller E, Patwardhan R, Rappaport EB, Salzman A, Rosiglitazone Clinical Trials Study Group (2001) Once- and twice-daily dosing with rosiglitazone improves glycaemic control in patients with type 2 diabetes. Rosiglitazone Clinical Trial Study Group. Diabetes Care 24:308–315
110. Stafford JM, Elasy T (2007) Treatment update: thiazolinediones in combination with metformin for the treatment of type 2 diabetes. Vasc Health Risk Manag 3:503–510
111. Kipnes MS, Krosnick A, Rendell MS, Egan JW, Mathisen AL, Schneider RL (2001) Pioglitazone hydrochloride in combination with sulfonylurea therapy improves glycemic control in patients with type 2 diabetes mellitus: a randomized, placebo-controlled study. Am J Med 111:10–17

112. Raskin P, Rendell M, Riddle MC, Dole JF, Freed MI, Rosenstock J (2001) A randomized trial of rosiglitazone therapy in patients with inadequately controlled insulin-treated type 2 diabetes. Diabetes Care 24:1226–1232
113. Waksman JC (2008) Cardiovascular risk of rosiglitazone: another perspective. J Pharm Pharmacol 60:1573–1582
114. Sarafidis PA, Pantelis A (2008) Thiazolidinedione derivatives in diabetes and cardiovascular disease: an update. Fundam Clin Pharmacol 22:247–264
115. Meriden T (2004) Progress with thiazolidinediones in the management of type 2 diabetes mellitus. Clin Ther 26:177–190
116. Einhorn D, Aroda VR, Henry RR (2004) Glitazones and the management of insulin resistance: what they do and how might they be used. Endocrinol Metab Clin North Am 33:595–616
117. Guan Y, Hao C, Cha DR, Rao R, Lu W, Kohan DE, Magnuson MA, Redha R, Zhang Y, Breyer MD (2005) Thiazolidinediones expand body fluid volume through PPAR gamma stimulation of EnaC-mediated renal salt absorption. Nat Med 11:861–866
118. Nesto RW, Bell D, Bonow RO, Fonseca V, Grundy SM, Horton ES, Le Winter M, Porte D, Semenkovich CF, Smith S, Young LH, Kahn R, American Heart Association; American Diabetes Association (2003) Thiazolidinedione use, fluid retention, and congestive heart failure: a consensus statement from the American Heart Association and American Diabetes Association. Circulation 108:2941–2948
119. Marcy TR, Britton ML, Blevins SM (2004) Second-generation thiazolinediones and hepatotoxicity. Ann Pharmacother 38:1419–1423
120. Selvin E, Bolen S, Yeh HC, Wiley C, Wilson LM, Marinopoulos SS, Feldman L, Vassy J, Wilson R, Bass EB, Brancati FL (2008) Cardiovascular outcomes in trials of oral diabetes medications: a systematic review. Arch Intern Med 168:2070–2080
121. Singh S, Loke YK (2008) The safety of rosiglitazone in the treatment of type 2 diabetes. Expert Opin Drug Saf 7:579–585
122. Grey A (2008) Skeletal consequences of thiazolidinedione therapy. Osteoporos Int 19:129–137
123. Meier C, Kraenzlin ME, Bodmer M, Jick SS, Jick H, Meier CR (2008) Use of thiazolidinediones and fracture risk. Arch Intern Med 168:820–825
124. Liazos E, Broadbent DM, Beare N, Kumar N (2008) Spontaneous resolution of diabetic macular oedema after discontinuation of thiazolidenediones. Diabet Med 25:860–862
125. Scheen AJ (1998) Clinical efficacy of acarbose in diabetes mellitus: a critical review of controlled trials. Diabetes Metab 24:311–320
126. Krentz AJ, Bailey CJ (2005) Oral antidiabetic agents: current role in type 2 diabetes mellitus. Drugs 65:385–411
127. Chiasson JL, Josse RG, Gomis R, Hanefeld M, Karasik A, Laakso M, STOP-NIDDM Trail Research Group (2002) Acarbose for prevention of type 2 diabetes mellitus: the STOP-NIDDM randomised trial. Lancet 359:2072–2077
128. van de Laar FA, Lucassen PL, Akkermans RP, van de Lisdonk EH, Rutten GE, van Weel C (2005) Alpha-glucosidase inhibitors for patients with type 2 diabetes: results from a Cochrane systematic review and meta-analysis. Diabetes Care 28:154–163
129. Crasto W, Jarvis J, Khunti K, Davies MJ (2009) New insulins and new insulin regimens: a review of their role in improving glycaemic control in patients with diabetes. Postgrad Med J 85:257–267
130. Henry RR, Gumbiner B, Ditzler T, Wallace P, Lyon R, Glauber HS (1993) Intensive conventional insulin therapy for type II diabetes. Metabolic effects during a 6-mo outpatient trial. Diabetes Care 16:21–31
131. Holst JJ, Vilsbøll T, Deacon CF (2009) The incretin system and its role in type 2 diabetes mellitus. Mol Cell Endocrinol 297:127–136
132. American Association of Clinical Endocrinologists (2007) American Association of Clinical Endocrinologists medical guidelines for clinical practice for the management of diabetes mellitus. Endocr Pract 13:1–68

133. Baggio LL, Drucker DJ (2007) Biology of incretins: GLP-1 and GIP. Gastroenterology 132:2131–2157
134. Sheffield CA, Kane MP, Busch RS, Bakst G, Abelseth JM, Hamilton RA (2008) Safety and efficacy of exenatide in combination with insulin in patients with type 2 diabetes mellitus. Endocr Pract 14:285–292
135. Drucker DJ (2005) Biologic actions and therapeutic potential of the proglucagon-derived peptides. Nat Clin Pract Endocrinol Metab 1:22–31
136. Barnett A (2007) Exenatide. Expert Opin Pharmacother 8:2593–2608
137. Lambeir AM, Durinx C, Scharpé S, De Meester I (2003) Dipeptidyl-peptidase IV from bench to bedside: an update on structural properties, functions, and clinical aspects of the enzyme DPP IV. Crit Rev Clin Lab Sci 40:209–294
138. Krentz AJ (2008) Management of type 2 diabetes in the obese patient: current concerns and emerging therapies. Curr Med Res Opin 24:401–417
139. Doyle ME, Egan JM (2007) Mechanisms of action of glucagons like peptide 1 in the pancreas. Pharmacol Ther 113:546–593
140. Croom KF, McCormack PL (2009) Liraglutide: a review of its use in type 2 diabetes mellitus. Drugs 69:1985–2004
141. Heine RJ, Van Gaal LF, Johns D, Mihm MJ, Widel MH, Brodows RG, GWAA Study Group (2005) Exenatide versus insulin glargine in patients with suboptimally controlled type 2 diabetes: a randomized trial. Ann Intern Med 143:559–569
142. Kendall DM, Riddle MC, Rosenstock J, Zhuang D, Kim DD, Fineman MS, Baron AD (2005) Effects of exenatide (exendin-4) on glycaemic control over 30 weeks in patients with type 2 diabetes treated with metformin and a sulfonylurea. Diabetes Care 28:1083–1091
143. U.S. Food and Drug Administration (2008) Information for healthcare professionals: exenatide (marketed as Byetta). http://www.fda.gov/Drugs/DrugSafety/PostmarketDrugSafety InformationforPatientsandProviders/ucm124713.htm. Accessed 10 Jan 2013
144. Ahren B (2007) Dipeptidyl peptidase-4 inhibitors: clinical data and clinical implications. Diabetes Care 30:1344–1350
145. Hale C, Wang M (2008) Development of 11beta-HSD1 inhibitors for the treatment of type 2 diabetes. Mini Rev Med Chem 8:702–710
146. Nishimura T, Nakatake Y, Konishi M, Itoh N (2000) Identification of a novel FGF, FGF-21, preferentially expressed in the liver. Biochim Biophys Acta 1492:203–206
147. Kharitonenkov A, Shiyanova TL, Koester A, Ford AM, Micanovic R, Galbreath EJ, Sandusky GE, Hammond LJ, Moyers JS, Owens RA, Gromada J, Brozinick JT, Hawkins ED, Wroblewski VJ, Li DS, Mehrbod F, Jaskunas SR, Shanafelt AB (2005) FGF-21 as a novel metabolic regulator. J Clin Invest 115:1627–1635
148. Kharitonenkov A, Wroblewski VJ, Koester A, Chen YF, Clutinger CK, Tigno XT, Hansen BC, Shanafelt AB, Etgen GJ (2007) The metabolic state of diabetic monkeys is regulated by fibroblast growth factor-21. Endocrinology 148:774–781
149. Coskun T, Bina HA, Schneider MA, Dunbar JD, Hu CC, Chen Y, Moller DE, Kharitonenkov A (2008) Fibroblast growth factor 21 corrects obesity in mice. Endocrinology 149:6018–6027
150. Fisher FM, Chui PC, Antonellis PJ, Bina HA, Kharitonenkov A, Flier JS, Maratos-Flier E (2010) Obesity is a fibroblast growth factor 21 (FGF21)-resistant state. Diabetes 59:2781–2789
151. Edelman S, Maier H, Wilhelm K (2008) Pramlintide in the treatment of diabetes mellitus. Bio Drugs 22:375–386
152. Wysham C, Lush C, Zhang B, Maier H, Wilhelm K (2008) Effect of pramlintide as an adjunct to basal insulin on markers of cardiovascular risk in patients with type 2 diabetes. Curr Med Res Opin 24:79–85
153. Defronzo RA (2011) Bromocriptine: a sympatholytic, d2-dopamine agonist for the treatment of type 2 diabetes. Diabetes Care 34:789–794
154. Scranton RE, Gaziano JM, Rutty D, Ezrokhi M, Cincotta A (2007) A randomized, double-blind, placebo-controlled trial to assess safety and tolerability during treatment of type 2 diabetes with usual diabetes therapy and either Cycloset or placebo. BMC Endocr Disord 7:3

155. Zieve FJ, Kalin MF, Schwartz SL, Jones MR, Bailey WL (2007) Results of the glucose-lowering effect of WelChol study (GLOWS): a randomized, double-blind placebo-controlled pilot study evaluating the effect of colesevelam hydrochloride on glycaemic control in subjects with type 2 diabetes. Clin Ther 29:74–83
156. Couzin J (2008) Medicine: bypassing medicine to treat diabetes. Science 320:438–440
157. Buchwald H, Estok R, Fahrbach K, Banel D, Jensen MD, Pories WJ, Bantle JP, Sledge I (2009) Weight and type 2 diabetes after bariatric surgery: systematic review and meta-analysis. Am J Med 122:248–256.e5
158. Rubino F, Schauer PR, Kaplan LM, Cummings DE (2010) Metabolic surgery to treat type 2 diabetes: clinical outcomes and mechanisms of action. Annu Rev Med 61:393–411
159. Frachettia KJ, Goldfine AB (2009) Bariatric surgery for diabetes management. Curr Opin Endocrinol Diabetes Obes 16:119–124
160. Buchwald H (2008) Introduction and current status of bariatric procedures. Surg Obes Relat Dis 4:S1–S6
161. Vetter ML, Cardillo S, Rickels MR, Iqbal N (2009) Narrative review: effect of bariatric surgery on type 2 diabetes mellitus. Ann Intern Med 150:94–103
162. Ballantyne GH, Gumbs A, Modlin IM (2005) Changes in insulin resistance following bariatric surgery and the adipoinsular axis: role of the adipocytokines, leptin, adiponectin and resistin. Obes Surg 15:692–699
163. Rubino F, Marescaux J (2004) Effect of duodenal-jejunal exclusion in a non-obese animal model of T2DM: a new perspective for an old disease. Ann Surg 239:1–11
164. Rubino F (2008) Is type 2 diabetes an operable intestinal disease? A provocative yet reasonable hypothesis. Diabetes Care 31:S290–S296
165. Knop FK (2009) Resolution of type 2 diabetes following gastric bypass surgery: involvement of gut-derived glucagon and glucagonotropic signalling? Diabetologia 52: 2270–2276
166. Strader AD, Vahl TP, Jandacek RJ, Woods SC, D'Alessio DA, Seeley RJ (2005) Weight loss through ileal transposition is accompanied by increased ileal hormone secretion and synthesis in rats. Am J Physiol Endocrinol Metab 288:E447–E453
167. Korner J, Bessler M, Cirilo LJ, Conwell IM, Daud A, Restuccia NL, Wardlaw SL (2005) Effects of Roux-en-Y gastric bypass surgery on fasting and postprandial concentrations of plasma ghrelin, peptide YY, and insulin. J Clin Endocrinol Metab 90:359–365
168. Morínigo R, Moizé V, Musri M, Lacy AM, Navarro S, Marín JL, Delgado S, Casamitjana R, Vidal J (2006) Glucagon-like peptide-1, peptide YY, hunger, and satiety after gastric bypass surgery in morbidly obese subjects. J Clin Endocrinol Metab 91:1735–1740
169. Isbell JM, Tamboli RA, Hansen EN, Saliba J, Dunn JP, Phillips SE, Marks-Shulman PA, Abumrad NN (2010) The importance of caloric restriction in the early improvements in insulin sensitivity after Roux-en-Y gastric bypass surgery. Diabetes Care 33:1438–1442
170. Furet JP, Kong LC, Tap J, Poitou C, Basdevant A, Bouillot JL, Mariat D, Corthier G, Doré J, Henegar C, Rizkalla S, Clément K (2010) Differential adaptation of human gut microbiota to bariatric surgery-induced weight loss: links with metabolic and low-grade inflammation markers. Diabetes 59:3049–3057
171. Houmard JA, Tanner CJ, Yu C, Cunningham PG, Pories WJ, MacDonald KG, Shulman GI (2002) Effect of weight loss on insulin sensitivity and intramuscular long-chain fatty acyl-CoAs in morbidly obese subjects. Diabetes 51:2959–2963
172. Henry RR (2008) Evolving concepts of type 2 diabetes management with oral medications: new approaches to an old disease. Curr Med Res Opin 24:2189–2202
173. Philippe J, Raccah D (2009) Treating type 2 diabetes: how safe are current therapeutic agents? Int J Clin Pract 63:321–332
174. Christensen M, Knop FK, Holst JJ, Vilsboll T (2009) Lixisenatide, a novel GLP-1 receptor agonist for the treatment of type 2 diabetes mellitus. IDrugs 12:503–513
175. Balakumar P, Rose M, Ganti SS, Krishan P, Singh M (2007) PPAR dual agonists: are they opening Pandora's Box? Pharmacol Res 56:91–98

176. Henry RR, Lincoff AM, Mudaliar S, Rabbia M, Chognot C, Herz M (2009) Effect of the dual peroxisome proliferator-activated receptor α/γ: results from SYNCHRONY, a phase II, randomised, dose-ranging study in patients with type 2 diabetes. Lancet 374:126–135
177. Lan H, Vassileva G, Corona A, Liu L, Baker H, Golovko A, Abbondanzo SJ, Hu W, Yang S, Ning Y, Del Vecchio RA, Poulet F, Laverty M, Gustafson EL, Hedrick JA, Kowalski TJ (2009) GPR119 is required for physiological regulation of glucagon-like peptide-1 secretion but not for metabolic homeostasis. J Endocrinol 201:219–230
177. Lauffer LM, Iakoubov R, Brubaker PL (2009) GPR119 Is essential for oleoylethanolamide-induced glucagon-like peptide-1 secretion from the intestinal enteroendocrine L-cell. Diabetes 58:1058–1066
179. Lauffer L, Iakoubov R, Brubaker PL (2008) GPR119: "double-dipping" for better glycaemic control. Endocrinology 149:2035–2037
180. Jones RM, Leonard JN, Buzard DJ, Lehmann J (2009) GPR119 agonists for the treatment of type 2 diabetes. Expert Opin Ther Pat 19:1339–1359
181. Positive results from Metabolex' Phase 1 MBX-2982 clinical trial for treating type 2 diabetes. News-Medical.Net. http://www.news-medical.net/news/20091014/Positive-results-from-Metabolex-Phase-1-MBX-2982-clinical-trial-for-treating-type-2-diabetes.aspx. Accessed 10 Jan 2013
182. Larsen CM, Faulenbach M, Vaag A, Vølund A, Ehses JA, Seifert B, Mandrup-Poulsen T, Donath MY (2007) Interleukin-1–receptor antagonist in type 2 diabetes mellitus. N Engl J Med 356:1517–1526
183. Donath MY, Mandrup-Poulsen T (2008) The use of interleukin- 1-receptor antagonists in the treatment of diabetes mellitus. Nat Rev Endocrinol 4:240–241
184. Donath MY, Weder C, Brunner A, Keller C, Whitmore J, Der K, Zayed H, Scannon PJ, Feldstein JD, Dinarello CA, Solinger AM (2009) XOMA 052, a potential disease modifying anti-IL-1beta antibody, shows sustained HbA1c reductions 3 months after a single injection with no increases in safety parameters in subjects with type 2 diabetes. Proc 69th scientific sessions meeting of the American diabetes association, New Orleans, (Abstract 113-OR)
185. Vaughn DE, Yocum RC, Muchmore DB, Sugarman BJ, Vick AM, Bilinsky IP, Frost GI (2009) Accelerated pharmacokinetics and glucodynamics of prandial insulins injected with recombinant human hyaluronidase. Diabetes Technol Ther 11:345–352
186. Hompesch M, Muchmore DB, Morrow L, Vaughn DE (2011) Accelerated insulin pharmacokinetics and improved postprandial glycaemic control in patients with type 1 diabetes after coadministration of prandial insulins with hyaluronidase. Diabetes Care 34:666–668
187. Halford JCG, Harrold JA (2008) Neuropharmacology of human appetite expression. Dev Disabil Res Rev 14:158–164
188. Smith SR, Weissman NJ, Anderson CM, Sanchez M, Chuang E, Stubbe S, Bays H, Shanahan WR, Behavioral Modification and Lorcaserin for Overweight and Obesity Management (BLOOM) Study Group (2010) Multicenter, placebo-controlled trial of lorcaserin for weight management. N Engl J Med 363:245–256
189. Cole P, Vicente M, Castaner R (2008) Dapagliflozin: SGLT2 inhibitor antidiabetic agent. Drug Future 33:745–751
190. Jabbour SA, Goldstein BJ (2008) Sodium glucose co-transporter 2 inhibitors: blocking renal tubular reabsorption of glucose to improve glycaemic control in patients with diabetes. Int J Clin Pract 62:1279–1284
191. Idris I, Donnelly R (2009) Sodium-glucose co-transporter-2 inhibitors: an emerging new class of oral antidiabetic drug. Diabetes Obes Metab 11:79–88
192. Ikari A, Nakano M, Ishibashi M, Kawano K, Suketa Y, Harada H, Takagi K (2003) Recovery from heat shock injury by activation of Na+-glucose cotransporter in renal epithelial cells. Biochim Biophys Acta 1643:47–53
193. Arakawa K, Ishihara T, Oku A, Nawano M, Ueta K, Kitamura K, Matsumoto M, Saito A (2001) Improved diabetic syndrome in C57BL/KsJ-db/db mice by oral administration of the Na+-glucose cotransporter inhibitor T-1095. Br J Pharmacol 132:578–586

194. Vemula S, Roder KE, Yang T, Bhat GJ, Thekkumkara TJ, Abbruscato TJ (2009) A functional role for sodium dependent glucose transport across the blood-brain barrier during oxygen glucose deprivation. J Pharmacol Exp Ther 328:487–495
195. Zhou L, Cryan EV, D'Andrea MR, Belkowski S, Conway BR, Demarest KT (2003) Human cardiomyocytes express high level of Na+/glucose cotransporter 1 (SGLT1). J Cell Biochem 90:339–346
196. Han S, Hagan DL, Taylor JR, Xin L, Meng W, Biller SA, Wetterau JR, Washburn WN, Whaley JM (2008) Dapagliflozin, a selective SGLT2 inhibitor, improves glucose homeostasis in normal and diabetic rats. Diabetes 57:1723–1729
197. Brooks AM, Thacker SM (2009) Dapagliflozin for the treatment of type 2 diabetes. Ann Pharmacother 43:1286–1293
198. Johnson & Johnson Pharmaceutical Research & Development, LLC (2010) CANagliflozin treatment and trial analysis-sulfonylurea (CANTATA-SU) SGLT2 add-on to metformin versus glimepiride. ClinicalTrials.gov Idenitifier: NCT00968812. http://clinicaltrials.gov/ct2/show/NCT00968812?term=canagliflozin&rank=5. Accessed 10 Jan 2013
199. Johnson & Johnson Pharmaceutical Research & Development, LLC (2010) The CANTATA-M (CANagliflozin treatment and trial analysis—monotherapy) trial. ClinicalTrials.gov Idenitifier: NCT01081834. http://clinicaltrials.gov/ct2/show/NCT01081834?term=canagliflozin&rank=2. Accessed 10 Jan 2013
200. Johnson & Johnson Pharmaceutical Research & Development, LLC (2010) The CANTATA-MSU trial (Canagliflozin treatment and trial analysis—metformin and sulphonylurea). ClinicalTrials.gov Idenitifier: NCT01106625. http://clinicaltrials.gov/ct2/show/NCT01106625?term=canagliflozin&rank=3. Accessed 10 Jan 2013
201. Johnson & Johnson Pharmaceutical Research & Development, LLC (2010) The CANTATA-D trial (Canagliflozin treatment and trial analysis—DPP-4 inhibitor comparator trial). ClinicalTrials.gov. Idenitifier: NCT01106677. http://clinicaltrials.gov/ct2/show/NCT01106677?term=canagliflozin&rank=4. Accessed 10 Jan 2013
202. Bhanot S, Murray SF, Booten SL, Chakravarty K, Zanardi T, Henry S, Watts LM, Wancewicz EV, Siwkowski A, Monia BP (2009) ISIS 388626, an SGLT2 antisense drug, causes robust and sustained glucosuria in multiple species and is safe and well-tolerated. Proc 69th scientific sessions meeting of the American diabetes association, New Orleans, (Abstract 328-OR)
203. Price SA, Dent C, Duran-Jimenez B, Liang Y, Zhang L, Rebar EJ, Case CC, Gregory PD, Martin TJ, Spratt SK, Tomlinson DR (2006) Gene transfer of an engineered transcription factor promoting expression of VEGF-A protects against experimental diabetic neuropathy. Diabetes 55:1847–1854
204. Dent CL, Lau G, Drake EA, Yoon A, Case CC, Gregory PD (2007) Regulation of endogenous gene expression using small molecule-controlled engineered zinc-finger protein transcription factors. Gene Ther 14:1362–1369
205. Benaim E, Hamilton S, Spratt K, Giedlin M, Kim H, Ando D; Sangamo DN Study Group (2009) Vascular endothelial growth factor zinc finger protein activator (SB-509) in mild to moderate diabetic peripheral neuropathy patients. Interim phase 2 results (SB-509-0601 study). Proc 69th scientific sessions meeting of the American diabetes association, New Orleans, (Abstract 859-P)
206. Alberti L, Girola A, Gilardini L, Conti A, Cattaldo S, Micheletto G, Invitti C (2007) Type 2 diabetes and metabolic syndrome are associated with increased expression of 11-β-hydroxysteroid dehydrogenase 1 in obese subjects. Int J Obes 31:1826–1831
207. Rosenstock J, Banarer S, Fonseca VA, Inzucchi SE, Sun W, Yao W, Hollis G, Flores R, Levy R, Williams WV, Seckl JR, Huber R, INCB13739–202 Principal Investigators (2010) The 11-β-hydroxysteroid dehydrogenase type 1 inhibitor INCB-13739 improves hyperglycaemia in patients with type 2 diabetes inadequately controlled by metformin monotherapy. Diabetes Care 33:1516–1522
208. Schuster D, Maurer EM, Laggner C, Nashev LG, Wilckens T, Langer T, Odermatt A (2006) The discovery of new 11β-hydroxysteroid dehydrogenase type 1 inhibitors by common feature pharmacophore modelling and virtual screening. J Med Chem 49:3454–3466

209. Dostálová I, Haluzíková D, Haluzík M (2009) Fibroblast growth factor 21: a novel metabolic regulator with potential therapeutic properties in obesity/type 2 diabetes mellitus. Physiol Res 58:1–7
210. Potthoff MJ, Inagaki T, Satapati S, Ding X, He T, Goetz R, Mohammadi M, Finck BN, Mangelsdorf DJ, Kliewer SA, Burgess SC (2009) FGF21 induces PGC-1alpha and regulates carbohydrate and fatty acid metabolism during the adaptive starvation response. Proc Natl Acad Sci USA 106:10853–10858
211. Mai K, Andres J, Biedasek K, Weicht J, Bobbert T, Sabath M, Meinus S, Reinecke F, Möhlig M, Weickert MO, Clemenz M, Pfeiffer AF, Kintscher U, Spuler S, Spranger J (2009) Free fatty acids link metabolism and regulation of the insulin-sensitizing fibroblast growth factor–21. Diabetes 58:1532–1538
212. Zhang BB, Zhou G, Li C (2009) AMPK: an emerging drug target for diabetes and the metabolic syndrome. Cell Metab 9:407–416
213. Kadowaki T, Yamauchi T (2005) Adiponectin and adiponectin receptors. Endocr Rev 26:439–451
214. Snijder MB, Heine RJ, Seidell JC, Bouter LM, Stehouwer CD, Nijpels G, Funahashi T, Matsuzawa Y, Shimomura I, Dekker JM (2006) Associations of adiponectin with incident impaired glucose metabolism and type 2 diabetes in older men and women. Diabetes Care 29:2498–2503
215. Fang X, Sweeney G (2006) Mechanisms regulating energy metabolism by adiponectin in obesity and diabetes. Biochem Soc Trans 34:798–801
216. Blackburn GL (2010) From bench to bedside: novel mechanisms and therapeutic advances through the development of selective peroxisome proliferator-activated receptor c modulators. Am J Clin Nutr 91:251S–253S
217. Cheatham WW (2010) Peroxisome proliferator-activated receptor translational research and clinical experience. Am J Clin Nutr 91:262S–266S
218. Higgins LS, DePaoli AM (2010) Selective peroxisome proliferator-activated receptor c (PPARc) modulation as a strategy for safer therapeutic PPARc activation. Am J Clin Nutr 91:267S–272S
219. Ziemke F, Mantzoros CS (2010) Adiponectin in insulin resistance: lessons from translational research. Am J Clin Nutr 91:258S–261S
220. Billyard T, McTernan P, Kumar S (2007) Potential therapies based on antidiabetic peptides. Clin Endocrinol Metab 21:641–655
221. Cool B, Zinker B, Chiou W, Kifle L, Cao N, Perham M, Dickinson R, Adler A, Gagne G, Iyengar R, Zhao G, Marsh K, Kym P, Jung P, Camp HS, Frevert E (2006) Identification and characterization of a small molecule AMPK activator that treats key components of type 2 diabetes and the metabolic syndrome. Cell Metab 3:403–416
222. Goransson O, McBride A, Hawley SA, Ross FA, Shpiro N, Foretz M, Viollet B, Hardie DG, Sakamoto K (2007) Mechanism of action of A-769662, a valuable tool for activation of AMP-activated protein kinase. J Biol Chem 282:32549–32560
223. Sanders MJ, Ali ZS, Hegarty BD, Heath R, Snowden MA, Carling D (2007) Defining the mechanism of activation of AMP-activated protein kinase by the small molecule A-769662, a member of the thienopyridone family. J Biol Chem 282:32539–32548
224. Moreno D, Knecht E, Viollet B, Sanz P (2008) A769662, a novel activator of AMP-activated protein kinase, inhibits non-proteolytic components of the 26S proteasome by an AMPK-independent mechanism. FEBS Lett 582:2650–2654
225. Pang T, Zhang ZS, Gu M, Qiu BY, Yu LF, Cao PR, Shao W, Su MB, Li JY, Nan FJ, Li J (2008) Small molecule antagonizes autoinhibition and activates AMP-activated protein kinase in cells. J Biol Chem 283:16051–16060
226. Haigis MC, Guarente LP (2006) Mammalian sirtuins —emerging roles in physiology, aging, and calorie restriction. Genes Dev 20:2913–2921
227. Milne JC, Lambert PD, Schenk S, Carney DP, Smith JJ, Gagne DJ, Jin L, Boss O, Perni RB, Vu CB, Bemis JE, Xie R, Disch JS, Ng PY, Nunes JJ, Lynch AV, Yang H, Galonek H, Israelian K, Choy W, Iffland A, Lavu S, Medvedik O, Sinclair DA, Olefsky JM, Jirousek MR, Elliott

PJ, Westphal CH (2007) Small molecule activators of SIRT1 as therapeutics for the treatment of type 2 diabetes. Nature 450:712–716
228. Yoshizaki T, Milne JC, Imamura T, Schenk S, Sonoda N, Babendure JL, Lu JC, Smith JJ, Jirousek MR, Olefsky JM (2008) SIRT1 exerts anti-inflammatory effects and improves insulin sensitivity in adipocytes. Mol Cell Biol 29:1363–1374
229. Ota H, Akishita M, Eto M, Iijima K, Kaneki M, Ouchi Y (2007) Sirt1 modulates premature senescencelike phenotype in human endothelial cells. J Mol Cell Cardiol 43:571–579
230. Zhang QJ, Wang Z, Chen HZ, Zhou S, Zheng W, Liu G, Wei YS, Cai H, Liu DP, Liang CC (2008) Endothelium-specific overexpression of class III deacetylase SIRT1 decreases atherosclerosis in apolipoprotein E-deficient mice. Cardiovasc Res 80:191–199
231. Rush JW, Quadrilatero J, Levy AS, Ford RJ (2007) Chronic resveratrol enhances endothelium-dependent relaxation but does not alter eNOS levels in aorta of spontaneously hypertensive rats. Exp Biol Med 232:814–822
232. Potente M, Dimmeler S (2008) Emerging roles of SIRT1 in vascular endothelial homeostasis. Cell Cycle 7:2117–2122
233. Matschinsky FM, Magnuson MA, Zelent D, Jetton TL, Doliba N, Han Y, Taub R, Grimsby J (2006) The network of glucokinase-expressing cells in glucose homeostasis and the potential of glucokinase activators for diabetes therapy. Diabetes 55:1–12
234. Leighton B, Atkinson A, Coghlan MP (2005) Small molecule glucokinase activators as novel anti-diabetic agents. Biochem Soc Trans 33:371–374
235. Matschinsky FM (2009) Assessing the potential of glucokinase activators in diabetes therapy. Nat Rev Drug Discov 8:399–416
236. Grimsby J, Sarabu R, Corbett WL, Haynes NE, Bizzarro FT, Coffey JW, Guertin KR, Hilliard DW, Kester RF, Mahaney PE, Marcus L, Qi L, Spence CL, Tengi J, Magnuson MA, Chu CA, Dvorozniak MT, Matschinsky FM, Grippo JF (2003) Allosteric activators of glucokinase: potential role in diabetes therapy. Science 301:370–373
237. Drug candidates in development by phase. DiabesityDigest.com. http://diabesitydigest.com/development.php?order=devel. Accessed 10 Jan 2013
238. Broca C, Breil V, Cruciani-Guglielmacci C, Manteghetti M, Rouault C, Derouet M, Rizkalla S, Pau B, Petit P, Ribes G, Ktorza A, Gross R, Reach G, Taouis M (2004) Insulinotropic agent ID-1101 (4-hydroxyisoleucine) activates insulin signaling in rat. Am J Physiol Endocrinol Metab 287:E463–E471
239. Boyd SA (2008) Arry-588 and Arry-403: novel glucokinase activators with potent glucose-lowering activity in animals of type 2 diabetes mellitus. SMI Metabolic Diseases Conference http://www.arraybiopharma.com/_documents/Publication/PubAttachment297.pdf. Accessed 10 Jan 2013
240. Aicher T, Anderson D, Boyd S, Chicarelli M, Condroski K, DeWolf W Jr, Fell JB, Fischer J, Frank M, Galbraith S, Garrey R, Gunawardana I, Hinklin R, Hirsch J, Lee P, Neitzel N, Pratt S, Singh A, Thomas L, Turner T, Voegtli W, Wallace E, Williams L (2008) Discovery of ARRY-588, a novel glucokinase activator with potent glucose lowering activity in animal models of type 2 diabetes mellitus. www.arraybiopharma.com/_documents/Publication/PubAttachment288.pdf. Accessed 10 Jan 2013
241. Edgerton DS, Johnson KM, Cherrington AD (2009) Current strategies for the inhibition of hepatic glucose production in type 2 diabetes. Front Biosci 14:1169–1181
242. Kaal MJH, Collet JT (1993) Inhibition of gluconeogenesis an option for the treatment of type 2 diabetes mellitus? Pharmaceut Week 128:507–511
243. van Poelje PD, Dang Q, Erion MD (2007) Fructose-1,6-bisphosphatase as a therapeutic target for type 2 diabetes. Drug Discov Today 4:103–109
244. Erion MD, van Poelje PD, Dang Q (2005) MB06322 (CS-917): a potent and selective inhibitor of fructose 1,6-bisphosphatase for controlling gluconeogenesis in type 2 diabetes. Proc Natl Acad Sci USA 102:7970–7975
245. Freemantle N, Blonde L, Duhot D, Hompesch M, Eggertsen R, Hobbs FD, Martinez L, Ross S, Bolinder B, Stridde E (2005) Availability of inhaled insulin promotes greater perceived acceptance of insulin therapy in patients with type 2 diabetes. Diabetes Care 28:427–428

246. Arbit E, Kidron M (2009) Oral insulin: the rationale for this approach and current developments. J Diabetes Sci Technol 3:562–567
247. Amin A, Shah T, Patel J, Gajjar A (2008) Annual update on non-invasive insulin delivery technologies. Drug Deliv Technol 8:43–48
248. Clement S, Dandona P, Still JG, Kosutic G (2004) Oral modified insulin (HIM2) in patients with type 1 diabetes mellitus: results from a phase I/II clinical trial. Metabolism 53:54–58
249. Maher S, Leonard TW, Jacobsen J, Brayden DJ (2009) Safety and efficacy of sodium caprate in promoting oral drug absorption: from in vitro to the clinic. Adv Drug Deliv Rev 61:1427–1449
250. Johnson TO, Ermolieff J, Jirousek MR (2002) Protein tyrosine phosphatase 1B inhibitors for diabetes. Nat Rev Drug Discov 1:696–709
251. Pei Z, Liu G, Lubben TH, Szczepankiewicz BG (2004) Inhibition of protein tyrosine phosphatase 1B as a potential treatment of diabetes and obesity. Curr Pharm Des 10:3481–3504
252. Mehdi MZ, Pandey SK, Théberge JF, Srivastava AK (2006) Insulin signal mimicry as a mechanism for the insulin-like effects of vanadium. Cell Biochem Biophys 44:73–81
253. Srivastava AK, Mehdi MZ (2005) Insulino-mimetic and anti-diabetic effects of vanadium compounds. Diabet Med 22:2–13
254. Vardatsikos G, Mehdi MZ, Srivastava AK (2009) Bis(maltolato)-oxovanadium (IV)-induced phosphorylation of PKB, GSK-3 and FOXO1 contributes to its glucoregulatory responses. Int J Mol Med 24:303–309
255. Bertolaso M, Olsson J, Picardi A, Rakela J (2010) Gene therapy and enhancement for diabetes (and other diseases): the multiplicity of considerations. Diabetes Metab Res Rev 26:520–524
256. Levine F, Leibowitz G (1999) Towards gene therapy of diabetes mellitus. Mol Med Today 5:165–171
257. Smeekens SP, Montag AG, Thomas G, Albiges-Rizo C, Carroll R, Benig M, Phillips LA, Martin S, Ohagi S, Gardner P, Swift HH, Steiner DF (1992) Proinsulin processing by the subtilisin-related proprotein convertases furin, PC2, and PC3. Proc Natl Acad Sci USA 89:8822–8826
258. Won JC, Rhee BD, Ko KS (2009) Glucose-responsive gene expression system for gene therapy. Adv Drug Deliv Rev 61:633–640
259. Wong MS, Hawthorne WJ, Manolios N (2010) Gene therapy in diabetes. Self Nonself 1:165–175
260. Han J, McLane B, Kim EH, Yoon JW, Jun HS (2011) Remission of diabetes by insulin gene therapy using a hepatocyte-specific and glucose-responsive synthetic promoter. Mol Ther 19:470–478
261. Limbert C, Päth G, Ebert R, Rothhammer V, Kassem M, Jakob F, Seufert J (2011) PDX1- and NGN3-mediated in vitro reprogramming of human bone marrow-derived mesenchymal stromal cells into pancreatic endocrine lineages. Cytotherapy 13:802–813
262. Kajiyama H, Hamazaki TS, Tokuhara M, Masui S, Okabayashi K, Ohnuma K, Yabe S, Yasuda K, Ishiura S, Okochi H, Asashima M (2009) Pdx1-transfected adipose tissue-derived stem cells differentiate into insulin-producing cells in vivo and reduce hyperglycaemia in diabetic mice. Int J Dev Biol 54:699–705
263. Kojima H, Fujimiya M, Matsumura K, Younan P, Imaeda H, Maeda M, Chan L (2003) NeuroD-betacellulin gene therapy induces islet neogenesis in the liver and reverses diabetes in mice. Nat Med 9:596–603
264. Gefen-Halevi S, Rachmut IH, Molakandov K, Berneman D, Mor E, Meivar-Levy I, Ferber S (2010) NKX6.1 promotes PDX-1-induced liver to pancreatic β-cells reprogramming. Cell Reprogram 12:655–664
265. Cunha-Vaz JG (1978) Pathophysiology of diabetic retinopathy. Br J Ophthalmol 62:351–355
266. Cooke A, Phillips JM, Parish NM (2011) Tolerogenis strategies to halt or prevent type 1 diabetes. Nat Immunol 2:810–815
267. Long SA, Cerosaletti K, Bollyky PL, Tatum M, Shilling H, Zhang S, Zhang ZY, Pihoker C, Sanda S, Greenbaum C, Buckner JH (2010) Defects in IL-2R signaling contribute to

diminished maintenance of FOXP3 expression in CD4(+)CD5(+) regulatory T-cells of type 1 diabetes subjects. Diabetes 59:407–415
268. Bennett CL, Christie J, Ramsdell F, Brunkow ME, Ferguson PJ, Whitesell L, Kelly TE, Saulsbury FT, Chance PF, Ochs HD (2001) The immune dysregulation, polyendocrinopathy, enteropathy, X-linked syndrome (IPEX) is caused by mutations in FOXP3. Nat Genet 27:20–21
269. Goudy KS, Johnson MC, Garland A, Li C, Samulski RJ, Wang B, Tisch R (2011) Reduced IL-2 expression in NOD mice leads to a temporal increase in CD62LIo FoxP3+ CD4+ T cells with limited suppressor activity. Eur J Immunol 41:1480–1490
270. Rehman KK, Trucco M, Wang Z, Xiao X, Robbins PD (2008) AAV8-mediated gene transfer of interleukin-4 to endogenous beta-cells prevents the onset of diabetes in NOD mice. Mol Ther 16:1409–1416
271. Jean M, Alameh M, Buschmann MD, Merzouki A (2011) Effective and safe gene-based delivery of GLP-1 using chitosan/plasmid-DNA therapeutic nanocomplexes in an animal model of type 2 diabetes. Gene Ther 18:807–816
272. Riedel MJ, Gaddy DF, Asadi A, Robbins PD, Kieffer TJ (2010) DsAAV8-mediated expression of glucagon-like peptide-1 in pancreatic beta-cells ameliorates streptozotocin-induced diabetes. Gene Ther 17:171–180
273. Bindom SM, Hans CP, Xia H, Boulares AH, Lazartigues E (2010) Angiotensin I-converting enzyme type 2 (ACE2) gene therapy improves glycemic control in diabetic mice. Diabetes 59:2540–2548
274. Engerman RL (1989) Pathogenesis of diabetic retinopathy. Diabetes 38:1203–1206
275. Stratton IM, Adler AI, Neil HA, Matthews DR, Manley SE, Cull CA, Hadden D, Turner RC, Holman RR (2000) Association of glycaemia with macrovascular and microvascular complications of type 2 diabetes (UKPDS 35): prospective observational study. BMJ 321:405–412
276. Campochiaro PA (2006) Potential applications for RNAi to probe pathogenesis and develop new treatments for ocular disorders. Gene Ther 13:559–562
277. Sarkar K, Fox-Talbot K, Steenbergen C, Bosch-Marcé M, Semenza GL (2009) Adenoviral transfer of HIF-1alpha enhances vascular responses to critical limb ischemia in diabetic mice. Proc Natl Acad Sci USA 106:18769–18774
278. Samuel SM, Akita Y, Paul D, Thirunavukkarasu M, Zhan L, Sudhakaran PR, Li C, Maulik N (2010) Coadministration of adenoviral vascular endothelial growth factor and angiopoietin-1 enhances vascularization and reduces ventricular remodeling in the infarcted myocardium of type 1 diabetic rats. Diabetes 59:51–60
279. Catrina SB, Okamoto K, Pereira T, Brismar K, Poellinger L (2004) Hyperglycemia regulates hypoxia-inducible factor-1alpha protein stability and function. Diabetes 53:3226–3232
280. Mace KA, Yu DH, Paydar KZ, Boudreau N, Young DM (2007) Sustained expression of Hif-1alpha in the diabetic environment promotes angiogenesis and cutaneous wound repair. Wound Repair Regen 15:636–645
281. Liu L, Marti GP, Wei X, Zhang X, Zhang H, Liu YV, Nastai M, Semenza GL, Harmon JW (2008) Age-dependent impairment of HIF-1alpha expression in diabetic mice: correction with electroporation-facilitated gene therapy increases wound healing, angiogenesis, and circulating angiogenic cells. J Cell Physiol 217:319–327
282. Bento CF, Pereira P (2011) Regulation of hypoxia-inducible factor 1 and the loss of the cellular response to hypoxia in diabetes. Diabetologia 54:1946–1956
283. Ceradini DJ, Yao D, Grogan RH, Callaghan MJ, Edelstein D, Brownlee M, Gurtner GC (2008) Decreasing intracellular superoxide corrects defective ischemia-induced new vessel formation in diabetic mice. J Biol Chem 283:10930–10938
284. Bento CF, Fernandes R, Ramalho J, Marques C, Shang F, Taylor A, Pereira P (2010) The chaperone-dependent ubiquitin ligase CHIP targets HIF-1α for degradation in the presence of methylglyoxal. PLoS One 5:e15062
285. Wang J, Jiao H, Stewart TL, Lyons MV, Shankowsky HA, Scott PG, Tredget EE (2007) Accelerated wound healing in leukocyte-specific, protein 1-deficient mouse is associated with increased infiltration of leukocytes and fibrocytes. J Leukoc Biol 82:1554–1563

286. Brem H, Kodra A, Golinko MS, Entero H, Stojadinovic O, Wang VM, Sheahan CM, Weinberg AD, Woo SL, Ehrlich HP, Tomic-Canic M (2009) Mechanism of sustained release of vascular endothelial growth factor in accelerating experimental diabetic healing. J Invest Dermatol 129:2275–2287
287. Prud'homme GJ, Draghia-Akli R, Wang Q (2007) Plasmid-based gene therapy of diabetes mellitus. Gene Ther 14:553–564
288. Dube MG, Beretta E, Dhillon H, Ueno N, Kalra PS, Kalra SP (2002) Central leptin gene therapy blocks high-fat diet-induced weight gain, hyperleptinemia, and hyperinsulinemia: increase in serum ghrelin levels. Diabetes 51:1729–1736
289. Prud'homme GJ, Glinka Y, Khan AS, Draghia-Akli R (2006) Electroporation-enhanced nonviral gene transfer for the prevention or treatment of immunological, endocrine and neoplastic diseases. Curr Gene Ther 6:243–273
290. Kalra SP (2011) Pivotal role of leptin-hypothalamus signaling in the aetiology of diabetes uncovered by gene therapy: a new therapeutic intervention? Gene Ther 18:319–325
291. Kalra SP (2009) Central leptin gene therapy ameliorates diabetes type 1 and 2 through two independent hypothalamic relays; a benefit beyond weight and appetite regulation. Peptides 30:1957–1963
292. Orellana-Gavaldà JM, Herrero L, Malandrino MI, Pañeda A, Sol Rodríguez-Peña M, Petry H, Asins G, Van Deventer S, Hegardt FG, Serra D (2011) Molecular therapy for obesity and diabetes based on a long-term increase in hepatic fatty-acid oxidation. Hepatology 53:821–832
293. Wild S, Roglic G, Green A, Sicree R, King H (2004) Global prevalence of diabetes: estimates for the year 2000 and projections for 2030. Diabetes Care 27:1047–1053
294. Gastaldelli A, Ferrannini E, Miyazaki Y, Matsuda M, DeFronzo RA (2004) Beta-cell dysfunction and glucose intolerance: results from the San Antonio metabolism (SAM) study. Diabetologia 47:31–39
295. Hughes TE (2009) Emerging therapies for metabolic diseases – the focus is on diabetes and obesity. Curr Opin Chem Biol 13:332–337

Identification of Biomarkers for Beta Cell Failure as Detection Tools for Predictive Screening in Type 2 Pre-diabetes

Jadranka Koehn and Kurt Krapfenbauer

Abstract It has been suggested that a more precise selection of predictive biomarkers may prove useful in the early diagnosis of type 2 diabetes (T2D), even when glucose tolerance is normal. This is vital since many T2D cases may be preventable by avoiding those factors that trigger the disease process (primary prevention) or by use of therapy that modulates the disease process before the onset of clinical symptoms (secondary prevention) occurs. The selection of predictive markers must be carefully assessed and depends mainly on three important parameters: sensitivity, specificity and positive predictive value. Unfortunately, biomarkers with ideal specificity and sensitivity are difficult to find. One potential solution is to use the combinatorial power of a large number of biomarkers, each of which alone may not offer satisfactory specificity and sensitivity. Recent technological advances in proteomics and bioinformatics offer a great opportunity for discovery of potential predictive markers. In this chapter, we described a cellular T2D model as an example with the intent of providing specific enrichment and new identification strategies, which might have the potential to improve predictive biomarker identification and to bring accuracy in disease diagnosis and classification, as well as therapeutic monitoring in the early phase of T2D.

Keywords Glycotoxicity • Lipotoxicity • Beta-cell dysfunction • Impaired glucose tolerance • Insulin fasting glucose • Type 2 diabetes • Type 1 diabetes • Pancreas specific biomarker

---

J. Koehn (✉)
Department of Cranio-Maxillofacial and Oral Surgery, Medical University of Vienna, Waehringer Guertel 18-20, A-1090 Vienna, Austria
e-mail: j.koehn@gmx.net

K. Krapfenbauer
Department of Otorhinolaryngology, Head and Neck Surgery, Medical University of Vienna, Waehringer Guertel 18-20, A-1090 Vienna, Austria

M.S. Mozaffari (ed.), *New Strategies to Advance Pre/Diabetes Care: Integrative Approach by PPPM*, Advances in Predictive, Preventive and Personalised Medicine 3, DOI 10.1007/978-94-007-5971-8_4,

1 General Introduction

Type 2 diabetes (T2D) is one of the fastest growing health concerns worldwide with a prevalence of approximately 5% in the United States. Due to sedentary lifestyle, unhealthy nutrition and a prolonged lifespan, the prevalence in the developed and developing countries is still increasing. Health experts have warned of a global epidemic of diabetes caused by a rise in overweight and obesity. There are currently 120–140 million people worldwide with T2D, and if trends continue, this number is predicted to double in the next 25 years. In the Western world, around 90% of T2D cases are attributable to weight gain. Because of the severe health and cost implications of this disease, organizations such as the International Diabetes Federation (IDF) have called for increased efforts to prevent its development. The IDF estimates that 314 million people worldwide, or 8.2% of the global population, have impaired glucose tolerance (IGT), a state that often precedes T2D. In addition, T2D is a heavily under-diagnosed disease and is usually only discovered at a late stage, when the first complications occur. Detection in an early state would be highly desirable, as the progression of T2D can then be slowed down by simple, nonpharmacologic means, i.e. weight loss and exercise. The existing diagnostic tests, fasting plasma glucose (FPG), the oral glucose tolerance (OGT) test or the euglycemic hyperinsulinemic clamp are either relatively insensitive and inaccurate or so difficult to perform that they do not lend themselves to mass screening.

The identification of a reliable, accurate, cost-effective and non-invasive test for T2D that could be applied in the prediabetic state would present a sizeable business opportunity for diagnostics and would be valuable as a monitoring marker for clinical studies in T2D. Although significant progress has been made in the treatment of T2D, new insulin sensitizing drugs such as the thiazolidinediones display significant side-effects and major health concerns still prevail for this class of compounds. Most importantly, there is no medication on the market that slows down the progressive beta-cell (β-cell) failure, a hallmark of late-stage T2D. The identification of pathways and factors involved in either insulin-resistance of skeletal muscle or β-cell failure could lead to new drug targets and the development of novel pharmaceuticals for the treatment of impaired glucose tolerance (IGT) and for the prevention of β-cell failure, respectively.

Beta-cell failure is explained by both a relative loss of β-cell mass as well as secretory defects including enhanced basal insulin secretion by the β-cells and a selective loss of sensitivity to insulin mainly in skeletal muscle but also in other organs. Beta-cell failure is the last step in the progression of type 2 diabetics towards dependence on external insulin. The loss of β-cell function is believed to be triggered by long-term exposure to enhanced levels of glucose and lipids (glycol- and lipotoxicity). However, the molecular mechanisms of this phenomenon are poorly understood. Although insulin sensitizers hold some promise, there is currently no proven treatment for T2D patients that would prevent progression towards β-cell failure. It would also be useful to identify better targets and/or predictive markers for β-cell failure or function that are more sensitive or more reliable than the targets/markers commonly used, such as proinsulin or its processing products insulin or C-peptide. This is vital since many T2D cases may be preventable by avoiding those factors that trigger the disease process (primary prevention) or by use of therapy that modulates the disease process before the onset of clinical symptoms (secondary prevention) occurs. For a better understanding of the molecular events leading to the loss of β-cell function and to define possible targets for a function-preserving treatment a cellular model for β-cell failure is thus required. Furthermore, there is currently no good test available to assess β-cell functionality. Such a test would be of great value in potential clinical trials for pharmaceuticals that preserve β-cell function. For this reason, several labs systematically analyzed beta-cell failure to identify significant changes that occur after exposure to both palmitate and high glucose.

The rat insulinoma cell line INS-1 displays β-cell-like features making it a good in vitro model for this cell type. It has preserved many β-cell characteristics, including glucose-stimulated insulin secretion. Furthermore, study of these β-cells with decreased insulin secretion as compared to cells with normal function could lead to a better understanding of the biology of β-cell failure and could highlight possible targets for therapeutic intervention. Study of these cells may also lead to the identification of markers for β-cell failure or function that are more reliable or sensitive than the C-peptide.

Using intracellular insulin content and insulin mRNA synthesis as endpoints, conditions were established where insulin production was decreased as a consequence of elevated glucose levels (glucotox), high palmitate levels (lipotox), and a combination of both. Cell cultures propagated under these conditions were analyzed for their mRNA and protein composition. The semiquantitative protein expression data obtained by proteomics analysis was systematically analyzed to identify significant changes that occur only after exposure to both palmitate and high glucose but not by either one of these treatments alone.

In conclusion, important steps were made towards the identification of β-cell specific proteins that might serve in the future as predictive markers for β-cell function or failure. Furthermore, it would be an advantage to identify such markers that can be detected in plasma of humans. Follow-up experiments using animal models and human blood and plasma samples were designed to validate opportunities to pharmaceutically target β-cell failure and to generate novel medicines for T2D.

2 Current Status of Diabetes Diagnostics

Blood glucose testing is the most commonly used method for the diagnosis and monitoring of diabetes mellitus. FPG levels and plasma glucose levels 2 h after the uptake of a 75 g dose of glucose are two possible methods to detect the disease. Two independently obtained positive test results are required to establish the diagnosis. As it is easier and less time-consuming to perform, the FPG test is recommended by the American Diabetes Association. However, FPG levels will only be increased, when an increased insulin secretion fails to compensate for insulin resistance, which will only be the case in late stages of T2D. The OGT test and the euglycemic hyperinsulinemic clamp [1] are two methodologies that have the potential to detect insulin resistance at an earlier stage, but are relatively time-consuming to perform and are not amenable to mass testing. Once the diagnosis is established, the efficacy of a treatment regimen can be monitored by measuring glycated hemoglobin (HbA1c) which reflects the plasma glucose levels over the last 4–6 weeks [2]. For an assessment of β-cell function, the measurement of the concentration of C-peptide is, at this time, the best method available [3, 4]. Measurement of this parameter is however not recommended for clinical practice as it suffers from a large variability [5].

3 Biomarkers for Diabetes

The classical risk factors for type 1 diabetes (T1D) are, apart from familiar predisposition, genetic markers, like certain HLA-DQ or CTLA4 genotypes associate with increased risk of developing T1D. The detection of auto antibodies to islet cell antigens (ICA), insulin, glutamic acid decarboxylase (GAD-65), or IA-2 protein tyrosine phosphatase marks the onset of the autoimmune disease and can be detected years before clinical symptoms of diabetes appear. While certain variants in genes encoding a β-cell specific ion-channel (SUR8/Kir6.2) are associated with transient or permanent neonatal diabetes and mutations in the ABCA1 gene have recently been associated with T2D (reviewed in [6]), obesity is still the greatest known risk factor for T2D. Though familiar clustering of T2D implies the existence of genetic risk factors, these are still unknown.

While for T1D early diagnostic markers exist in the form of auto antibodies, T2D can so far only be diagnosed when the clinical symptoms appear at the later stages of the disease. Metabolic markers like hyperglycemia, insulin or C-peptide level, and HbA1c are used to diagnose diabetes and monitor the disease state in all types of diabetes. Since an early diagnosis and an early therapy hold the greatest promises to postpone full disease development and to delay secondary morbidities like cardiovascular diseases, reliable, sensitive markers for early T2D are a key concern.

Several novel biomarkers for T2D have been reported and reviewed recently [7]. Among the T2D marker candidates are markers of fatty liver routinely tested like alanine aminotransferase (ALT) and γ-glutamyl transferase (γ-GT), both well correlated with diabetes risk [8]. Low levels of sex homone binding globuline (SHBG), a hepatically secreted hormone negatively regulated by insulin, can indicate high insulin levels, insulin resistance and risk of T2D, a correlation that is stronger in women than in men [9]. Low SHBG has been reported to be predictive of gestational diabetes [10]. Inflammatory markers interleukin 6 (IL-6) and

C-reactive protein (CRP) induced by IL-6 have also been associated with T2D [11], as has another acute phase marker ferritin [12]. While high CRP and IL-6 levels show a robust correlation to T2D incidences, also independent of obesity, serum ferritin, initially a marker of cellular iron stores, is not fully validated as a indicator of T2D risk [13]. Adiponectin, a hormone produced by adipocytes and involved in the regulation of fat and glucose metabolism, has an inverse correlation to diabetes, i.e. low adiponectin levels correlate with high T2D risk [14] and genetic variants of adiponectin have been associated with increased risk of T2D [15]. Adiponectin exerts insulin-sensitizing and anti-infamatory functions through two different seven-transmembrane receptors expressed predominantly in β-cells and skeletal muscles (ADIPOR1) or the liver (ADIPOR2) [16]. Leptin, another adipocyte hormone regulates appetite and has been associated with T2D but the significance is unclear, as leptin levels can be deregulated in leptin resistant obese patients and a positive and a negative association with T2D can be found [17]. Tissue plasminogen activator (tPA) and its inhibitor PAI-1 have also been linked to T2D [18–20]. Both of these proteins can be secreted form endothelial cells and high plasma levels usually reflect endothelial stress or inflammation. PAI-1 can also be released from adipocytes and hepatocytes and its plasma levels increase with adiposity, weakening the association with T2D. Like tPA and PAI-1 von Willebrand factor (vWF) is a maker of endothelial dysfunction associated with risk of T2D in the literature [21] and like the former it is induced by proinflammatory cytokines. The association of vWF with T2D is less robust and significance is lost after adjustment for obesity or inflammatory cytokines in some studies [20]. Other markers or risk factors for T2D have been described, many of these are also associated with obesity or inflammation and thus might not be useful to discern an obese patient at risk to develop T2D form one not at risk. Furthermore, many risk factors have been analyzed in small study groups. All markers identified so far are present in healthy populations as well though at a different levels. But it is not only T2D that will influence their expression level, age, sex, ethnicity might also play a role increasing the variability and reducing the specificity.

4 Current Treatment Scheme for T2D

In the early stages of the disease, weight loss and an increase in physical activity are often sufficient to restore insulin sensitivity. If this regimen fails, non-insulinotropic agents such as acarbose or metformin (and glitazones) are the first-line therapeutics. When insulin levels drop, insulin secretagogues (sulfonylureas, repaglinide) need to be added to the therapeutic regimen. Eventually, when this regimen has become insufficient, insulin injection becomes necessary.

5 Scientific Rationale for β-Cell Failure as a Disease Model for T2D

Beta-cell failure occurs at a late stage in the progression of T2D and leads to overt diabetes. The mechanisms by which β-cell function is impaired are largely unknown but have been shown to involve hyperglycemia as well as hyperlipidemia [22].

Interestingly, β-cell failure does not appear to be an inevitable consequence of hyperinsulinemia. Some animal models of hyperinsulinemia/diabetes show persistent life-long hyperinsulinemia (e.g. ob/ob mice, fa/fa rats) while others show deteriorating β-cell function and apoptosis of β-cells (e.g. db/db mice, *Psammomys obesus*, and rhesus monkeys) [23]. The reasons for these differences are unclear and include genetic factors [24]. The situation in humans may be similar, as insulin resistance is an almost inevitable consequence of obesity but only some individuals have frank diabetes and progress into insulin dependence. The fact that β-cell failure is avoidable in some animal models encourages research towards the prevention of β-cell failure. In vitro studies have demonstrated a glucotoxic effect that was only partially reversible [23, 25, 26]. Lipotoxicity plays an equally important role [27, 28]. A profound effect on β-cell function is observed in models combining both factors [22]. Some scientists speculate that these two mechanisms are independent of each other and involve distinct pathways (M.Y. Donath, Zuerich, personal communication 2004). Apoptosis plays a critical role in β-cell failure and T2D [29] and is induced by Fas-FasL interaction where the latter is constitutively expressed by β-cells and the former induced by cytokines and glucose (M.Y. Donath, personal communication 2004). FLIP blocks the cascade from Fas-FasL interaction → caspase 8 → 3 and finally apoptosis and even reverses the effect of high glucose, thus resulting in proliferation instead of apoptosis. This suggests that prevention of β-cell failure is feasible. The apoptotic effect of lipids, mediated by Bcl-2 and ceramide, is less well documented. Importantly, although apoptotic pathways have been explored extensively, the present study addresses functional β-cell failure outside apoptosis. A decrease in β-cell mass was observed in patients with advanced T2D but not in the early stages of the disease [30]. The metabolic disturbances of most patients will go unnoticed until frank hyperglycemia is detected. However, the vast majority of patients will remain in a stage of partial compensation before moving on to dependence on exogenous insulin. At this particular stage, compounds preventing β-cell failure could be applied. The goals of the proteomics study are twofold: (i) The first line of experiments will be directed towards the identification and validation of a potential monitoring marker for clinical studies. Indeed, the current gold standard for the measurement of β-cell function, the measurement of the insulin C-peptide, is not sufficiently reliable for this purpose. To ensure its broad acceptance, a marker would have to be established well in advance of a clinical program aimed at demonstrating disease modifying (i.e., β-cell failure preventing) efficacy of a novel compound. (ii) Secondly, the changes in protein expression patterns will be studied in order to improve our understanding of the biological processes involved in β-cell failure, ultimately leading to the identification of pathways and novel targets for therapeutic intervention.

6 Methodology Leading to the Identification of Potential Biomarkers

6.1 Cellular Model System for the Identification of Potential Markers of β-Cell Failure

Isolated islets and insulin secreting cell lines become desensitized upon exposure to high levels of glucose and palmitate [22]. The rat insulinoma cell line INS-1 and the mouse cell line Min6 appear to be the best available model cell lines at this time [31, 32]. They have retained many characteristics of primary β-cells, including the susceptibility to the combined toxic effect of glucose and lipids.

Two classes of proteins could fulfill the criteria for a marker of β-cell failure: (i) Highly expressed β-cell specific proteins could become detectable in the bloodstream upon cell death, and (ii) the level of β-cell specific secreted proteins could significantly drop as a sign of impaired β-cell function. Ideally the latter would be constitutively secreted and not depend on glucose concentration as shown in the results of the OGTT (see Fig. 9). For this purpose, a catalogue of proteins expressed in β-cell lines were established to identify β-cell specific proteins that have the potential to become detectable in human blood and that could serve as a marker for β-cell failure. Beta cell specific proteins identified by this strategy will require to be first validated in an animal model for β-cell failure, for example in blood samples collected from rat strains rapidly developing full-blown diabetes (e.g. ZDF rats), before human blood samples can be tested.

6.2 New Enrichment and Identification Strategies of Differentially Regulated Proteins in Failing β-Cells

Beta-cell failure can be mimicked in several cell culture models by adding a combination of fatty acids and increased concentrations of glucose to the culture medium [22, 28]. Comparison of expression patterns from untreated cells, cells treated with either a fatty acid or high glucose, and cells treated with both fatty acid and high glucose should highlight gene products whose regulation correlates with β-cell failure in a cell culture model. The results from this study directed the proteome comparison towards the compartments that appeared most affected based on the transcript analysis. In a second step these changes were validated using animal models such as ZDF rats or db/db mice. Finally, findings from the primary β-cells were validated by using human plasma samples from five different donor groups (impaired glucose tolerance (IGT), impaired fasting glucose (IFG), type 2 diabetes (T2D), type 1 diabetes (T1D), and healthy donors) (Fig. 8).

6.3 *Critical Points of the Beta Cell Failure Model*

- A reliable, reproducible model of β-cell failure induced by gluco- and lipotoxicity after a reasonable exposure time of 35 h but not days was established first.
- Triggering failure of the cultured cells to secrete insulin is a time-dependent process that was carefully monitored (see Fig. 2). Establishing the appropriate monitoring readouts was therefore crucial to this project. Selecting the appropriate time point based on the measurement of the selected parameters was difficult and required kinetic measurements.
- Up-scaling of the cell culture was necessary for obtaining sufficient quantities of starting material for a proteome analysis (see Fig. 4). This may lead to variations. Careful monitoring by measurement of the parameters defined by the pilot experiments was therefore mandatory.
- Beta-cell specific proteins, even if highly abundant intracellularly, or secreted proteins may not be present in blood in detectable quantities. Therefore highly sensitive ELISA were used to validate our findings in human blood samples
- A fractionation scheme that efficiently enriches for secretory vesicles has to be established.
- The concentration of secreted proteins in the culture supernatant is likely to be very low. The accessibility of these proteins by mass spectrometric methods was therefore assessed.

6.4 *Cellular Study Design*

Features of β-cell failure can be performed by chronic exposure of β-cells to a combination of high glucose/fatty acids (FAs), suggesting that hyperlipidemia as well as hyperglycemia may contribute to decompensation of β-cells. INS-1 cells pretreated for 30 h with a combination of 10 mM glucose and 0.5 mM palmitate can be used for these experiments. Establishment of β-cell culture conditions for proteomics analysis was carried out in parallel (see Figs. 1 and 3). Screen INS-1 cell cultures under treatment with glucose/fatty acids and identification of different expressed protein levels between glyco/lipotoxicity and combination. Focus on secreted proteins in the supernatant as possible markers for β-cell failure/function. Enrichment and identification of secreted INS-1 peptides/proteins in the medium by heparin chromatography followed by LC-MS were applied.

6.5 *General Introduction and Strategies for Enrichment and Purification of Secreted Proteins*

In order to identify proteins secreted by INS-1 [31, 33] two methods were applied: (i) Fractionation of the cells by differential sedimentation into subcellular

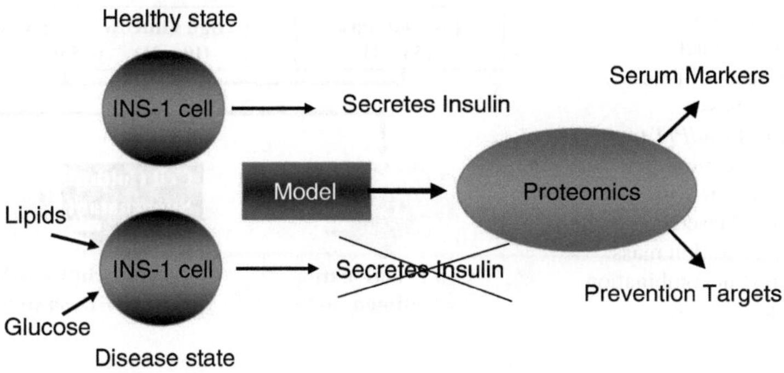

Fig. 1 *In vitro* model of β-cell failure: concept of gluco/lipotoxicity

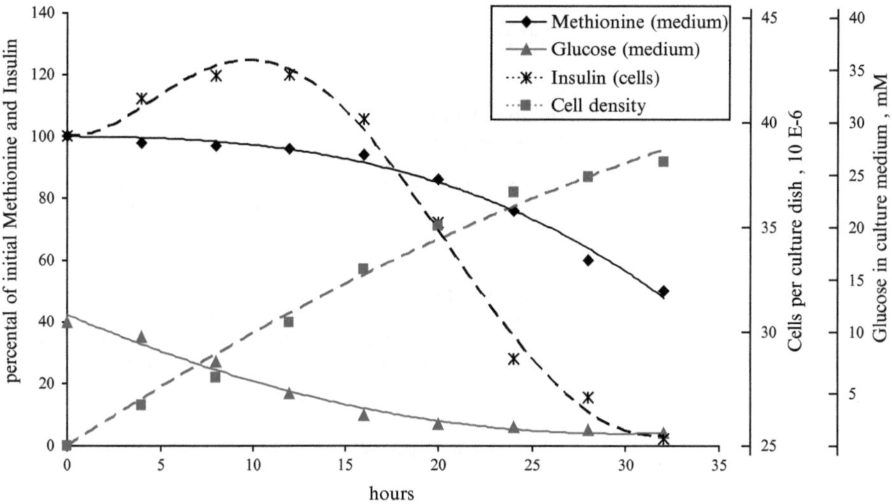

Fig. 2 Characterization and whole protein profiling of INS-1 cells: model of β-cell failure induced by gluco- and lipotoxicity after 35 h exposure time

compartments with subsequent identification of the proteins based on their peptide mass fingerprint using MALDI-TOF mass spectrometry, and (ii) enrichment of glycoprotein's by heparin chromatography followed by one-dimensional SDS-PAGE and identification of proteins by analysis of the tryptic peptides resulting from protein digest by liquid chromatography coupled to tandem mass spectrometry resulting in identification based on protein sequence tags (see Fig. 4). The combination of these two purification strategies allowed us to increase the efficiency of protein identification in the cellular compartments as well as in the medium of cultured cells.

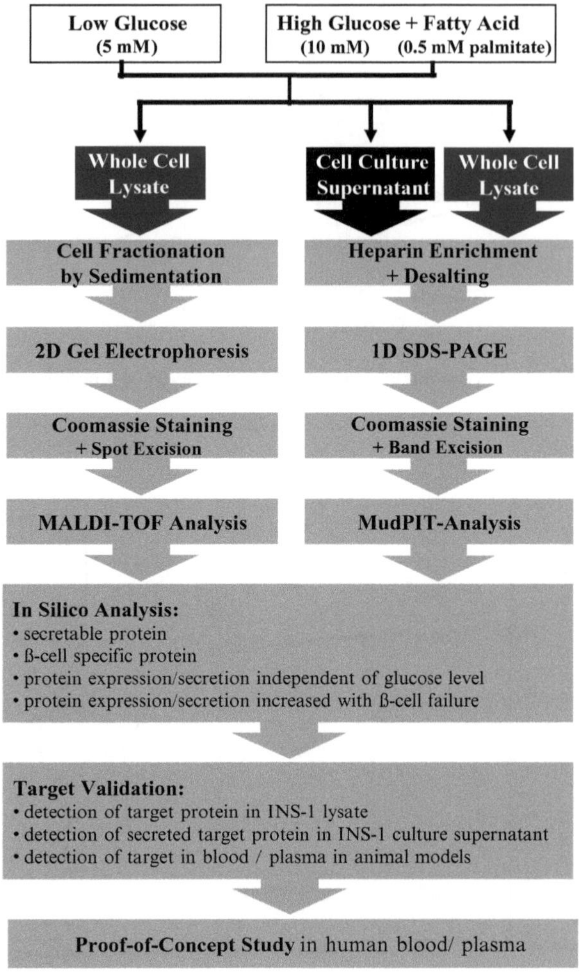

Fig. 3 Flowchart of proteomics approach to identify markers of β-cell failure using mass spectrometry (*MudPIT* multi protein identification technology and *MALDI* matrix assisted laser desorption/ionization mass spectrometry) in combination with in Silico analysis

6.6 Enrichment of Putative Secreted Proteins by Heparin Columns from the Medium and Identification by LC-MS

Visualization and identification of secreted proteins may facilitate the identification of novel drug targets and different biomarkers including predictive markers. However, not all proteins of an organism are expressed in amounts sufficient for detection by two-dimensional or one dimensional polyacrylamide gel electrophoresis. In order to visualize and identify low-copy gene products with signaling function, we enriched low-abundance cytosolic proteins by applying heparin chromatography. This step was chosen because heparin has a high protein binding capacity and can discriminate and enrich proteins with minor differences in their pI-values and glycosylation patterns [34, 35]. As most secreted proteins are glycosylated [36],

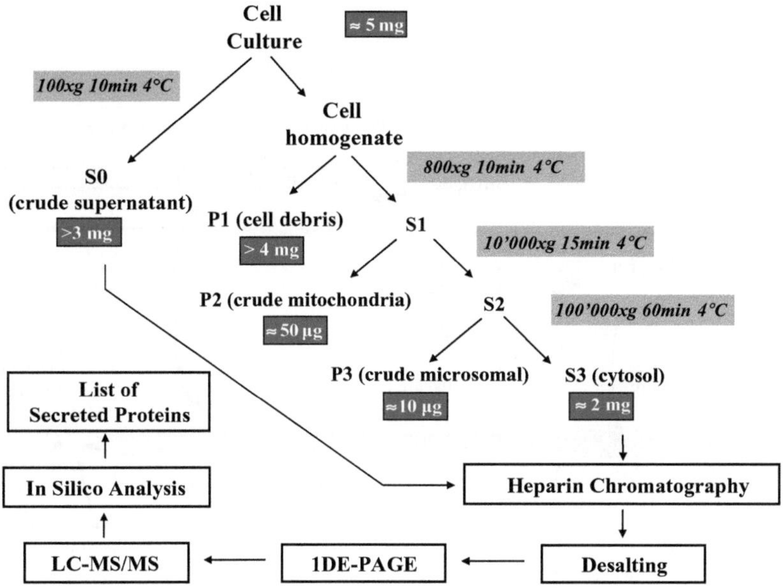

Fig. 4 Scheme of centrifugal prefractionation of rat INS-1 proteins. Different centrifugal force leads to enrichment of cellular components such as mitochondria, microsomal and cytosolic proteins. The cytosolic fraction was subjected to further fractionation by heparin chromatography followed by separation on a 10% homogenous polyacrylamide gel and protein identification by LC-MS/MS

heparin sepharose and/or lectin columns are versatile tools for the enrichment of many classes of glycosylated proteins such as proteins with signaling functions, growth factors, coagulation factors and steroid receptors [37, 38]. The ligand in a heparin sepharose column is a naturally occurring sulfated glycosaminoglycan, which is extracted from the native proteoglycan of porcine intestinal mucosa. Heparin consists of alternating units of uronic acid and D-glucosamine, most of which are substituted with one or two sulphate groups. Immobilized heparin has two main modes of interaction with proteins: It can operate as an affinity ligand; e.g., in its interaction with coagulation factors and it can also function as a high capacity cation exchanger because of its anionic sulphate groups, leading thus to an additional enrichment of positively charged proteins.

To enrich for secreted rat insulinoma proteins with signaling function heparin chromatography was applied prior to proteomic analysis. The column was operated with a syringe instead of a liquid chromatography pump. Elution was performed by increasing the ionic strength with 2 M NaCl. Separation of the eluted proteins was carried out by one-dimensional (1D-) polyacrylamide gel electrophoresis (PAGE) and the proteins were identified by multi dimensional protein identification technology (MudPIT) tandem mass spectrometry [39, 40] in combination with *in silico* analysis (see Fig. 5).

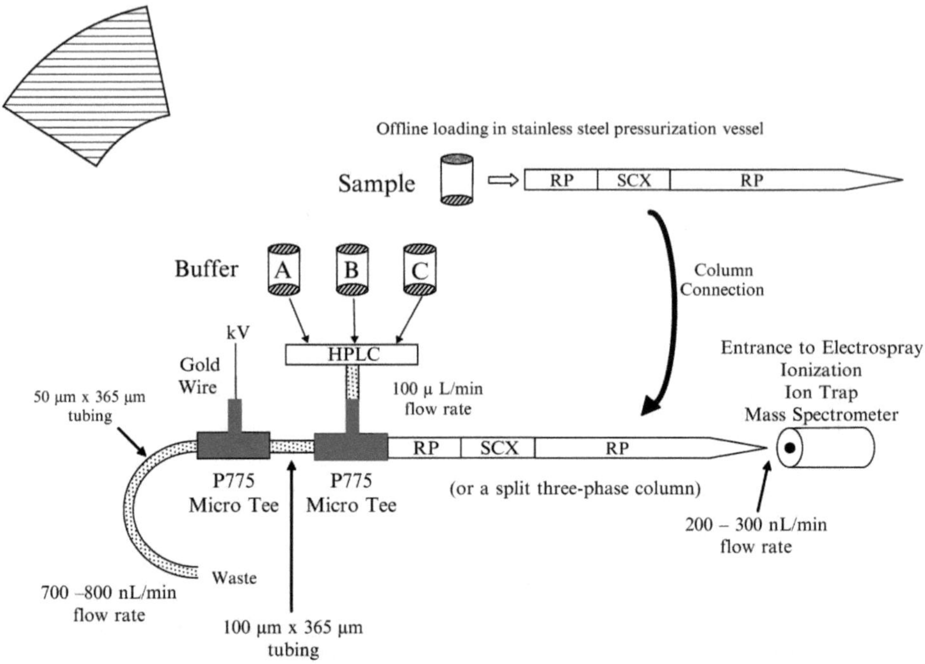

Fig. 5 Three-phase multidimensional protein identification technology (MudPIT) column setup. In the MudPIT system, a triphasic microcapillary column with both strong cation exchange (SCX) and reversed-phase (RP) packing material is prepared and loaded off-line. Upon insertion of the system, a fully automated analysis can be run with salt pumps moving fraction from the RP to the SCX by a first step reversed-phase gradient and to the RP using a series of salt pumps. Upon the application of RP gradients, peptides elute into the mass spectrometer

Moreover, heparin chromatography may be useful in the depletion of albumin from body fluids, such as plasma and cerebrospinal fluid, in which it represents more than 50% of total proteins [34, 41]. For example, serum albumin, which is represented by a strong band in the starting material (see Fig. 6), was completely recovered in the flow-through fraction. Specific removal of albumin, as well as of other high abundance proteins allowed the visualization and identification of minor components of the samples, whose levels may change in certain disorders [42]. In addition to the easier design of protein purification steps, use of selected chromatography steps prior to the LC-MS/MS, can significantly facilitate the analysis of complex protein mixtures.

6.7 *Sample Preparation*

Twenty-five milliliter of the medium were centrifuged at 10,000 g, for 10 min at 4 °C in order to remove apoptotic cells and other unsolved materials. The sample solution was adjusted to the composition of the binding buffer. This was done by

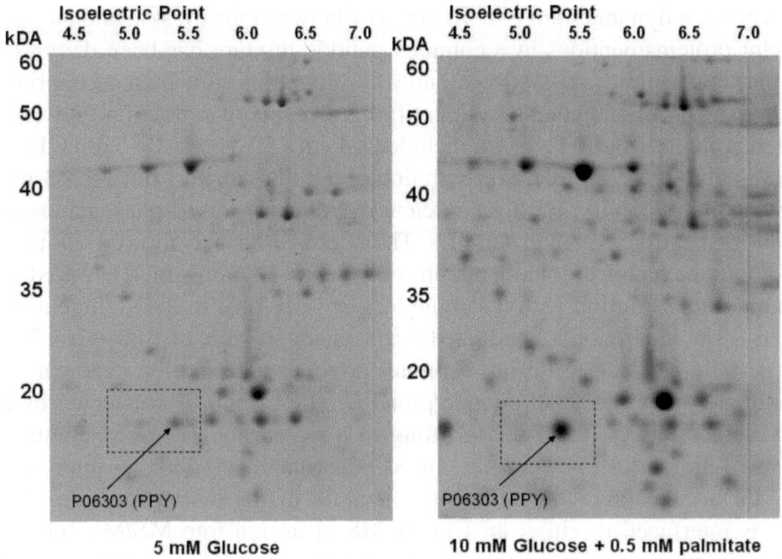

Fig. 6 2D-PAGE images of differential protein expression patterns in INS-1 cells incubated under lipo/glucotoxic conditions (10 mM glucose and 0.5 palimate) *versus* control (5 mM glucose). The selected spot of protein (P06303) was identified by MALDI-TOF as pancreatic polypeptide Y

diluting the sample by adding 25 ml of a 20 mM sodium phosphate buffer solution (pH = 7). The sample was centrifuged immediately before applying it on the column. The flow rate of the heparin Column (HiTrap Heparin HP, 1 ml, Cat. Nr. 17-0406-01, Amersham) was 5 ml for 1 ml column.

Sample fractions eluted from heparin column were desalted by using reversed phase chromatography (POROS R2, PerSeptive Biosytems), and dried by using a speed vac. In addition to this, samples were dissolved in sample buffer mentioned below and the protein content was determined by Bradford reaction.

LC-MS: Analysis of tryptic digests is performed by the LC-MS/MS approach termed multidimensional protein identification technology (MudPIT) as described [39, 40] with minor modifications. It combines multidimensional liquid chromatography with electrospray ionization tandem mass spectrometry, integrating strong cation-exchange (SCX) resin and reversed-phase resin in a biphasic column. Analysis is usually performed in duplicate and the separation should be reproducible to within 0.5% between two analyses. Analysis of replicates demonstrated that the fraction of identifications shared between replicates (expressed as: [number of proteins identified in the replicate run identifying fewer proteins]/[number of proteins identified in the replicate run identifying more proteins]) was approximately 0.8. For values less than 0.8 more replicates are recommended. In order to separate the digested proteins enriched by heparin columns, our multidimensional liquid chromatography method integrates a strong cation-exchange (SCX) resin and reversed-phase resin in a biphasic column. Each MudPIT analysis was done in duplicate and separation was reproducible within 0.5% between two analyses.

Furthermore, a dynamic range of 10,000 to 1 between the most abundant and least abundant proteins/peptides in a complex peptide mixture has been demonstrated. By improving sample preparation along with separations, the method improved the overall analysis of proteomes by identifying proteins of a fraction enriched with secreted proteins. MudPIT system included a 4 cm × 50 – μm i.d. × 5 μm C18 micro-SPE pre-column for sample concentration and an 85 cm × 15-μm i.d. × 3 μm C18 packed capillary column for high efficiency gradient reversed-phase nanoscale LC separation of extremely small samples. The micro-SPE stage allowed solution to be loaded onto the nanoLC column at approximately 8 μL/min which required <2 min to load a 10 μL solution with a sample loss of <5% (due to the syringe and valve adapters). The separation is conducted at a constant pressure of 10,000 psi. The long 3-μm particle packed 15-μm-i.d. capillary provides a separation peak capacity of approximately few thousand peptide/proteins. The column is connected by a zero dead volume stainless steel union fitting to a replaceable nanoESI emitter made from a 10-μm -i.d. × 150- μm-o.d. fused silica capillary with an approximately 2-μm- i.d orifice for highly efficient ionization of the eluting peptide. The ESI source is interfaced to either an FTICR MS or an ion trap MS/MS for peptide/protein detection and identification. An FTICR mass spectrometer was used for single-stage MS based upon high-accuracy mass measurements and the use of relative retention time (RRT) information, and a Finnigan ion trap mass spectrometer (LCQ XP, ThermoQuest Corp., San Jose, CA) was used for MS/MS.

6.8 Mapping and Identification of Signal Proteins in Cell Compartment by 2D-Electrophoresis and Identification by MALDI-MS

Samples prepared from each cell line were subjected to 2-DE as described elsewhere [43, 44] and 2-DE was performed essentially as reported [45]. Samples were desalted by using membrane filter tubes (Millipore, Art. No. UFV4BGC25) and 2.0 mg were applied on immobilized pH 3–10 non linear gradient strips (Amersham, Pharmacia Biotechnology, Uppsala, Sweden) at both the basic and acid ends of the strips. The proteins were focused at 200 V after which the voltage is gradually increasing to 5,000 V with 2 V/min. Focusing was continued at 5,000 V for 24 h. The second-dimensional separation was performed on a 12% polyacrylamide gel (Biosolve, Walkinswaard, and Netherland). The gels (180 × 200 × 1.5 mm) were run at 50 mA/gels, in an Ettan DALT II system (Amersham, Pharmacia Biotechnology, Uppsala, Sweden) accommodating 12 gels. After protein fixation for 12 h in 50% methanol containing 5% phosphoric acid, the gels were stained with colloidal Coomassie blue (Novex, San Diego, CA) for 24 h. Molecular masses were determined by running standard protein markers (Gibco, Basel, Switzerland), covering the range of 10–200 kDa. PI values were used as given by supplier of the IPG strips

(Amersham Pharmacia, Uppsala, Sweden). Gels were destained with H_2O and scanned in an AGFA DUOSCAN densitometer. Electronic images of the gels were recorded using Photoshop (Adobe) and PowerPoint (Microsoft).

MALDI-MS: MS analysis was performed as described [43] with minor modifications. Briefly, spots were excised, destained with 30% (v/v) acetonitrile in 0.1 M ammonium bicarbonate and dried in a Speed Vac evaporator. The dried gel pieces were reswollen with 5 μl of 5 mM ammonium bicarbonate, (pH 8.8) containing 50 ng trypsin (Promega, Madison, WI, USA) were added, centrifuged for 1 min and left at room temperature for about 12 h. After digestion, 5 μl of water was added, followed 10 min later by 10 μl 75% acetonitrile, containing 0.3% trifluoroacetic acid, was added, centrifuged for 1 min and the content was vortexed for 20 min. For MALDI-MS 1.5 μl from the separated liquid was mixed with 1 μl saturated alpha-cyano cinnamic acid in 50% acetonitril, 0.1% TFA in water and applied to the MALDI target. The samples were analyzed in a time-of-flight Bruker mass spectrometer (Reflex III, Reflex and Autoflex) equipped with a reflector and delayed extraction. An accelerating voltage of 20 kV was used. Des-Arg-1 Bradykinin (Sigma) and ACTH (18–38) (Sigma) were used as standard peptides. Calibration was internal to the samples. The peptide masses were matched with the theoretical peptide masses of all proteins from all species of the SWISS-Prot database.

Peptide matching and protein searches were performed automatically. The peptide masses were compared with the theoretical peptide masses of all available proteins from all species. Monoisotopic masses were used and a mass tolerance of 0.0025% was allowed.

Spectra were analyzed and protein sequence databases were searched using the programs Fragment 21 and MSROFIT, respectively. Databases queried were SWISS-PROT (http://www.expasy.ch) and PIR (http://www.nbrf.georgetown.edu.pir).

The algorithm used for determining the probability of a false positive match with a given MS-spectrum published [45] and [46].

6.9 *Identification of Putative Secreted Proteins by in Silico Analysis*

Secreted proteins are characterized by a signal peptide sequence that helps in threading the protein's N terminus through a membrane before it is cleaved off by a signal peptidase. Signal peptides are only loosely defined, often with a positively charged, polar section, followed by a hydrophobic stretch, and a short pattern around the cleavage site.

The web tool that we used in order to identified or predicted signal peptide from SwissProt, was an house developed software tool based on a set of specialized, manually curated Hidden Markov Models (HMMs) that attempt to recognize the (sparse) sequence features common to signal peptides or anchors, respectively. As these sequence signals cannot be reliably predicted, the "signal" and "anchor" scores that any input sequence is assigned are fed into a Support Vector Machine (SVM) in a second analysis step. The SVM was trained on a set of *bona fide* examples for both classes. On this training set, the SVM obtained the following results on three training sets (signal – anchor – neither):

signal 96% sensitivity and 96% specificity
anchor 87% sensitivity and 98% specificity
neither 96% sensitivity and 97% specificity

These results are as good as the best claims in the literature (note, however, that most tools only look at the signal peptides and do not attempt to predict membrane anchors).

For the default setting of this web interface, a list is produced that simply classifies each input sequence in either of the three categories "Signal", "Anchor", or "Normal" (the latter indicating that neither signal peptide nor membrane anchor are predicted).

For the "verbose" mode, there is a much more detailed table that has, for every input sequence, the following columns:

seqID, Classification – these are the same as in the default output discussed above.

Confidence – this qualifier can take the values "Unique", "Ambiguous", or "No Class", depending on the SVM results. The SVM determines how well the sequence scores fit its model of either of the three classes. In the "Unique" case, the scores match only one of the three classes, to which the sequence is then assigned. In the "Ambiguous" case, the sequence scores match more than one class – the sequence is assigned to that class into which it fits best. In the "No class" case, the scores do not match the expectations for any of the three classes – the sequence is assigned to that class for which the discrepancy is smallest.

D (Anchor), D (Normal), D (Signal) – these numbers specify the "distances" to a line separating the three classes (to use an intuitive metaphor). For a "Unique" classification, only one of these "distances" is positive (indicating that the sequence is "on the right side of the fence"), whereas the other two are negative ("on the wrong side of the fence" for the two other classes). The larger the numbers, the clearer the classification. For an "Ambiguous" classification, more than one of the D's is positive, for a "No class" classification, all three D's are negative.

SW-classification, SW-note – finally, sequences from the SwissProt database already come with an annotation regarding signal peptides and membrane anchors. For them, we show the SwissProt classifications and any additional comments on them for reference and comparison. Sequences from the HUMANGP, MOUSEGP, and RATGP databases have "none" for both fields.

The *HUMANGP*, *MOUSEGP*, and *RATGP* databases have been completely annotated with this prediction algorithm, the results are displayed in the *XR SIGNAL/ ANCHOR* annotation lines.

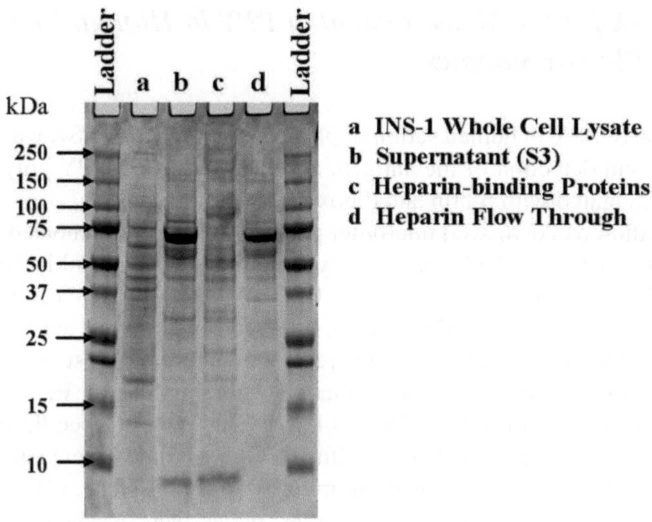

Fig. 7 1DE gel analysis of fractions eluted from heparin chromatography. Fraction enriched with cytosolic rat INS-1 proteins was prepared as stated in Sect. 6, further separated as stated in Sect. 6.5. and analyzed by 1DE as stated in Sect. 6.6. Gels were stained with colloidal Coomassie blue and protein bands were identified by MudPIT

7 Visualization and Identification of Putative Secreted Proteins by MudPIT Analysis Enriched by Heparin Chromatography

Heparin binding fractions of cell culture supernatants were analyzed by LC-MS and MALDI-MS and lead to the validation of the putative markers. We confirmed five secreted proteins: Chromogranin b, Pancreatic Polypeptide (PPY), Phospholipase A2, Fibronectin and Vinculin were clearly shown to be secreted by INS-1. The amount of secretion into the medium correlated strongly to the insulin content of the INS-1 β-cells. Based on the organ specify, PPY seemed to be one of the most interesting candidates and was selected for further validation by ELISA. PPY is a secretory glycoprotein with a widespread distribution mostly expressed in endocrine pancreas. These data were underlined by additional OGTT studies which corroborate our findings and show clearly, that PPY is secreted from the endocrine pancreas independent from the glucose level in blood (see Fig. 9). Furthermore it has been suggested that PPY has an autocrine inhibitory effect on iAPP and insulin secretion by different intra- and extracellular mechanism [46]. Ten proteins were selected for further evaluation. Evaluation/validation by immunoassays in a well-characterized patient population was performed only with PPY. Pancreatic Polypeptide levels are higher in diabetics than controls. A more profound analysis has to demonstrate that the levels correlate with beta cell failure and not other diabetes-related phenomena (see Fig. 8). A priority application was made for all candidates before a full validation was done (Fig. 7).

7.1 ELISA for the Measurement of PPY in Human Serum and Plasma Samples

For detection of PPY in human serum or plasma, a sandwich ELISA was developed. For capture and detection of the antigen, aliquots of the anti-PPY polyclonal antibody are conjugated with biotin and digoxygenin, respectively.

Streptavidin-coated 96-well microtiter plates (Nunc) were incubated with 100 μl biotinylated anti-PPY polyclonal antibody for 60 min at 10 μg/ml in 10 mM phosphate, pH 7.4, 1% BSA, 0.9% NaCl and 0.1% Tween 20. After incubation, plates were washed three times with 0.9% NaCl, 0.1% Tween 20. Wells are then incubated for 2 h with either a serial dilution of the recombinant protein as standard antigen or with diluted plasma samples from patients. After binding of PPY, plates were washed three times with 0.9% NaCl, 0.1% Tween 20. For specific detection of bound PPY, wells were incubated with 100 μl of digoxygenylated anti-PPY polyclonal antibody for 60 min at 10 μg/ml in 10 mM phosphate, pH 7.4, 1% BSA, 0.9% NaCl and 0.1% Tween 20. Thereafter, plates were washed three times to remove unbound antibody. In a next step, wells were incubated with 20 mU/ml anti-digoxigenin-POD conjugates (Roche Diagnostics GmbH, Mannheim, Germany, Catalog No. 1633716) for 60 min in 10 mM phosphate, pH 7.4, 1% BSA, 0.9% NaCl and 0.1% Tween 20. Plates were subsequently washed three times with the same buffer. For detection of antigen-antibody complexes, wells were incubated with 100 μl ABTS solution (Roche Diagnostics GmbH, Mannheim, Germany, Catalog No. 11685767) and OD was measured after 30–60 min at 405 nm with an ELISA reader.

7.2 Detection of PPY in Human Plasma: Proof-of-Concept

Clinical utility of the novel marker PPY was assessed by measuring its levels in ten diabetic patients depending on injections of exogenous insulin and comparing the levels with those measured in ten patients with demonstrated normal β-cell function. The studies have been approved by the ethic committee and have therefore been performed in accordance with the ethical standards laid down in the 1964 Declaration of Helsinki. All persons gave their informed consent prior to their inclusion in the study.

The OGTT has fallen into disfavor as a tool to examine the mechanism responsible for impaired glucose metabolism in T2D subjects. The rate of glucose absorption varies considerable from one subject to another and cannot be quantified easily. Consequently, the rise in plasma glucose concentration differently markedly from one subject to another and is constantly changing. This present an ever-fluctuating glycemic stimulus that, when combined with intrinsic differences in β-cell function from one individual to another, leads to large variations in the plasma insulin profile between subjects as shown in [47]. Because the two primary

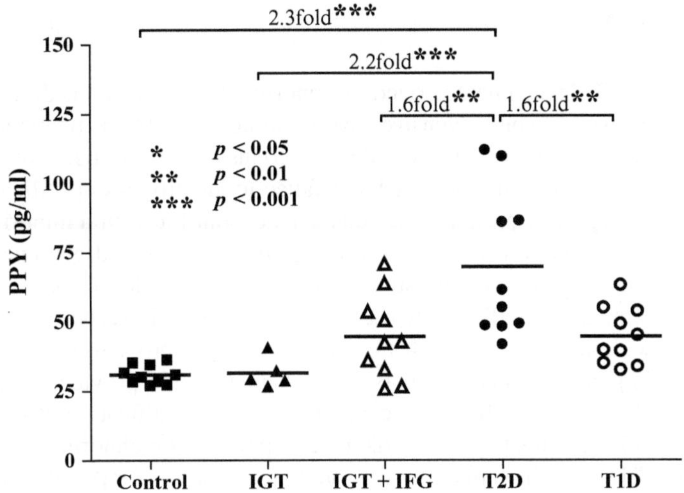

Fig. 8 Primary results: validation of PPY in human blood/plasma samples in five different donor groups by ELISA

variables of interest – plasma glucose and plasma insulin concentration – are changing simultaneously, it is difficult to draw any conclusion about insulin secretion or insulin sensitivity. Lastly, it is difficult to quantitate changes in hepatic glucose production after glucose ingestion because the rate of entry of glucose into circulation is unknown. In order to investigate the influence of the plasma glucose concentration on the PPY level, an OGTT with four healthy subjects were performed and all have revealed very similar results as shown in Fig. 9. Our investigations indicate that early insulin secretion (absolute insulin concentration at 30 min, the incremental insulin concentration (ΔI) at 30 min, or the incremental insulin concentration factored by the incremental glucose concentration ($\Delta I/\Delta G$) at 30 min) had less predictive value for the β-cell function than the determination of PPY. PPY absolute concentration doesn't varied so much between the healthy individuals and it was shown that the secretion rate to be independent of the plasma glucose and insulin concentration. Furthermore, the PPY plasma concentration between the healthy individuals was very constant and only significantly elevated in T2D and IGT + IFG individuals as shown in Fig. 8 proved to be a better predictor of progression to T2D as OGTT. Since the plasma glucose concentration is the primary determinant of insulin secretion, it is obvious that fasting insulin and glucose concentration alone can not be used to predict the progression of normal-glucose-tolerant subjects to T2D. It is anticipated that our results and methods are able to establish a framework for the identification of additional marker candidates that need to be examined and provide a sound scientific basis for the identification and more precise definition of predictive biomarkers in T2D.

8 Summary

Individuals with T2D are characterized by abnormalities in insulin action and insulin secretion. However, despite intensive investigation, the proteins/peptides or genes responsible for the insulin resistance and impaired insulin secretion remain undefined. The candidate-proteins/genes approach has failed to identify any specific proteins or genes or combination of protein/genes that can account for even a minority of adult cases of T2D. Although a number of laboratories have initiated genome- and proteome-wide search to identify potential susceptibility targets for T2D, consistent and reproducible linkage to predictive markers has yet to merge. It has been suggested that a more precise definition of the diabetic phenototype may prove useful in delineating diabetogenetic genes or gene products. This chapter should provides an inside into new analytical strategy for the specific enrichment and identification of secreted, and pancreas specific proteins/peptides, which are responsible for abnormalities of insulin secretion in T2D and which may serve as predictive markers for β-cell function.

All together the analysis resulted in the identification of 391 different gene products derived from the INS cells. 188 proteins bound to the heparin matrix, 101 of which were identified as secreted proteins with signaling functions and are listed in Table 1. Forty-four of the enriched proteins had not been detected by 2-DE without previous enrichment by combination of subcellular prefractionation and heparin chromatography. Heparin chromatography specifically enriched several enzymes that had not been identified before [48], like peptidyl-glycine alpha-amidating monooxygenase, sodium/hydrogen exchanger 5, etc. The identified proteins can be divided into five classes:

- Hormones and related molecules (amylin, pancreatic polypeptide, neuroendocrine convertase)
- Protease inhibitors (e.g. cystatin)
- Secretory vesicle proteins (e.g. chromogranin, secretogranin)
- Cell adhesion or extracellular matrix proteins (e.g. protocadherin)
- Secreted enzymes (e.g. phospholipase)

From all identified secreted proteins only pancreatic polypeptide Y (PPY) was identified as pancreas specific, fulfill the criteria of a predictive marker and hence was selected for further validation.

PPY (pancreatic polypeptide Y), a 78-amino acid peptide is synthesized in pancreatic islets of Langerhans and acts as a regulator of pancreatic and gastrointestinal functions. The peptide is structurally similar to other pancreatic hormone family consisting of neuropeptide Y (NPY) b, and peptide YY (PYY) which are closely related to each other. The propeptide is enzymatically cleaved to yield the mature active peptide with amidated C-terminal ends; receptor binding and activation functions may reside in the N- and C-termini respectively; occurs in neurons, intestinal endocrine cells, and pancreas; exist as monomers and dimers. The physiological role for the icosapeptide has not yet been elucidated. PPY was identified as specifically secreted by the endocrine pancreas and it is mainly produced by β- and f-cells [46]. Its plasma concentration has been used as a marker of parasympathetic activity. Recent work in rodents suggests that there is both sympathetic and parasympathetic innervation of white adipose tissue and that parasympathetic activity is anabolic

Table 1 List of identified secreted INS-1 proteins enriched by heparin chromatography

Fraction	SW.-Nr.	Access. Nr.	Status	Full protein name	kDa
Sup.	S106_RAT	P05964	Sec.	Calcyclin (Prolactin receptor associated protein)	10.1
Hep.	KV5C_MOUSE	P01635	Sec.	Ig kappa chain V-V region K2 precursor (Fragment)	12.6
Sup.	TL19_MOUSE	Q9CQU0	Sec.	Thioredoxin-like protein p19 precursor	19.0
Sup.	CYPC_MOUSE	P30412	Sec.	Peptidyl-prolyl cis-trans isomerase C (EC 5.2.1.8)	22.8
Hep.	CYPB_RAT	P24368	Sec.	Peptidyl-prolyl cis-trans isomerase B prec. (EC 5.2.1.8)	23.0
Hep.	TIM2_RAT	P30121	Sec.	Metalloproteinase inhibitor 2 precursor (TIMP-2)	24.3
Hep.; Sup.	TRY1_RAT	P00762	Sec.	Trypsin I, anionic precursor (EC 3.4.21.4)	25.9
Hep.; Sup.	TRY2_MOUSE	P07146	Sec.	Trypsin II, anionic precursor (EC 3.4.21.4)	26.2
Hep.	PAHO_RAT	P06303	Sec.	Pancreatic prohormone prec. (Pancreatic polypeptide)	27.1
Hep.; Sup.	IBP4_RAT	P21744	Sec.	Insulin-like growth factor binding protein 4 (IGFBP-4)	27.7
Hep.	ER29_RAT	P52555; P80749	Sec.	Endoplasmic reticulum protein ERp29 prec. (ERp31)	28.6
Sup.	PDX4_MOUSE	O08807;	Sec.	Peroxiredoxin 4 (EC 1.11.1.-) (Prx-IV)	31.0
Hep.; Sup.	SPRC_RAT	P16975; O08953	Sec.	SPARC precursor	34.4
Hep.	FSL1_RAT	Q62632	Sec.	Follistatin-related protein 1 prec. (Follistatin-like 1)	34.6
Hep.; Sup.	OSTP_RAT	P08721	Sec.	Osteopontin precursor (Secreted phosphoprotein 1)	34.9
Hep.; Sup.	DRNG_MOUSE	O55070	Sec.	Deoxyribonuclease gamma precursor (EC 3.1.21.-)	35.7
Sup.	CALU_RAT	O35783	Sec.	Calumenin precursor (Crocalbin)	37.0
Sup.	CATM_MOUSE	Q9JL96; Q91Z75	Sec.	Cathepsin M precursor (EC 3.4.22.-)	37.4
Sup.	CATB_RAT	P00787;	Sec.	Cathepsin B precursor (EC 3.4.22.1) (Cathepsin B1)	37.4
Hep.	CATL_RAT	P07154	Sec.	Cathepsin L prec. (EC 3.4.22.15)	37.6
Hep.	IF35_MOUSE	Q9DCH4	Sec.	Eukaryotic translation initiation factor 3 subunit 5	38.0
Sup.	RCN1_MOUSE	Q05186	Sec.	Reticulocalbin 1 precursor	38.1
Sup.	DKK3_MOUSE	Q9QUN9	Sec.	Dickkopf related protein-3 precursor (Dkk-3)	38.4
Hep.	PGS2_RAT	Q01129	Sec.	Decorin precursor (Bone proteoglycan II)	39.8

(continued)

Table 1 (continued)

Fraction	SW.-Nr.	Access. Nr.	Status	Full protein name	kDa
Sup.	CB45_RAT	Q91ZS3	Sec.	45 kDa calcium-binding protein precursor (Cab45)	42.0
Hep.	SELP_RAT	P25236	Sec.	Selenoprotein P precursor (SeP)	42.6
Hep.	PAI1_RAT	P20961	Sec.	Plasminogen activator inhibitor-1 precursor (PAI-1)	45.0
Sup.	T11B_RAT	O08727	Sec.	Tumor necrosis factor recept. superfamily member11B	46.2
Hep.; Sup.	PEDF_MOUSE	P97298	Sec.	Pigment epithelium-derived factor precursor (PEDF)	46.2
Hep.	HS47_RAT	P29457	Sec.	47 kDa heat shock protein prec.	46.5
Sup.	MFGM_RAT	P70490	Sec.	Lactadherin precursor (Milk fat globule-EGF factor 8)	47.4
Hep.; Sup.	CRTC_RAT	P18418; P10452	Sec.	Calreticulin precursor (CRP55) (Calregulin) (HACBP)	48.0
Hep.	LGMN_RAT	Q9R0J8	Sec.	Legumain precursor (EC 3.4.22.34)	49.4
Sup.	NCB2_RAT	Q9JI85	Sec.	Nucleobindin 2 precursor (DNA-binding protein NEFA)	50.1
Hep.	PCO1_RAT	O08628	Sec.	Procollagen C-proteinase enhancer protein precursor	50.2
Sup.	VAS1_RAT	O54715	Sec.	Vacuolar ATP synthase subunit S1 prec. (EC 3.6.3.14)	51.1
Hep.	HRA1_MOUSE	Q9R118	Sec.	Serine protease HTRA1 precursor (EC 3.4.21.-)	51.2
Hep.	CLUS_RAT	P05371	Sec.	Clusterin precursor (Sulfated glycoprotein 2) (SGP-2)	51.3
Hep.; Sup.	SRPX_RAT	Q63769	Sec.	Sushi repeat-containing protein SRPX prec.	51.5
Hep.	ANT3_MOUSE	P32261	Sec.	Antithrombin-III precursor (ATIII)	52.0
Hep.	LIPL_RAT	Q06000	Sec.	Lipoprotein lipase precursor (EC 3.1.1.34) (LPL)	53.0
Hep.	SG3_MOUSE	P47867	Sec.	Secretogranin III precursor (SgIII)	53.3
Hep.	NCB1_RAT	Q63083; P97623	Sec.	Nucleobindin 1 precursor (CALNUC)	53.5
Hep.	ARSA_MOUSE	P50428	Sec.	Arylsulfatase A precursor (EC 3.1.6.8) (ASA)	53.7
Hep.; Sup.	PRTP_MOUSE	P16675	Sec.	Lysosomal protective protein precursor (EC 3.4.16.5)	53.8
Hep.; Sup.	PLTP_MOUSE	P55065; Q99L70	Sec.	Phospholipid transfer protein precursor	54.4
Sup.	DPP2_RAT	Q9EPB1	Sec.	Dipeptidyl-peptidase II precursor (EC 3.4.14.2)	55.1
Hep.	CPS1_MOUSE	P03940	Sec.	Cytochrome P450 XXIA1 (EC 1.14.99.10)	55.3

(continued)

Table 1 (continued)

Fraction	SW.-Nr.	Access. Nr.	Status	Full protein name	kDa
Sup.	PDA3_RAT	P11598	Sec.	Protein disulfide isomerase A3 precursor (EC 5.3.4.1)	56.6
Sup.	PDI_RAT	P04785; P13700	Sec.	Protein disulfide isomerase precursor (PDI)	56.9
Hep.	MM19_MOUSE	Q9JHI0	Sec.	Matrix metalloproteinase-19 precursor (EC 3.4.24.-)	59.1
Hep.; Sup.	HEXA_MOUSE	P29416; Q64246	Sec.	Beta-hexosaminidase alpha chain prec. (EC 3.2.1.52)	60.6
Sup.	TPP1_RAT	Q9EQV6	Sec.	Tripeptidyl-peptidase I precursor (EC 3.4.14.9) (TPP-I)	61.3
Hep.	GPC1_RAT	P35053	Sec.	Glypican-1 precursor (HSPG M12)	61.7
Sup.	ALBU_RAT	P02770; P11382	Sec.	Serum albumin precursor	68.7
Sup.	ASM_MOUSE	Q04519	Sec.	Sphingomyelin phosphodiester-ase prec. (EC 3.1.4.12)	69.9
Hep.	SPL1_RAT	P24054	Sec.	SPARC-like protein 1 prec. (Matrix glycoprotein Sc1).	70.6
Hep.; Sup.	KNG_RAT	P08934; P08933	Sec.	Kininogen precursor	70.9
Sup.	IDUA_MOUSE	P48441	Sec.	Alpha-L-iduronidase precursor (EC 3.2.1.76)	71.1
Hep.	GR78_RAT	P06761	Sec.	78 kDa glucose-regulated protein precursor (GRP 78)	72.3
Hep.	PDA4_RAT	P38659	Sec.	Protein disulfide isomerase A4 precursor (EC 5.3.4.1)	72.8
Hep.; Sup.	BGLR_MOUSE	P12265; Q61601	Sec.	Beta-glucuronidase precursor (EC 3.2.1.31)	74.2
Hep.	FBL1_MOUSE	Q08879; Q08878	Sec.	Fibulin-1 prec. (Basement-membrane protein 90)	78.0
Hep.	SM3B_MOUSE	Q62177	Sec.	Semaphorin 3B precursor (Semaphorin A)	82.8
Sup.	PLO1_RAT	Q63321	Sec.	Procollagen-lysine, 2-oxoglutarate 5-dioxygenase 1	83.6
Hep.	LOL3_MOUSE	Q9Z175; Q9JJ39	Sec.	Lysyl oxidase homolog 3 precursor (EC 1.4.3.-)	83.6
Hep.; Sup.	AD10_MOUSE	O35598	Sec.	ADAM 10 precursor (EC 3.4.24.-)	83.9
Sup.	MEPA_MOUSE	P28825	Sec.	Meprin A alpha-subunit precursor (EC 3.4.24.18)	84.1
Hep.	PLO3_MOUSE	Q9R0E1	Sec.	Procollagen-lysine, 2-oxoglutarate 5-dioxygenase 3	84.9
Hep.; Sup.	A4_RAT	P08592	Sec.	Alzheimer's disease amyloid A4 protein homolog	86.6
Sup.	APP2_RAT	P15943	Sec.	Amyloid-like protein 2 precursor	86.8

(continued)

Table 1 (continued)

Fraction	SW.-Nr.	Access. Nr.	Status	Full protein name	kDa
Hep.	ENPL_MOUSE	P08113; P11427	Sec.	Endoplasmin prec. (Endoplasmic reticulum protein 99)	92.4
Sup.	DAG1_MOUSE	Q62165; Q61094	Sec.	Dystroglycan precursor	96.8
Hep.	FPRP_RAT	Q62786	Sec.	Prostaglandin F2 receptor negative regulator precursor	98.7
Hep.	NAH5_RAT	Q9Z0X2	Sec.	Sodium/hydrogen exchanger 5 (Na(+)/H(+)) exchanger	99.0
Hep.; Sup.	DSC3_MOUSE	P55850; O55110	Sec.	Desmocollin 3 precursor	100.6
Hep.	TSP3_MOUSE	Q05895	Sec.	Thrombospondin 3 precursor	103.9
Hep.; Sup.	PEX6_RAT	P54777; O55097	Sec.	Peroxisome assembly factor-2 (PAF-2)	104.4
Hep.	ITH2_MOUSE	Q61703	Sec.	Inter-alpha-trypsin inhibitor heavy chain H2 precursor	105.9
Hep.	CA16_MOUSE	Q04857	Sec.	Collagen alpha 1(VI) chain precursor	108.4
Hep.	AMD_RAT	P14925; P70710	Sec.	Peptidyl-glycine alpha-amidating monooxygenase	108.6
Sup.	CLS1_MOUSE	Q9EPL2	Sec.	Calsyntenin-1 precursor	108.8
Sup.	OXRP_RAT	Q63617	Sec.	150 kDa oxygen-regulated protein precursor (Orp150)	111.2
Hep.	M2B1_MOUSE	O09159; O55037	Sec.	Lysosomal alpha-mannosidase precursor (EC 3.2.1.2)	114.5
Sup.	M2B2_MOUSE	O54782	Sec.	Epididymis-specific alpha-mannosidase (EC 3.2.1.24)	115.6
Hep.; Sup.	TSP1_MOUSE	P35441	Sec.	Thrombospondin 1 precursor	129.6
Hep.; Sup.	LMG2_MOUSE	Q61092	Sec.	Laminin gamma-2 chain precursor	130.1
Hep.	DIA1_MOUSE	O08808	Sec.	Diaphanous protein homolog 1	139.3
Sup.	RGSC_RAT	O08774; O88383	Sec.	Regulator of G-protein signaling 12 (RGS12)	150.4
Hep.; Sup.	A2MG_RAT	P06238	Sec.	Alpha-2-macroglobulin precursor (Alpha-2-M)	163.6
Sup.	A1I3_RAT	P14046	Sec.	Alpha-1-inhibitor III precursor	163.7
Hep.; Sup.	LMG3_MOUSE	Q9R0B6	Sec.	Laminin gamma-3 chain precursor	172.2
Hep.	LMG1_MOUSE	P02468	Sec.	Laminin gamma-1 chain precursor (Laminin B2 chain)	177.2
Sup.	CO3_RAT	P01026	Sec.	Complement C3 precursor	186.3
Sup.	LTB1_RAT	Q00918	Sec.	Latent transforming growth factor beta binding prot. 1	186.5
Hep.	LMB1_MOUSE	P02469	Sec.	Laminin beta-1 chain precursor (Laminin B1 chain)	196.8
Hep.	LMA4_MOUSE	P97927; O88785	Sec.	Laminin alpha-4 chain precursor	201.7
Hep.	PGG2_RAT	Q00657	Sec.	Chondroitin sulfate proteoglycan NG2 precursor	251.8
Sup.	FINC_RAT	P04937	Sec.	Fibronectin precursor (FN)	272.3

(continued)

Table 1 (continued)

Fraction	SW.-Nr.	Access. Nr.	Status	Full protein name	kDa
Hep.	PGBM_MOUSE	Q05793	Sec.	Basement memb.-spec. heparan sulfate proteoglycan	398.0
Hep.; Sup.	LMA5_MOUSE	Q61001	Sec.	Laminin alpha-5 chain precursor	403.7
Hep.; Sup.	LRP2_RAT	P98158	Sec.	Low-density lipoprotein receptor-related protein 2	518.9

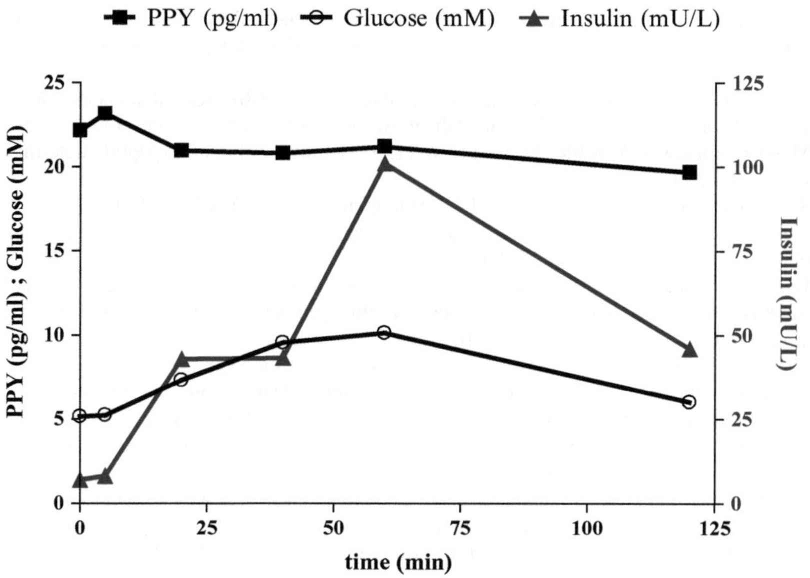

Fig. 9 Means of plasma glucose, insulin and PPY concentration during a 100 g OGTT in 4 healthy normal weight individuals. The results indicate that early insulin secretion has a less predictive value for the β-cell function than the determination of PPY levels. The absolute PPY concentration varies much less between healthy individuals and our results show that the secretion rate of PPY from the endocrine pancreas is independent of the plasma glucose and insulin concentration

resulting in lipid accumulation [49]. We have examined whether in humans increased PP levels are associated with increased β-cell dysfunctions, and thereby early β-cell failure in T2D. Together with amylin, the peptide has been shown to be the precursor for the amyloid deposits frequently observed in patients with T2D [49–52]. At very high doses PPY inhibits insulin secretion by the perfused rat pancreas in vitro [52] and pancreatic PPY and amylin deposits have been shown to precede the appearance of glucose intolerance in spontaneously diabetic monkeys [53]. In the course of our study we were able to demonstrate that elevated plasma PPY levels can be observed in IGT and IFG subjects and in glucose-intolerance first degree relatives of type 2 diabetic patients (see Fig. 8), and in animal models of diabetes. Although pancreatic and plasma PPY levels were significantly elevated, hyperglycemia and hyperinsulinemia were not observed (see Fig. 9). In summary, based on our findings/observation, the evidence that PPY is responsible for some of the defects in pancreatic islets is very strong and it seems that PPY plays an important role in impaired β-cell function in T2D.

References

1. Garvey WT, Olefsky JM, Griffin J, Hamman RF, Kolterman OG (1985) The effect of insulin treatment on insulin secretion and insulin action in type II diabetes mellitus. Diabetes 34:222–234
2. Little RR, Rohlfing CL, Wiedmeyer HM, Myers GL, Sacks DB, Goldstein DE (2001) The national glycohemoglobin standardization program: a five-year progress report. Clin Chem 47:1985–1992
3. Landin-Olsson M, Nilsson KO, Lernmark A, Sundkvist G (1990) Islet cell antibodies and fasting C-peptide predict insulin requirement at diagnosis of diabetes mellitus. Diabetologia 33:561–568
4. Meier C, Ladewig A, Keller U, Berger W (1997) Clinical value of C-peptide determination. Praxis 86:1289–1295
5. Sacks DB, Bruns DE, Goldstein DE, Maclaren NK, McDonald JM, Parrott M (2002) Guidelines and recommendations for laboratory analysis in the diagnosis and management of diabetes mellitus. Clin Chem 48:436–472
6. Koehn J, Fountoulakis M, Krapfenbauer K (2008) Multiple drug resistance associated with function of ABC-transporters in diabetes mellitus: molecular mechanism and clinical relevance. Infect Disord Drug Targets 8:109–118
7. Sattar N, Wannamethee SG, Forouhi NG (2008) Novel biochemical risk factors for type 2 diabetes: pathogenic insights or prediction possibilities? Diabetologia 51:926–940
8. Wannamethee SG, Shaper AG, Lennon L, Whincup PH (2005) Hepatic enzymes, the metabolic syndrome, and the risk of type 2 diabetes in older men. Diabetes Care 28:2913–2918
9. Ding EL, Song Y, Malik VS, Liu S (2006) Sex differences of endogenous sex hormones and risk of type 2 diabetes: a systematic review and meta-analysis. JAMA 295(11):1288–1299
10. Smirnakis KV, Plati A, Wolf M, Thadhani R, Ecker JL (2007) Predicting gestational diabetes: choosing the optimal early serum marker. Am J Obstet Gynecol 196:410 e411–416; discussion 410 e416–417
11. Pradhan AD, Manson JE, Rifai N, Buring JE, Ridker PM (2001) C-reactive protein, interleukin 6, and risk of developing type 2 diabetes mellitus. JAMA 286:327–334
12. Kaye TB, Guay AT, Simonson DC (1993) Non-insulin-dependent diabetes mellitus and elevated serum ferritin level. J Diabetes Complications 7:246–249
13. Eshed I, Elis A, Lishner M (2001) Plasma ferritin and type 2 diabetes mellitus: a critical review. Endocr Res 27:91–97
14. Spranger J, Kroke A, Mohlig M, Bergmann MM, Ristow M, Boeing H, Pfeiffer AF (2003) Adiponectin and protection against type 2 diabetes mellitus. Lancet 361:226–228
15. Vasseur F, Lepretre F, Lacquemant C, Froguel P (2003) The genetics of adiponectin. Curr Diab Rep 3:151–158
16. Yamauchi T, Kamon J, Ito Y, Tsuchida A, Yokomizo T, Kita S, Sugiyama T, Miyagishi M, Hara K, Tsunoda M, Murakami K, Ohteki T, Uchida S, Takekawa S, Waki H, Tsuno NH, Shibata Y, Terauchi Y, Froguel P, Tobe K, Koyasu S, Taira K, Kitamura T, Shimizu T, Nagai R, Kadowaki T (2003) Cloning of adiponectin receptors that mediate antidiabetic metabolic effects. Nature 423:762–769
17. Schmidt MI, Duncan BB, Vigo A, Pankow JS, Couper D, Ballantyne CM, Hoogeveen RC, Heiss G (2006) Leptin and incident type 2 diabetes: risk or protection? Diabetologia 49:2086–2096
18. Festa A, D'Agostino R Jr, Tracy RP, Haffner SM (2002) Elevated levels of acute-phase proteins and plasminogen activator inhibitor-1 predict the development of type 2 diabetes: the insulin resistance atherosclerosis study. Diabetes 51:1131–1137

19. Eliasson MC, Jansson JH, Lindahl B, Stegmayr B (2003) High levels of tissue plasminogen activator (tPA) antigen precede the development of type 2 diabetes in a longitudinal population study. The Northern Sweden MONICA study. Cardiovasc Diabetol 2:19
20. Wannamethee SG, Sattar N, Rumley A, Whincup PH, Lennon L, Lowe GD (2008) Tissue plasminogen activator, von Willebrand factor, and risk of type 2 diabetes in older men. Diabetes Care 31:995–1000
21. Meigs JB, O'Donnell CJ, Tofler GH, Benjamin EJ, Fox CS, Lipinska I, Nathan DM, Sullivan LM, D'Agostino RB, Wilson PW (2006) Hemostatic markers of endothelial dysfunction and risk of incident type 2 diabetes: the Framingham Offspring Study. Diabetes 55:530–537
22. Jacqueminet S, Briaud I, Rouault C, Reach G, Poitout V (2000) Inhibition of insulin gene expression by long-term exposure of pancreatic beta cells to palmitate is dependent on the presence of a stimulatory glucose concentration. Metabolism 49:532–536
23. Donath MY, Gross DJ, Cerasi E, Kaiser N (1999) Hyperglycemia-induced beta-cell apoptosis in pancreatic islets of Psammomys obesus during development of diabetes. Diabetes 48:738–744
24. Shafrir E, Ziv E, Mosthaf L (1999) Nutritionally induced insulin resistance and receptor defect leading to beta-cell failure in animal models. Ann N Y Acad Sci 892:223–246
25. Anello M, Rabuazzo AM, Degano C, Caltabiano V, Patane G, Vigneri R, Purrello F (1996) Fast reversibility of glucose-induced desensitization in rat pancreatic islets. Evidence for an involvement of ionic fluxes. Diabetes 45:502–506
26. Marshak S, Leibowitz G, Bertuzzi F, Socci C, Kaiser N, Gross DJ, Cerasi E, Melloul D (1999) Impaired beta-cell functions induced by chronic exposure of cultured human pancreatic islets to high glucose. Diabetes 48:1230–1236
27. Busch AK, Cordery D, Denyer GS, Biden TJ (2002) Expression profiling of palmitate- and oleate-regulated genes provides novel insights into the effects of chronic lipid exposure on pancreatic beta-cell function. Diabetes 51:977–987
28. Carlsson C, Borg LA, Welsh N (1999) Sodium palmitate induces partial mitochondrial uncoupling and reactive oxygen species in rat pancreatic islets in vitro. Endocrinology 140:3422–3428
29. Shimabukuro M, Zhou YT, Levi M, Unger RH (1998) Fatty acid-induced beta cell apoptosis: a link between obesity and diabetes. Proc Natl Acad Sci U S A 95:2498–2502
30. Mandrup-Poulsen T (2001) beta-cell apoptosis: stimuli and signaling. Diabetes 50(Suppl 1):S58–S63
31. Asfari M, Janjic D, Meda P, Li G, Halban PA, Wollheim CB (1992) Establishment of 2-mercaptoethanol-dependent differentiated insulin-secreting cell lines. Endocrinology 130:167–178
32. Miyazaki J, Araki K, Yamato E, Ikegami H, Asano T, Shibasaki Y, Oka Y, Yamamura K (1990) Establishment of a pancreatic beta cell line that retains glucose-inducible insulin secretion: special reference to expression of glucose transporter isoforms. Endocrinology 127(1):126–132
33. Praz GA, Halban PA, Wollheim CB, Blondel B, Strauss AJ, Renold AE (1983) Regulation of immunoreactive-insulin release from a rat cell line (RINm5F). Biochem J 210:345–352
34. Karlsson K, Cairns N, Lubec G, Fountoulakis M (1999) Enrichment of human brain proteins by heparin chromatography. Electrophoresis 20:2970–2976
35. Shin JH, Krapfenbauer K, Lubec G (2006) Large-scale identification of cytosolic mouse brain proteins by chromatographic prefractionation. Electrophoresis 27:2799–2813
36. Conrad EH (ed) (1998) Heparin-binding proteins. Academic Press, Inc., San Diego
37. Pollak D, Krapfenbauer K, Fountoulakis M, Peyrl A, Lubec G (2004) Expressional pattern of known and predicted signaling proteins in seven human cell lines. J Chromatogr B Analyt Technol Biomed Life Sci 808:185–208
38. Shin JH, Krapfenbauer K, Lubec G (2005) Column chromatographic prefractionation leads to the detection of 543 different gene products in human fetal brain. Electrophoresis 26:2759–2778
39. Cagney G, Park S, Chung C, Tong B, O'Dushlaine C, Shields DC, Emili A (2005) Human tissue profiling with multidimensional protein identification technology. J Proteome Res 4:1757–1767

40. Florens L, Washburn MP (2006) Proteomic analysis by multidimensional protein identification technology. Methods Mol Biol 328:159–175
41. Fountoulakis M, Juranville JF, Jiang L, Avila D, Roder D, Jakob P, Berndt P, Evers S, Langen H (2004) Depletion of the high-abundance plasma proteins. Amino Acids 27:249–259
42. Lubec G, Krapfenbauer K, Fountoulakis M (2003) Proteomics in brain research: potentials and limitations. Prog Neurobiol 69:193–211
43. Fountoulakis M, Langen H (1997) Identification of proteins by matrix-assisted laser desorption ionization-mass spectrometry following in-gel digestion in low-salt, nonvolatile buffer and simplified peptide recovery. Anal Biochem 250:153–156
44. Peyrl A, Krapfenbauer K, Slavc I, Yang JW, Strobel T, Lubec G (2003) Protein profiles of medulloblastoma cell lines DAOY and D283: identification of tumor-related proteins and principles. Proteomics 3:1781–1800
45. Langen H, Roder D, Juranville JF, Fountoulakis M (1997) Effect of protein application mode and acrylamide concentration on the resolution of protein spots separated by two-dimensional gel electrophoresis. Electrophoresis 18:2085–2090
46. Rockwood AL, Haimi P (2006) Efficient calculation of accurate masses of isotopic peaks. J Am Soc Mass Spectrom 17:415–419
47. DeFronzo RA (1997) Pathogenesis of type 2 diabetes: metabolic and molecular implications for identifying diabetes genes. Diabetes Rev 5:177–269
48. Krapfenbauer K, Fountoulakis M (2009) Improved enrichment and proteomic analysis of brain proteins with signaling function by heparin chromatography. Methods Mol Biol 566:165–180
49. Clark A (1989) Islet amyloid and type 2 diabetes. Diabet Med 6(7):561–567
50. Nishi M, Sanke T, Nagamatsu S, Bell GI, Steiner DF (1990) Islet amyloid polypeptide. A new beta cell secretory product related to islet amyloid deposits. J Biol Chem 265:4173–4176
51. Westermark P, Wernstedt C, Wilander E, Hayden DW, O'Brien TD, Johnson KH (1987) Amyloid fibrils in human insulinoma and islets of Langerhans of the diabetic cat are derived from a neuropeptide-like protein also present in normal islet cells. Proc Natl Acad Sci U S A 84:3881–3885
52. Ohsawa H, Kanatsuka A, Yamaguchi T, Makino H, Yoshida S (1989) Islet amyloid polypeptide inhibits glucose-stimulated insulin secretion from isolated rat pancreatic islets. Biochem Biophys Res Commun 160:961–967
53. Howard CF Jr (1986) Longitudinal studies on the development of diabetes in individual Macaca nigra. Diabetologia 29:301–306

Endothelial Dysfunction in Diabetes: Role of Circulating Biomarkers as Potential Diagnostic and Prognostic Tools

Worku Abebe, Babak Baban, and Mahmood S. Mozaffari

Abstract Diabetes is a major health problem worldwide, type 2 diabetes accounting for about 90% of the diagnosed cases. Vascular disorder is a prominent feature of diabetes and nearly 80% of diabetic mortality is a consequence of this disorder. Endothelial dysfunction is a predisposing factor for vascular disease in diabetes. The methods currently utilized for assessing endothelial function/dysfunction in humans are associated with various limitations making them less applicable for routine tests. Circulating endothelial-linked biomarkers have been proposed to serve as alternative/additional options for evaluating the status of the endothelium in diabetes and related conditions affecting the vasculature. These indicators include von Willebrand factor, thrombomodulin, E-selectin, tissue plasminogen activator, endothelial microparticles, circulating endothelial cells and circulating endothelial progenitor cells. While tentative evidence provides support that most of these biomarkers have the potential to serve as reliable indicators of endothelial status, their usefulness for routine clinical applications has not yet been established. Thus, there is presently intense research interest in circulating endothelial markers and it is anticipated that as more information becomes available, the improved quantification of these substances can provide a more suitable predictive, diagnostic and prognostic means for endothelial-associated events in diabetes and related disorders. The possibility of application of these biomarkers for early detection of endothelial dysfunction can have the added advantage of preventing the vascular complications of diabetes, particularly of type 2 diabetes.

W. Abebe (✉)
Department of Oral Biology, CL 2140, College of Dental Medicine,
Georgia Regents University, Augusta, GA 30912-1128, USA
e-mail: wabebe@GRU.edu

B. Baban • M.S. Mozaffari
Department of Oral Biology, College of Dental Medicine,
Georgia Regents University, Augusta, GA 30912-1128, USA
e-mail: mmozaffa@gru.edu

M.S. Mozaffari (ed.), *New Strategies to Advance Pre/Diabetes Care: Integrative Approach by PPPM*, Advances in Predictive, Preventive and Personalised Medicine 3,
DOI 10.1007/978-94-007-5971-8_5,

Keywords Microangiopathy • Macroangiopathy • Insulin resistance • Prediabetes • Improved technology limitations • Endothelial-associated diagnostic • Prognostic tools

1 Introduction

Over the last several decades it has become evident that diabetes mellitus is a major non-infectious chronic disease with increasing prevalence rate worldwide. The World Health Organization (WHO) reports that, in the year 2000, more than 171 million people worldwide had diabetes and the number of diabetic patients is now risen to 346 million [1]. About 3.4 million deaths were attributed to diabetes in 2004 and it is projected that this figure will be double by 2030. Depending upon the prevalence of the disease and the treatments available locally, health care costs for diabetes are estimated to cover 2.5–15% of annual health care budgets [1]. Additionally, the costs of lost production due to diabetes are reported to reach five times the health care costs. The United States (US) is considered among the top ten countries in the world concerning the number of people suffering from diabetes. According to the American Diabetes Association (ADA), currently, there are nearly 26 million (8.3%) Americans with diabetes and compared to a decade ago, this number is significantly higher [2]. Of the total number of diabetic patients, about one-quarter are undiagnosed and are not aware that they have the disease. In addition, there are now 79 Americans with prediabetes, who are likely to progress to a stage of type 2 diabetes within 10–30 years [2].

Diabetes mellitus is known to exist in several forms but two are considered major categories: type 1 and type 2 diabetes. Type 1 diabetes is characterized by an absolute insulin deficiency attributable to pancreatic insufficiency caused by autoimmune destruction of β cells [3]. It often starts in childhood and accounts for 5–10% of all diagnosed cases of diabetes. On the other hand, type 2 diabetes is characterized by insulin resistance associated with a reduction in glucose uptake by peripheral tissues (e.g., skeletal muscle, adipose tissues) during insulin exposure, thus representing a deficiency in the effectiveness of insulin despite high plasma hormone level. This condition eventually can lead to an absolute insulin deficiency due to the progressive dysfunction and failure of the β cells [4]. Genetic predisposition is considered to be a key factor for developing type 2 diabetes, but it may progress with additional risk factors such as obesity [5, 6] and sedentary lifestyle [7, 8]. Type 2 diabetes usually manifests itself later in life in adulthood following slowly developing metabolic processes that include a phase of glucose intolerance/prediabetes. Type 2 diabetes accounts for approximately 90–95% of all diagnosed cases of diabetes. Increasing incidences of type 2 diabetes, which progressively inflicts children primarily due to the epidemic of overweight and obesity, is a major contributory factor to the overall rise in the number of diabetic patients worldwide.

Being associated with hyperglycemia and related metabolic disturbances, both type 1 and type 2 diabetes can lead to serious secondary complications, such as cardiovascular disorders, nephropathy, retinopathy and neuropathy [9]. Heart disease and stroke are major manifestations of cardiovascular disorders in both types of diabetes. Diabetes-induced impairment of blood flow and neuropathy in foot leads to increased risk of foot ulcers and eventual amputation. Diabetic retinopathy occurring as a consequence of damage to the small blood vessels (microvessels) in the retina is a major cause of blindness. The effect of diabetes on the kidney (i.e., nephropathy) results in kidney failure and death in 10–20% of people with diabetes [10].

Whereas diabetes is recognized to be associated with a number of pathophysiological complications, vascular disorder is a prominent feature of both type 1 and type 2 diabetes. It is reported that as high as 80% of the diabetes mortality is a consequence of this disorder [11]. Although the reason(s) for diabetic vasculopathy (angiopathy) is incompletely understood, it is now established that the vascular endothelium plays a central role [12]. As discussed in some depth below, the endothelium is critical for proper functioning of blood vessels and, hence, target organs. Accordingly, endothelial dysfunction or injury predisposes diabetic patients to vascular disorder manifested by both microangiopathy and macroangiopathy [12]. Microangiopathy in diabetes is a hallmark feature of retinopathy, neuropathy and nephropathy whereas macroangiopathy is more closely linked to accelerated atherosclerosis. Therefore, maintaining the function of the endothelium is of prime importance for healthy vasculature and proper target organ functions. Measuring the functionality of the endothelium using appropriate methodologies can be a valuable step for this assessment.

Even though strict glycemic control can lower the incidence of diabetic complications, this is often difficult to achieve and is not also sufficient to avert the vascular problem [13]. Moreover, the reliability or sensitivity of existing diagnostic tests for glucose has been questioned [14]. Besides contributing to the development of vascular problems, evidence is also available indicating that endothelial dysfunction is an early event preceding the clinical manifestation of type 2 diabetes [15]. On the other hand, in type 1 diabetes, impairment of the endothelium occurs much later after a period of exposure to hyperglycemia [16]. Accordingly, while endothelial dysfunction may reflect a 'postdiabetes' event in type 1 diabetes, it appears to occur prior to the clinical development of type 2 diabetes, making it a viable target for early detection of type 2 diabetes or the vascular complications associated with it. Experimentally, diabetes-associated endothelial impairment has been examined by measuring various outcome indicators in different animal models of the disease. However, assessing the status of the endothelium in humans using the currently utilized functional assay techniques has been more problematic both for routine determination and for detecting subtle changes (see below for details). On the other hand, there are emerging assay methods based on the measurements of circulating endothelial markers which have the potential to serve as suitable alternatives/additions for routine human application. However, presently, these assays are utilized primarily for research purposes while at the same time they are under consideration to develop clinically-applicable procedures for routine work.

This chapter presents a synopsis of the currently available endothelial-derived biomarkers in the circulation, with the potential to be used as sensitive, convenient, reliable and cost effective predictors of endothelial dysfunction and the associated disorders in humans, particularly with type 2 diabetes/prediabetes. As background information, we have briefly reviewed the function of the endothelium under normal and diabetic/prediabetic conditions, and also summarized the methods that have been commonly used for evaluating endothelial function/dysfunction in humans.

2 Endothelial Function and Its Impairment in Diabetes

2.1 Overview of Endothelial Function

The endothelium is a mono-layer of special epithelial cells that cover the lumen of all blood vessels. It provides an interface between circulating blood and the underlying vascular smooth muscle. Besides serving as a physical barrier between the blood and vascular tissue, the endothelium is involved in multiple other functions, including regulation/maintenance of vascular tone, blood flow, blood coagulation, fibrinolysis, cell growth, synthesis and release of vasoactive substances, platelet function and leukocyte adherence [17, 18]. The type and location of the blood vessel can determine the interplay between the endothelium and the rest of the vessel. For instance, while endothelial cells in conduit arteries limit activation of clotting and pro-inflammatory factors, they contribute to the regulation of blood flow in resistance vessels and facilitate the transport and distribution of nutrients and hormones in precapillary arterioles [17, 18].

The role of the endothelium as a regulator of vascular tone and blood flow was demonstrated experimentally about three decades ago by the 1998 Nobel Laureate Dr. Robert Furchgott [19]. The vasodilatory function was shown to be mediated by what was then known as endothelium-derived relaxing factor (EDRF), which was subsequently identified as nitric oxide (NO). Nitric oxide is now considered a key vasodilator substance generated in the vascular endothelium, although other released substances including the arachidonic acid metabolic product prostacyclin (PGI_2) and endothelium-derived hyperpolarizing factor (EDHF) also play a similar role (albeit to a lesser extent) [20] (Fig. 1). To induce this vascular response, endothelial cells constitutively express the enzyme NO synthase, more specifically NO synthase-type III (eNOS). For NO formation, the initial step is activation of the endothelial cells which results in increased intracellular free Ca^{2+}concentration. Calcium in turn binds to calmodulin forming a complex. The eNOS enzyme, with Ca^{2+}/calmodulin binding, then generates NO using l-arginine as a substrate together with reduced nicotinamide adenine dinucleotide phosphate (NADPH), flavin adenine dinucleotide/mononucleotide (FAD/FMN), and tetrahydrobiopterin (BH_4) as cofactors [20]. In addition, more recently it was shown that NO itself can activate eNOS in endothelial cells, suggesting its involvement in a positive feedback process to further increase its own release [21]. The NO produced in the endothelial cells rapidly diffuses into underlying vascular smooth muscle cells and binds to a heme group of soluble guanylyl cyclase (sGC), resulting in activation of the enzyme

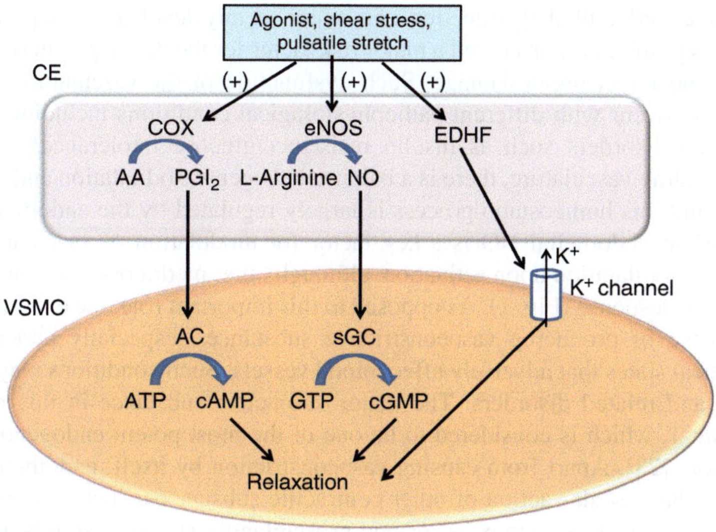

Fig. 1 Effects of endothelial cell mediators on vascular smooth muscle tone. Endothelial cells synthesize and release NO, PGI_2 and EDHF which diffuse into adjacent smooth muscle cells to activate respective signaling mechanisms that lead to a reduction in intracellular calcium (Ca_i^{2+}), and cause smooth muscle relaxation. *Abbreviations*: *NO* nitric oxide, *sGS* soluble guanylate cyclase, *AA* arachidonic acid, *COX* cyclooxygenase, *AC* adenylyl cyclase, *GTP* guanosine triphosphate, *cGPM* cyclic GMP, *AMP* adenine monophosphate, *cAMP* cyclic AMP, *eNOS* NO synthase-type III, *PGI_2* prostaglandin I_2, *EDHF* endothelial-derived hyperpolarizing factor, *EC* endothelial cell, *VSMC* vascular smooth muscle cell

and formation of cyclic GMP (cGMP) [22]. Cyclic GMP in turn activates a cGMP-dependent protein kinase, leading to a reduction in cytosolic Ca^{2+} in vascular smooth muscle cells and followed by inhibition of the contractile machinery and vasodilatation [23] (Fig. 1).

Various agents activating receptor-operated mechanisms in endothelial cells are known to cause the release of NO. Examples include acetylcholine (Ach), bradykinin, serotonin and substance P [23, 24]. Nitric oxide is also released by receptor-independent mechanisms, such as physical stimuli (e.g., shear stress, pulsatile stretch), most likely independent of the involvement of Ca^{2+}/calmodulin [25]. The release of NO by endothelial cells takes place seemingly under baseline conditions and this process is believed to play a role in the regulation of blood pressure. Consequently, many studies have demonstrated that eNOS inhibitors or mechanisms that impair NO generation elevate blood pressure. Besides causing vasodilation, NO also mediates many other functions, which are predominantly protective. For example, NO by inhibiting expression of pro-inflammatory cytokines, chemokines and leukocyte adhesion molecules, limits vascular recruitment of leukocytes and platelets, thus limiting such vascular disorders as atherosclerosis and thrombosis [26]. Nitric oxide also inhibits vascular smooth muscle cell proliferation, which is an early sign of vasculopathy such as atherosclerosis [26].

Vascular endothelial dysfunction can occur at any level and, as pointed out above, this problem is considered a major risk factor for the development of vascular diseases and target organ damage. Such dysfunction of the vasculature has been observed to occur with different pathophysiological conditions including diabetes and related disorders such as insulin resistance/glucose intolerance/prediabetes [27]. In healthy vasculature, there is a balance between vasodilatation and vasoconstriction and this homeostatic process is largely regulated by the endothelium. As noted earlier, endothelial NO is a key factor for modulation of vascular tone by inducing vasodilatation upon activation, although other mediators also contribute to the vascular response (Fig. 1). As opposed to this important role, the endothelium is also capable of producing vasoconstrictive substances, especially under certain pathological states that adversely affect blood vessels. Such conditions may include diabetes and related disorders. The major vasoactive substance in this regard is endothelin-1, which is considered to be one of the most potent endogenous vasoconstrictors [28]. Apart from causing vasoconstriction by itself, endothelin-1 also augments the vascular action of other contractile substances, such as norepinephrine, angiotensin II, thromboxane A_2 and prostaglandin H_2. Accordingly, the chain of events that lead from endothelial dysfunction to overt vasculopathy in diabetes and related diseases are suggested to be due to an imbalance in the production of these contrasting vasoactive mediators. It should also be noted that besides causing vasoconstriction, the contractile agents also have other effects on the vasculature such as causing vascular smooth muscle cell proliferation. This abnormality during vascular disease conditions may affect the critical role of the endothelium in overall homeostasis, predisposing it to pro-angiopathic factors [24, 28].

Numerous studies have identified impairment of NO production and/or loss of its biological activity as the main immediate culprit lesion for the dysfunction of the endothelium [29]. With the availability of suboptimal concentrations of substrate and/or cofactors for NO synthesis, eNOS becomes uncoupled thereby impairing the synthesis of NO. This situation leads to the generation of reactive oxygen species (ROS), such as superoxide anion and hydrogen peroxide, a condition collectively defined as oxidative stress [30]. Excessive ROS generation during oxidative stress cannot be completely scavenged by the antioxidant mechanisms which function efficiently under normal conditions. With excess ROS, the scavenger system becomes overwhelmed and thus ineffective, enabling NO to rapidly react with superoxide to form peroxynitrite, which further exacerbates tissue injury [31].

2.2 *Influence of Diabetes on Endothelial Function*

2.2.1 Overview

There is ample clinical and experimental evidence indicating that both type 1 and type 2 diabetes cause dysfunction of the endothelium, with or without overt vasculopathy [12, 24, 27]. The degree of the dysfunction depends upon the severity and

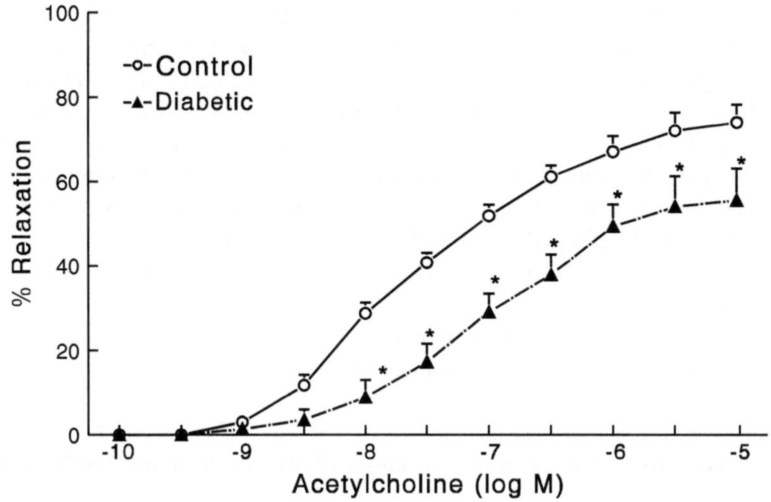

Fig. 2 Effect of streptozotocin (STZ)-induced type 1-like diabetes of 14 weeks duration on endothelium-dependent relaxant responses of aortas from male rats to acetylcholine (n=6 rats/group). *Significantly different from corresponding control responses ($P<0.05$)

duration of the disease. Figure 2 presents data from our laboratory demonstrating impairment of endothelium-dependent relaxation of aortas from streptozotocin (STZ)-treated diabetic rats of 14 weeks of duration. This animal model resembles type 1-like diabetes, with hyperglycemia, hypoinsulinimia and other diabetes-associated characteristics. In this animal model, hyperglycemia develops within a few days but endothelial dysfunction occurs later. This observation suggests that in type 1 diabetes, endothelium dysfunction is manifested after a period of diabetes state or hyperglycemia. On the other hand, as stated earlier, endothelial dysfunction precedes the clinical development of type 2 diabetes. This can occur during prediabetes or even prior to the manifestation of this condition. This contention is supported, for example, by the finding of impaired vascular endothelial responsiveness in subjects with normal glucose tolerance, but with family history of type 2 diabetes [32]. In this regard, endothelial dysfunction may be a critical early target for preventing overt vascular disease that may subsequently develop as diabetes progresses. From the aforementioned discussion, such a consideration is particularly relevant for type 2 diabetes. Despite differences in onset time, in both cases of type 1 and type 2 diabetes, endothelial dysfunction, once initiated, can cause vascular alterations in various aspects, which may further lead to target organ damage.

Why is the endothelium so susceptible to the effects of diabetes or prediabetes? Several reasons can be forwarded as likely answers to this question. The endothelium is directly exposed to hyperglycemia in the circulation, but it is unable to autoregulate glucose transport, thereby accumulating excess glucose intracellularly, which can result in adverse biological effects (e.g., cell damage due to excess ROS production). The endothelium is located at a favorable site for interactions with the immune system, inflammatory mediators and other vasoactive circulating substances, making it responsive to these stimuli almost on a continuous basis. The endothelium is also in a strategic location to be constantly exposed to blood flow-induced shear stress which can potentially cause cell perturbations. Depending upon the magnitude of these effects, the consequences may at least include increased generation of ROS, osmolarity changes, enhanced expression of adhesion molecules, augmented formation of proinflammatory cytokines and chemokines, increased membrane permeability and impairment of functional responsiveness [33]. An important aspect of endothelial cell alterations/damage is the release of chemical substances or cellular structures into the circulation which can serve as biomarkers of dysfunction of the endothelium.

The major basic mechanisms associated with endothelial dysfunction in diabetes/prediabetes are discussed below in some detail.

2.2.2 Mechanisms Underlying Diabetes-Induced Endothelial Dysfunction

The overall mechanisms underlying endothelial dysfunction in diabetes and associated complications are very complex and incompletely understood. However, the mechanisms described below have been shown to be involved in this dysfunction. While only a partial account of these mechanisms is provided, detailed information is available from other sources in the scientific literature [15, 22, 30].

Mechanisms Linked to Hyperglycemia

Both type 1 and type 2 diabetes are characterized by hyperglycemia, among many other things. Hyperglycemia plays a more important role in diabetes-associated microvascular complications. Hyperglycemia and its metabolic products have been shown to directly affect endothelial function. To verify this, our laboratory examined the effect of high glucose medium on endothelium-mediated vasorelaxation in rat aortas in response to acetylcholine (Fig. 3). It was shown that incubation of the rat aorta with 45 mM for 3 h causes inhibition of endothelium-dependent relaxation [34]. It is likely that similar effects are produced under in vivo situation in diabetes.

The transport of glucose into endothelial cells occurs by facilitated diffusion that is not dependent on insulin. This transport is not autoregulated in endothelial cells, in which case an increase in blood glucose concentration in the circulation leads to increased intracellular accumulation of glucose and its metabolites [15, 30, 35–37]. As mentioned above, this situation makes the endothelium relatively highly venerable to the adverse effects of hyperglycemia.

A number of investigators have proposed various possibilities to explain how hyperglycemia causes vascular endothelial dysfunction in diabetes. Under normal conditions, glucose undergoes cellular metabolism through the glycolytic pathway. With hyperglycemia four major pathways are proposed to be involved in dysfunction of the endothelium (Fig. 4). Accordingly, hyperglycemia-induced increased generation of reactive oxygen species (e.g., mitochondrial superoxide)

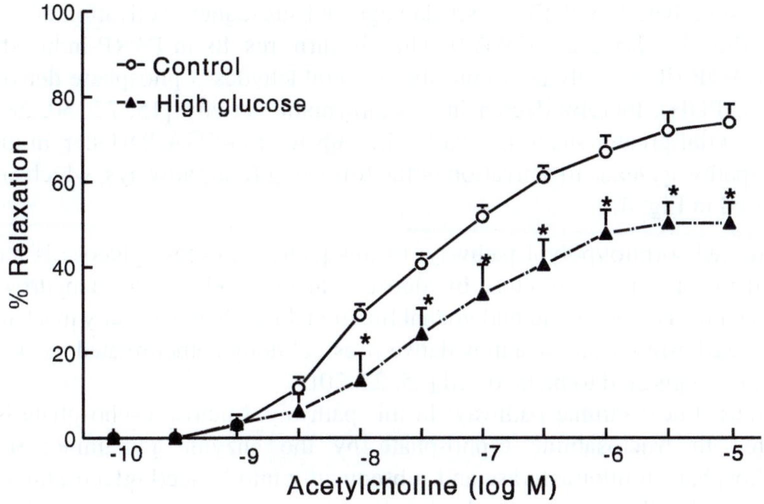

Fig. 3 Effect of high glucose concentration (45 mM for 3 h) on endothelium-dependent relaxant responses of aortas from male rats to acetylcholine (n=6 rats/group). *Significantly different from corresponding control responses ($P<0.05$)

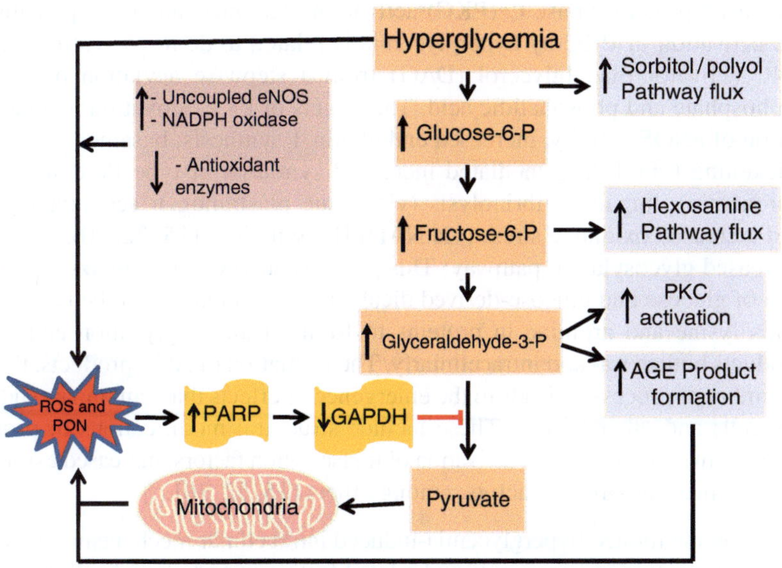

Fig. 4 Proposed hyperglycemia-induced intracellular mechanisms in endothelial cells. Overproduction of reactive oxygen species such as superoxide (primarily from mitochondria) induced by hyperglycemia causes oxidative DNA damage and activation of PARP. In turn, increased PARP activity causes addition of ADP-ribose polymer onto the GAPDH resulting in its inhibition. Inhibition of GAPDH will result in build-up of glycolytic intermediates and their subsequent diversion into and augmentation of the following four pathways: the sorbitol/polyol, hexosamine, PKC and glycosylation pathways. *Abbreviations*: *NADPH* reduced nicotinamide adenine dinucleotide phosphate, *eNOS* nitric oxide synthase-type III, *PON* peroxynitrite, *PKC* protein kinase C, *ROS* reactive oxygen species, *PARP* poly(ADP-ribose) polymerase, *GAPDH* glyceraldehydes-3-phosphate dehydrogenase, *AGEs* advanced glycation end (products)

causes exacerbated oxidative DNA damage and subsequent activation of the poly (ADP-ribose) polymerase (PARP). This, in turn, results in PARP-induced addition of ADP-ribose polymers onto the glyceraldehydes-3 phosphate dehydrogenase (GAPDH), thereby decreasing its enzymatic activity [15, 22, 30, 38]. The consequential greater substrate availability, upstream of GAPDH step in the glycolytic pathway, leads to activation of the following four pathways, which are also illustrated in Fig. 4.

- Increased sorbitol/polyol pathway: In this pathway, excess glucose is metabolized to sorbitol and fructose by aldose reductase and sorbitol dehydrogenase. This pathway may impair endothelial function through two primary mechanisms: increased osmotic stress and oxidative stress, although other related mechanisms are also suggested to be involved [15, 22, 30].
- Increased hexosamine pathway: In this pathway, fructose-6-phosphate is converted to fructosamine 6-phosphate by the enzyme glutamine: fructose 6-phosphate amidotrasferase and subsequently into N-acetlyglucosamine. This pathway has been shown to be involved in augmentation of transcription of plasminogen activator inhibitor-1 (PAI-1) and transforming growth factor β1 (TGF-β1) and modification of protein kinase C (PKC) and eNOS [15, 22, 30]. The consequence of this process is enhancement of activities of the different pathways associated with the factors described.
- Increased protein kinase C (PKC) activation: The mechanism responsible for the activation of PKC by hyperglycemia is related to de novo synthesis of the PKC activator diacylglycerol (DAG) from a stepwise acylation of glycerol 3-phosphate and phosphatidic acid. The effects of PKC activation include inhibition of eNOS activity, increased endothelin-1 synthesis, basement membrane thickening (via TGF-β-mediated increased synthesis of type IV collagen and fibronectin), impaired fibrinolysis (via tissue plasminogen activator (PAI-1)) and increased oxidative stress (via NADPH regulation) [15, 22, 30].
- Increased glycosylation pathway: This pathway involves a non-enzymatic reaction of glucose and glucose-derived dicabonyl compounds with the basic amino acids lysine and arginine in proteins to form advanced glycation end (AGE) products both extra- and intracellularly. The formation of AGE products, through a number of processes, leads to the emergence of effects that can impair endothelial cells and other tissues. These include structural modifications of proteins, increase in oxidative stress, activation of transcription factors, increased expression of cytokines and growth factors, among others [15, 22, 30].

The aforementioned hyperglycemia-induced intracellular mechanisms associated with overproduction of superoxide and related factors in endothelial cells have been proposed as a possible cause of what is known as metabolic memory of hyperglycemic stress after glucose normalization [39]. More recently, this phenomenon has also been described as glucose memory or glucose legacy and shown to be caused by epigenetic changes that involve the histone H3 lysine 4 (H3K4) methyltransferase enzyme, Set7 [39, 40]. The accumulation of this enzyme in the nucleus under hyperglycemic environment appears to catalyze histone methylation/modification and regulate chromatin changes and gene expression (e.g., interleukin-8 gene) via both mono-methylation of histone H3 lysine 4 (H3K4m1)-dependent and independent pathways. Thus, it appears that epigenetic changes leave stable footprints which exert their detrimental impact on target organs even after normalization of blood glucose level. Although this epigenetic mechanism of glucose memory is proposed to occur both in type 1 and type 2 diabetes, all diabetic patients do not develop the debilitating consequences of the disease associated with exposure to hyperglycemia [39]. This observation indicates that genetic background of patients could be an important determinant of the glucose memory phenomenon. The phenomenon glucose memory may also have implications in relation to predictive, preventive and personalized medicine since existing indices for determining glycemic control primarily rely on whole blood (e.g., hemoglobin A1c), plasma (e.g., glucose and/or insulin) or urine (e.g., glucose excretion) assays. Since glucose memory manifests itself primarily at cellular /tissue level, it would be of great interest to identify novel biomarkers whose changes at tissue or cell level would mirror those occurring in the blood (e.g., endothelial progenitor cells). As research continues in this new area, future possibility exists for developing protocols to identify susceptible patients affected by this condition for predictive, preventive and consideration of individualized treatment measures.

Mechanisms Linked to Insulin Action

Besides its metabolic effects, insulin is a vasoactive hormone which acutely can cause vasodilation and increased blood flow. This effect involves various mechanisms that follow binding of insulin to its receptors on endothelial cell plasma membrane. By this process, insulin increases the production of NO in endothelial cells which subsequently causes relaxation of the underlying vascular smooth muscle, although the molecular pathways involved have not yet been fully elucidated. However, among other processes, it has been suggested to involve the activation of the intracellular enzymes phosphatidylinositol-3-kinase (PI3K) and Akt, which ultimately lead to phosphorylation and activation of eNOS. In addition to the generation of NO, insulin, via another pathway (mitogen-activated protein kinase (MAPK) pathway), causes the release of endothelin-1 by endothelial cells, and the growth and migration of vascular smooth muscle cells [15, 22, 27, 41, 42] (Fig. 5).

The action of insulin as a vasodilator is impaired under conditions of insulin-resistance, which is observed more notably in prediabetes, type 2 diabetes, obesity and hypertension. It is also decreased by chemical mediators closely associated with insulin-resistance, namely tumor necrosis factor-α (TNF-α), non-esterified fatty acids (NEFAs) and an impaired NO-dependent action [15, 22]. Similarly, when there is an absolute lack or impairment of insulin production as in type 1 diabetes, the effects normally mediated by insulin in endothelial cells (and elsewhere) will be absent, potentially shifting the vascular tone balance in favor of

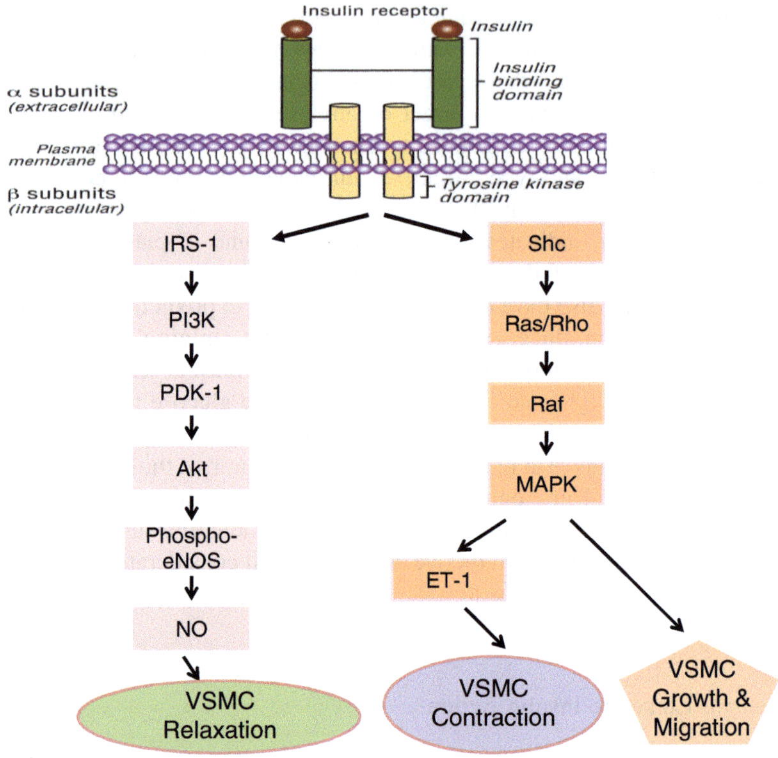

Fig. 5 Proposed mechanisms of effect of insulin in endothelial cells. The PI3K/Akt branch regulates NO production in endothelial cells and vasodilation in vascular smooth muscle cells. The MAPK branch controls secretion of endothelin-1 in the endothelium, and also regulates growth and mitogenesis. Under normal conditions, there is an optimum balance between these pathways. *Abbreviations*: *IRS-1* insulin receptor substrate protein-1, *PI3K* phosphatidylinositol-3-kinase, *PDK-1* phospho-3′-phosphoinositide-dependent kinase-1, *NO* nitric oxide, *eNOS* NO synthase-type III, *MAPK* mitogen-activated protein kinase, *ET-1* endothelin-1, *VSMC* vascular smooth muscle cell

vasoconstriction. Supporting the impairment of vascular relaxation under conditions of insulin resistance, endothelin-1 released from the endothelium by plasma insulin has been suggested to play a contributory role [22, 41].

Of additional consideration of insulin resistance/type 2 diabetes is that the disorder is also associated with elevated plasma free fatty acids which are capable of generating ROS via mitochondrial oxidation [18, 23, 26]. The ROS produced subsequently undergo similar metabolic pathways as noted for hyperglycemia (Fig. 4). However, unlike the role of hyperglycemia (which contributes more to microangiopathy), this mechanism is proposed to be involved more in diabetic macrovascular dysfunction [30].

Mechanisms Linked to Growth Factor and Cytokine Overproduction

Diabetes is associated with overproduction of growth factors and cytokines which have effects on endothelial cell function. Hyperglycemia is known to contribute to the effects of growth factors and cytokines by inducing increased synthesis of these factors in different types of cells [15, 27, 43]. Among other effects, the mechanisms involve hyperglycemia-induced PKC activation, AGE production and cytokine-mediated activation of endothelial cells. Hyperglycemia being a common factor, theses cellular processes take place in both type 1 and type 2 diabetes.

Excessive generation of growth factors has been implicated as a link between diabetes and proliferation of endothelial cells together with vascular smooth muscle cells, which possibly promote neovascularization [15, 27, 43]. Among the growth factors that play important roles in diabetes complications, TGF-β1 and vascular endothelial growth factor (VEGF) have greater influence on endothelial cells. The receptors for both growth hormones are expressed in endothelial cells [15, 27]. Both growth factors regulate multiple cellular functions in different tissues as well. For TGF-β1, these include cell growth, cell death/apoptosis, cellular differentiation and extracellular matrix protein synthesis. Similarly, VEGF stimulates cell differentiation, survival, migration, and proliferation, tubulogenesis and vascular permeability [15, 27]. However, although the levels of these growth factors have been shown to increase in animal models of diabetes, the temporal sequence of activity is not well defined, implying that the roles of growth factors in diabetes require further investigation.

With regard to cytokines, although some of them appear to play important roles in endothelial functions, many cytokines produce effects to which endothelial cells respond variably. In short, inflammatory cytokines enhance vascular permeability, alter vasoregulatory responses, increase leukocyte adhesion to the endothelium, facilitate procoagulant activity, and regulate gene expression in endothelial cells, among several other effects [15, 27]. Of the many known cytokines, TNF-α is clearly implicated as a link between insulin resistance, diabetes, and endothelial dysfunction. In view of the observed association between obesity and insulin resistance, increased expression of TNF-α (and other cytokines) in humans with obesity provides support for the hypothesis that elevation of this cytokine contributes to insulin resistance. TNF-α also induces the synthesis of other cytokines, such as interleukin (IL)-1, which in turn regulate the expression of c-reactive protein (CRP), an important downstream marker of inflammation [15, 27]. These mediators, alone or in concert with each others, contribute to atherothrombosis by altering, at least in part, endothelial function.

More recent clinical and experimental observations have indicated that there is a strong association between obesity and insulin resistance/type 2 diabetes. Accordingly, endothelial dysfunction that occurs in obese subjects or animal models of obesity can be related to insulin resistance/type 2 diabetes, although obesity can also be an independent risk factor for endothelial dysfunction and the associated disorders. This relationship is believed to be, at least partly, the result of overproduction of chemical mediators from adipose tissue in the obese subject. Evidence has been accumulating that visceral adipose tissue is specially an active endocrine organ that releases relatively high amounts of various substances, including certain hormones, cytokines and enzymes that are capable of attenuating insulin sensitivity [32, 42, 44]. Of these substances, cytokines (e.g., TNF-α, IL-6, PAI-1) and adipokines (e.g., diponectin, lepin) have been proposed to play greater roles in inflammation, insulin resistance and associated vascular diseases [27, 32, 42]. Studies have also shown that while some of these mediators are protective, others cause a variety of adverse effects, as mentioned earlier. On balance, however, visceral obesity gives rise to insulin resistance and endothelial dysfunction primarily through a cascade of secreted proinflammatory substances. Furthermore, insulin resistance per se by causing hyperglycemia perpetuates inflammation via augmented oxidative stress [30, 32, 42].

Of recent interest in relation to inflammation is the identification of endogenous inhibitors of signaling cascades involved in leukocyte adhesion to endothelial cells. These endogenous inhibitors include pentraxin-3, growth differentiation factor-15 (GDF-15) and developmental endothelial locus-1 (Del-1) [45, 46]. While pentraxin-3 inhibits rolling of leukocytes, GDF-15 blocks chemokine-induced activation of intergrins. On the other hand, Del-1 antagonizes LFA-1 (∝Lβ2)-dependent leukocyte adhesion and recruitment; LFA-1 is a member of β2-integrin family expressed on lymphocytes and neutrophils. Importantly, Del-1 is secreted by endothelial cells during inflammation and its subsequent anti-inflammatory actions may represent an endogenous homeostatic mechanism to curb inflammatory outcomes. Del-1 is a 52 kDa glycoprotein associated with endothelial cell surface and extracellular matrix [47]. Given the increasing recognition that inflammation plays a pivotal role in the pathogenesis of diabetes mellitus and associated pathologies, unraveling the potential role of Del-1 in this condition represents a fertile ground for future investigations. Also, the determination of Del-1 in quantitative terms during the development and progression of diabetes is an avenue to be explored for assessment of diabetes based on endothelial responsiveness

3 Methods of Endothelial Assessment in Humans and Their Limitations

Ever since the initial seminal reports of Furchgott and collaborators [19] demonstrating the vasodilatory effect the endothelium, determination of the role of the endothelium in vascular diseases has become a major consideration. Subsequent to the observation that derangement in NO formation contributes to the development of atherosclerosis, various techniques of assessment of endothelial function/dysfunction in humans in relation to NO release have been developed for both coronary and peripheral circulation. It should be noted that most of the tests performed in this regard were designed to evaluate abnormalities in the regulation of the lumen sizes of blood vessels [22, 48–51]. Since this approach has been realized to be associated with certain limitations, its application in combination with other assessment methods, such as use of biomarkers (e.g., nitrite assay) or measurement of blood vessel structural changes (e.g., carotid intima media thickness) has recently become a subject of consideration.

Following is a brief description of the methods that are commonly utilized presently for assessing endothelial function/dysfunction in humans. As such, the parameter measured is considered to be an outcome of endothelial cell activity.

3.1 Plethysmography

This technique is based on strain-gauge venous impedance plethysmography for studying endothelium-dependent as well as independent vasodilatation in the peripheral circulation [22, 48, 50, 51]. It is especially designed for determining dose–response relationships of stimulators and inhibitors of endothelial cell activity. This method often uses direct infusion of Ach or other muscarinic receptor agonists (as a selective activator of endothelial cells) in the brachial artery and determines vasodilator responses in the forearm resistance vessels. Sodium nitroprusside is administered as a control agent (for comparison) in order to assess endothelium-independent vasodilatation. The determination of changes in blood flow (reflective of vasodilation/lumen size), contributed by the whole limb portion is believed to represent a measure of endothelial function. However, the invasive nature of the technique and adverse effects of the drugs administered limit its repeatability, prohibiting its use in large scale studies or for routine application. In addition, the results are also difficult to standardize because resistance vessel tone is variable [22, 48–51].

3.2 Ultrasound and Doppler

Before the development of most of the currently used methods, various types of non-invasive techniques were used to assess endothelial function/dysfunction, particularly in the coronary circulation. These included Doppler echocardiography, positron emission tomography and phase-contrast magnetic resonance imaging [48, 49, 51, 52]. However, due to certain shortcomings associated with these techniques, especially in relation to technical difficulties and high-costs, their uses have remained very limited. Subsequently, quantitative coronary angiography with an intra-coronary ultrasound and Doppler transducer has been developed and used more often to assess endothelial function in the coronary blood vessels [22, 48–52]. By a graded administration of Ach and sodium nitroprusside, endothelium-dependent and independent vasodilatation, respectively, can be quantified using this approach. Nonetheless, in addition to being invasive and associated with drug administration, this technique is also technically complicated for routine application. This led to the emergence of a simpler and a less invasive method based on flow-mediated vasodilatation (FMD) in brachial artery. Although it does not directly involve coronary vasculature, this method has been found to be more suitable for repeated studies and for evaluating large groups of patients [22, 48, 50, 52, 53]. The method uses ultrasound Doppler for detection of changes in artery diameter (lumen size) and blood flow. Flow-mediated vasodilatation thus measured in brachial artery has been correlated with endothelial function in coronary circulation and other proximal sites [54]. Because of the advantages stated, this technique has been more widely used to test endothelial functions in different parts of the vascular system. In fact, some consider this method the gold standard for clinical research on conduit artery endothelial biology [55]. Flow-mediated vasodilatation is based on the theory of shear stress, whereby a short period (for example, 5 min) of arterial occlusion increases flow in the artery upon release. This procedure is proposed to induce the endothelium to release NO with post-occlusive vasodilatation. Nitroglycerin (or another nitrate) is usually administered sublingually/orally to evaluate endothelium-independent vasorelaxant responses to be used for reference. In shear stress, no specific receptor signaling pathway has been reported to be activated or inhibited. Overall, the precise mechanisms how shear stress and subsequently occurring signaling messages modulate vascular tone are not well understood [56].

Brachial artery FMD has been studied widely in clinical research because, among other things, it has been found useful for serial assessment of young subjects for endothelial functionality, such as in the testing of lifestyle and pharmacological interventions at an early stage of vascular diseases [22, 48, 51]. However, besides the need for occasional administration of drugs, the equipment used in this technique are still costly and the procedures involved are also technically demanding, although the protocol can be reliably standardized to yield reproducible results that correlate with coronary vascular endothelial function as well [48, 51].

3.3 Limitations of Above Techniques

In spite of improvements in recent years, the above techniques of assessing endothelial functions in humans are associated with certain limitations for routine applications. These limitations include invasiveness, low degree sensitivity, lack of clarity of specificity for endothelium, technical challenges, high cost and variability of measurement outcomes. As a result, currently, these methods have been used primarily for research purposes. This implies that there is a need for further improvement of the methods being utilized or for developing alternative methods which can be used to test the endothelium for routine applications.

4 Circulating Endothelial Biomarkers in Diabetes

Given the limitations of the above techniques to human applications, more recent efforts have been directed towards exploring other approaches of endothelial evaluation based on measurement of circulating endothelial markers. The advantages of measuring these indicators may include (a) possibility of earlier detection of

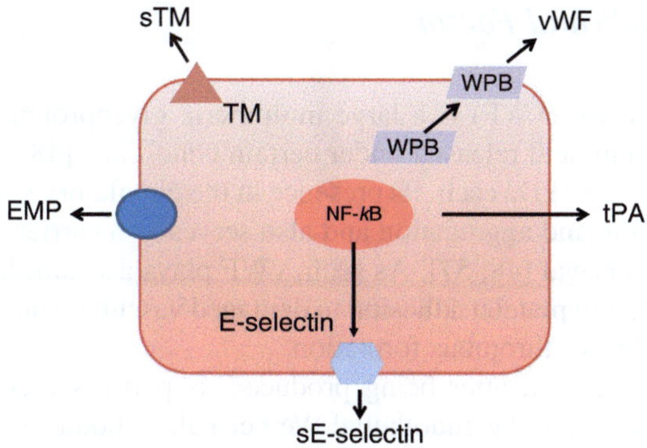

Fig. 6 Endothelial-derived circulating biomarkers with the potential for detecting endothelial function/dysfunction. These biomarkers and endothelin-1 are released into the circulation by endothelial cells. However, the assay method available for endothelin-1 is not sensitive enough to detect subtle changes in the endothelium. There are also other biomarkers that are released by endothelial cells and other types of cells, making them less specific to reflect the status of the endothelium. These widely dispersed biomarkers include nitric oxide, soluble vascular cell adhesion molecule-1, soluble intracellular adhesion molecule and prostaglandin I_2. *Abbreviations*: *NF-kB* nuclear factor *k*B, *EMP* endothelial microparticle, *TM* thrombomodulin, *sTM* soluble TM, *vWF* von Willebrand factor, *WPB* Weibel-Palade body, *tPA* tissue plasminogen activator

endothelial impairment (b) simplicity of quantification with current immunological techniques for most of them (e.g., using ELISA), (c) release of the biomarkers specifically from the vascular endothelium and provision of information on status of the endothelium, and (d) greater sensitivity, rapidity, simplicity and cost effectiveness of the assay methods. Because of these possible advantages, there appears to be a general consensus that circulating endothelial biomarkers have the potential to be used for routine clinical testing of endothelial dysfunction in patients with vascular disorders. However, it is noteworthy that the usefulness of this approach has not yet been fully verified for clinical applications and some of the procedures involved, as used presently, may also require delicate conditions of sampling and/or analysis [48, 49, 51].

Below is a brief review of the biomarkers which have been reported relevant for assessing endothelial function/dysfunction in diabetes (and other similar conditions) (Fig. 6). With the exception of endothelial progenitor cells (EPCs), these biomarkers have the common property of being released specifically from the vascular endothelium. Endothelial progenitor cells are also closely associated with the endothelium as they are involved in the maintenance and repair of the endothelium, as detailed below. It should be noted that there are also several other circulating biomarkers which are released from the endothelium as well as from other tissues, and endothelial-derived substance whose assays are less sensitive to detect subtle changes in the endothelium. Examples of such biomarkers include some adhesion molecules [e.g., P-selectin, vascular cell adhesion molecule-1 (VCAM-1), intracellular adhesion molecule-1 (ICAM-1)], endothelin-1 and NO/nitrite [48–52, 57] (Fig. 6).

4.1 Von Willebrand Factor

Von Willebrand factor (vWF) is a large multimeric glycoprotein produced in the vascular endothelium and released under certain conditions [48, 57]. It consists of over 80 subunits of 250 kDa each. Its presence in the circulatory system is important for platelet adhesion and aggregation and also serves as a carrier protein for factor VIII, helping its survival [48, 57]. As such, vWF plays a vital role in hemostatasis by specially mediating platelet adhesion to damaged vascular walls, thereby causing localized blood clots or thrombus formation.

Von Willebrand factor, after being produced, is partly secreted constitutively, while the rest is stored in the endothelial Weibel-Palade bodies to be released later into blood on demand (e.g., endothelial cell injury or stimulation) (Fig. 6). In the circulation, the multimeric vWF is in a folded form with important epitopes hidden. However, under certain conditions (e.g., shear stress) the molecule's binding sites becomes exposed making it hemostatically functional. The molecule also becomes susceptible to the effect of the metalloprotease, ADAMTS13, limiting its hemostatic activity. It is reported that deficiency in the function of ADMT13 leads to increased levels of the large circulating multimers. Measurement of circulating levels of vWF is considered a gold standard in the assessment of endothelial damage or associated vascular disorders. Its plasma concentration has been shown to increase in type 2 diabetes particularly in association with cardiovascular disorders manifested by both macro- and microangiopathies [48, 57]. The amount of vWF measured in the blood has been found to be correlated with the severity of diabetes and the attendant cardiovascular events. From these observations, vWF is considered potentially a valuable diagnostic as well as prognostic indicator for endothelial dysfunction or vascular diseases at least in type 2 diabetes [48, 57]. However, plasma concentration of glycoprotein in not altered by conditions of nondiabetic insulin resistance, indicating that vWF is not associated with risk of type 2 diabetes. It is, thus, less useful to serve as a predictive tool. The mechanism(s) behind the increased vWF levels with endothelial/vascular dysfunction in diabetes is unclear but it is hypothesized to be related at least in part to impairment of ADMT13 activity. Plasma vWF can be measured by enzyme-linked immunosorbent assay (ELISA) kits which makes use of a monoclonal antibody.

4.2 Thrombomodulin

Thrombomodulin (TM) (CD141/BDCA-3) is a membrane glycoprotein expressed abundantly on the surface of endothelial cells (primarily in microvasculature) [48, 57]. It has a molecular mass of 74 KDa, consisting of a single chain with five distinct domains. It serves as a cofactor for protein C activation in anticoagulant pathways involving thrombin, thus neutralizing the thrombotic potential of thrombin.

Thrombomodulin can be released from endothelial cells with injury or cleaved enzymatically to yield a soluble form, (soluble thrombomodulin, sTM) which has been demonstrated to be a specific circulating biomarker of endothelial cell damage (Fig. 6). While there are not many studies reported in the literature, in one study the concentrations of sTM have been found to be high in type 2 diabetic patients associated with endothelial damage or vascular complications [58]. Plasma sTM concentration can be measured reliably by ELISA kits [48, 57].

4.3 E-Selectin

E-selectin is a member of the selectin family adhesion molecules, which is specifically expressed on the surface of cytokine-activated endothelial cells [48, 49, 57]. In earlier studies, it was observed that maximal expression of E-selectin on the surface of human umbilical vein endothelial cells (HUVECs) occurred 6 h after stimulation by TNF-α or IL-1 and the level remained sustained for at least 24 h [48, 49]. E-selectin plays an important part in inflammation by enabling the recruitment of leukocytes to the site of cell injury.

A circulating (soluble) form of E-selectin (sE-selectin) is released from endothelial cells by enzymatic cleavage or as a result of shedding of damaged or highly activated cells [48, 49, 57] (Fig. 6). Plasma concentration of sE-selectin is correlated to its expression on the surface of HUVECs [48, 49]. Hence, plasma level of sE-selectin has been proposed to serve as a marker of endothelial cell damage and/or activation. The role of E-selectin in vascular disease has been extensively investigated in recent years and it appears that there is the possibility of developing an E-selectin assay system for routine clinical application [48, 49]. With regard to diabetes, Meigs et al. [59] reported that elevated plasma concentrations of sE-selectin could be measured in nondiabetic women who subsequently developed type 2 diabetes, with or without cardiovascular complications. This finding indicates that the blood level of this endothelial-derived biomarker is increased before the clinical development of type 2 diabetes, suggesting its predictive potential for this disease condition or the vascular disorder that may be associated with it. Predictably, enhanced blood levels of E-selectin were also assayed in both men and women with established type 2 diabetes irrespective of overt cardiovascular events [59]. Overall, the reported findings suggest that circulation E-selectin can serve as a predictive, diagnostic and/or prognostic tool for endothelial dysfunction at least in type 2 diabetes. Soluble E-selectin in the plasma can be detected by a sensitive ELISA kit [48, 49].

4.4 Tissue Plasminogen Activator

Tissue plasminogen activator (tPA) is a protein (serine protease) released by vascular endothelial cells [48, 49]. As a protease, it is involved in breakdown of blood clots by catalyzing the conversion of plasminogen to plasmin. It also plays a role in cell migration and tissue remodeling. Being specifically released by endothelial cells, circulating tPA has been suggested to serve as a suitable biomarker of endothelial cells (Fig. 6). It has been shown to be elevated in hypercholestrolemia and subclinical conditions of stroke, myocardial infarction and atheosclerososis [48, 49]. Studies also indicate that the production of tPA is higher in atheromatous arteries than in normal blood vessels [48, 49]. From these observations, it has been suggested that increased tPA levels may reflect excessive endothelial cell activation and/or damage. There is only one study in the literature demonstrating that tAP level in serum is elevated in patients with either prediabetes or type 2 diabetes [60]. These results suggest the potential for tAP to serve as an early indicator of endothelial dysfunction in type 2 diabetes.

Circulating tPA levels have been shown to increase rapidly with local adrenergic or cholinergic receptor stimulation [61]. This observation suggests that assessment of tPA release after adrenergic or cholinergic receptor stimulation may also be a valuable tool for evaluating endothelial function.

Tissue plasminogen activator (tPA) can be assayed by ELISA and/or, less commonly, by measuring its activity, both assays providing closely similar results [48].

4.5 Endothelial Microparticles

Endothelial microparticles (EMPs) are circulating phospholipids-rich submicron vesicles that are released from membranes of endothelial cells as a consequence of cellular damage, activation or apoptosis, although some studies suggest that similar particles are also released from platelets and T lymphocytes under certain conditions [48, 51, 57, 62, 63] (Fig. 6). These microparticles consist of plasma membrane structures containing receptors and other cell surface molecules which enable the identification of their endothelial origin. This allows them to be distinguished from microparticles originating from other possible cell sources. Although circulating EMPs can be found in the blood of normal subjects, increased number of the microparticles have been measured in individuals with diseases affecting blood vessels including diabetes [48, 51, 57, 62, 63]. As a result, EMPs have gained greater considerations to be utilized as an index of changes in endothelial function/structure for such types of disease states. Accordingly, endothelial changes that may be correlated with levels of circulating EMPs can serve to indicate the severity of these diseases.

Concerning diabetes, in a number of studies increased EMP release has been reported along with reductions in endothelium-dependent vasorelaxation and NO production in type 2 diabetes [62, 63]. Corroborating this observation, a correlation has also been found between EMPs and reduced FMD of arteries in diabetic patients with renal failure. The effect of type 2 diabetes has been observed to cause even greater release of EMPs when associated with peripheral vascular diseases [64]. In another group of patients with type 2 diabetes, augmented circulating EMPs was measured together with an increase in ICAM-1 [62, 63]. From these studies, it appears that the more severe the diabetes state, the greater is the amount of EMP in the circulation, both of which correlate with the degree of dysfunction of the endothelium. To our knowledge, although there are not many reported studies on the effect of type 1 diabetes on EMP levels, one study has clearly shown increased circulating levels of EMPs in patients with type 1 diabetes [65]. On the other hand, no significant changes in serum EMPs concentrations were found in nondiabetic patients with insulin resistance [66]. This finding together with the aforementioned results provide evidence that while blood levels of EMPs have the potential to serve as diagnostic and/or prognostic tools for type 2 diabetes (and possibly type 1 diabetes), their alterations do not seem to precede the clinical development of the disease to serve as predictors or risk indicators of this diabetes type.

More recent investigations have revealed that EMPs per se impair endothelium-dependent vasorelaxation and NO production in the rat aorta [62, 63]. This effect may be related to an increase in superoxide anion production, which might reduce the bioavailability of NO. Therefore, microparticles of endothelial origin may cause cellular injury by themselves, and, as such, may serve as a strong circulating marker for endothelial/vascular dysfunction that can readily be measured in the blood [62, 63].

Endothelial microparticles can be assayed by analyzing the expression of $CD31^{+}$/$CD41^{-}$, $CD144^{+}$ and $CD62e^{+}$ on circulating cell surface by use of flow cytometry technique [62, 63]. Among these surface markers, $CD62e^{+}$ is believed to be more specific to endothelial cells; the expression of this marker is enhanced by cytokine. Another less widely utilized method is functional assay performed using the procoagulant property of EMPs [62, 63]. However, due to lack of standardization, the measurement of EMPs by both methods remains a technical challenge.

4.6 Circulating Endothelial Cells

Circulating endothelial cells (CEC) are mature cells expressing the membrane glycoproteins CD146, CD31 and CD144 [51, 57, 63, 67]. They are rarely found in the blood of healthy subjects but are elevated in disease conditions associated with cardiovascular and inflammatory disorders. They present a heterogeneous size ranging from 10 to 50 μm. These cells are negative for leukocyte markers and, unlike EPCs, do not express immature markers such as CD133 and do not rise with high proliferative potential [63, 67]. Circulating endothelial cells may arise from damaged endothelium, although the mechanisms of endothelial cell detachment are multiple and not exclusive. Denudation of the vessels may be triggered by mechanical injury, defective adhesion, protease- or cytokine-mediated detachment or activation of apoptosis, among other possible reasons. A major problem reported to be associated with CEC quantification is their low number in blood. To circumvent this limitation, CECs first have to be enriched (using CD146-coated magnetic microbeads) and then phenotyped using monoclonal antibodies to cell surface glycoproteins [57, 63, 67]. This process is known as immuno-magnetic separation (IMS) assay, which has now become a consensus protocol for CEC quantification. Alternative method for

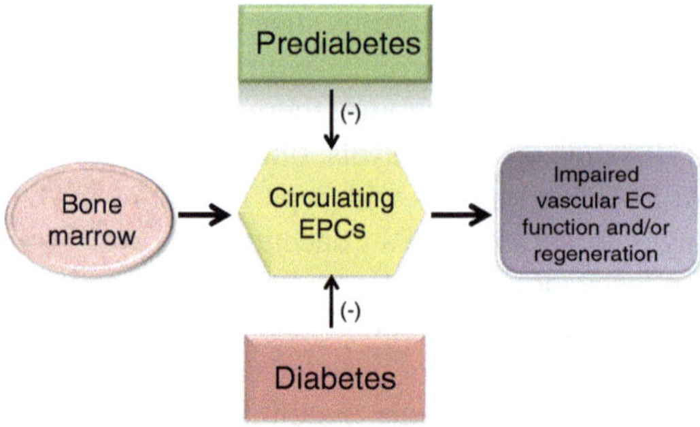

Fig. 7 Proposed effects of diabetes/prediabetes on circulating EPCs and their relation to vascular endothelial cells. EPCs released from bone marrow are involved in the repair of endothelial cells under normal conditions and circulatory levels of these cells have been shown to decrease significantly and this is associated with impairment/damage to the endothelium. The sign (–) indicates a negative effect. *Abbreviations*: *EPCs* endothelial progenitor cells, *EC* endothelial cells

CEC counting includes direct flow cytometry; however, due to heterogeneity of number of cells measured, in most cases, this method provides inconsistent results relative to the IMS assay [65].

The use of CECs as a biomarker of vascular endothelial dysfunction and the associated risk factors has recently been reviewed in relation to cardiovascular diseases [63, 67]. In fact, an inverse correlation was found between CEC levels and FMD, indicating a positive relationship between endothelial dysfunction and increased CEC number. A positive correlation was also shown between CECs and vWF, suggesting that CECs may reflect endothelial damage [63, 67]. Furthermore, CECs levels have been observed to positively relate to the degree of coronary syndrome and peripheral arterial disease. With regard to diabetes, there is a report which demonstrates elevated CECs in patients with unstable diabetes [68]. The cell number tended to normalize with treatment of the diabetic patients. However, since the protocol adopted for this particular study appears to be questionable (due to low number of sample sizes and large variability among sample population), the reported results need to be confirmed.

4.7 *Circulating Endothelial Progenitor Cells*

Endothelial progenitor cells (EPCs) are a group of heterogeneous immature (stem) cells released from bone marrow into the circulation. They have been found to be critical for maintenance and repair of the vascular endothelium [69] (Fig. 7). In this process, EPCs undergo differentiation into mature ECs following their recruitment. It is therefore plausible that endothelial dysfunction that occurs in diabetes may be linked to alterations of EPCs. This observation has been explored to find out if EPCs can be used as a viable biomarker for risk assessment/prediction as well as for diagnosis and prognosis of endothelial dysfunction/vascular disorders in diabetes [49, 57, 63, 70] (Fig. 7).

Endothelial progenitor cells (EPCs) can be characterized by expression of surface markers, more commonly CD34, DC133 and vascular endothelial growth factor receptor-2 (VEGF-2, also known as KDR), by acetylated low density lipoprotein (LDL) uptake and by lecitin binding [63–65]. With cultivation of mononuclear cells (MNCs) isolated from human peripheral blood, these cells attach and develop an endothelium-like phenotype in blood vessels. Under in vivo situation, after being transported, the cells can incorporate into sites of active neovascularization

Cardiovascular risk factors, including diabetes, that cause endothelial dysfunction, have been shown to be associated with decreased number and functional ability of circulating EPCs [71–73]. Such an inverse association between circulating EPCs and endothelial dysfunction has been observed in human patients as well as animal models of both type 1 and type 2 diabetes [69, 71, 72]. When diabetes is accompanied by a significant condition of vasculopathy, the reduction in ECPc number becomes even more marked [74]. The magnitudes of EPC alterations observed, both in number and function, have been found to be related to glycemic states, as assessed by blood glucose and glycosylated hemoglobin concentrations [75]. In EPC colonies from culture of MNCs, diabetes and hyperglycemia-induced functional impairments have been reported to be mediated at least partly via mechanisms involving alterations in chemokines (e.g., growth factors, stromal-derived factor (SDF-1), granulocyte-monocyte stimulating factor (GM-SCF)), eNOS/NO, and matrix metalloproteinase-9 (MMP-9) [71–76]. It is noteworthy that while most studies on EPCs reported to date have been conducted under overt conditions of diabetes, there are limited studies demonstrating the effects of insulin resistance or prediabetes per se. The results of these studies have also shown that insulin resistance/prediabetes without obvious vascular effects causes decreased circulating EPCs. This observation implies that circulating EPCs has the potential to serve as a risk predictor for type 2 diabetes. On the other hand, the altered number and functionality of EPCs associated with diabetes are indicative of their possible usefulness as diagnostic and/or prognostic tools. However, in both cases, a lot more remains to be done before these possibilities are put into practice.

A number of chemokines and related factors appear to regulate the release and function of EPCs [71–73]. Vascular endothelial growth factor (VEGF), which regulates vasculogenesis and angiogenesis, is involved in the mobilization of EPCs from bone marrow. Nitric oxide, in addition to modulating various aspects of endothelial function has been shown to have a significant role in the activities of EPCs [72]. Impaired MMP-9 function has been demonstrated to contribute to reduced EPC mobilization in such cardiovascular conditions as post-myocardial infarction [73], but this has not been determined specifically for blood vessels. Furthermore, SDF-1 has been reported to be involved in the pathophysiology of EPCs and promotion of angiogenesis by enhancing EPC release and its incorporation into new vessels [77, 78]. Granulocyte-monocyte stimulating factor (GM-SCF) is another substance with significant ability to stimulate the mobilization of EPCs from bone marrow, thereby leading to homage to regions of vascular damage for regeneration [71–76]. Thus, the consideration of the above factors may be relevant when studying the role of EPCs in relation to endothelial function/dysfunction in diabetes/prediabetes.

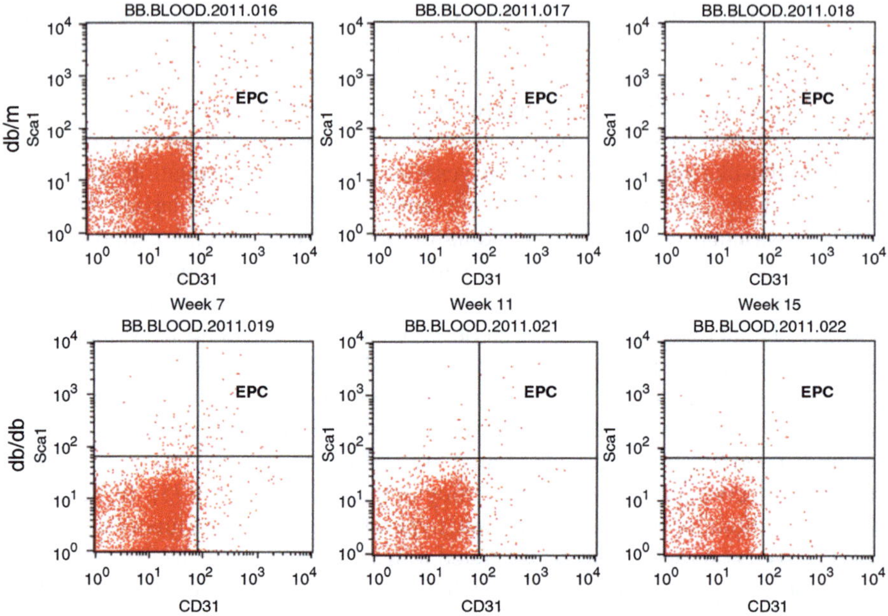

Fig. 8 Representative flow cytometry analysis of whole blood EPCs from 7, 11 and 15 weeks old db/db diabetic mice and corresponding controls (db/m) for expression of CD31 and Sca1

Circulating EPCs can be quantified in whole blood using the technique of flow cytometry [79]. Different laboratories have utilized slightly different cell surface markers to identify EPCs, including CD34/KDR, CD133/KDR and CD31/Sca1, double positive identification being the commonly utilized techniques.

To further investigate the effect of diabetes, our laboratory determined EPCs in whole blood from db/db and db/m mice of 7–15 weeks of age using $CD31^+/Sca1^+$ in flow cytometry assay (Fig. 8). The db/db mouse is one of the best characterized and widely used animal model for type 2 diabetes. With the development of the disease, db/db mouse becomes significantly obese, insulin resistant and hyperglycemic, reflecting a full-fledged diabetes sate. The db/m mouse is its heterozygous littermate which serves as a control. Our experiments showed that EPCs were significantly lower in peripheral blood of db/db mice at 7, 11 and 15 weeks of age (Fig. 8). These results are generally consistent with those reported in the literature. However, at the age of 7 weeks, the blood glucose level of the db/db mice was not significantly greater than that of the control mice. This suggests that changes in EPCs can occur without a change in blood glucose level. Future experiments will investigate EPCs in the diabetic mice even at younger ages to identify the starting time for EPC reduction in the absence of hyperglycemia.

The functional abilities of EPCs can be determined in some detail ex vivo using expanded cell culture by the application of different assay methods. For culturing, the MNCs are usually separated in histopaque – 1,083 density gradient medium, and then expanded using appropriate media and conditions [79]. The number of cell colonies obtained from a sufficient number of wells is determined manually under a microscope by a trained observer. Commonly performed assays to evaluate the functionalities of EPCs include chemotaxis, cell proliferation, eNOS activity/expression, senescence-associated β-galactosidase activity, apoptosis/cell survival rate and matrigel tube formation [73–76, 80–82]. The measurement of these parameters can enable to determine if there is a correlation between the number and functions of EPCs and ultimately the status of the endothelium in health and disease.

5 Summary and Concluding Remarks

Diabetes is associated with vascular complications, which contribute to the death of as high as 80% of diabetic patients. Endothelial dysfunction plays a critical and an initiating role in vascular disorders in both type 1 diabetes, type 2 diabetes and insulin resistance/prediabetes. Assessing the status of the endothelium can serve as a valuable predictive, diagnostic and prognostic tool for vascular disease in diabetes, particularly type 2 diabetes. Endothelial function/dysfunction has been assessed in humans by techniques that utilize measurement of the functional consequences of endothelial activity (i.e., relaxation/lumen size changes of blood vessels). The commonly utilized techniques include plethysmography, ultrasound and/or Doppler which are mostly used to measure blood flow or lumen size in coronary arteries and peripheral blood vessels, such as brachial arteries. While these techniques have been used primarily for research purposes, some of them appear to be clinically relevant for a limited scope of endothelial evaluation. Certain limitations associated to these methods have precluded their wider application and these include technical difficulties and cost factors, and lack of desired sensitivity, specificity, simplicity and speed of operation.

More recently, circulating biomarkers derived from the vascular endothelium have been suggested for consideration as potential indicators of endothelial status. Based on research evidence so far, it is predicted that the use of such indicators will have several advantages over the application of plethysmography, ultrasound and/or Doppler. These biomarkers include vWF, sTM, sE-selectin, tPA and EMPs. While the relative significance of each of these substances (or their combinations) for endothelial evaluation still remains to be defined or established, in most cases, these biomarkers provide good indications of endothelial dysfunction, as it occurs in diabetes and related disorders. Other potential indicators of endothelial function/dysfunction under consideration are CECs and circulating EPCs. Although reported data on these cells provide encouraging evidence, only very few studies have investigated their potential clinical applications as biomarkers, indicating the limitation of information on their usefulness for diagnostic and/or prognostic purposes. The different types of circulating markers discussed and their relation to the vascular endothelium and the possible effects of diabetes and prediabetes are summarized in Fig. 9.

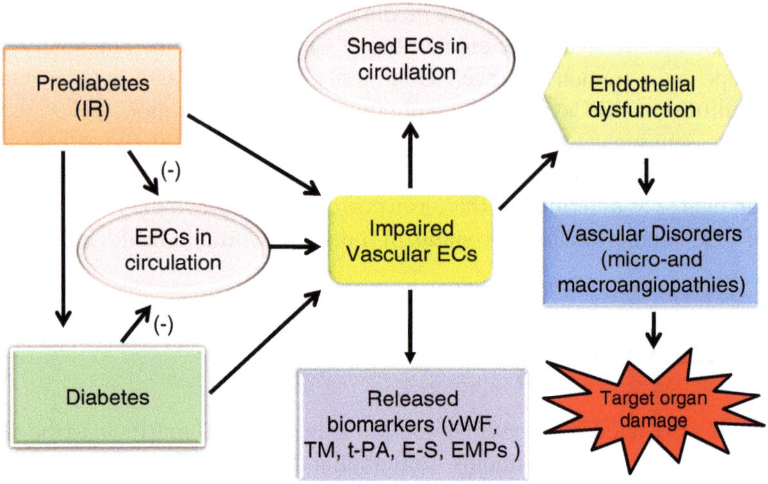

Fig. 9 Summary of proposed effects of diabetes/prediabetes on circulating EPCs, ECs and endothelial-derived biomarkers, and on functions of the endothelium, vasculature and target organs. *Abbreviations*: The sign (–) indicates a negative effect (as for Figs. 6 and 7)

Taken together, currently, the issue of circulating endothelial markers is an area of intense research interest with the potential to lead to the development of clinically relevant assessment techniques of the endothelium. It is hoped that as more data become available, the improved measurement of circulating endothelial indicators will make it possible to obtain valuable diagnostic and prognostic information on vascular events in diabetes and related conditions (e.g., prediabetes) for routine clinical testing.

Acknowledgements EPC data reported for db/db mice in the manuscript and data shown in Figs. 2 and 3 are from the laboratories of Drs. Babak Baban and Worku Abebe, respectively, whose studies were supported by Georgia Regents University, Augusta, GA, USA.

References

1. World Health Organization www.who.int/mediacentre/factsheets/fs312/en/print.html. Accessed 22 Oct 2011
2. American Diabetes Association www.diabetes.org. Accessed 22 Oct 2011
3. Korc M (2004) Update on diabetes mellitus. Dis Markers 20:161–165
4. Kumar V, Abbas A, fausto N (2003) The endocrine pancreas. In: Robbins and Cotran pathological basis of disease, 7th ed. Elsevier Saunders, Philadelphia
5. Neel JV (1962) Diabetes mellitus: a "thrifty" genotype rendered detrimental by "progress"? Am J Hum Genet 14:353–362
6. Hartz AJ, Rupley DC, Kalkhoff RD, Rimm AA (1983) Relationship of obesity to diabetes: influence of obesity level and body fat distribution. Prev Med 12:351–357
7. Helmrich SP, Ragland DR, Leung RW, Paffenbarger RS Jr (1991) Physical activity and reduced occurrence of non-insulin-dependent diabetes mellitus. N Engl J Med 325:147–152

8. Kriska AM, Bennett PH (1992) An epidemiological perspective of the relationship between physical activity and NIDDM: from activity assessment to intervention. Diabetes Metab Rev 8:355–372
9. Martin JH, Mangiafico S, Kelly DJ (2009) Role of statins in diabetes complications. Curr Diabetes Rev 5:165–170
10. Ulrich B (2009) Diabetes and kidney disease. Nephrol Nurs J 36:461
11. Cooper ME, Bonnet F, Oldfield M, Jandeleit-Dahm K (2001) Mechanisms of diabetic vasculopathy: an overview. Am J Hypertens 14:475–486
12. Potenza MA, Gagliardi S, Nacci C, Carratu' MR, Montagnani M (2009) Endothelial dysfunction in diabetes: from mechanisms to therapeutic targets. Curr Med Chem 16:94–112
13. UK prospective Diabetes Study (UKPDS) Group (1998) Intensive blood glucose control with sulphonylureas or insulin compared with conventional treatment and risk of complications in patients with type 2 diabetes. Lancet 352:837–853
14. Koehn J, Turhani D, Krapfenbauer K (2009) Potential predictive biomarkers for the dysfunction of β-cell in type 2 diabetes. In: Golubnitschaja O (ed) Predictive diagnostics and personalized treatment: dream or reality. Nova Science Publishers Inc, New York
15. Schalkwijk CG, Stehouwer CD (2005) Vascular complications in diabetes mellitus: the role of endothelial dysfunction. Clin Sci (Lond) 109:143–159
16. Fitzgerald SM, Brands MW (2000) Nitric oxide may be required to prevent hypertension at the onset of diabetes. Am J Physiol Endocrinol Metab 279:E762–E768
17. Szmitko PE, Wang CH, Weisel RD, de Almeida JR, Anderson TJ, Verma S (2003) New markers of inflammation and endothelial cell activation: part I. Circulation 108:1917–1923
18. Szmitko PE, Wang CH, Weisel RD, Jeffries GA, Anderson TJ, Verma S (2003) Biomarkers of vascular disease linking inflammation to endothelial activation: part II. Circulation 108:2041–2048
19. Furchgott RF, Zawadzki JV (1980) The obligatory role of endothelial cells in the relaxation of arterial smooth muscle by acetylcholine. Nature 288:373–376
20. Furchgott RF, Vanhoutte PM (1989) Endothelium-derived relaxing and contracting factors. FASEB J 3:2007–2018
21. Zhang J, Xie Z, Dong Y, Wang S, Liu C, Zou MH (2008) Identification of nitric oxide as an endogenous activator of the AMP-activated protein kinase in vascular endothelial cells. J Biol Chem 283:27452–27461
22. Nathanson D, Nyström T (2009) Hypoglycemic pharmacological treatment of type 2 diabetes: targeting the endothelium. Mol Cell Endocrinol 297(1–2):112–126
23. Mombouli JV, Vanhoutte PM (1999) Endothelial dysfunction: from physiology to therapy. J Mol Cell Cardiol 31(1):61–74
24. Verma S, Anderson TJ (2002) Fundamentals of endothelial function for the clinical cardiologist. Circulation 105(5):546–549
25. Fleming I, Busse R (1999) Signal transduction of eNOS activation. Cardiovasc Res 43:532–541
26. Furchgott RF (1996) The 1996 Albert Lasker Medical Research Awards. The discovery of endothelium-derived relaxing factor and its importance in the identification of nitric oxide. JAMA 276:1186–1188
27. Calles-Escandon J, Cipolla M (2001) Diabetes and endothelial dysfunction: a clinical perspective. Endocr Rev 22:36–52
28. Tesfamariam B, Jakubowski JA, Cohen RA (1989) Contraction of diabetic rabbit aorta caused by endothelium-derived PGH2-TxA2. Am J Physiol 257:H1327–H1333
29. Harrison DG (1997) Endothelial function and oxidant stress. Clin Cardiol 20(11 Suppl 2):11–17
30. Giacco F, Brownlee M (2010) Oxidative stress and diabetic complications. Circ Res 107:579–591
31. Tesfamariam B, Cohen RA (1992) Free radicals mediate endothelial cell dysfunction caused by elevated glucose. Am J Physiol 263:H321–H326
32. Caballero AE (2003) Endothelial dysfunction in obesity and insulin resistance: a road to diabetes and heart disease. Obes Res 11:1278–1289

33. De Vriese AS, Verbeuren TJ, Van de Voorde J, Lameire NH, Vanhoutte PM (2000) Endothelial dysfunction in diabetes. Br J Pharmacol 130:963–974
34. Abebe W (2008) Effect of taurine on the reactivity of aortas from diabetic rats. Life Sci 82:279–289
35. Baumgartner-Parzer SM, Wagner L, Pettermann M, Grillari J, Gessl A, Waldhäusl W (1995) High-glucose-triggered apoptosis in cultured endothelial cells. Diabetes 44:1323–1327
36. McGinn S, Saad S, Poronnik P, Pollock CA (2003) High-glucose-mediated effects on endothelial cell proliferation occur via p38 MAP kinase. Am J Physiol 285:E708–E717
37. Graier WF, Grubenthal I, Dittrich P, Wascher TC, Kostner GM (1995) Intracellular mechanism of high D-glucose-induced modulation of vascular cell proliferation. Eur J Pharmacol 294:221–229
38. Mozaffari MS, Abdelsayed R, Schaffer SW (2009) Diabetic complications: pathogenic mechanisms and prognostic indicators. In: Golubnitschaja O (ed) Predictive diagnostics and personalized treatment: dream or reality. Nova Science Publishers Inc, New York
39. Jayaraman S (2012) Epigenetic mechanisms of metabolic memory in diabetes. Circ Res 110:1039–1041
40. Okabe J, Orlowaki C, Balcerczyk A, Tikellis C, Thomas MC, Cooper ME, El-Osta A (2012) Distinguishing hyperglycemic changes by Set7 in vascular endothelial cells. Circ Res 110:1067–1076
41. Kim JA, Montagnani M, Koh KK, Quon MJ (2006) Reciprocal relationships between insulin resistance and endothelial dysfunction: molecular and pathophysiological mechanisms. Circulation 113:1888–7904
42. Bakker W, Eringa EC, Sipkema P, van Hinsbergh VW (2009) Endothelial dysfunction and diabetes: roles of hyperglycemia, impaired insulin signaling and obesity. Cell Tissue Res 335:165–189
43. Kofler S, Nickel T, Weis M (2005) The role of cytokines in cardiovascular diseases: a focus on endothelial responses to inflammation. Clin Sci (Lond) 108:205–213
44. Arner P (2003) The adipocyte in insulin resistance: key molecules and the impact of the thiazolidinediones. Trends Endocrinol Metab 14:137–145
45. Choi EY (2009) Inhibition of leukocyte adhesion by developmental endothelial locus-1 (del-1). Immune Netw 9(5):153–157
46. Chavakis T (2012) Leucocyte recruitment in inflammation and novel endogenous negative regulators thereof. Eur J Clin Invest 42(6):686–691
47. Hidai C, Zupancic T, Penta K, Mikhail A, Kawana M, Quertermous EE, Aoka Y, Fukagawa M, Matsui Y, Platika D, Auerbach R, Hogan BL, Snodgrass R, Quertermous T (1998) Cloning and characterization of developmental endothelial locus-1: an embryonic endothelial cell protein that binds the alphavbeta3 integrin receptor. Genes Dev 12(1):21–33
48. Raitakari OT, Celermajer DS (2000) Testing for endothelial dysfunction. Ann Med 32:293–304
49. Deanfield JE, Halcox JP, Rabelink TJ (2007) Endothelial function and dysfunction: testing and clinical relevance. Circulation 115:1285–1295
50. Goligorsky MS (2006) Clinical assessment of endothelial dysfunction: combine and rule. Curr Opin Nephrol Hypertens 15:617–624
51. Münzel T, Sinning C, Post F, Warnholtz A, Schulz E (2008) Pathophysiology diagnosis and prognostic implications of endothelial dysfunction. Ann Med 40:180–196
52. Hadi HA, Suwaidi JA (2007) Endothelial dysfunction in diabetes mellitus. Vasc Health Risk Manag 3:853–876
53. Celermajer DS, Sorensen KE, Gooch VM, Celermajer DS, Sorensen KE, Gooch VM (1992) Non-invasive detection of endothelial dysfunction in children and adults at risk of atherosclerosis. Lancet 340:1111–1115
54. Anderson TJ, Uehata A, Gerhard MD, Meredith IT, Knab S, Delagrange D, Lieberman EH, Ganz P, Creager MA, Yeung AC, Selwyn AP (1995) Close relation of endothelial function in the human coronary and peripheral circulations. J Am Coll Cardiol 26:1235–1241
55. Bots ML, Westerink J, Rabelink TJ, de Koning EJ (2005) Assessment of flow-mediated vasodilatation (FMD) of the brachial artery: effects of technical aspects of the FMD measurement on the FMD response. Eur Heart J 26:363–368

56. Corretti MC, Anderson TJ, Benjamin EJ, Celermajer D, Charbonneau F, Creager MA, Deanfield J, Drexler H, Gerhard-Herman M, Herrington D, Vallance P, Vita J, Vogel R, International Brachial Artery Reactivity Task Force (2002) Guidelines for the ultrasound assessment of endothelial-dependent flow-mediated vasodilation of the brachial artery: a report of the International Brachial Artery Reactivity Task Force. J Am Coll Cardiol 39:257–265, Erratum in: J Am Coll Cardiol 39:1082
57. Constans J, Conri C (2006) Circulating markers of endothelial function in cardiovascular disease. Clin Chim Acta 368:33–47
58. Aso Y, Inukai T, Takemura Y (1998) Mechanisms of elevation of serum and urinary concentrations of soluble thrombomodulin in diabetic patients: possible application as a marker for vascular endothelial injury. Metabolism 47:362–365
59. Meigs JB, Hu FB, Rifai N, Manson JE (2004) Biomarkers of endothelial dysfunction and risk of type 2 diabetes mellitus. JAMA 291:1978–1986
60. Al-Hamodi Z, Ismail IS, Saif-Ali R, Ahmed KA, Muniandy S (2011) Association of plasminogen activator inhibitor-1 and tissue plasminogen activator with type 2 diabetes and metabolic syndrome in Malaysian subjects. Cardiovasc Diabetol 10:23
61. Salomaa V, Stinson V, Kark JD, Folsom AR, Davis CE, Wu KK (1995) Association of fibrinolytic parameters with early atherosclerosis. The ARIC Study. Atherosclerosis Risk in Communities Study. Circulation 91:284–290
62. Benameur T, Andriantsitohaina R, Martínez MC (2009) Therapeutic potential of plasma membrane-derived microparticles. Pharmacol Rep 61:49–57
63. Sabatier F, Camoin-Jau L, Anfosso F, Sampol J, Dignat-George F (2009) Circulating endothelial cells, microparticles and progenitors: key players towards the definition of vascular competence. J Cell Mol Med 13:454–471
64. Koga H, Sugiyama S, Kugiyama K, Watanabe K, Fukushima H, Tanaka T, Sakamoto T, Yoshimura M, Jinnouchi H, Ogawa H (2005) Elevated levels of VE-cadherin-positive endothelial microparticles in patients with type 2 diabetes mellitus and coronary artery disease. J Am Coll Cardiol 45:1622–1630
65. Sabatier F, Darmon P, Hugel B, Combes V, Sanmarco M, Velut JG, Arnoux D, Charpiot P, Freyssinet JM, Oliver C, Sampol J, Dignat-George F (2002) Type 1 and type 2 diabetic patients display different patterns of cellular microparticles. Diabetes 51:2840–2845
66. Jung C, Shemyakin A, Böhm F, Pernow J (2009) Endothelial microparticles in patients with insulin resistance. Diabetes Metab 35:71–73
67. Erdbruegger U, Haubitz M, Woywodt A (2006) Circulating endothelial cells: a novel marker of endothelial damage. Clin Chim Acta 373:17–26
68. El Amine M, Sohawon S, Lagneau L, Gaham N, Noordally S (2010) Plasma levels of ICAM-1 and circulating endothelial cells are elevated in unstable types 1 and 2 diabetes. Endocr Regul 44:17–24
69. Werner N, Kosiol S, Schiegl T, Ahlers P, Walenta K, Link A, Böhm M, Nickenig G (2005) Circulating endothelial progenitor cells and cardiovascular outcomes. N Engl J Med 353:999–1007
70. Ding H, Triggle CR (2005) Endothelial cell dysfunction and the vascular complications associated with type 2 diabetes: assessing the health of the endothelium. Vasc Health Risk Manag 1:55–71
71. Kalka C, Masuda H, Takahashi T, Gordon R, Tepper O, Gravereaux E, Pieczek A, Iwaguro H, Hayashi SI, Isner JM, Asahara T (2000) Vascular endothelial growth factor(165) gene transfer augments circulating endothelial progenitor cells in human subjects. Circ Res 86:1198–1202
72. Balligand JL, Feron O, Dessy C (2009) eNOS activation by physical forces: from short-term regulation of contraction to chronic remodeling of cardiovascular tissues. Physiol Rev 89:481–534
73. Maclauchlan S, Skokos EA, Agah A, Zeng J, Tian W, Davidson JM, Bornstein P, Kyriakides TR (2009) Enhanced angiogenesis and reduced contraction in thrombospondin-2-null wounds is associated with increased levels of matrix metalloproteinases-2 and –9, and soluble VEGF. J Histochem Cytochem 57:301–313

74. Lee LC, Chen CS, Choong PF, Low A, Tan HC, Poh KK (2009) Time-dependent dynamic mobilization of circulating progenitor cells during percutaneous coronary intervention in diabetics. Int J Cardiol 142:199–201
75. Vöö S, Dunaeva M, Eggermann J, Stadler N, Waltenberger J (2009) Diabetes mellitus impairs CD133+ progenitor cell function after myocardial infarction. J Intern Med 265:238–249
76. Hu CH, Li ZM, Du ZM, Zhang AX, Yang DY, Wu GF (2009) Human umbilical cord-derived EPCs promote growth cytokines-mediated neorevascularization in rat myocardial infarction. Chin Med J (Engl) 122:548–555
77. Seandel M, Butler J, Lyden D, Rafii S (2008) Catalytic role for proangiogenic marrow-derived cells in tumor neovascularization. Cancer Cell 13:181–183
78. Yao EH, Fukuda N, Matsumoto T, Katakawa M, Yamamoto C, Han Y, Ueno T, Kobayashi N, Matsumoto K (2008) Effects of the antioxidative beta-blocker celiprolol on endothelial progenitor cells in hypertensive rats. Am J Hypertens 21:1062–1068
79. Wu X, Wang K, Cui L, Wang Y, Wang X, Meng L, Cheng Y (2008) Effect of granulocyte-colony stimulating factor on the repair of balloon-injured arteries. Pathology 40:513–519
80. Rousseau A, Ayoubi F, Deveaux C, Charbit B, Delmau C, Christin-Maitre S, Jaillon P, Uzan G, Simon T (2008) Impact of age and gender interaction on circulating endothelial progenitor cells in healthy subjects. Fertil Steril 93:843–846
81. Xiao Q, Xu Q (2009) Caspase-8: a double-edged sword for EPC functioning. Arterioscler Thromb Vasc Biol 29:444–446
82. Tramontano AF, Lyubarova R, Tsiakos J, Palaia T, Deleon JR, Ragolia L (2010) Circulating endothelial microparticles in diabetes mellitus. Mediators Inflamm 2010:250476

Correction of Diabetes-Induced Endothelial Progenitor Dysfunction to Promote Retinal Vascular Repair

Sergio Li Calzi, Matthew Neu, Lynn C. Shaw, and Maria B. Grant

Abstract Progressive obliteration of the retinal microvessels is a characteristic of diabetic retinopathy and the resultant retinal ischemia that can lead to sight-threatening macular edema, macular ischemia and ultimately preretinal neovascularization. Bone marrow derived endothelial progenitor cells (EPCs) play a critical role in vascular maintenance and repair. There is still great debate about the most appropriate markers that define an EPC and much deliberation occurs regarding the plasticity of monocytes and their ability to promote repair as well as contribute to angiogenesis. EPC populations can be isolated using cell sorting by surface phenotype selection or *in vitro* cell culture. For freshly isolated cells, EPC cell sorting is heavily dependent on the surface markers used; EPC populations can also be isolated by *in vitro* propagation of heterogeneous mixtures of cells in culture using adhesion to specific substrates and cell growth characteristics. *In vitro* isolation enables consistent reproducibility and with this approach at least two distinct types of populations with different angiogenic properties have been identified from adult peripheral and umbilical cord blood. These groups are early EPCs (eEPCs) and "endothelial colony forming cells" (ECFCs), initially termed "late outgrowth endothelial progenitor cells" (OECs).

Emerging studies demonstrate the potential of these cells in revascularization of ischemic/injured tissues including the retina. Since ischemic retinopathies are the leading causes of blindness, they are a potential disease target for cell based therapy. In this chapter, we summarize the current knowledge about the most well described EPC populations and discuss the possibility of cell therapy for treatment of diabetic macular ischemia and the vasodegenerative phase of diabetic retinopathy. We also report on current pharmacological options that can be utilized to correct diabetes

---

S. Li Calzi • M. Neu • L.C. Shaw • M.B. Grant (✉)
Department of Pharmacology and Therapeutics, College of Medicine, University of Florida, 1600 SW Archer Road, Gainesville, FL 32610, USA
e-mail: grantma@ufl.edu

M.S. Mozaffari (ed.), *New Strategies to Advance Pre/Diabetes Care: Integrative Approach by PPPM*, Advances in Predictive, Preventive and Personalised Medicine 3,
DOI 10.1007/978-94-007-5971-8_6,

associated defects in one particular EPC population, $CD34^+$ cells, so that these cells can potentially be used therapeutically.

Keywords Angiogenesis • Endothelial progenitor cells • Adult stem cells • Diabetic retinopathy • Ischemia • Targeted treatment

1 Introduction

Diabetic retinopathy (DR), the leading cause of visual impairment in the Western world [1], will develop in the majority of type 1 diabetics and about 20–30% of these individuals will advance to the blinding stage of the disease, proliferative diabetic retinopathy (PDR). Greater than 60% of patients with type 2 diabetes will develop retinopathy. With the global epidemic of obesity and subsequently of type 2 diabetes, the number of individuals developing DR will also greatly increase. Over 360 million people are projected to suffer from diabetes and its complications by 2030.

Hyperglycemia damages the retinal microvasculature, which results in increased permeability (loss of the functional integrity of the blood retinal barrier), leakage of fluid into the extra vascular space, a progressive decline in retinal blood flow, and damaging reductions in blood supply to the neural retina. The level of damage to the microvasculature is impacted by the duration of diabetes, the degree of blood sugar and hypertension control, and yet to be determined patient-specific differences in the ability to repair the damaged endothelium. In the later stages of DR, the formation of acellular capillaries leads to areas of retinal nonperfusion and ischemia. Capillary leakage in the region of the macula is called diabetic macular edema (DME), the principal cause of vision loss in type 2 diabetes. When capillary closure occurs in the macula, it results in diabetic macular ischemia (DMI) and causes loss of vision. Bilateral DMI causes legal blindness and unfortunately there is no therapy for DMI. Moreover, patients commonly have mixed DME and DMI.

Strict metabolic and blood pressure control can lower the risk of developing retinopathy and reducing disease progression [1]. Newer therapies targeted at treating either DMI or PDR include corticosteroids and anti-VEGF strategies. These approaches represent promising alternatives to control retinopathy progression. Laser therapy, the standard of care for DME, can be destructive to various neurovascular layers of the retina and choroid, but results in modest preservation of visual acuity [2, 3]. Laser is less effective in the presence of diffuse or chronic DME and in the presence of "mixed" DME and capillary closure [3]. When retinopathy progresses to the late stages, chronic hypoxia/nonperfusion leads to compensatory neovascularization which occurs in an aberrant manner mostly on the surface of the retina. For this stage, PDR, laser treatment is the most cost-effective approach. However, anti VEGF approaches similar to those used for choroidal neovascularization show considerable promise [4]. However long term anti-VEGF treatments may have limitations as VEGF is an important survival factor for neurons and neuronal damage/dysfunction exists early in DR.

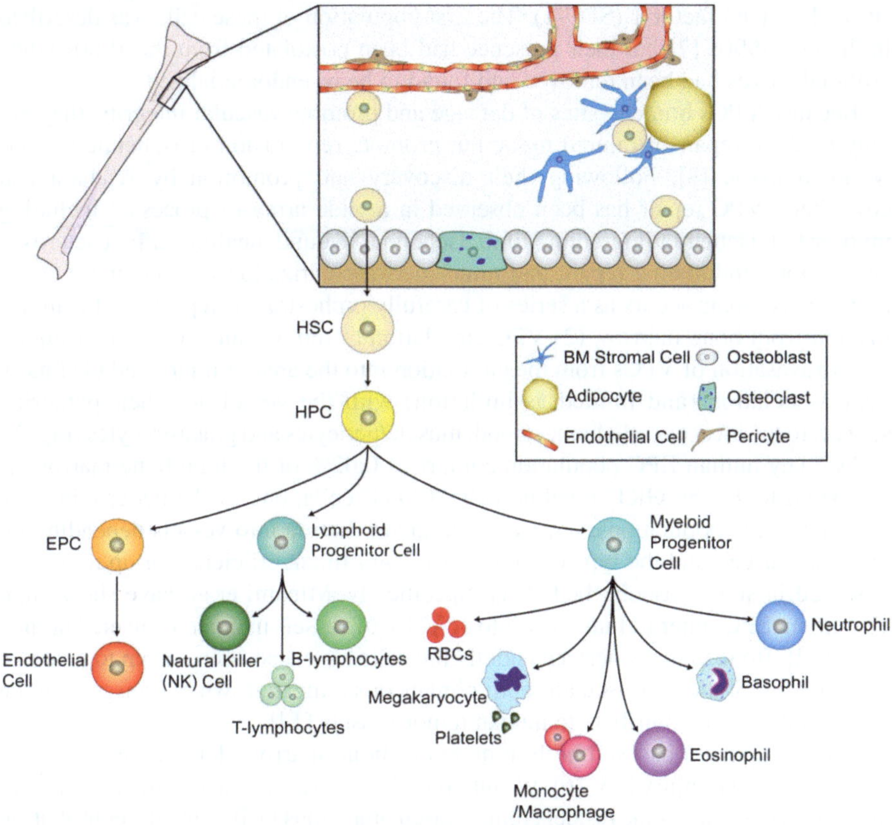

Fig. 1 Adult stem cells of the bone marrow. The bone marrow hosts at least two known types of adult stem cells, the mesenchymal stem cells (MSC) and the hematopoietic stem cells (HSC); the most prominent adult stem cell in the bone marrow. The HSCs can give rise to the hematopoietic progenitor cells (HPCs) which in turn give rise to the lymphoid progenitor cell, the myeloid progenitor cells, and likely the EPCs. The precise origin of the EPC is under debate as this cell may directly arise from HSCs or from HPCs. The bone marrow microenvironment is composed of bone marrow stromal cells (which are the source of SDF-1), adipocytes, and cells of the bone matrix, osteoblasts and osteoclasts. The vessels within the bone marrow, composed of pericytes and endothelium, function to provide a barrier between the hematopoietic compartment and the circulatory system (figure adapted from Domen et al. [5])

The populations of endothelial progenitor cells may be better named or described as "vascular progenitor cells" (VPCs) as these cells provide mostly paracrine support to the vasculature and very limited "direct" incorporation into the endothelium. These cells also provide paracrine support to ablumenal populations, such as pericytes and astrocytes. VPCs are believed to be derived from hematopoietic stem cells (HSCs, Fig. 1) or in the case of ECFCs from the endothelium itself [6]. In the last decade, it has been recognized that VPCs are recruited to sites requiring repair responding to hypoxic signals released by ischemic tissue, typically VEGF and stromal derived factor-1 (SDF-1). The first population of these cells was described in the late 1990s [7] but their presence had been postulated from the 1960s when artificial valves had been removed and found to be re-endothelialized.

Because VPCs home to sites of damage and promote vascular integrity, they not only mediate repair of injured tissue but promote reperfusion of ischemic regions within a tissue [8]. Following their discovery and promotion by Asahara and coworkers, VPC repair has been observed in a wide array of processes including myocardial ischemia/infarction, limb ischemia, wound healing, atherosclerosis, endogenous endothelial repair, and tumor neovascularization in mice and humans [7, 9]. This repair occurs as a series of carefully orchestrated steps: (1) VPC mobilization from bone marrow, (2) VPC circulation to remote sites of vascular injury, (3) extravasation of VPCs from the circulation into the area of injury and (4) finally their association (and limited assimilation) with the vessel and their paracrine support to the local endothelium, smooth muscle/pericytes and glia/astrocytes (Fig. 2).

One key human EPC population comprises 0.02% of the total bone marrow as compared to 4% by $GR1^+$ myeloid cells. $CD34^+$ cells, one of the better clinically described EPC populations, have variable incorporation into vessels depending on the vascular bed and the type of injury. An engraftment efficiency of up to 95% is observed in some vascular beds [11]. Specifically, Minami et al. have shown that circulating EPCs engraft luminally into 15–29% of vessels in the transplanted human heart [12]. Bone marrow derived endothelial cells have also been shown to give rise to up to 16% of the neovasculature in spontaneous tumors grown in transgenic mice [13] and they also contribute to human tumor vessels [14].

Our group has observed high numbers of bone marrow derived cells homing to areas of ocular injury [15, 16]. We interpret this high percentage to be secondary to the quiescent nature of the resident retinal vasculature, thus facilitating the contribution of circulating VPCs to the resident vessel.

2 Characterization of EPCs

Characterization of the various EPC subpopulations is an area of heated debate but there exist two main approaches to define and/or isolate cells with endothelial characteristics and vascular regenerative capacity: culture selection from mononuclear cells (either BM- or PB-derived) or the use of a set of antibodies to various "identifying" markers of these cells.

3 Identification of Endothelial Precursors Based on Culture

Asahara and colleagues first characterized the endothelial precursors in 1997 as a subset of $CD34^+$ hematopoietic progenitor cells [7]. They reported that peripheral blood mononuclear cells (PB-MNCs) enriched for $CD34^+$ cells could differentiate into endothelial-like cells following culture on fibronectin in the presence of growth factors.

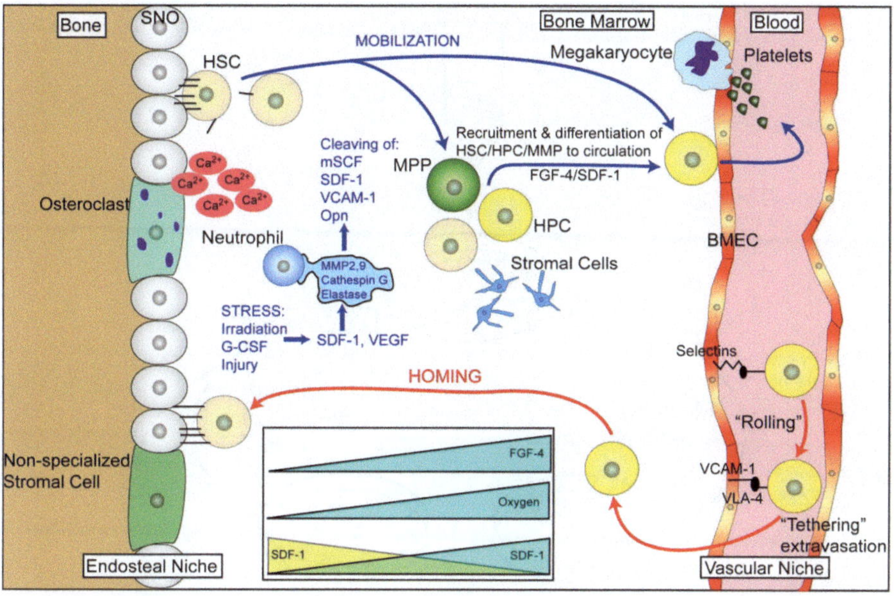

Fig. 2 Mobilization and homing are two closely related processes. Mobilization involves the exodus of HSC/HPC from the bone marrow into the circulation while homing is the "opposite" of this event. HSCs mobilize from the endosteal niche, move to the vascular niche, and ultimately into the circulation. This normally occurs when stress induces changes of SDF-1 levels in the bone marrow. The mechanism of stress-induced mobilization as occurs following irradiation or G-CSF-induced mobilization is not fully known, but is in part accomplished by the upregulation of proteases such as MMP-2, MMP-9, cathepsin-G and elastase. These proteases cleave niche retention signals like membrane-bound stem cell factor (mSCF), SDF-1, VCAM-1 and osteopontin (Opn). Gradients of fibroblast growth factor 4 (FGF-4) also regulate mobilization. For homing events, key steps are needed. Upon reaching the bone marrow vasculature, SDF-1-stimulated circulating HSC/HPC express integrins such as very late antigen 4 (VLA-4) and hyaluronan binding–cellular adhesion molecule (CD44). These integrins, in turn, interact with vascular cell adhesion molecule 1 (VCAM-1), intercellular adhesion molecule 1 (ICAM-1), E-and P-selectins expressed on bone marrow endothelial cells which slow down the circulating HSCs/HSPs in a process known as "rolling". Following rolling, firm adhesion and subsequent endothelia trans- migration into the hematopoietic compartment is mainly accomplished by VLA-4 interactions. Once extravasated, the cells migrate along extravascular hematopoietic cords toward specific niches through a SDF-1 gradient or receding oxygen gradient originating from the supporting osteoblastic or endothelia niches. ***BMEC*** Bone marrow microvascular endothelial cell (figure adapted from Wilson and Trumpp [10])

After 7 days in culture the fraction of cells co-expressing CD34 and VEGFR-2 increased. These cells also expressed other endothelial markers such as CD31 (platelet endothelial cell adhesion molecule-1, PECAM-1), tyrosine kinase with immunoglobulin-like and EGF-like domains 2 (Tie-2) and eNOS, incorporated acetylated low-density lipoprotein (acLDL), bound Ulex (UEA-1) lectin and formed tube-like structures *in vitro*, supporting the contention that they possess the ability to differentiate into endothelial cells. These findings were corroborated by Shi et al. [17]. Currently, culture selection involves the growth of PB-MNCs in selective medium

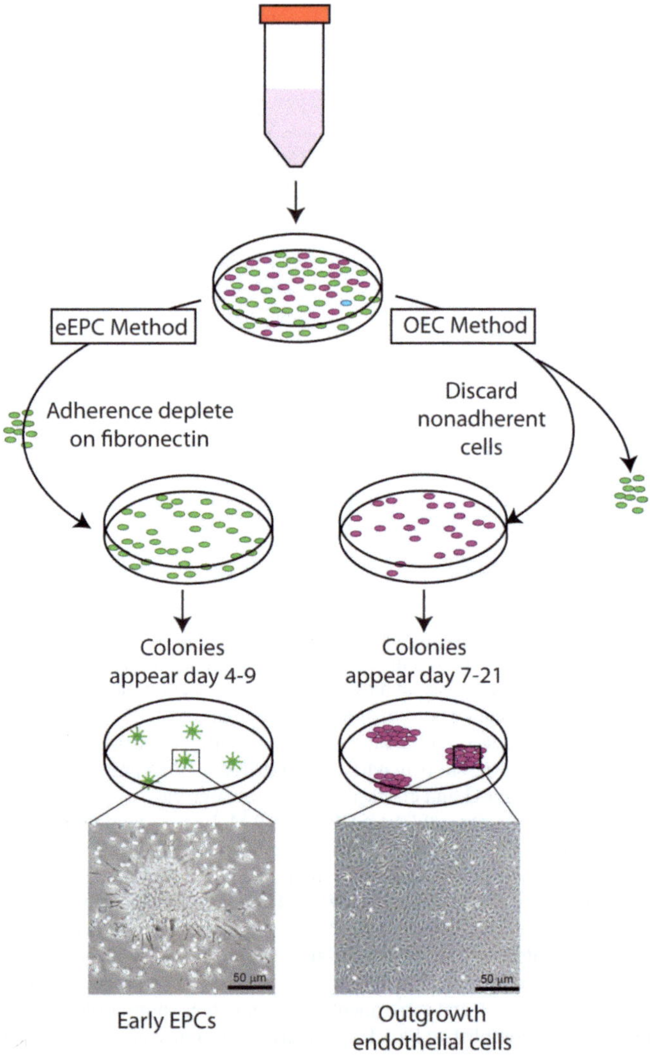

Fig. 3 Common methods of precursor isolation. Culture of eEPCs includes a 5-day process wherein nonadherent MNCs give rise to the EPC colony. OECs are derived from adherent MNCs cultured on collagen for 21 days in endothelial growth conditions and demonstrate typical cobblestone morphology

and on either fibronectin- or collagen-coated dishes [17]. The medium typically contains a cocktail of endothelial growth factors, such as VEGF, FGF-2, IGF-1, PDGF-BB, as well as ascorbic acid and hydrocortisone. The medium is changed regularly (usually every 24–28 h). Culture-selected cells can be classified into two distinct phenotypes as shown in Fig. 3 [18].

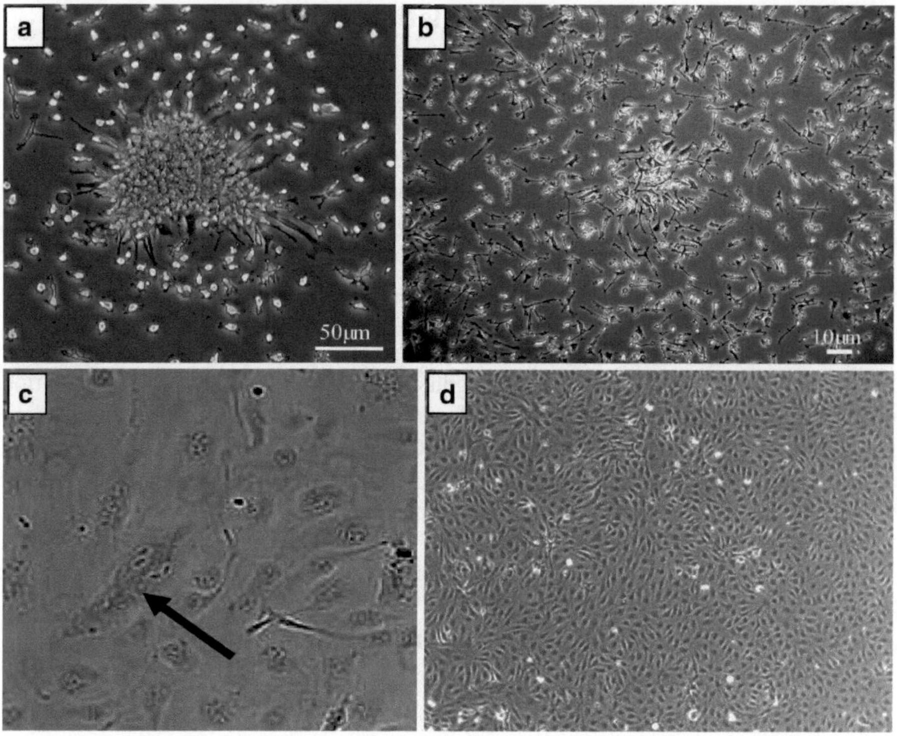

Fig. 4 Morphology of early and late outgrowth EPCs. (**a**) Typical colony of early EPCs which develop into spindle shaped cells by day 7 (**b**). Early EPCs do not proliferate readily, which leads to the typical sub-confluent appearance. In contrast, late outgrowth ECs (**c**) show rapid growth as indicated by the dividing cell (*black arrow*) and display the typical endothelial cobblestone morphology and proliferate abundantly (**d**)

eEPCs, which are the cells originally identified by Asahara, have been most studied to date. eEPCs are mainly derived from monocytes and do not proliferate, but rather begin to gradually die after a few weeks in culture [18, 19]. In contrast, OECs/ECFCs usually begin growing after 2 weeks following isolation, eventually proliferate very rapidly, resemble microvascular ECs (with a cobblestone morphology), and seem to be expandable nearly indefinitely (Fig. 4) [20, 21]. The distinction between these two cell types is reinforced by their expression of different markers. ECFCs do not express CD14 and have a low expression of CD45, the pan-leukocyte marker [22]. The precise origin of the ECFC is still in question yet, Yoder's group believes that the endothelium rather than the bone marrow is the source of this progenitor and that these cells are derived from a $CD34^+$ $VEGFR\text{-}2^+$ cell found in the circulation [6, 23, 24]. The ECFC represents one cell in a million MNCs found in the circulation and thus can only be "identified" once it has grown out in culture [25].

4 Cell Therapy with EPC Populations

Most early studies of cell therapy in the heart and limbs employed eEPCs; thus, there is abundant evidence that these transplanted cells can enhance revascularization [26], although there is much debate about the mechanism of their effects. In contrast, ECFCs have been less well studied *in vivo* [24]. Still their higher proliferative potential may be important when a large supply of regenerative cells is required. Thus, ECFCs may have great therapeutic utility as a cellular therapy [8]. It is our belief that both the CD34^{+} cell and the ECFC population are impaired in diabetes and may be responsible in part for the vasodegenerative phase of DR. We speculate that when this population reappears, it may do so as a more aggressive and proliferative phenotype which triggers the "angiogenic switch" in PDR. The eEPC population never disappears entirely but rather shifts in its level of activity, sometimes being more inflammatory than other times. The EPC phenotype can change from reparative to non-reparative. Moreover, it has been shown by Yoon et al. that early EPCs and ECFC can act together to stimulate vascular repair [24].

5 Identification of EPCs Based on FACS or Magnetic Bead Selection

In contrast to differential culture, endothelial precursors can be isolated using a panel of immunological and non-immunological markers and then used for either *in vitro* or *in vivo* studies. This is the approach commonly used for cell therapy. The cells are isolated from a patient's own bone marrow or peripheral blood following mobilization (using GM-CSF). The original combination of surface markers, CD34 and VEGFR-2, remains the most common, but several other markers such as CD133 (prominin) have been used to refine the identification of a therapeutic cell type [27].

Bone marrow-derived CD14^{+} monocytes also demonstrate potential to differentiate into endothelial cells [18]. Interestingly, although several independent groups have shown a clear development of an endothelial phenotype after selective culture, they also found that the expression of monocytic antigens persisted [18, 28]. Several additional studies confirmed the overlap between endothelial and monocyte phenotypes, suggesting that using traditional endothelial cell markers such as acLDL uptake, Ulex binding, CD31, CD105 (endoglin), CD144 (vascular endothelial cadherin), VEGFR-2, CD34 and Tie-2 may not be enough to distinguish between endothelial cells and monocytes [29–31]. Similarly, Harraz et al. showed that cultured CD34^{-}CD14^{+} peripheral blood derived mononuclear cells express Tie-2, VEGFR-2, CD144, von Willebrand factor (vWF), CD146, CD105 and eNOS [28]. Interestingly, another report suggested that monocytes (CD14^{+}), which express low levels of CD34 (undetectable with conventional techniques), may in fact contain the true progenitor cell population, with the greatest regenerative capacity [32]. Considering the uncertainty as to the most reliable method to identify circulating EPCs, several studies have employed CD34/VEGFR-2 or CD34/CD133 double positivity to quantify circulating EPCs and to correlate their concentration to clinical conditions [33, 34]. This is the strategy that we have selected for our studies using human cells.

A study by Loomans et al. [35] and an excellent review by Schatteman et al. [8] have suggested that despite phenotypic overlap between various bone marrow-derived or circulating cell types, the endothelial phenotype can best be characterized by three features: eNOS expression, integration into tube-like structures formed by human umbilical vein ECs (HUVECs), and stimulation of tube formation by HUVECs. It is suggested that these criteria might be the best *in vitro* method of judging the endothelial phenotype, regardless of the presence or absence of particular cell surface antigens. This simplified interpretation is important, as it acknowledges the limitation of surface markers and affirms the importance of the identification of EPCs according to their potential to acquire functional properties of ECs.

The failure to reliably characterize EPCs has generated skepticism, confusion and tarnished their "image" in regenerative medicine. This has lead us and others to re-label these cells simply as vascular progenitors allowing for angiogenic monocytes to be included with $CD34^{+}$cells and cells that incorporate into the actual endothelium to be included with cells that directly associate vessels. Furthermore, cells that durably engraft and cells that provide paracrine support can all be considered.

6 Diabetic EPCs are Dysfunctional

It is our contention that for durable repair and sustained correction of retinal ischemia the use of both freshly isolated cells and *ex vivo* expanded populations may be needed. Moreover, an extended *ex vivo* period allows time for correction of diabetes-induced dysfunction. In diabetes bone marrow-derived progenitors are dysfunctional, producing fewer endothelial cells with reduced proliferative and migratory potential [36]. Enhanced oxidative stress in diabetes contributes to progenitor dysfunction [37]. Accumulation of ROS increases cellular/replicative senescence in these progenitors as does increased angiotensin II, oxidized low-density lipoprotein (ox-LDL) and homocysteine. EPCs of diabetic origin show a reduced ability to integrate into EC tubes *in vitro* compared to EPCs of non-diabetic origin [38–40]. Vascularization is depressed when EPCs from STZ treated mice are injected into normal mice [41]. Recently, we showed that activation of the HDL receptor is protective to EPCs by increasing eNOS [42], whereas activation of ox-LDL receptor down-regulates eNOS, supporting a key role for NO in the function of progenitors [43].

The development of acellular capillaries may be due to failed attempts at repair of injured capillaries and persistence of ischemia. The basic mechanisms responsible for the diabetes-associated defect in EPC function remains incompletely understood. Correcting this defect may allow the use of a diabetic patient's own cells for repair of injured retinal and systemic vasculature. Specifically in the retina, correction of this dysfunction could treat early and intermediate stages of vasodegeneration to enhance vessel repair, reverse ischemia, and prevent progression to the late stages of DR (Fig. 5).

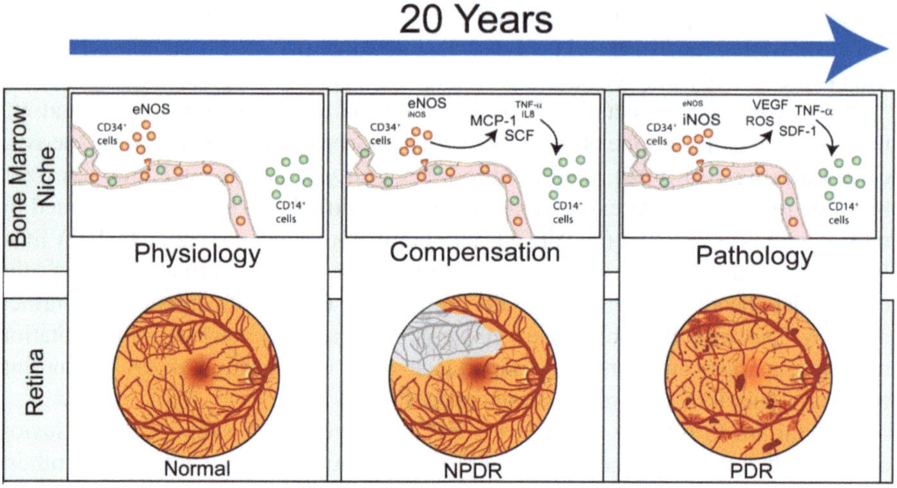

Fig. 5 Schematic of the hypothetic diabetic retinopathy progress. In physiological conditions, CD34+ EPCs contribute to routine blood vessel maintenance through eNOS activation and NO-mediated stimulation of CD14+ EPCs. In diabetes, initially, cytokines like stem cell factor (SCF), monocyte chemoattractant protein-1 (MCP-1), interleukin-8 (IL-8), and tumor necrosis factor alpha (TNF-α) released by dysfunctional CD34+ EPCs initiate CD14+ EPC-mediated aberrant vascular repair resulting in retinal ischemia. This phase is referred to as non proliferative diabetic retinopathy (NPDR). The vasodegenerative phase of diabetic retinopathy associated with reduced reparative function of EPCs evolves in the proliferative diabetic retinopathy (PDR). This phase is characterized by pathological neovascularization seen in the diabetic retina

7 Nitric Oxide (NO) in the Pathogenesis of DR and EPC Dysfunction

NO generation arises from the guanidino group of l-arginine and is a NADPH-dependent reaction catalyzed by a family of NOS. Three distinct isoforms of NOS, consisting of endothelial NOS (eNOS), neuronal NOS (nNOS), and inducible NOS (iNOS) are similar in structure and function, utilizing l-arginine, oxygen and NADPH as substrates, and requiring FAD, FMN, calmodulin, and tetrahydrobiopterin as co-factors. The catalytic mechanism of NOS involves flavin-mediated electron transport from NADPH to the terminal heme, where oxygen is bound and incorporated into NO and citrulline [44]. Maintaining an adequate cellular supply of l-arginine is critical for normal vascular function. Deficiencies in l-arginine supply have been strongly implicated in vascular diseases, including diabetes [45]. If the supply of l-arginine or any cofactor does not meet the needs of active NOS, NOS becomes "uncoupled" and uses molecular oxygen as a substrate to form O_2^- instead of NO. An imbalance between l-arginine availability and NOS activity can occur when cellular transport of l-arginine is inhibited [46] or when there is reduced recycling of l-citrulline back to l-arginine, or when an elevated catabolism of l-arginine by arginase exists [47, 48].

We demonstrated in a spontaneous rat model of type 2 diabetes that eNOS was decreased while NADH oxidase, a major source of superoxide in the vascular endothelium, iNOS, and $ONOO^-$ were all increased in the retinal vasculature [49]. The shift in redox state in diabetes with increased ROS causes increased interaction of NO with O_2^-, resulting in loss of bioavailable NO and formation of $ONOO^-$. The reduced NO and elevated $ONOO^-$ can lead to microvascular dysfunction in diabetes [50–52]. $ONOO^-$ is a potent oxidant that can attack many types of biological molecules; a high level of $ONOO^-$ initiates lipid peroxidation, hydroperoxides, nitration of amino acids such as tyrosine, oxidation of antioxidants such as ascorbic acid and α-tocopherol, and direct DNA damage. The increased oxidative stress associated with diabetes can limit tetrahydrobiopterin availability. Moreover, $ONOO^-$ formation affects tetrahydrobiopterin interaction with NOS [53], leading to this "switch" or uncoupling of NOS from producing NO to producing O_2^-. Advanced glycation end products (AGEs) seen in diabetic individuals can also directly react with NO, diminishing NO bioavailability both *in vitro* and *in vivo* [54].

In type 1 diabetes, uncoupled eNOS was found to be the major source of ROS generation and, when blocked, there was reduced retinal leukostasis, blood retinal breakdown and reduced retinopathy [55]. Whereas nNOS is highly expressed in the retina, its role in DR is still being determined. Interestingly, nNOS knock-out mice still develop retinal leukostasis and blood retinal barrier breakdown suggesting that iNOS or eNOS may have a more central role in these particular aspects of DR. In a model of diabetic ketoacidosis, in which acetone is administered to a mouse, tissue damage is mediated by free radical generation via iNOS overexpression [56]. Diabetes-induced vascular dysfunction in rats was prevented by administration of a specific inhibitor of iNOS, 1,400 W [57, 58].

8 Is Reduced Bioavailable NO Responsible for the Endothelial Progenitor Dysfunction Observed in DR?

NO-mediated signaling pathways are essential for EPC mobilization from the bone marrow [59–61]. NO activates MMP-9, releasing soluble Kit ligand, which shifts EPCs and hematopoietic stem cells from a quiescent to a proliferative niche and stimulates rapid cell mobilization to the PB [61, 62]. NO regulates migration of EPCs into ischemic sites [59, 63–65] and promotes their survival [66]. Diabetic EPCs have decreased eNOS activity; more importantly exogenous NO can correct the migratory defect in these cells (Fig. 6) [67, 68]. The impaired *in vivo* re-endothelialization capacity of diabetic EPCs from human subjects was restored by small interfering RNA silencing of NAD(P)H oxidase subunit p47(phox) [64]. Oxidative stress impairs *in vivo* reendothelialization capacity of EPCs [69]. We show in the preliminary data that pretreatment of diabetic $CD34^+VEGFR\text{-}2^+$ EPCs with apocynin or gp91ds-tat decreased superoxide production and increased NO availability. Pretreatment with apocynin or gp91ds-tat reversed the impaired migration of diabetic EPCs in response to SDF-1 and VEGF *in vitro* and enhanced their

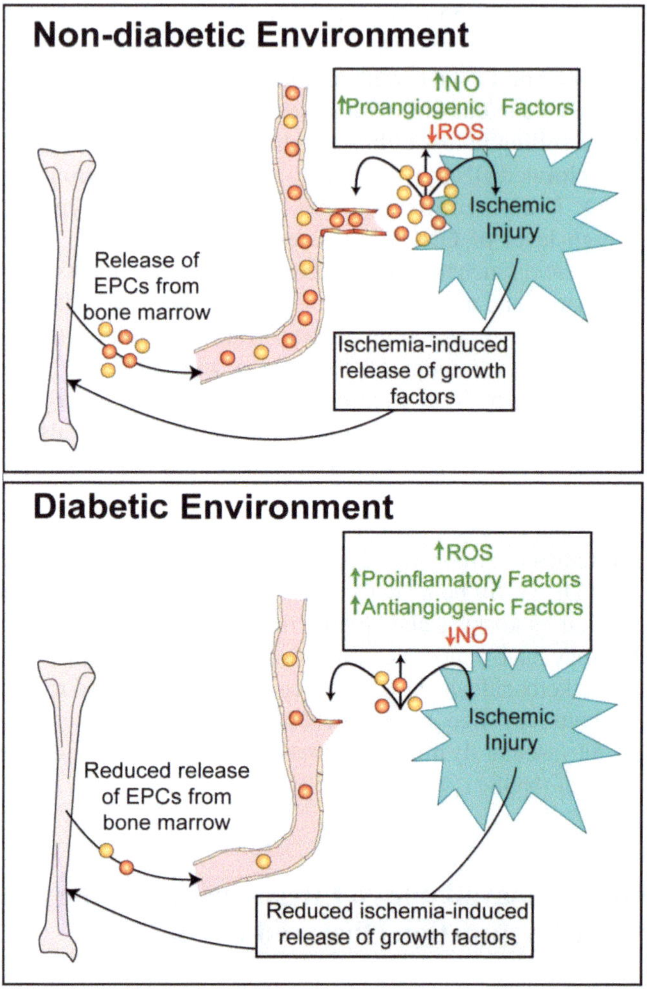

Fig. 6 Release of EPCs is reduced in a diabetic environment. In a non-diabetic environment ischemic injury results in the release of growth factors at the site of injury which stimulate the release of EPCs from bone marrow. EPCs then migrate to the site of injury and initiate blood vessel repair (angiogenesis). In a diabetic environment, ROS, proinflammatory and antiangiogenic factors are increased above non-diabetic levels while NO is reduced. This results in a blunted response to ischemic injury and marginal repair at the site of injury

integration into ischemic retinal vasculature *in vivo* [70]. Propofol, a peroxynitrite scavenger, inhibits NF-kB activation, increases NO production and protects EPCs from apoptosis [38, 71, 72]. High glucose enhances EPC senescence, impairs migration and tube formation. These effects can be ameliorated by co-incubation with the NO donor sodium nitroprusside and worsened by eNOS inhibition [73]. Diabetic EPC dysfunction was improved by the eNOS activator AVE 9488 and with PPARγ-agonists [74, 75]. Specifically, these agonists promote differentiation, increase EPC number and migratory activity, and reduce EPC apoptotic rate [76]. Decreased NO can be due to increased oxidative degradation as well as reduced synthesis. ADMA, an analogue of the amino acid l-arginine, inhibits the formation of NO and augments vascular oxidative stress, partly via eNOS uncoupling, resulting in higher superoxide generation. ADMA levels are increased in diabetes [76] and ADMA represses EPC proliferation, differentiation, and function, in a concentration-dependent manner [77]. The enzyme dimethylarginine dimethylaminohydrolase (DDAH) inactivates ADMA and regulates NO production [78]. Overexpression of DDAH *in vitro* and *in vivo* increased NO production [79].

9 BM Microenvironment is Dysfunctional in Diabetes

The bone marrow stem cell niche is a specialized microenvironment that nurtures and regulates the stem cell pool. We showed in a type 2 diabetes model that diabetic bone marrow has large amounts of fat compared to bone marrow from healthy controls [80]. Adipocyte-rich marrow harbors a decreased number of progenitors and relatively quiescent stem cells. Mice that are genetically deficient in adipogenesis show accelerated hematopoietic recovery after bone marrow ablation, a phenomenon that can be reproduced pharmacologically in wild-type mice through PPAR-γ inhibition [81]. These results suggest a novel therapeutic approach to enhance hematopoietic engraftment after marrow or cord blood transplantation, or to ameliorate aplasia in genetic bone marrow failure syndromes. Furthermore, this is a mechanism for the myelosuppression observed in patients treated with the PPAR- γ agonist rosiglitazone [82–84], a diabetes drug known to increase marrow adiposity [85].

Bone marrow adipocytes have a primarily suppressive effect on hematopoiesis within the BM microenvironment. Bone marrow adipocytes are less supportive of hematopoiesis *in vitro* than their undifferentiated stromal or pre-adipocytic counterparts, in part due to reduced production of growth factors such as granulocyte-macrophage colony-stimulating factor (GM-CSF) and granulocyte colony-stimulating factor (G-CSF) [86, 87]. Moreover, adipose tissue secretes neuropillin-1, [88] lipocalin [89, 90] adiponectin, [91] and TNF-α [92], each of which can impair hematopoietic proliferation. Of note, TNF-α and adiponectin inhibit progenitor activity [24, 92, 93] suggesting that adipocytes prevent hematopoietic progenitor expansion. Adipocytes and osteoblasts originate from mesenchymal stem cells within the bone marrow, where both compartments hold a reciprocal relationship [93]. Balancing the supportive role of the osteoblast in the HSC niche, adipocytes act as negative regulators of hematopoiesis and the diabetic fatty marrow likely has a deleterious effect on EPC function.

Based on the studies summarized above, we conclude that both freshly isolated and expanded EPCs can potentially be used for restoration of injured retinal vasculature and correction of retinal ischemia by restoring proper blood flow to the retina. Whether $CD34^+$ cells, angiogenic monocytes, such as $CD14^+$ cells, eEPCs or ECFCs represent the cell population with the greatest therapeutic utility is currently unknown. If there are unique characteristics of the retinal vasculature that make one of these populations better suited for retinal repair is also unknown. Only future studies will provide the needed information for determining the most effective cell therapy for diabetes induced vasodegenerative disease of the retina. Further complicating the situation, we believe that ECFCs may be markedly reduced in numbers in the blood of individuals with NPDR and may even be "transiently lost" from the circulation; thus are not available for repair of acellular capillaries.

Ex vivo correction of the diabetes-induced defects in EPCs, or systemic modulation of the BM to correct the dysfunction of these cells will reduce the development of retinopathy. We showed that healthy but not diabetic $CD34^+$ cells robustly home to injured retinal vasculature [15]. These studies support the potential of EPC therapy for DR and other ischemic retinopathies but they also highlight the need for a more thorough characterization of EPC subsets so that the precise fate and utility of delivered cells can be determined without the potential to evoke unwanted responses.

This is especially important in the context of a complex milieu such as diabetes which is known to alter EPC phenotype. For example, sca-1^+ bone marrow-derived EPCs transplanted into diabetic mice convert to a pro-inflammatory and anti-angiogenic phenotype and can exacerbate rather than improve limb ischemia [94].

10 The Role of Bone Marrow Derived Cells in Ocular Health and Disease

Retinopathy is the most common diabetic complication with almost all diabetic patients developing background retinopathy [95]. Both type 1 and type 2 diabetes is associated with widespread EPC dysfunction [96]. Several studies have shown that circulating EPC number is reduced in patients with NPDR [97] but increased in patients with PDR [98]. These observations were further confirmed by Brunner et al. who conducted a case-control study that compared 90 patients with type 1 diabetes with and without retinopathy. They also demonstrated that in type 1 patients with retinopathy, EPCs underwent stage-related regulation. In NPDR, a reduction of EPCs was observed, while in proliferative retinopathy, a dramatic increase of mature EPCs was observed [99]. These findings are consistent with the hypothesis that the vasodegenerative phase of diabetic retinopathy is associated with reduced reparative function of EPCs and that in PDR EPCs may contribute to pathological neovascularization.

Astrocytes, as mentioned earlier, are known to play a pivotal role in the normal developmental retinal vascularization. Otani et al. showed that intravitreally injected Lin^- bone marrow cells selectively target retinal astrocytes, cells that serve as a template for both developmental and injury-associated retinal angiogenesis [100]. When Lin^- bone marrow cells were injected into neonatal mouse eyes, they participated in normal developmental angiogenesis. When Lin^- cells were injected into the eyes of neonatal *rd*/*rd* mice, whose vasculature ordinarily degenerates with age, the cells prevented retinal vascular degeneration as this vascular rescue resulted in neuronal rescue [100]. Normal developmental vessels and the targeting of exogenous progenitor cells to astrocytes were both found to be dependent on R-cadherin as confirmed by disruption of Lin^- cell targeting to three distinct retinal vascular plexuses following R-cadherin blockade [101].

The role of glial cells during pathological retinal neovascularization is still under investigation. Loss of astrocytes and microglia directly correlates with the development of pathological neovascularization in a mouse model of oxygen-induced retinopathy (OIR). These two distinct glial cell populations were found to have cooperative survival effects *in vitro* and *in vivo*. The intravitreal injection of myeloid progenitor cells, astrocytes, or astrocyte-conditioned media rescued endogenous astrocytes from degeneration that normally occurs within the hypoxic, vaso-obliterated retina following return to normoxia. Protection of the retinal astrocytes and microglia was directly correlated with accelerated revascularization of the normal retinal plexuses and reduction of pathological intravitreal neovascularization normally associated with OIR. Using astrocyte-conditioned media, several factors were identified that may contribute to the observed astrocytic protection and subsequent normalization of the retinal vasculature, including VEGF and basic fibroblast growth factor (bFGF). Injection of VEGF or bFGF at specific doses rescued the retinas from developing OIR-associated pathology, an effect that was also preceded by protection of endogenous glia from hypoxia-induced degeneration. These data suggest that vascular-associated glia are required for normalized revascularization of the hypoxic retina and strategies designed to protect glial cells may provide a novel approach by which normalized revascularization can be promoted and abnormal neovascularization can be prevented [102].

Another condition associated with pathological angiogensis is age related macular degeneration.

The Csaky group demonstrated that the EPC population, ECFCs can be isolated and expanded successfully from the peripheral blood of elderly control and AMD-affected patients. Individuals with AMD or at risk for neovascular AMD demonstrated a significantly higher number of initial ECFCs clusters and increased expansion potential. The group postulated that ECFCs may be used for further phenotypic, genetic, and functional analyses in AMD patients [103].

Proliferative sickle cell retinopathy (PSR) is an uncommon complication in individuals with sickle cell trait, also called AS hemoglobinopathy (HAS), occurring more frequently in patients with SC hemoglobinopathy, S-thalassemia and SS hemoglobinopathy [104]. Interestingly, $KDR^+/CD34^+/Cd45^{dim}$ cells were found to be significantly higher during painful crisis [105], suggesting that patients during sickle crisis may be more at risk for development of ocular neovascularization.

11 Predictive Medicine and DR

Predictive medicine is a rapidly emerging field of medicine that involves anticipating disease and implementing effective measures in order to either prevent the disease altogether or significantly reduce morbidity and mortality. While the most fundamental way to predict future disease is based on genetics (genomics), other prediction methodologies, such as proteomics, and cytomics, allow for the early detection. However, most of the time, the detected biological markers exist because a disease process has already started. Diabetes, costing the United States more than $174 billion per year in 2007, is expected to take an increasingly large financial toll in subsequent years. Accurate projections of diabetes burden are essential to policymakers planning for future health care needs and costs [106]. DR is the most prevalent of all diabetic complications. The field of predictive medicine associated with DR is still in its early stages. HbA1c, high fasting plasma glucose (FPG) levels [107–109], and duration of diabetes [110] are predictors of DR. The DESIR (**D**ata from an **E**pidemiological **S**tudy on the **I**nsulin **R**esistance Syndrome) study was performed on a cohort of 700 men and women [109]. In this study, the positive predictive values for retinopathy increased sharply from 108 mg/dL (FPG) and from 6.0% (HbA1c) levels. These thresholds were proposed to identify patients at risk of developing retinopathy in 10 years [109]. There are conflicting reports about the importance of a lipid profile as predictor for retinopathy. Heydari and coworkers reported no difference in lipid profile between patients with and without retinopathy [110]. However, other studies have indicated increased serum triglycerides as predictive factors for the development and the progression of renal and retinal complications in patients with type 1 diabetes [111]. Nevertheless, all these factors, including blood pressure [107, 110], likely play a role.

12 Prevention and Treatment Options for DR

Despite advances in diabetes management, DR remains a leading cause of preventable vision loss. To date, 1 in 10 U.S. adults has diabetes. According to a new analysis from the Centers for Disease Control and Prevention [106], if current trends continue, by 2050 as many as 1 in 3 U.S. adults will likely have diabetes. The causes of this predicted sharp increase are an aging population more likely to develop type 2 diabetes, an increase in minority populations which are at high risk of developing type 2 diabetes and increased life expectancy for people with diabetes. Along with the evolving pandemic of type 2 diabetes, the occurrence of DR is expected to increase. Tight control of glycemia and blood pressure, together with improved life style, represent critical means for preventing or arresting the development of DR, but they are often difficult to achieve as mentioned earlier in this chapter. Current therapeutic strategies include laser photocoagulation and intravitreous injections of corticosteroids or anti-VEGF agents have an application limited to the advanced stage of the disease and are invasive. There is, therefore, a need to develop new therapeutic approaches that are less invasive, less destructive and aimed at early stages of DR.

Lipid-modifying therapy has recently gained great support as an effective strategy to reduce the diabetic retinopathy endpoints. The Fenofibrate Intervention and Event Lowering in Diabetes (FIELD) study, although inconclusive for statin therapy, show beneficial effects of fenofibrate in reducing the requirement for laser therapy, and particularly in preventing disease progression in patients with pre-existing DR [112].

13 Personalized Intervention Options

Most of today's therapies in use in clinical practice are aimed toward the general population and can, therefore, be ineffective in certain subgroups of patients or even cause, in such groups, adverse effects. On the contrary, personalized medicine is based on targeted therapeutic approaches that allow for patient-specific care [113] and aim to maximize therapy success while minimizing adverse effects.

The successful use of stem cells for therapeutic interventions requires taking into consideration several critical factors involved in the overall process including selection of the ideal stem cell population and optimizing the host microenvironment. While more relevant to replacement of retinal pigment epithelial cells and photoreceptors than endothelial cells, embryonic stem cells (ESCs) are now in clinical trials for AMD [114]. However, while ESCs can generate an unlimited supply of cells for therapy, this property also means they have the potential to generate teratomas. The use of ESCs raises unavoidable ethical issues. All the problems associated with the use of ESCs have been solved by the introduction of induced pluripotent stem cells (iPS) which are somatic cells reprogrammed into pluripotent stem cells by introducing four transcription factors (Sox2, Oct3/4, c-Myc and Klf4)[115–117]. While these cells escape the ethical controversy associated with the ESCs, their clinical application is also problematic. Their generation, being accomplished by retroviral transduction, makes them at high risk for tumor formation and recently they have been shown to be rejected from the host even if from autologous origin [118]. However, nonviral strategies to introduce these four transcription factors are now available, potentially decreasing the risks of use of these cells.

Among adult stem cells, EPCs, isolated from either peripheral blood or bone marrow, have been extensively and successfully used for revascularization of various ischemic tissues. Therapeutic neovascularization was first reported by Asahara's group for the intracardiac transplantation of human peripheral blood-derived culture-expanded EPCs into immunodeficient mice with hind limb ischemia [119]. Later, a similar study in which human EPCs were transplanted in a nude rat myocardial ischemia model demonstrated that transplanted EPCs were recruited to the ischemic tissue and differentiated into endothelial cells [120]. More studies followed in other animal models and clinical trials are in progress [119].

Bone marrow mesenchymal stem cells (BMSCs) have great potential for repair [121]. BMSCs are adult stem cells with important functional features including the potential for rapid self-renewal that give rise to connective tissues, including bone, cartilage, muscle, fat, hemopoiesis-supporting stroma and neuronal cells. BMSCs occur at very low frequencies of $2–5/1 \times 10^6$ mononuclear cells in bone marrow harvests. However, these cells have rapid *in vitro* potential for expansion by a factor of 10^3 within 14–21 days of culture. Unless cultured for prolonged periods, BMSCs do not show senescence and apoptosis. BMSCs may have a strong capability of penetrating the myocardium from the coronary artery [122] and can facilitate the regeneration of damaged heart tissue [123].

14 Autologous Therapy Using EPCs

Our group has shown that human EPCs from peripheral blood of healthy human donors, when intravitreally injected in diabetic rats, migrate to the area of ischemic injury and some cells integrate into the vessel wall. However, when EPCs were isolated from diabetic donors, they were unable to migrate and formed a monolayer on the surface of the retina [15] confirming the earlier reported defects observed in diabetic EPCs. Thus, while autologous cell therapy holds great promise for cell-based regenerative medicine, its application in DR patients faces great challenges. As discussed earlier in the chapter, the diabetic EPCs themselves are known to be dysfunctional, and the diabetic milieu poses additional challenges that need to be addressed. Cell-based therapies may be ineffective or may actually result in worsening of existing ocular pathologies. *In vitro* expansion, while increasing the number of available cells, does not correct the defects observed. Therefore, to have a successful outcome, the EPCs need to be manipulated *ex vivo* to correct their defects before being injected.

We and others have shown that the migratory defect of diabetic EPCs can be corrected *ex vivo*. Segal et al. have shown that the preconditioning of $CD34^+$ EPCs from peripheral blood of diabetic individuals with the nitric oxide (NO) donor DETA/NO resulted in restoration of migration in the diabetic EPCs [68]. Expression of NOX2, NADPH oxidase activity, and superoxide levels are higher in diabetic EPCs than in nondiabetic EPCs. *Ex vivo* NADPH oxidase inhibition with either apocynin or gp91ds-tat increased NO bioavailability without increasing eNOS activity and resulted in enhanced migratory responsiveness to SDF-1α in diabetic EPCs and enhanced their homing to ischemic retinal vasculature *in vivo* [124]. Treatment of human diabetic EPCs with the peroxisome proliferator-activated receptor (PPAR)-gamma agonist rosiglitazone has been shown to improve EPC angiogenic function in a mouse model of carotid injury by reducing NADPH oxidase–dependent ROS production and enhancing bioavailable NO levels [64]. *Ex* vivo treatment with the p38 MAPK inhibitor SB203580 has been shown to increase EPC proliferation [117]. However, not all types of adult stem cells are equal, and *ex vivo* treatments may affect each cell type differently. Ischemic cardiomyopathy patients (ICMP) show reduced NO bioavailability. The *ex vivo* treatment of bone marrow derived cells from ICMP patients with AVE9488, a molecule that increases eNOS expression, has been shown to enhance migratory capacity *in vitro* and to improve neovascularization capacity of the infused BMC in an ischemic hind limb model *in vivo* [125]. The same treatment on $CD34^+$ EPCs from PB of diabetic

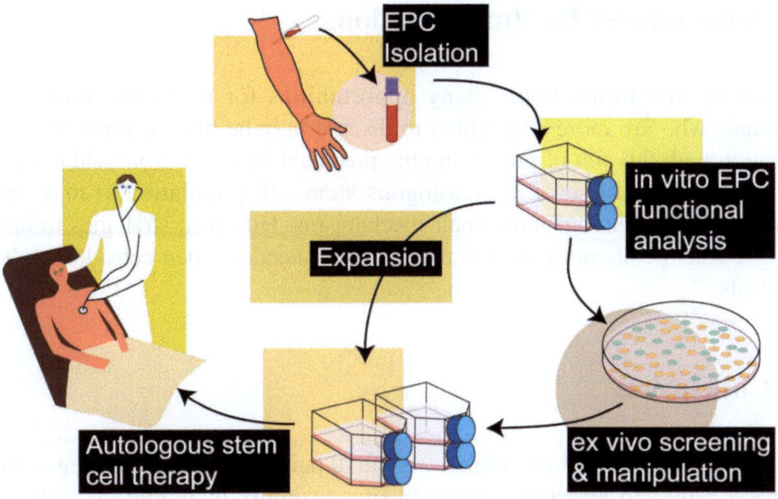

Fig. 7 Personalized intervention. When the individual is admitted to the hospital/outpatient facility for cell therapy. His/her own progenitor cells are isolated from the peripheral blood/bone marrow aspiration and first subjected to functional assays; such as, migration, tube formation, proliferation, nitric oxide generation, etc. After any functional defects are identified, the appropriate *ex vivo* corrective treatment is applied. While this testing of progenitors is occurring, the individual will undergo careful medical evaluation (glycemia, blood pressure, etc.) and proper measures will be taken to optimize the environment in which the modified cells are introduced

patients resulted in increased peroxynitrite levels and, therefore, failed to enhance NO-mediated functions *in vitro* and *in vivo* [124].

In addition to dealing with the differences found among adult stem cell types, we must consider the differences in the same cell type found between individuals. We believe that the stem cells of each diabetic individual with complications represents a unique "dysfunction identity" and this points to the need for the development of personalized intervention as the most effective way to treat each patient. While one treatment may work on EPCs from one individual and provide effective outcome, the same treatment strategy may not work on another individual.

In conclusion, we propose the following scenario for personalized intervention in the near future: when the individual is admitted to the hospital for treatment or if cells are isolated in the clinic setting, these cells, once isolated, are tested in a battery of functional assays such as migration, tube formation, proliferation, NO generation, etc. After the dysfunction is uniquely determined, the appropriate *ex vivo* corrective treatments are applied. However, while all this takes place, the patient has to undergo careful medical evaluation (glycemia, blood pressure, etc.) to optimize the environment in which the cells are reintroduced (Fig. 7). Furthermore, if the cells are *ex vivo* labeled with a contrast agent for MRI, their fate after transplantation can be easily followed overtime *in vivo* [126, 127].

15 Personalized Treatment Options

In summary, the future holds many opportunities for predictive medicine and individuals who are more susceptible to disease may be offered lifestyle advice or medication with the aim of preventing the predicted illness. For individuals that fail these strategies, the use of an autologous stem cell population(s) may serve to augment defective endogenous repair mechanisms. However, strategies to optimize autologous stem cells prior to their use are likely needed when considering diabetic individuals.

16 Conclusion

Even though EPCs constitute a relatively small percentage of circulating cells, they can specifically and effectively home to sites of injury including the retina. What exactly is the promise of cellular therapy for diabetic retinopathy? While peripheral blood or bone marrow $CD34^+$ cells show promise, we believe that the *ex vivo* expanded eEPCs and ECFCs may equally represent cell populations with great therapeutic utility and may provide the basis for an effective cell therapy for vasodegenerative disease in the retina. In NPDR, EPC populations have reduced function as they cannot be recruited into the retina to repair the acellular capillaries, while in PDR the EPCs may take on a pro-inflammatory, proangiogenic phenotype and perhaps too many EPCs lead to pathological neovascularization. While this hypothesis remains to be proven, the concept of identifying the "best" reparative cell and the ideal time for its administration is critically important to the derivation of cellular therapies. In diabetes, dysregulation of NO is central to phenotypic transitions of these EPC to cells with reduced reparative function and this cellular change may be further influenced by the shifting bone marrow microenvironment associated with diabetes. *Ex vivo* correction of the diabetes-induced defects in these cells prior to administration to the patient or systemic modulation of the bone marrow to correct the dysfunction of these cells will likely be one of the therapeutic modalities available to individuals with diabetic retinopathy.

Acknowledgements This work was supported by grants from the National Institutes of Health (NIH1R01 EY07739, NIH R01EY12601), American Heart Association (Beginning Grant-in-Aid # 0865213E to S. L.C.) and the Juvenile Diabetes Research Foundation (Grant 4-2000-847).

References

1. Fong DS, Sharza M, Chen W, Paschal JF, Ariyasu RG, Lee PP (2002) Vision loss among diabetics in a group model Health Maintenance Organization (HMO). Am J Ophthalmol 133:236–241
2. ETDRS (1991) Early photocoagulation for diabetic retinopathy. ETDRS report number 9. Early Treatment Diabetic Retinopathy Study Research Group. Ophthalmology 98:766–785

3. ETDRS (1995) Focal photocoagulation treatment of diabetic macular edema. ETDRS report number 19. Arch Ophthalmol 113:1144–1155
4. Avery RB, Diener-West M, Reynolds SM, Grossniklaus HE, Green WR, Albert DM (2008) Histopathologic characteristics of choroidal melanoma in eyes enucleated after iodine 125 brachytherapy in the collaborative ocular melanoma study. Arch Ophthalmol 126:207–212
5. Domen J, Wagers A, Weissman IL Bone marrow (hematiopoietic) stem cells. http://stemcells.nih.gov/staticresources/info/scireport/PDFs/D.%20Chapter%202.pdf. Accessed 20 Oct 2012
6. Yoder MC, Mead LE, Prater D, Krier TR, Mroueh KN, Li F, Krasich R, Temm CJ, Prchal JT, Ingram DA (2007) Redefining endothelial progenitor cells via clonal analysis and hematopoietic stem/progenitor cell principals. Blood 109:1801–1809
7. Asahara T, Murohara T, Sullivan A, Silver M, van der Zee R, Li T, Witzenbichler B, Schatteman G, Isner JM (1997) Isolation of putative progenitor endothelial cells for angiogenesis. Science 275:964–967
8. Schatteman GC, Dunnwald M, Jiao C (2007) Biology of bone marrow-derived endothelial cell precursors. Am J Physiol Heart Circ Physiol 292:H1–H18
9. Spring H, Schuler T, Arnold B, Hämmerling GJ, Ganss R (2005) Chemokines direct endothelial progenitors into tumor neovessels. Proc Natl Acad Sci U S A 102:18111–18116
10. Wilson A, Trumpp A (2006) Bone-marrow haematopoietic-stem-cell niches. Nat Rev Immunol 6:93–106
11. Rafii S, Lyden D (2003) Therapeutic stem and progenitor cell transplantation for organ vascularization and regeneration. Nat Med 9:702–712
12. Minami E, Laflamme MA, Saffitz JE, Murry CE (2005) Extracardiac progenitor cells repopulate most major cell types in the transplanted human heart. Circulation 112:2951–2958
13. Ruzinova MB, Schoer RA, Gerald W, Egan JE, Pandolfi PP, Rafii S, Manova K, Mittal V, Benezra R (2003) Effect of angiogenesis inhibition by Id loss and the contribution of bone-marrow-derived endothelial cells in spontaneous murine tumors. Cancer Cell 4:277–289
14. Peters BA, Diaz LA, Polyak K, Meszler L, Romans K, Guinan EC, Antin JH, Myerson D, Hamilton SR, Vogelstein B, Kinzler KW, Lengauer C (2005) Contribution of bone marrow-derived endothelial cells to human tumor vasculature. Nat Med 11:261–262
15. Caballero S, Sengupta N, Afzal A, Chang KH, Li Calzi S, Guberski DL, Kern TS, Grant MB (2007) Ischemic vascular damage can be repaired by healthy, but not diabetic, endothelial progenitor cells. Diabetes 56:960–967
16. Grant MB, May WS, Caballero S, Brown GA, Guthrie SM, Mames RN, Byrne BJ, Vaught T, Spoerri PE, Peck AB, Scott EW (2002) Adult hematopoietic stem cells provide functional hemangioblast activity during retinal neovascularization. Nat Med 8:607–612
17. Shi Q, Rafii S, Wu MH, Wijelath ES, Yu C, Ishida A, Fujita Y, Kothari S, Mohle R, Sauvage LR, Moore MA, Storb RF, Hammond WP (1998) Evidence for circulating bone marrow-derived endothelial cells. Blood 92:362–367
18. Fernandez Pujol B, Lucibello FC, Gehling UM, Lindemann K, Weidner N, Zuzarte ML, Adamkiewicz J, Elsässer HP, Müller R, Havemann K (2000) Endothelial-like cells derived from human CD14 positive monocytes. Differentiation 65:287–300
19. Eggermann J, Kliche S, Jarmy G, Hoffmann K, Mayr-Beyrle U, Debatin KM, Waltenberger J, Beltinger C (2003) Endothelial progenitor cell culture and differentiation in vitro: a methodological comparison using human umbilical cord blood. Cardiovasc Res 58:478–486
20. Hur J, Yoon CH, Kim HS, Choi JH, Kang HJ, Hwang KK, Oh BH, Lee MM, Park YB (2004) Characterization of two types of endothelial progenitor cells and their different contributions to neovasculogenesis. Arterioscler Thromb Vasc Biol 24:288–293
21. Delorme B, Basire A, Gentile C, Sabatier F, Monsonis F, Desouches C, Blot-Chabaud M, Uzan G, Sampol J, Dignat-George F (2005) Presence of endothelial progenitor cells, distinct from mature endothelial cells, within human CD146+ blood cells. Thromb Haemost 94:1270–1279
22. Gehling UM, Ergun S, Schumacher U, Wagener C, Pantel K, Otte M, Schuch G, Schafhausen P, Mende T, Kilic N, Kluge K, Schäfer B, Hossfeld DK, Fiedler W (2000) In vitro differentiation of endothelial cells from AC133-positive progenitor cells. Blood 95:3106–3112

23. Case J, Mead LE, Bessler WK, Prater D, White HA, Saadatzadeh MR, Bhavsar JR, Yoder MC, Haneline LS, Ingram DA (2007) Human CD34+AC133+VEGFR-2+ cells are not endothelial progenitor cells but distinct, primitive hematopoietic progenitors. Exp Hematol 35:1109–1118
24. Yoon CH, Hur J, Park KW, Kim JH, Lee CS, Oh IY, Kim TY, Cho HJ, Kang HJ, Chae IH, Yang HK, Oh BH, Park YB, Kim HS (2005) Synergistic neovascularization by mixed transplantation of early endothelial progenitor cells and late outgrowth endothelial cells: the role of angiogenic cytokines and matrix metalloproteinases. Circulation 112:1618–1627
25. Gulati R, Jevremovic D, Peterson TE, Chatterjee S, Shah V, Vile RG, Simari RD (2003) Diverse origin and function of cells with endothelial phenotype obtained from adult human blood. Circ Res 93:1023–1025
26. Crosby JR, Kaminski WE, Schatteman G, Martin PJ, Raines EW, Seifert RA, Bowen-Pope DF (2000) Endothelial cells of hematopoietic origin make a significant contribution to adult blood vessel formation. Circ Res 87:728–730
27. Kalka C, Masuda H, Takahashi T, Gordon R, Tepper O, Gravereaux E, Pieczek A, Iwaguro H, Hayashi SI, Isner JM, Asahara T (2000) Vascular endothelial growth factor(165) gene transfer augments circulating endothelial progenitor cells in human subjects. Circ Res 86:1198–1202
28. Harraz M, Jiao C, Hanlon HD, Hartley RS, Schatteman GC (2001) CD34- blood-derived human endothelial cell progenitors. Stem Cells 19:304–312
29. Rohde E, Malischnik C, Thaler D, Maierhofer T, Linkesch W, Lanzer G, Guelly C, Strunk D (2006) Blood monocytes mimic endothelial progenitor cells. Stem Cells 24:357–367
30. Schmeisser A, Garlichs CD, Zhang H, Eskafi S, Graffy C, Ludwig J, Strasser RH, Daniel WG (2001) Monocytes coexpress endothelial and macrophagocytic lineage markers and form cord-like structures in Matrigel under angiogenic conditions. Cardiovasc Res 49:671–680
31. Schmeisser A, Strasser RH (2002) Phenotypic overlap between hematopoietic cells with suggested angioblastic potential and vascular endothelial cells. J Hematother Stem Cell Res 11:69–79
32. Romagnani P, Annunziato F, Liotta F, Lazzeri E, Mazzinghi B, Frosali F, Cosmi L, Maggi L, Lasagni L, Scheffold A, Kruger M, Dimmeler S, Marra F, Gensini G, Maggi E, Romagnani S (2005) CD14+CD34low cells with stem cell phenotypic and functional features are the major source of circulating endothelial progenitors. Circ Res 97:314–322
33. George J, Herz I, Goldstein E, Abashidze S, Deutch V, Finkelstein A, Michowitz Y, Miller H, Keren G (2003) Number and adhesive properties of circulating endothelial progenitor cells in patients with in-stent restenosis. Arterioscler Thromb Vasc Biol 23:e57–e60
34. Ghani U, Shuaib A, Salam A, Nasir A, Shuaib U, Jeerakathil T, Sher F, O'Rourke F, Nasser AM, Schwindt B, Todd K (2005) Endothelial progenitor cells during cerebrovascular disease. Stroke 36:151–153
35. Loomans CJ, Wan H, de Crom R, van Haperen R, de Boer HC, Leenen PJ, Drexhage HA, Rabelink TJ, van Zonneveld AJ, Staal FJ (2006) Angiogenic murine endothelial progenitor cells are derived from a myeloid bone marrow fraction and can be identified by endothelial NO synthase expression. Arterioscler Thromb Vasc Biol 26:1760–1767
36. Loomans CJ, van Haperen R, Duijs JM, Verseyden C, de Crom R, Leenen PJ, Drexhage HA, de Boer HC, de Koning EJ, Rabelink TJ, Staal FJ, van Zonneveld AJ (2009) Differentiation of bone marrow-derived endothelial progenitor cells is shifted into a proinflammatory phenotype by hyperglycemia. Mol Med 15:152–159
37. Donato AJ, Eskurza I, Silver AE, Levy AS, Pierce GL, Gates PE, Seals DR (2007) Direct evidence of endothelial oxidative stress with aging in humans: relation to impaired endothelium-dependent dilation and upregulation of nuclear factor-kappaB. Circ Res 100:1659–1666
38. Imanishi T, Hano T, Sawamura T, Nishio I (2004) Oxidized low-density lipoprotein induces endothelial progenitor cell senescence, leading to cellular dysfunction. Clin Exp Pharmacol Physiol 31:407–413

39. Loomans CJ, de Koning EJ, Staal FJ, Rookmaaker MB, Verseyden C, de Boer HC, Verhaar MC, Braam B, Rabelink TJ, van Zonneveld AJ (2004) Endothelial progenitor cell dysfunction: a novel concept in the pathogenesis of vascular complications of type 1 diabetes. Diabetes 53:195–199
40. Zhu J, Xu ZK, Miao Y, Liu XL, Zhang H (2006) Changes of inducible protein-10 and regulated upon activation, normal T cell expressed and secreted protein in acute rejection of pancreas transplantation in rats. World J Gastroenterol 12:4156–4160
41. Tamarat R, Silvestre JS, Le Ricousse-Roussanne S, Barateau V, Lecomte-Raclet L, Clergue M, Duriez M, Tobelem G, Lévy BI (2004) Impairment in ischemia-induced neovascularization in diabetes: bone marrow mononuclear cell dysfunction and therapeutic potential of placenta growth factor treatment. Am J Pathol 164:457–466
42. Kielczewski JL, Jarajapu YP, McFarland EL, Cai J, Afzal A, Li Calzi S, Chang KH, Lydic T, Shaw LC, Busik J, Hughes J, Cardounel AJ, Wilson K, Lyons TJ, Boulton ME, Mames RN, Chan-Ling T, Grant MB (2009) Insulin-like growth factor binding protein-3 mediates vascular repair by enhancing nitric oxide generation. Circ Res 105:897–905
43. Ma FX, Zhou B, Chen Z, Ren Q, Lu SH, Sawamura T, Han ZC (2006) Oxidized low density lipoprotein impairs endothelial progenitor cells by regulation of endothelial nitric oxide synthase. J Lipid Res 47:1227–1237
44. Sobrevia L, Mann GE (1997) Dysfunction of the endothelial nitric oxide signalling pathway in diabetes and hyperglycaemia. Exp Physiol 82:423–452
45. Weyrich AS, Ma XL, Lefer AM (1992) The role of L-arginine in ameliorating reperfusion injury after myocardial ischemia in the cat. Circulation 86:279–288
46. Kaesemeyer WH, Ogonowski AA, Jin L, Caldwell RB, Caldwell RW (2000) Endothelial nitric oxide synthase is a site of superoxide synthesis in endothelial cells treated with glyceryl trinitrate. Br J Pharmacol 131:1019–1023
47. Berkowitz DE, White R, Li D, Minhas KM, Cernetich A, Kim S, Burke S, Shoukas AA, Nyhan D, Champion HC, Hare JM (2003) Arginase reciprocally regulates nitric oxide synthase activity and contributes to endothelial dysfunction in aging blood vessels. Circulation 108:2000–2006
48. Jean C, Fromentin G, Tome D, Larue-Achagiotis C (2002) Wistar rats allowed to self-select macronutrients from weaning to maturity choose a high-protein, high-lipid diet. Physiol Behav 76:65–73
49. Ellis EA, Guberski DL, Hutson B, Grant MB (2002) Time course of NADH oxidase, inducible nitric oxide synthase and peroxynitrite in diabetic retinopathy in the BBZ/WOR rat. Nitric Oxide 6:295–304
50. Camici GG, Schiavoni M, Francia P, Bachschmid M, Martin-Padura I, Hersberger M, Tanner FC, Pelicci P, Volpe M, Anversa P, Lüscher TF, Cosentino F (2007) Genetic deletion of p66(Shc) adaptor protein prevents hyperglycemia-induced endothelial dysfunction and oxidative stress. Proc Natl Acad Sci U S A 104:5217–5222
51. Kobayashi T, Taguchi K, Takenouchi Y, Matsumoto T, Kamata K (2007) Insulin-induced impairment via peroxynitrite production of endothelium-dependent relaxation and sarco/endoplasmic reticulum Ca(2+)-ATPase function in aortas from diabetic rats. Free Radic Biol Med 43:431–443
52. Szabo C, Mabley JG, Moeller SM, Shimanovich R, Pacher P, Virag L, Soriano FG, Van Duzer JH, Williams W, Salzman AL, Groves JT (2002) Part I: pathogenetic role of peroxynitrite in the development of diabetes and diabetic vascular complications: studies with FP15, a novel potent peroxynitrite decomposition catalyst. Mol Med 8:571–580
53. Munzel T, Sayegh H, Freeman BA, Tarpey MM, Harrison DG (1995) Evidence for enhanced vascular superoxide anion production in nitrate tolerance. A novel mechanism underlying tolerance and cross-tolerance. J Clin Invest 95:187–194
54. Bucala R, Tracey KJ, Cerami A (1991) Advanced glycosylation products quench nitric oxide and mediate defective endothelium-dependent vasodilatation in experimental diabetes. J Clin Invest 87:432–438

55. Leal EC, Manivannan A, Hosoya K, Terasaki T, Cunha-Vaz J, Ambrósio AF, Forrester JV (2007) Inducible nitric oxide synthase isoform is a key mediator of leukostasis and blood-retinal barrier breakdown in diabetic retinopathy. Invest Ophthalmol Vis Sci 48:5257–5265
56. Stadler K, Bonini MG, Dallas S, Duma D, Mason RP, Kadiiska MB (2008) Direct evidence of iNOS-mediated in vivo free radical production and protein oxidation in acetone-induced ketosis. Am J Physiol Endocrinol Metab 295:E456–E462
57. Nagareddy PR, Xia Z, McNeill JH, MacLeod KM (2005) Increased expression of iNOS is associated with endothelial dysfunction and impaired pressor responsiveness in streptozotocin-induced diabetes. Am J Physiol Heart Circ Physiol 289:H2144–H2152
58. Cheng X, Cheng XS, Kuo KH, Pang CC (2004) Inhibition of iNOS augments cardiovascular action of noradrenaline in streptozotocin-induced diabetes. Cardiovasc Res 64:298–307
59. Aicher A, Zeiher AM, Dimmeler S (2005) Mobilizing endothelial progenitor cells. Hypertension 45:321–325
60. Heissig B, Hattori K, Dias S, Friedrich M, Ferris B, Hackett NR, Crystal RG, Besmer P, Lyden D, Moore MA, Werb Z, Rafii S (2002) Recruitment of stem and progenitor cells from the bone marrow niche requires MMP-9 mediated release of kit-ligand. Cell 109:625–637
61. Heissig B, Werb Z, Rafii S, Hattori K (2003) Role of c-kit/Kit ligand signaling in regulating vasculogenesis. Thromb Haemost 90:570–576
62. Shintani S, Murohara T, Ikeda H, Ueno T, Sasaki K, Duan J, Imaizumi T (2001) Augmentation of postnatal neovascularization with autologous bone marrow transplantation. Circulation 103:897–903
63. Dimmeler S, Fleming I, Fisslthaler B, Hermann C, Busse R, Zeiher AM (1999) Activation of nitric oxide synthase in endothelial cells by Akt- dependent phosphorylation. Nature 399:601–605
64. Sorrentino SA, Bahlmann FH, Besler C, Müller M, Schulz S, Kirchhoff N, Doerries C, Horváth T, Limbourg A, Limbourg F, Fliser D, Haller H, Drexler H, Landmesser U (2007) Oxidant stress impairs in vivo reendothelialization capacity of endothelial progenitor cells from patients with type 2 diabetes mellitus: restoration by the peroxisome proliferator-activated receptor-gamma agonist rosiglitazone. Circulation 116:163–173
65. Thum T, Fraccarollo D, Schultheiss M, Froese S, Galuppo P, Widder JD, Tsikas D, Ertl G, Bauersachs J (2007) Endothelial nitric oxide synthase uncoupling impairs endothelial progenitor cell mobilization and function in diabetes. Diabetes 56:666–674
66. Vasa M, Breitschopf K, Zeiher AM, Dimmeler S (2000) Nitric oxide activates telomerase and delays endothelial cell senescence. Circ Res 87:540–542
67. Li Calzi S, Purich DL, Chang KH, Afzal A, Nakagawa T, Busik JV, Agarwal A, Segal MS, Grant MB (2008) Carbon monoxide and nitric oxide mediate cytoskeletal reorganization in microvascular cells via vasodilator-stimulated phosphoprotein (VASP) phosphorylation: evidence for blunted responsiveness in diabetes. Diabetes 57:2488–2494
68. Segal MS, Shah R, Afzal A, Perrault CM, Chang K, Schuler A, Beem E, Shaw LC, Li Calzi S, Harrison JK, Tran-Son-Tay R, Grant MB (2006) Nitric oxide cytoskeletal-induced alterations reverse the endothelial progenitor cell migratory defect associated with diabetes. Diabetes 55:102–109
69. Grunewald M, Avraham I, Dor Y, Bachar-Lustig E, Itin A, Jung S, Chimenti S, Landsman L, Abramovitch R, Keshet E (2006) VEGF-induced adult neovascularization: recruitment, retention, and role of accessory cells. Cell 124:175–189
70. Jarajapu YP, Thampi P, Caballero S, Boulton ME, Grant MB (2009) Differential paracrine dysfunction and hypoxic desensitization in diabetic CD34+ and CD14+ bone marrow derived cells (BMDC). Circulation 120:S1124
71. Gao J, Zhao WX, Zhou LJ, Zeng BX, Yao SL, Liu D, Chen ZQ (2006) Protective effects of propofol on lipopolysaccharide-activated endothelial cell barrier dysfunction. Inflamm Res 55:385–392
72. Luo T, Xia Z, Ansley DM, Ouyang J, Granville DJ, Li Y, Xia ZY, Zhou QS, Liu XY (2005) Propofol dose-dependently reduces tumor necrosis factor-alpha-induced human umbilical vein endothelial cell apoptosis: effects on Bcl-2 and Bax expression and nitric oxide generation. Anesth Analg 100:1653–1659

73. Chen YS, Chen KH, Liu CC, Lee CT, Yang CH, Chuang KC, Lin CR (2007) Propofol-induced vascular permeability change is related to the nitric oxide signaling pathway and occludin phosphorylation. J Biomed Sci 14:629–636
74. Dimmeler S, Aicher A, Vasa M, Mildner-Rihm C, Adler K, Tiemann M, Rütten H, Fichtlscherer S, Martin H, Zeiher AM (2001) HMG-CoA reductase inhibitors (statins) increase endothelial progenitor cells via the PI 3-kinase/Akt pathway. J Clin Invest 108:391–397
75. Landmesser U, Engberding N, Bahlmann FH, Schaefer A, Wiencke A, Heineke A, Spiekermann S, Hilfiker-Kleiner D, Templin C, Kotlarz D, Mueller M, Fuchs M, Hornig B, Haller H, Drexler H (2004) Statin-induced improvement of endothelial progenitor cell mobilization, myocardial neovascularization, left ventricular function, and survival after experimental myocardial infarction requires endothelial nitric oxide synthase. Circulation 110:1933–1939
76. Verma S, Szmitko PE, Anderson TJ (2004) Endothelial function: ready for prime time? Can J Cardiol 20:1335–1339
77. Thum T, Fraccarollo D, Galuppo P, Tsikas D, Frantz S, Ertl G, Bauersachs J (2006) Bone marrow molecular alterations after myocardial infarction: impact on endothelial progenitor cells. Cardiovasc Res 70:50–60
78. Achan V, Broadhead M, Malaki M, Whitley G, Leiper J, MacAllister R, Vallance P (2003) Asymmetric dimethylarginine causes hypertension and cardiac dysfunction in humans and is actively metabolized by dimethylarginine dimethylaminohydrolase. Arterioscler Thromb Vasc Biol 23:1455–1459
79. Dayoub H, Achan V, Adimoolam S, Jacobi J, Stuehlinger MC, Wang BY, Tsao PS, Kimoto M, Vallance P, Patterson AJ, Cooke JP (2003) Dimethylarginine dimethylaminohydrolase regulates nitric oxide synthesis: genetic and physiological evidence. Circulation 108: 3042–3047
80. Busik J, Tikhonenko M, Bhatwadekar A, Opreanu M, Yakubova N, Caballero S, Player D, Nakagawa T, Afzal A, Kielczewski J, Sochacki A, Hasty S, Li Calzi S, Kim S, Duclas SK, Segal MS, Guberski DL, Esselman WJ, Boulton ME, Grant MB (2009) Diabetic retinopahty is associated with bone marrow neuropathy and a depressed peripheral clock. J Exp Med 206:2897–2906
81. Naveiras O, Nardi V, Wenzel PL, Hauschka PV, Fahey F, Daley GQ (2009) Bone-marrow adipocytes as negative regulators of the haematopoietic microenvironment. Nature 460: 259–263
82. Berria R, Glass L, Mahankali A, Miyazaki Y, Monroy A, De Filippis E, Cusi K, Cersosimo E, Defronzo RA, Gastaldelli A (2007) Reduction in hematocrit and hemoglobin following pioglitazone treatment is not hemodilutional in Type II diabetes mellitus. Clin Pharmacol Ther 82:275–281
83. Digman C, Borto D, Nasraway SA Jr (2005) Hyperglycemia in the critically ill. Nutr Clin Care 8:93–101
84. Maaravi Y, Stessman J (2005) Mild, reversible pancytopenia induced by rosiglitazone. Diabetes Care 28:1536
85. Lazarenko OP, Rzonca SO, Hogue WR, Swain FL, Suva LJ, Lecka-Czernik B (2007) Rosiglitazone induces decreases in bone mass and strength that are reminiscent of aged bone. Endocrinology 148:2669–2680
86. Corre J, Barreau C, Cousin B, Chavoin JP, Caton D, Fournial G, Penicaud L, Casteilla L, Laharrague P (2006) Human subcutaneous adipose cells support complete differentiation but not self-renewal of hematopoietic progenitors. J Cell Physiol 208:282–288
87. Nishikawa S, Kumar PK, Jeoung YH, Kawakami J, Nishikawa F, Suh Y, Ohtsuka E, Taira K (1993) Chemical probing studies of the hepatitis delta virus (HDV) genomic ribozyme. Nucleic Acids Symp Ser 29:119–120
88. Belaid-Choucair Z, Lepelletier Y, Poncin G, Thiry A, Humblet C, Maachi M, Beaulieu A, Schneider E, Briquet A, Mineur P, Lambert C, Mendes-Da-Cruz D, Ahui ML, Asnafi V, Dy M, Boniver J, Nusgens BV, Hermine O, Defresne MP (2008) Human bone marrow adipocytes block granulopoiesis through neuropilin-1-induced granulocyte colony-stimulating factor inhibition. Stem Cells 26:1556–1564

89. Miharada K, Hiroyama T, Sudo K, Danjo I, Nagasawa T, Nakamura Y (2008) Lipocalin 2-mediated growth suppression is evident in human erythroid and monocyte/macrophage lineage cells. J Cell Physiol 215:526–537
90. Yan QW, Yang Q, Mody N, Graham TE, Hsu CH, Xu Z, Houstis NE, Kahn BB, Rosen ED (2007) The adipokine lipocalin 2 is regulated by obesity and promotes insulin resistance. Diabetes 56:2533–2540
91. Yokota T, Oritani K, Takahashi I, Ishikawa J, Matsuyama A, Ouchi N, Kihara S, Funahashi T, Tenner AJ, Tomiyama Y, Matsuzawa Y (2000) Adiponectin, a new member of the family of soluble defense collagens, negatively regulates the growth of myelomonocytic progenitors and the functions of macrophages. Blood 96:1723–1732
92. Zhang Y, Harada A, Bluethmann H, Wang JB, Nakao S, Mukaida N, Matsushima K (1995) Tumor necrosis factor (TNF) is a physiologic regulator of hematopoietic progenitor cells: increase of early hematopoietic progenitor cells in TNF receptor p55-deficient mice in vivo and potent inhibition of progenitor cell proliferation by TNF alpha in vitro. Blood 86:2930–2937
93. DiMascio L, Voermans C, Uqoezwa M, Duncan A, Lu D, Wu J, Sankar U, Reya T (2007) Identification of adiponectin as a novel hemopoietic stem cell growth factor. J Immunol 178:3511–3520
94. Awad O, Jiao C, Ma N, Dunnwald M, Schatteman GC (2005) Obese diabetic mouse environment differentially affects primitive and monocytic endothelial cell progenitors. Stem Cells 23:575–583
95. Klein R, Knudtson MD, Lee KE, Gangnon R, Klein BE (2008) The Wisconsin Epidemiologic Study of Diabetic Retinopathy: XXII the twenty-five-year progression of retinopathy in persons with type 1 diabetes. Ophthalmology 115:1859–1868
96. Fadini GP, Miorin M, Facco M, Bonamico S, Baesso I, Grego F, Menegolo M, de Kreutzenberg SV, Tiengo A, Agostini C, Avogaro A (2005) Circulating endothelial progenitor cells are reduced in peripheral vascular complications of type 2 diabetes mellitus. J Am Coll Cardiol 45:1449–1457
97. Kusuyama T, Omura T, Nishiya D, Enomoto S, Matsumoto R, Takeuchi K, Yoshikawa J, Yoshiyama M (2006) Effects of treatment for diabetes mellitus on circulating vascular progenitor cells. J Pharmacol Sci 102:96–102
98. Lee IG, Chae SL, Kim JC (2006) Involvement of circulating endothelial progenitor cells and vasculogenic factors in the pathogenesis of diabetic retinopathy. Eye (Lond) 20:546–552
99. Brunner S, Schernthaner GH, Satler M, Elhenicky M, Hoellerl F, Schmid-Kubista KE, Zeiler F, Binder S, Schernthaner G (2009) Correlation of different circulating endothelial progenitor cells to stages of diabetic retinopathy: first in vivo data. Invest Ophthalmol Vis Sci 50:392–398
100. Otani A, Kinder K, Ewalt K, Otero FJ, Schimmel P, Friedlander M (2002) Bone marrow-derived stem cells target retinal astrocytes and can promote or inhibit retinal angiogenesis. Nat Med 8:1004–1010
101. Dorrell MI, Otani A, Aguilar E, Moreno SK, Friedlander M (2004) Adult bone marrow-derived stem cells use R-cadherin to target sites of neovascularization in the developing retina. Blood 103:3420–3427
102. Dorrell MI, Friedlander M (2006) Mechanisms of endothelial cell guidance and vascular patterning in the developing mouse retina. Prog Retin Eye Res 25:277–295
103. Thill M, Strunnikova NV, Berna MJ, Gordiyenko N, Schmid K, Cousins SW, Thompson DJ, Csaky KG (2008) Late outgrowth endothelial progenitor cells in patients with age-related macular degeneration. Invest Ophthalmol Vis Sci 49:2696–2708
104. Livrea MA, Tesoriere L, Pintaudi AM, Calabrese A, Maggio A, Freisleben HJ, D'Arpa D, D'Anna R, Bongiorno A (1996) Oxidative stress and antioxidant status in beta-thalassemia major: iron overload and depletion of lipid-soluble antioxidants. Blood 88:3608–3614
105. Van Beem RT, Nur E, Zwaginga JJ, Landburg PP, van Beers EJ, Duits AJ, Brandjes DP, Lommerse I, de Boer HC, van der Schoot CE, Schnog JJ, Biemond BJ, CURAMA Study

Group (2009) Elevated endothelial progenitor cells during painful sickle cell crisis. Exp Hematol 37:1054–1059
106. Boyle JP, Thompson TJ, Gregg EW, Barker LE, Williamson DF (2010) Projection of the year 2050 burden of diabetes in the US adult population: dynamic modeling of incidence, mortality, and prediabetes prevalence. Popul Health Metr 8:29
107. Cardoso CR, Saller GF (2008) Predictors of development and progression of microvascular complications in a cholort of Brazilian type 2 diabetic patients. J Diabetes Complications 22:167–170
108. Conway BN, Miller RG, Klein R, Orchard TJ (2009) Prediction of prolilferative diabetic retinopathy with hemoglobin level. Arch Ophthalmol 127:1494–1499
109. Massin P, Lange C, Tichet J, Vol S, Erginay A, Cailleau M, Eschwège E, Balkau B, DESIR (Data From an Epidemiological Study on the Insulin Resistance Syndrome) Study Group (2011) Hemoglobin A1c and fasting plasma glucose levels as predictors of retinopathy at 10 years: the French DESIR study. Arch Ophthalmol 129:188–195
110. Heydari B, Yaghoubi G, Yaghoubi MA, Miri MR (2012) Prevalence and risk factors for diabetic retinopathy: an Iranian eye study. Eur J Ophthalmol 22:393–397
111. Hadjadj S, Duly-Bouhanick B, Bekherraz A, Br Idoux F, Gallois Y, Mauco G, Ebran J, Marre M (2004) Serum triglycerides are a predictive factor for the development and the progression of renal and retinal complications in patients with type 1 diabetes. Diabetes Metab 30:43–51
112. Simo R, Hernandez C (2012) Prevention and treatment of diabetic retinopathy: evidence from large. Randomized trials. The emerging role of fenofibrate. Rev Recent Clin Trials 7:71–80
113. Patel SA, King CC, Lim PK, Habiba U, Dave M, Porecha R, Rameshwar P (2010) Personalizing stem cell research and therapy: the arduous road ahead or missed opportunity? Curr Pharmacogenomics Person Med 8:25–36
114. Tibbetts MD, Samuel MA, Chang TS, Ho AC (2012) Stem cell therapy for retinal disease. Curr Opin Ophthalmol 23:226–234
115. Masui S, Nakatake Y, Toyooka Y, Shimosato D, Yagi R, Takahashi K, Okochi H, Okuda A, Matoba R, Sharov AA, Ko MS, Niwa H (2007) Pluripotency governed by Sox2 via regulation of Oct3/4 expression in mouse embryonic stem cells. Nat Cell Biol 9:625–635
116. Nakatake Y, Fukui N, Iwamatsu Y, Masui S, Takahashi K, Yagi R, Yagi K, Miyazaki J, Matoba R, Ko MS, Niwa H (2006) Klf4 cooperates with Oct3/4 and Sox2 to activate the Lefty1 core promoter in embryonic stem cells. Mol Cell Biol 26:7772–7782
117. Seeger FH, Haendeler J, Walter DH, Rochwalsky U, Reinhold J, Urbich C, Rössig L, Corbaz A, Chvatchko Y, Zeiher AM, Dimmeler S (2005) p38 mitogen-activated protein kinase downregulates endothelial progenitor cells. Circulation 111:1184–1191
118. Zhao T, Zhang ZN, Rong Z, Xu Y (2012) Immunogenicity of induced pluripotent stem cells. Nature 474:212–215
119. Kalka C, Masuda H, Takahashi T, Kalka-Moll WM, Silver M, Kearney M, Li T, Isner JM, Asahara T (2000) Transplantation of ex vivo expanded endothelial progenitor cells for therapeutic neovascularization. Proc Natl Acad Sci U S A 97:3422–3427
120. Senegaglia AC, Barboza LA, Dallagiovanna B, Aita CA, Hansen P, Rebelatto CL, Aguiar AM, Miyague NI, Shigunov P, Barchiki F, Correa A, Olandoski M, Krieger MA, Brofman PR (2010) Are purified or expanded cord blood-derived CD133+ cells better at improving cardiac function? Exp Biol Med (Maywood) 235:119–129
121. Tomita S, Li RK, Weisel RD, Mickle DA, Kim EJ, Sakai T, Jia ZQ (1990) Autologous transplantation of bone marrow cells improves damaged heart function. Circulation 100:II247–II256
122. Strauer BE, Brehm M, Zeus T, Gattermann N, Hernandez A, Sorg RV, Kögler G, Wernet P (2001) Intracoronary, human autologous stem cell transplantation for myocardial regeneration following myocardial infarction. Dtsch Med Wochenschr 126:932–938
123. Emerson CP, Geng YT, James TW (2003) Adults stem cell therapy in perspective. Circulation 107:935–938

124. Jarajapu YP, Caballero S, Verma A, Nakagawa T, Lo MC, Li Q, Grant MB (2011) Blockade of NADPH oxidase restores vasoreparative function in diabetic CD34+ cells. Invest Ophthalmol Vis Sci 52:5093–5104
125. Sasaki K, Heeschen C, Aicher A, Ziebart T, Honold J, Urbich C, Rossig L, Koehl U, Koyanagi M, Mohamed A, Brandes RP, Martin H, Zeiher AM, Dimmeler S (2006) Ex vivo pretreatment of bone marrow mononuclear cells with endothelial NO synthase enhancer AVE9488 enhances their functional activity for cell therapy. Proc Natl Acad Sci U S A 103:14537–14541
126. Li Calzi S, Kent DL, Chang KH, Padgett KR, Afzal A, Chandra SB, Caballero S, English D, Garlington W, Hiscott PS, Sheridan CM, Grant MB, Forder JR (2009) Labeling of stem cells with monocrystalline iron oxide for tracking and localization by magnetic resonance imaging. Microvasc Res 78:132–139
127. Yao Y, Li Y, Ma G, Liu N, Ju S, Jin J, Chen Z, Shen C, Teng G (2011) In vivo magnetic resonance imaging of injected endothelial progenitor cells after myocardial infarction in rats. Mol Imaging Biol 13:303–313

Oxidative Stress and Apoptotic Biomarkers in Diabetic Retinopathy

Mohamed Al-Shabrawey and Sylvia Smith

Abstract Diabetic retinopathy (DR) is the foremost cause of blindness in working-aged worldwide; it is characterized by vascular and neural degeneration. Features of DR include leukocyte adhesion, increased vascular permeability, neovascularization and neuronal cell death. Early diagnosis and intervention are important to prevent or at least ameliorate the development of DR. Recent reports indicate that pathophysiological mechanisms leading to diabetic retinopathy include oxidative stress and retina l cell death cascades. Circulating biomarkers of oxidative stress such as malondialdehyde (MDA), thiobarbituric acid reacting substances (TBARS), conjugated diene, advanced oxidation protein products (AOPP), protein carbonyl, 8-hydroxy-deoxyguanosin (8-OHdG), nitrotyrosine, and F(2) isoprostanes and pro-apoptosis molecules (caspase-3, Fas, and Bax) are associated with increased susceptibility to develop DR in diabetic subjects. Thus, identification of oxidative stress and cell death biomarkers in diabetic patients could be in favor of predicting, diagnosis, and prevention of DR, and to target for novel therapeutic interventions.

Keywords Diabetic Retinopathy • Oxidative stress • NADPH Oxidase • Apoptosis • Biomarkers

M. Al-Shabrawey (✉)
Oral Biology and Anatomy, College of Dental Medicine, Georgia Regents University (GRU), Augusta, GA 30912, USA

Ophthalmology and Vision Discovery Institute, Medical College of Georgia, GRU, Augusta, GA 30912, USA

Vascular Biology Center, Medical College of Georgia, GRU, Augusta, GA 30912, USA
e-mail: malshabrawey@gru.edu

S. Smith
Ophthalmology and Vision Discovery Institute, Medical College of Georgia, GRU, Augusta, GA 30912, USA

Cellular Biology and Anatomy, GRU, Medical College of Georgia, Augusta, GA 30912, USA
e-mail: sbsmith@gru.edu

M.S. Mozaffari (ed.), *New Strategies to Advance Pre/Diabetes Care: Integrative Approach by PPPM*, Advances in Predictive, Preventive and Personalised Medicine 3,
DOI 10.1007/978-94-007-5971-8_7,

Diabetic retinopathy (DR) is the most frequent cause of blindness among adults aged 20–74 years. It is characterized by neurodegenerative features associated with extensive vascular changes. Nearly all patients with type 1 diabetes and >60% of patients with type 2 diabetes have retinopathy during the first two decades of disease [1]. Hyperglycemia triggers endothelial cell dysfunction and increases leukostasis leading to breakdown of the blood retina barrier (BRB) and vascular leakage. This leakage results in diabetic macular edema, the most common cause of decreased visual acuity in diabetic patients. Later, capillary degeneration and ischemia develop which leads to uncontrolled neovascularization in attempt to compensate for the lack of blood flow [2, 3].The current treatment modalities include timely laser photocoagulation, vitrectomy and repeated intraocular injections of anti-VEGF or steroids, are invasive and limited by considerable side effects and they may prevent progression of DR but do not restore the visual acuity. Many patients have evidence of retinal subclinical inflammation at the time they are diagnosed with diabetes [4]. This means that pro-DR mechanisms arise early during the course of diabetes, although the nature of these mechanisms is not clearly evident. Hence, it is important to shift investigative efforts to focus on elucidating the mechanisms initiating and promoting the early changes in retinal function so that more effective therapeutic approaches may be developed with the goal of preventing (or at least minimizing) the development/progression of DR.

It remains unclear how the vascular and neuronal injury relate to each other and their net contribution to retinal damage. There are multiple biochemical pathways which contribute to the development of the neurovascular injury in DR. As a result, biomarkers which reflect different pathways are released locally and into the circulation. Early identification of these biomarkers could be in favor of predicting and efficient management of DR. Among these biomarkers are the ones related to inflammatory response, oxidative stress and retinal cell death. Diabetes increases oxidative stress, which plays a key regulatory role in the development of its complications [5, 6]. Hyperglycemia-induced reactive oxygen species (ROS) generation is considered a causal link between elevated glucose and the pathways of development of diabetic complications [7]. Oxidative stress may lead to cell death [8] via apoptotic means. Apoptosis of retinal neurons particularly ganglion cells [9–11] has been demonstrated in diabetic retinopathy as demonstrated by profound retinal abnormalities, evaluated by electroretinography, and potential visual changes evoked before the onset of the first vascular change is detectable in the diabetic retina [12, 13]. Retinal capillary cells also undergo accelerated apoptosis, which precedes the detection of any histopathological changes characteristic of diabetic retinopathy [14]. The current chapter discusses the markers of oxidative stress and retinal cell death associated with DR.

1 Oxidative Stress and Diabetic Retinopathy

Chronic hyperglycemia plays a crucial role in the pathogenesis of diabetic retinopathy (DR). The mechanism of hyperglycemia-induced retinal damage is still to be elucidated. However, the oxidative stress which represents an imbalance between excess generation and/or impaired removal of reactive oxygen species (ROS) have been suggested to be the key events in the pathogenesis of diabetic retinopathy. The high content of polyunsaturated fatty acids, oxygen uptake and glucose oxidation make the retina more susceptible to oxidative stress than other tissues [15]. ROS are produced continuously in all cells to support normal cellular functions. However, excess production of ROS, or inefficient removal of ROS, could result in pathological conditions. In addition to ROS, reactive nitrogen species [16] are also a part of normal physiological function, and have great potential to contribute to oxidative stress [17]. In the presence of superoxide, nitric oxide spontaneously forms peroxynitrite. Peroxynitrite is much more reactive than superoxide and nitric oxide and can exert direct oxidative modifications through one- or two-electron oxidation processes [18, 19]. Thus, excessive abundance of ROS and RNS with concurrent dysfunction of antioxidant defense systems, which includes reducing enzymes such as superoxide dismutase (SOD), catalase and glutathione peroxidase (GSH) contributes to oxidative stress in diabetic retina. Chronic oxidative stress causes damage to DNA, lipids, proteins, and carbohydrates and disruption in cellular homeostasis resulting in many disease processes of clinical interest [20]. Accumulation of damaged molecules and ROS that are not easily removed contributes not only to the pathogenesis of DR but also to the resistance of retinopathy to reverse even after good glycemic control is reestablished – the metabolic memory phenomenon [21].

1.1 Sources of ROS in Diabetic Retina

Chronic exposure of retinal cells to hyperglycemia causes overproduction of ROS by activation of different enzymatic pathways which more likely interact to create the retinal damage seen in DR (Fig. 1). These sources include but not limited to NADPH oxidase, mitochondrial electron transport chain (ETC), formation of advanced glycation end products (AGEs), aldose reductase/polyol pathway, protein kinas C (PKC) activation and uncoupling of endothelial nitric oxide synthase (eNOS).

1.1.1 NADPH Oxidase

Recent studies demonstrated that NADPH oxidase is a major source of ROS in diabetic patients and animals and retinal endothelial cells treated with high glucose [22–25]. ROS production by NADPH oxidase has been linked to the early signs of DR, including increases in ROS generation, vascular endothelial growth factor (VEGF) expression, leukostasis and vascular permeability [22, 26]. Additionally, increased NADPH oxidase activity and expression of its catalytic subunit NOX2 has been correlated to increased VEGF expression and retinal neovascularization in a mouse model for ischemic retinopathy [27]. Phagocyte NADPH oxidase consists of two membranous subunits, large catalytic subunit gp91phox (NOX2) and small

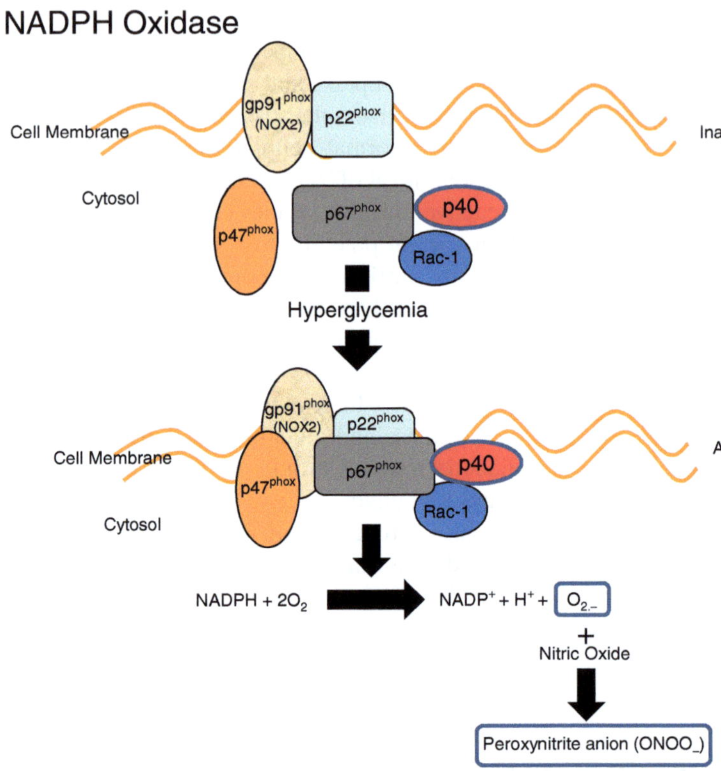

Fig. 1 NADPH oxidase is a major contributor to oxidative stress in diabetes. NADPH oxidase consists of two membranous subunits, large catalytic subunit gp91phox (NOX2) and small p22phox as well as cytosolic regulatory subunits, p40 phox, p47phox and p67phox, and the low molecular weight G protein Rac-1. Hyperglycemia causes phosphorylation of p47phox which in turn leads to the assembly of the complex and binding of the complex with activated Rac-1, both are critical for NADPH oxidase activation. Superoxide generation by the activated NADPH oxidase can further cause more oxidative stress by binding nitric oxide (NO) to form peroxynitrite ($ONOO^-$)

p22phox as well as cytosolic regulatory subunits, p40 phox, p47phox and p67phox, and the low molecular weight G protein Rac-1 [28, 29]. Phosphorylation of p47phox is crucial for the assembly of the complex [30], while P67phox subunit mediates binding of the complex with activated Rac [31], both are critical for NADPH oxidase activation. Vascular endothelial cells express the same subunits as well as two NOX2 homologues, NOX1 and NOX4 [25]. Previous studies in animal and tissue culture models have shown that NOX2 is expressed at low levels in normal retinas and in retinal endothelial cells maintained under control conditions, but is substantially increased in retinal vessels of animals with diabetic or ischemic retinopathy and in retinal endothelial cells exposed to high glucose or hypoxia [22, 27]. During diabetes other cells in addition to endothelial cells become activated and become sources of NADPH oxidase derived ROS formation. For example, leukocytes are suggested to regulate retinal edema, ischemia and angiogenesis [32] and shown to be activated [33–35] and to become source of oxidative stress [36, 37].

1.1.2 Mitochondrial Electron Transport Chain (ETC)

Mitochondria are the principal endogenous source of superoxide. Hyperglycemia disrupts the normal ETC electron flow leading to electron leak and superoxide generation which induces mutations in mitochondrial DNA resulting in defective subunits of the electron transport complexes causing increased superoxide production at physiological concentrations of glucose [6, 38]. Moreover, mitochondrial ROS suppresses the antioxidant defense mechanisms through inhibition of SOD and GSH leading to enhanced sensitivity of retinal cells to oxidative stress since they cannot scavenge ROS effectively [39–41]. The critical role of mitochondrial superoxide in DR has been demonstrated by recent studies using transgenic mice that overexpress mitochondrial SOD. Overexpression of mitochondrial SOD protected the diabetic retina from oxidative damage and capillary degeneration [41]. This indicates that hyperglycemia-induced oxidative stress can modulate mitochondria function, thus contributing to the development of DR.

1.1.3 Endothelial Nitric Oxide Synthase (eNOS)

eNOS uncoupling is an additional source of superoxide generation in diabetic retina. This occurs, when there is limited amount of its substrate L-arginine or its co-factor tetrahydrobiopterin. Uncoupled eNOS use molecular oxygen as a substrate producing superoxide instead of nitric oxide (NO) [42–44]. Several studies demonstrated the role of eNOS uncoupling in vascular dysfunctions associated with diabetes which reversed by supplemental L-arginine treatment [45, 46]. The role of eNOS in DR has been also demonstrated in human as certain eNOS polymorphism are associated with DR [47].

1.1.4 Polyol Pathway

The polyol pathway involves the conversion of glucose into sorbitol by aldose reductase. Sorbitol is then oxidized to form fructose by sorbitol dehydrogenase. Increased polyol pathway in diabetes causes depletion of NADPH. Since NADPH is required for regenerating the intracellular antioxidant, GSH, increased polyol pathway in diabetes enhances oxidative stress [48]. The role of this pathway in the development of DR has been reported in animal model of experimental diabetes and high glucose-treated endothelial cells [49, 50].

1.1.5 Advanced Glycation End Products

The AGEs are produced from strong glycating dicarbonyl compounds such as methylglyoxal and glyoxal [51]. Chronic hyperglycemia favors glycation reactions and nonenzymatic glycation leading to cross-linking of proteins and in turn alterations in their function, activity, and physical properties. Diabetes, increases production of

AGE and its receptor, RAGE, in the retinal microvasculature which becomes irreversible in the late stages of DR [52]. Interaction of AGEs with RAGE activates PKC-δ and subsequently NADPH oxidase to enhance oxidative stress in diabetic retina [53]. ROS generation by AGEs leads to the activation of nuclear transcriptional factor, NF-kB, [54] which has been linked to retinal capillary cell apoptosis [55] vascular inflammation [56, 57] and neovascularization [58, 59].

1.1.6 Protein Kinase C Activation

Activation of PKC is also considered as a major pathway implicated in the pathogenesis of diabetic retinopathy [60, 61]. High glucose levels increase the release of ROS and the synthesis of diacylglycerol (DAG) increasing the activity of PKC [62]. Activated PKC has been implicated in changes characteristic of diabetic retinopathy such as increased vascular permeability, neovascularization, endothelial proliferation and apoptosis, and regulating the action of VEGF [63–65]. There are cumulative evidences supporting the role of PKC in ROS-mediated diabetic complications. For example, diabetes-induced oxidative stress has been shown to be prevented by PKCβ specific inhibitor (LY53331), and in mice lacking PKCβ isoform [66–68].

1.2 Oxidative Stress and Vascular Changes in Diabetic Retinopathy

Diabetic retinopathy is associated with alterations in the structure and function of retinal microvasculature. For example, basement membrane thickening, pericyte and endothelial cell loss, acellular capillary formation [69, 70], microaneurysm, altered blood flow, leukocyte adhesion, hyperpermeability and neovascularization. Many of these changes have been linked to the oxidative stress-induced biochemical changes. ROS mediate these changes by both direct and indirect mechanisms which include increases in the transcriptional activation of VEGF expression [71], inflammatory mediators [4, 72], AGEs formation on collagen which results in basement membrane thickening [73] and modulation of vasoactive effector molecules such as endothelin-1 and NOS which contribute to changes in blood flow and vascular permeability [74, 75]. VEGF, a proangiogenic factor, is known to play a pivotal role in the development of vascular changes associated with diabetic retinopathy [76, 77]. Hyperglycemia-induced oxidative stress plays a role in mediating both the expression and pathological effect of VEGF [22, 78, 79]. For example, our subsequent studies demonstrated the implication of NADPH oxidase-derived ROS in the development of vascular dysfunctions in ischemic retinopathy. We reported abrogation in ROS formation, retinal expression of VEGF and new vessel formation in mouse model of ischemic retinopathy when treated by NADPH oxidase inhibitor [27]. Further studies on experimental diabetes demonstrated decreased ROS formation, VEGF expression leukostasis, and vascular permeability by NADPH oxidase inhibitor or deletion of its catalytic subunit, NOX2, [26]. Other studies have reported the beneficial effect of antioxidants such as N-acetylcysteine in preventing pericyte loss and endothelial/perivascular cell changes occur early in the pathogenesis of DR [80]. Moreover, PEDF, an essential antiangiogenic factor that produced by different retinal cells has been reported to prevent neuronal derangements, vascular hyperpermeability and leukostasis and VEGF expression in diabetic retinopathy via inhibition of ROS formation [81]. Taken together, these findings confirm the crucial role of ROS in the development of vascular dysfunction in DR.

1.3 Markers of Oxidative Stress

Oxidative stress induces biochemical changes in the serum and target organs of diabetic patients including severe lipid peroxidation, protein oxidation, and oxidative DNA damage and changes in their antioxidant status (Fig. 2). Thus, identification of the by-products induced by oxidative stress in serum, vitreous body, aqueous humor, fibrovascular tissue on the retinal surfaces and urine of diabetic patients could be in favor of predicting, diagnosis and prevention of microvascular complications of diabetes such as retinopathy and nephropathy. Oxidative stress parameters include malondialdehyde (MDA), thiobarbituric acid reacting substances (TBARS), conjugated diene (CD) [82], advanced oxidation protein products (AOPP), protein carbonyl, 8-hydroxydeoxyguanosin (8-OHdG), nitrotyrosine, and F(2) isoprostanes. Several studies demonstrated a marked increase in these products in serum/retinas of diabetic patients or experimental animals and significantly higher in patients or animals with DR than in diabetic patients/animals without DR. Additionally, amelioration of the antioxidant status plays a crucial role in development of DR.

1.3.1 Lipid Peroxidation

Lipid peroxidation of cellular structures, an oxidative stress-induced activity, is thought to play an important role in complications of diabetes mellitus, especially in diabetic retinopathy. Measurement of TBARS, MDA or CD provides a convenient assessment of lipid hydroperoxide (LHP) concentration in serum. CD is the initial formation of lipid peroxide while MDA is a decomposition product of peroxidised polyunsaturated fatty acids. The elevated levels of Poly Unsaturated Fatty Acids (PUFAs) and increased rate of glucose oxidation in diabetes make the retina susceptible for oxidative stress and lipid peroxidation. Increased lipid peroxides has been demonstrated in serum of diabetic patients with vascular complications [83–90]. Lipid peroxidation increases also in retinas of diabetic animals and mouse model of ischemic retinopathy [22, 27]. Compared with healthy subjects, diabetic patients demonstrate significantly higher levels of lipid peroxidation products [91–95]. Additionally, patients with features of DR are found to have much more lipid

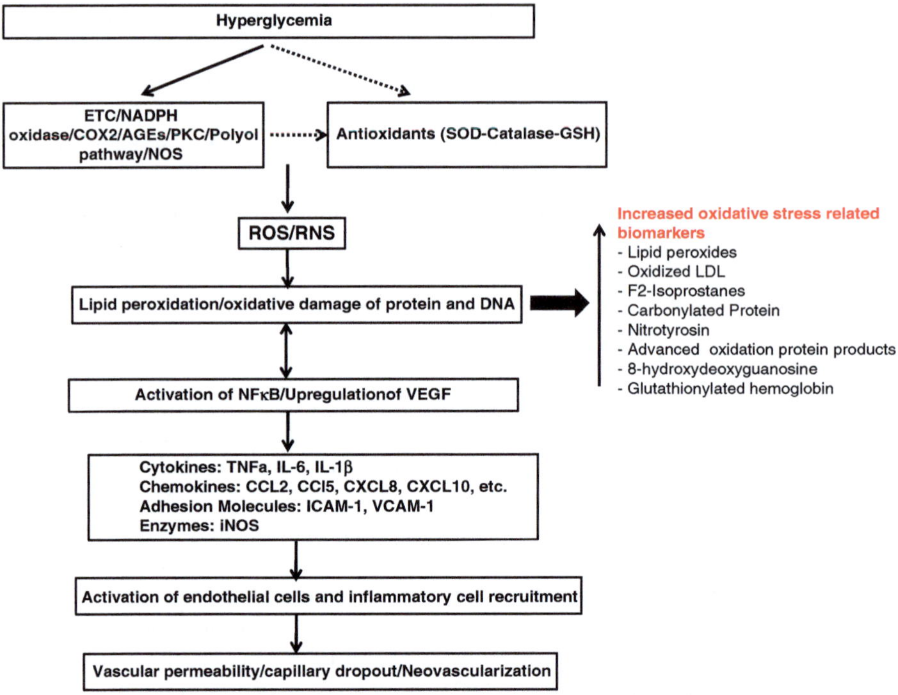

Fig. 2 Cascade of events that contribute to the oxidative stress and development of vascular injury in diabetic retinopathy. Hyperglycemia induces generation of reactive oxygen species (ROS) and reactive nitrogen species (RNS) via multiple sources, mainly nicotinamide adenine dinucleotide phosphate oxidase (NADPH) and mitochondrial electron transport chain (ETC.) and others such as cycloxygenase (COX2), advanced glycation end products (AGEs), protein kinase C (PKC), polyol pathway and nitric oxide synthase. These changes cause lipid peroxidation and oxidative damage of protein and DNA resulting in production of biomarkers such as lipid peroxides and nitrotyrosin. Lipid and protein modification by hyperglycemia-induced oxidative stress leads to upregulation of VEGF and activation of NFκB and in turn upregulation of cytokines, chemokines and iNOS. This leads to upregulation of adhesion molecules and subsequent leukocyte/endothelial interaction, hyperpermeability, capillary dropout and neovascularization

peroxidation products compared with those without DR [90, 91]. Studies by Armstrong et al. demonstrated a statistically significant correlation between the increasing serum lipid peroxidation levels and prevalence of retinopathy in diabetic patient [96] and reduction in retinal cell number in streptozotocin (STZ)-induced diabetic rat model [97]. Similar results have observed by Augustin et al. [98, 99] in patients with PDR where they demonstrated a significant increase in lipid peroxidation in the vitreous and epiretinal membranes.

In addition to the previous markers of lipid peroxidation several studies have demonstrated the role of oxidative stress-induced modification of low density lipoprotein in the pathogenesis of DR [100–102]. Fredrikson et al. [103] reported that patients affected by retinopathy had significantly higher levels of autoantibodies against apolipoprotein B peptides than patients without retinopathy. Thus, autoantibodies against oxidized LDL antigens might also reflect disease severity and the risk of developing vascular complication of diabetes including DR.

A major step regarding the measurement of lipid peroxidation products came with the discovery of F_2-isoprostanes [104, 105]. The F_2-isoprostanes are a family of PG (prostaglandin)-like compounds produced by non-enzymatic peroxidation of arachidonic acid. Of these, 8-epi-PGF_2^α (also known as 8-iso-PGF_2^α or 15-F_{2t}-isoprostane) has received most attention because it has been shown to possess certain adverse biological activities [106–109]. Circulating 8-epi-PGF_2^α is mainly present bound to phospholipids in situ and is released by the action of phospholipase A_2 [110] and has proved a reliable marker of oxidative stress in human body. However, the quantification of urinary 8-epi-PGF_2^α has been proposed as being superior to the measurement of circulating 8-epi-PGF_2^α levels, as it is believed to represent a more accurate systemic index of oxidative stress [111]. In their study on human retinas obtained from healthy donors Nourooz-Zadeh and Pereira provided the first direct evidence for free-radical-catalysed peroxidation of AA in retinal tissue by detecting esterified F(2)-IPs in human retina. Thus they proposed analysis of F(2)-IPs for studying the implications of oxidative injury in various diseases of retina, including diabetic retinopathy and age-related macular degeneration [112].

1.3.2 Biomarkers of DNA and Protein Oxidative Damage

8-Hydroxydeoxyguanosine (8-OHdG)

ROS can cause DNA damage and base modifications. 8-OHdG, an oxidized nucleoside of DNA, results from oxidation of guanine residues and is frequently used as DNA lesion marker. Higher levels of 8-OHdG were observed in the mononuclear cells, urine, pancreatic islet and mitochondrial DNA from diabetic patients [113–117] and in kidney tissues of streptozotocin-induced diabetic rats [118]. Moreover, the content of 8-OHdG in the urine, serum and mononuclear cells of patients with type 2 diabetes with either retinopathy or nephropathy were much higher than those in patients without complications [91, 119, 120]. Therefore, the contents of 8-OHdG in serum could act as a sensitive biomarker for the diabetic retinopathy (Fig. 2).

Carbonylated Proteins

Proteins can also be modified and become cytotoxic by ROS and RNS, thus contributing to the development of DR. There is association between the elevated biomarkers of protein oxidation with aging and different diseases such as Alzheimer's, diabetes mellitus and cancer [91, 121, 122]. Protein carbonylation and nitration have attracted a great deal of attention due to its irreversible and unrepairable nature which may serve as an important biomarker in DR.

Carbonylated proteins are marked for proteolysis by the proteasome but can escape degradation and form high-molecular-weight aggregates that accumulate with age. Such carbonylated aggregates can become cytotoxic and have been associated with a large number of diseases and age-related disorders, including Parkinson's disease, Alzheimer's disease, cancer, cataractogenesis, diabetes, and sepsis [123, 124], for review see [125]. The implication of protein carbonylation in reflecting the severity of diabetes and its complication has been investigated by many researchers [91, 126, 127]. There is a positive correlation between protein carbonylation and the clinical severity of DR. For example, compared to healthy individuals diabetic patients have significantly increased levels of carbonylated proteins. Furthermore, patients with DR show higher level of protein carbonylation more than in diabetic patients without DR [91, 128].

Protein Nitration

Nitrotyrosin is additional marker for oxidative damage of protein that disrupts its normal functions. Several studies have demonstrated that diabetes increases protein nitration in retinas of experimental animals [22, 129, 130]. Nitrotyrosine is an oxidative protein modification which formed from the reaction of free or protein-bound tyrosine with reactive nitrogen species such as free radical nitrogen dioxide [131] and peroxynitrite [132] and has been used as a marker of oxidative stress in retina of experimental animals [22, 129]. Although tyrosine nitration has been associated with development of diabetic and ischemic retinopathies little is known about the targets of tyrosine nitration in retina and the role nitration may play in the pathophysiology of diabetic retinopathy. Recently, Zhan et al. identified some proteins that are nitrated in diabetic rat retina and in rat retinal Müller cells grown in high glucose which could be linked to retinal neovascularization[130]. Additionally, subsequent studies by El-Remessy et al. have shown that peroxynitrite-induced tyrosine nitration correlates with accelerated retinal endothelial cell death, breakdown of the brain-retinal barrier, and accelerated neuronal cell death in models of experimental diabetes and neurotoxicity and human diabetes [129, 133–135]. Recent reports demonstrated that accumulation of nitrosylated protein in peripheral blood monocyte of diabetic patients is an emerging marker that reflects oxidative stress-induced protein damage. [136, 137]. Julius et al. 2009 reported that diabetic subjects accumulate significant amount of nitrosylated protein in peripheral blood monocyte and this is positively correlated with the severity of diabetic microangiopathy such as nephropathy and retinopathy [137]. Thus, targeting tyrosine nitration represents potentially effective therapeutic strategy in attenuating retinal vascular neurodegenerative diseases.

Advanced Oxidation Protein Products (AOPP)

Serum level of AOPP is also a novel protein oxidation marker which has been described by different investigators [138–141]. They are elevated in patients with renal insufficiency and diabetes mellitus [142–144]. Additionally, the level AOPP in diabetic patients with features of DR was reported to be significantly higher than in diabetic patients without DR [91, 141].

In addition to the previous markers of oxidative damage of protein and DNA in diabetic patient, mass spectrometric quantification of amino acid oxidation products and glutathionylated hemoglobin has been proposed to serve as markers for the oxidative stress in diabetic patients, in particular those with complications such as DR [145, 146].

1.3.3 Antioxidant Status

The antioxidant status of diabetic patient plays a crucial role in producing oxidative stress and development of vascular complications. ROS generation is inhibited by an intracelluar enzymatic system composed mainly of glutathione peroxidase (GPx), SOD and catalase (CAT). All serum antioxidants are usually measured together as total antioxidant status (TAS). The reports about the status of antioxidants and antioxidant enzymes in diabetic patients are very contradictory, both increases and decreases of antioxidant activity have been reported [147–150]. Generally, TAS has been shown to be significantly lower in patients with proliferative retinopathy than in diabetics not developing retinopathy [151]. Antioxidants SOD and vitamin C decrease with the progression of the disease, however GPx tends to increase in the later part of the disease [152]. Kesavulu et al., in their study they found that among the erythrocyte antioxidant enzymes, CAT activity was significantly increased in all the diabetics compared to controls, however, there was no difference in the CAT activity between the diabetic patients with or without microvascular complications (MVC) such as diabetic retinopathy or nephropathy. In contrast, GPx activity was significantly decreased in the diabetic patients compared to controls, and further decreased in the diabetics with MVC compared to those without the complications. The SOD activity in diabetic patients was not much different from that of non diabetic controls, but a significant decrease in SOD activity was observed in diabetic patients with MVC compared to those without these complications [90]. The total thiol level as a marker of antioxidant status in diabetics has also shown to be significantly decreased in patient with DR than those without DR [141]. An inverse relation between the level of GSH and HbA1c was also reported in diabetic patients indicating reduction in the antioxidant status in poorly controlled diabetics [146]. Additionally, nonenzymic antioxidants such as vitamin C, vitamin E, and β-carotene that exist biologically for the regulation of redox homeostasis are also depressed by diabetes [153]. Therefore, the level and activity of specific antioxidant enzymes compared to other enzymes that do not match to development of DR could be predictive for the susceptibility to develop DR in diabetic patients.

1.3.4 Oxidative Stress Related Gene Polymorphism

There are several studies showing that vascular complications of diabetes develop only in genetically susceptible patients [154–156] and hyperglycemia by itself is not sufficient for the development of diabetic vascular disease. There is tendency to have familial clustering of diabetic nephropathy and coronary vascular disease (CVD). Additionally, there is a significant ethnic difference in the incidence of CVD in subjects with diabetes that could not be explained by differences in conventional cardiovascular risk factors (UK Prospective Diabetes Study Group 1998). Various candidate genes that affect the oxidative stress such as manganese mitochondrial superoxide dismutase (Mn-SOD), endothelial nitric oxide synthase (eNOS), angiotensin-converting enzyme, glutathione peroxidase, superoxide dismutase and haptoglobin demonstrate functional polymorphisms which correlated with the increased susceptibility to vascular diseases and diabetes [157–161].

Manganese Superoxide Dismutase (MnSOD) Polymorphism

Hovnik et al. in 2009 investigated polymorphic markers in genes encoding enzymes regulating production of reactive oxygen species in association with diabetic retinopathy (DR) or diabetic nephropathy (DN) in 124 patients with type 1 diabetes and matched for sex, age and duration of diabetes. A positive association of MnSOD genotype Val/Val with diabetic retinopathy but not with diabetic nephropathy was demonstrated [162]. Moreover, the association of diabetic retinopathy with single nucleotide polymorphisms (SNPs) on MnSOD was investigated and a statistically significant association of MnSOD Ala16Val polymorphism with diabetic retinopathy was found compared with the diabetic control. Additionally, the Ala(−9)Val [161] and VV genotype of the V16A [160] polymorphism of the Mn-SOD gene are also reported to be significantly higher in patients with DR retinopathy in Chinese and Caucasians type 2 diabetic patients respectively.

Haptoglobin Polymorphism

Haptoglobin (Hp) is a hemoglobin-binding serum protein. It is found at levels of 30–300 mg/dL in normal human serum [163]. However, serum level of Hp shows three- to eight-fold increase during the acute phase reaction and in response to injury [164]. Free hemoglobin [165] is considered as an extremely potent pro-oxidant agent which catalyzes various oxidative and peroxidative reactions [166]. Haptoglobin (Hp) plays an essential role in capturing free Hb, thus preventing oxidative tissue damage mediated by free Hb. For example, Hp has been demonstrated to inhibit Hb-induced linolenic acid and low-density lipoprotein (LDL) oxidation [167]. Additionally, Hp knockout mice have shown to be more prone to oxidative tissue damage such as renal damage and endothelial dysfunction [168]. Recent reports demonstrated that a polymorphism in the haptoglobin (Hp) gene is an independent risk factor for diabetic vascular disease [158, 159]. In man, there are two alleles for Hp, denoted 1 and 2, giving rise to three major phenotypes. Individuals homozygous for the 1 allele express the Hp1-1 phenotype at the protein level. Individuals homozygous for the 2 allele express the Hp2-2 phenotype, whereas heterozygotes express the Hp2-1 phenotype. Nakhoul et al. in 2000 have observed a marked decrease in the prevalence of diabetic retinopathy in patients with Hp1-1 phenotype compared with patients with the Hp2-1 and Hp2-2 phenotypes [165]. Hp1-1 and Hp1-2 types have been found to have significantly higher levels of serum vitamin C [169], lower serum iron, and lower circulating oxidized LDL (oxLDL)/LDL ratios compared with those with the Hp2-2 [170, 171]. Thus, antioxidant therapy may be more beneficial in diabetic patients with the Hp2-2 phenotype as compared with those with the Hp1-1 phenotype. Furthermore, a highly significant increase in retinal capillary basement membrane thickness was observed in mice with streptozotocin-induced diabetes with the Hp 2 genotype. These data provide important support for association studies done in humans showing an increased prevalence of diabetic retinopathy in individuals with the Hp 2 genotype [158, 159].

eNOS Gene Polymorphism

Polymorphisms of eNOS gene have been reported to be associated with multiple health conditions including DR, hypertension, nephropathy, and cardiovascular diseases in several ethnic groups. Several studies demonstrated the association between certain eNOS polymorphism with DR in human [47, 172, 173]. For example, the eNOS 4b/b [47] or 4a/b [172] polymorphism demonstrate increased eNOS expression and activity and associated with sever DR. On the other hand, homozygous deletion of eNOS 4a/a causes reduction in eNOS expression and activity and susceptibility to DR [174].

1.3.5 Inflammation Related Markers

In addition to oxidative stress related biomarkers, generation of ROS in diabetic patients is known to initiate a cascade of signaling pathways responsible for the inflammatory component of DR. This includes activation of NFκB and up-regulation of genes responsive to inflammatory processes such as adhesion molecules [26, 57, 175, 176]. Golubnitschaja et al. in 2005 have studied the alteration in inflammatory related genes in mononuclear blood cells (MBCs) isolated from diabetic patients with and without active retinopathy. They found significant increase in the expression of recoverin in diabetic patients compared to the control group. "Real-Time"-PCR showed that recoverin levels were positively correlated with the stage of DR and with the activity of serum matrix metalloprotinases (MMPs) [177]. Thus, identification of inflammation related biomarkers in isolated MBCs and serum of diabetic subjects may be helpful for the prediction of active proliferative retinopathy in diabetics.

2 Neuronal Death and Diabetic Retinopathy

The preceding section of this chapter describes the complex vascular changes that accompany diabetic retinopathy. For many years, the disease was considered primarily vascular in nature. Beginning in the late 1990s, however, information about the neuronal components of diabetic retinopathy emerged. Several excellent reviews published in the last decade have summarized the evidence that neurons, particularly retinal ganglion cells, are vulnerable and die in diabetic retinopathy [9–11, 178, 179]. Loss of color [180, 181] and contrast [182, 183] sensitivity are early signs of neural retinal dysfunction in humans that occur within 2 years of diabetes onset. Bresnick proposed that diabetic retinopathy is a primary neurosensory disorder that precedes vasculopathy by many years [12, 184]. It is noteworthy that as early as the 1960's Wolter [185] and Bloodworth [186] described loss of retinal neurons in diabetic patients. However, owing to the more easily recognizable vascular features as viewed via ophthalmoscopy, these observations were largely overlooked. More sophisticated electrophysiological studies using macular or focal ERG [82], which can detect electrical responses of ganglion cells, showed ganglion cell dysfunction early in diabetes [187]. Though FERG is not used widely clinically, important information has been forthcoming from several studies. Greco et al. [188] used FERG in prepubescent diabetic children and found that ganglion cell function was significantly impaired, while other cells such as photoreceptors appeared unaffected. Others have reported similar findings, including ganglion cell dysfunction in IDDM patients without vasculopathy [189, 190]. Other investigations using ERG, dark adaptation, contrast sensitivity, and color vision tests have provided convincing functional evidence that neural retina function is compromised oftentimes before the onset of vascular lesions in humans [191–196].

2.1 *Markers for Apoptosis in Human Tissues*

Biomarkers of neuronal death in diabetic retinopathy are summarized in Fig. 3. The relevance of this functional evidence of neuronal degeneration in diabetic retinopathy has been borne out by molecular and biochemical analyses of the retinas of human patients. Barber et al. used post-mortem tissues from human patients with diabetic retinopathy and reported for the first time that ganglion cells die in this disease and do so via an apoptotic mechanism [197]. Subsequent studies by Abu El-Asrar used eyes from human diabetic patients and confirmed the expression of several pro-apoptosis molecules (caspase-3, Fas, and Bax) emphasizing the vulnerability of these cells [198]. They also found strong immunoreactivity for ERK1/2 in the inner nuclear layer of diabetic retinas as well as the nerve fiber and ganglion cell layers. This group extended their findings by examining additional anti- and proapoptotic markers in humans and showed that ganglion cells in diabetic and nondiabetic retinas express the anti-apoptotic molecules Cox-2, Akt, and Mcl-1. Retinal ganglion cells express the proapoptotic molecule Bad in response to diabetes-induced neuronal injury; diabetic retinas also show upregulation of the mitochondrial proteins cytochrome c, and AIF [199]. Human tissue from diabetic patients has also been shown to express connective tissue growth factor (CTGF), CD105 and gelatinase B, and ganglion cells and microglia were the sites of immunoreactivity [200]. Other investigators have provided additional support that the death of

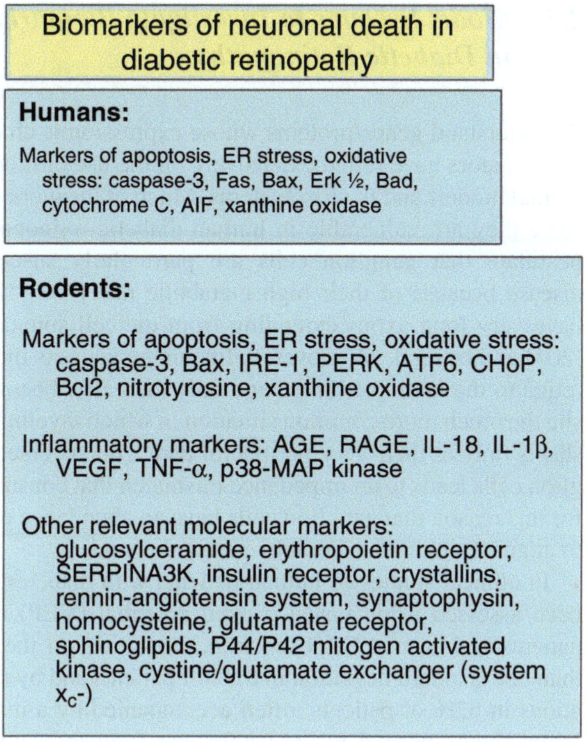

Fig. 3 Biomarkers of neuronal death in diabetic retinopathy. A number of studies performed using tissues from diabetic human patients and from rodent models of diabetic retinopathy reveal an alteration in expression of several genes and the proteins they encode including those involved in apoptosis, inflammation and other pathways. The data from rodent models form the foundation for the investigations in humans and to date several markers of ER stress and oxidative stress are observed in the human population. Whether they predict disease or are secondary is an area of intense investigation

neurons in human diabetic retinas is mediated via mitochondria- and caspase-dependent cell-death pathways [201].

These data collected from human tissues provide strong evidence that neurons die in diabetic retinopathy and that they do so via an apoptotic mechanism, however no comprehensive reports are available indicating specific markers that predict neuronal cell death in human subjects. By definition, biomarkers should reflect a change in expression or state of a protein that correlates with the risk or progression of disease, or with the susceptibility of the disease to a given treatment. The field of biomarker research is just emerging and the published reports on biomarkers for neuronal cell death, especially in humans with diabetic retinopathy, are limited. An additional consideration for any analysis of biomarkers for diabetic retinopathy is that non-diabetic retinal changes can occur in diabetic patients and may confound interpretation of studies. For example, the occurrence of chronic open angle glaucoma in patients with diabetes ranges from 4.5 to 6.5%, while the prevalence of age-related macular degeneration in patients with diabetes varies (ranging from 0.1 to 9.3%) [202]. Markers that are present in multiple diseases will not be as useful for distinguishing diabetic retinopathy as those that are specific.

2.2 Model Systems to Investigate Biomarkers of Neuronal Death in Diabetic Retinopathy

To understand genes/proteins whose expression is changed in diabetic retinopathy, investigators have relied extensively on the use of model systems including *in vivo* animal models and *in vitro* systems. Much of the work has focused on ganglion cells since they are vulnerable in human diabetic retinopathy. Schmidt and colleagues postulate that ganglion cells are particularly susceptible to neurodegenerative disease because of their high metabolic rate [203]. They note that ganglion cells have very long axons extending from the cell soma, through the lamina cribrosa [204] and beyond. Moreover, unlike many neurons they are unmyelinated from the retina to the lamina cribrosa and only thereafter become myelinated. This is opposite the much more common situation in which myelinated neurons lose their myelin sheets only at their very peripheral end. This reversed myelination pattern in ganglion cells leads to an impedance mismatch that consumes considerable energy. It is for this reason that ganglion cells have an abundance of mitochondria in their axons to augment their considerable energy demands.

In clinical studies of human patients with diabetes, optic nerve involvement has been assessed using visual evoked potential (VEP) responses [205]. While these patients had normal ERG function, the latency of the VEP was increased by more than one standard deviation in 81% of patients and by more than three standard deviations in 62% of patients, often accompanied by a marked reduction in amplitude. Interestingly, normal visual acuity was frequently noted in patients with severely slowed conduction reflecting subclinical affection of optic nerves in diabetes.

2.3 Studies in Diabetic Rats

Animal models of diabetic retinopathy include the well-characterized streptozotocin-induced diabetic rat. In rats, diabetes accelerates retinal neuronal cell death and inhibits neurite regeneration via an apoptotic mechanism that involves upregulation of Bax, a pro-apoptotic gene [206]. In the rat model, a number of inflammatory markers have been shown to be upregulated concomitant with ganglion cell death. These markers include interleukin (IL)-18 [207], IL-1β [129, 207], vascular endothelial growth factor (VEGF) [129], tumor necrosis factor-α (TNF-α) [129, 207], activated p38 MAP kinase [129]. Related to these observations are the data showing a significant ~30% increase in glucosylceramide levels in diabetic rats compared to controls [208]. These investigators were the first to demonstrate that augmented glycosphingolipid metabolism may contribute to the neuronal pathology of diabetic retinopathy and may serve as an important and relevant biomarker in human patients. Nakazawa et al. have shown that anti-inflammatory agents (statins) can prevent ganglion cell death in ischemic/reperfusion rat models [209]. Zhang and co-workers observed an increase in the receptor for erythropoietin, a glycoprotein hormone that controls erythropoiesis in diabetic rats [210]. Another interesting molecule whose expression is decreased in the diabetic rat retina is SERPINA3K, an extracellular serine proteinase inhibitor [211]. This observation is noteworthy because SERPINA3K prevents intracellular calcium overload and calcium is well known to trigger neuronal death. Several groups have demonstrated that insulin mediates a pro-survival pathway in retinal neurons. It has been demonstrated that insulin receptor signaling diminishes with increased diabetes-induced apoptosis [212]. Fort and colleagues have applied complementary proteomic and genomic techniques and discovered that the proteins of the crystallin superfamily are increased dramatically in early diabetic retinopathy [213]. Their data suggest that the changes observed in crystallins are greater than other changes described so far in diabetic retinopathy. The work also emphasizes the power of the 'omic' approaches (genomic, proteomic, etc.) to uncovering biomarkers for diabetic retinopathy.

As data are accumulated about molecules whose expression changes in diabetic retinopathy it is tempting to target them as a means to protect against deleterious consequences of diabetes. However, caution must be exercised regarding this strategy. A case in point is VEGF. VEGF plays key role in vasculogenesis and has been an anti-angiogenic target in treating a number of blinding diseases [214]. VEGF-A has been recognized as an important neuroprotectant in the central nervous system [215]. Nishijima and co-workers used a model of ischemia-reperfusion injury and found that VEGF-A exposure resulted in a dose-dependent reduction in retinal neuron apoptosis [216]. Indeed, chronic inhibition of VEGF-A function in normal adult animals led to a significant loss of retinal ganglion cells yet had no observable effect on several vascular parameters. Their findings have important implications for both neural pathologies and ocular vascular diseases, particularly diabetic retinopathy. It is important to note that the VEGF family consists of seven members (VEGF-A,B,C,D,E,F and PIGF), as well as alternative splicing variants, which include $VEGF_{121,165,189,206.}$ For an excellent comprehensive review of VEGF and its role in eye disease, the reader is referred to the work of Penn and colleagues [217].

2.4 Studies in Diabetic Mice

Mice constitute a powerful model system for understanding disease mechanisms, particularly the influence of genetic mutations on the progression of pathology. The field of diabetic retinopathy is no exception and a number of mouse models have proven informative with respect to gene/protein changes that occur in this disease. One of the most useful models is the *Ins2*$^{Akita/+}$ mouse model, which has proven relevant to and useful for studies of diabetic retinopathy [218–222]. The *Ins2*$^{Akita/+}$ mouse has a point mutation of the *Insulin2* gene leading to hypoinsulinemia and hyperglycemia in heterozygous mice by ~4 weeks [223]. The note: homozygosity for the mutation is embryonically lethal. Over a period of several weeks (~10–25 weeks), heterozygous mice demonstrate marked retinal alterations including ~20–25%

reduction inner plexiform layer thickness, a ~16% reduction of inner nuclear layer thickness and a ~25% reduction in the number of cell bodies in the RGC layer [219, 220, 222]. The cell death appears to occur via apoptosis as evidenced by data from TUNEL assays [222] and detection of active caspase 3 [219]. Oshitara and colleagues have utilized the *Ins2*$^{Akita/+}$ mouse to examine the role of ER stress in diabetic retinopathy [178]. ER stress refers to the phenomenon whereby misfolded proteins aggregate in the ER lumen activating the unfolded protein response (UPR). The aim of the UPR is to alleviate ER stress, restore ER homeostasis and prevent cell death. Activation of the response is mediated by GPR78 (BiP, binding protein), which dissociates from three resident proteins: inositol requiring ER-to-nucleus signal kinase (IRE)1, activating transcription factor (ATF) 6 and soluble-stranded RNA-activated kinase (PKR)-like ER kinase (PERK) [224]. Studies of the *Ins2*$^{Akita/+}$ mouse demonstrated that ER stress is involved in the death of both retinal neurons and vascular cells in diabetic eyes as manifested by increased expression of BiP as well as PERK, ATF6, IRE1 [178]. These data are particularly interesting to our group because of our work showing that (+)-pentazocine, a ligand for sigma receptor 1 (σR1), affords profound neuroprotection in the *Ins2*$^{Akita/+}$ mouse [222]. σR1 is a molecular chaperone protein that resides at the ER-mitochondrial membrane and interacts with BiP. We are currently examining the expression of ER-stress related genes and our preliminary data suggest that expression of PERK, ATF6 and IRE1 as well as downstream effectors CHOP, BCL2 are increased in the diabetic retina, but expression returns to normal upon sustained treatment with (+)-pentazocine (manuscript in preparation). ER stress genes/proteins may prove to be important additional biomarkers of neuronal death in diabetic retinopathy and reducing or blocking ER stress may be a potential therapy for preventing the onset and the progression of this disease.

Other mouse models, in which diabetes occurs endogenously, have demonstrated neuronal degeneration as well. The retinas of hyperglycemic, hyperlipidemic (HGHL, apolipoprotein E(−/−) db/db) have been examined for advanced glycation endproducts (AGE) and receptors for these endproducts [87, 225, 226]. The mice demonstrate early inner retinal neuronal dysfunction, manifested by prolonged latencies of the oscillatory potentials and b-wave as well as acellular capillaries and pericyte ghosts. The investigators found AGEs in the vitreous and associated with the inner limiting membrane formed by the Müller cell endfeet. They noted that attenuation of the RAGE axis with soluble RAGE ameliorated neuronal dysfunction and reduced the development of capillary lesions in these mice. The KKAY mouse is another model that is useful for studies of Type 2 diabetes. As with the *Ins2*$^{Akita/+}$ and the db/db mice, the KKAY mouse develops retinal neuropathy and microangiopathy [227], and hence should prove useful for analyses of biomarkers especially in determining whether there are differences in gene expression associated with this model of type 2 diabetes versus those that more closely mimic type 1. Other groups have induced diabetes in mice using streptozotocin and examined effects on neuronal function [228]. When the mice are not maintained on insulin, which is known for its neuroprotective properties [212, 229, 230], death of ganglion cells has been documented [228]. Using this model Kurihara and colleagues examined the relationship of the rennin-angiotensin system (RAS) with synaptophysin, a synaptic vesicle protein [230]. They examined the streptozotocin-induced mouse model have examined synergistic effects of diabetes neuronal activity in streptozotocin-induced diabetic mice by ERG. Using a variety of molecular and biochemical approaches they found that diabetes was associated with a significant increase in retinal production of angiotensin II and AT1R together with ERK activation. They observed ERG changes and a reduction in the level of synaptophysin protein (but not mRNA levels). Based on these observations, they treated the diabetic mice with the angiotensin II type 1 receptor (AT1R) blocker telimsartan or valsartan and found that they could reverse the ERG changes and the decreased synaptophysin.

Efforts to mitigate angiotensin II have led to the use of the angiotensin receptor blocker candesartan in the "Diabetic Retinopathy Candesartan Trials" (known as "DIRECT"). The findings of this compound in type 1 and type 2 patients have been reported in two papers [231, 232]. This multinational study with over 300 sites worldwide, involving a follow-up of subjects for 4 years, demonstrated some benefits of candesartan treatment in the various studies. However, some effects observed were borderline and of debatable clinical significance [233].

Other studies have used models with other genetic defects. For example, mice with a defect in the *cystathionine-β-synthase* gene develop hyperhomocysteinemia [234]. Increased levels of homocysteine are implicated in several ocular diseases including diabetic retinopathy [235–237] and glaucoma [238, 239]. In the *cbs+/–* mutant mice, retinal ganglion cells die by apoptosis [240] and diabetes accelerates the retinal neuronal loss in these mice [241].

2.5 Studies in Cell Lines and Purified Ganglion Cells

While *in vivo* models are critical for advancing our understanding of biomarkers in neuronal death associated with diabetic retinopathy, there is a place also for studies using a single cell type. Indeed initial screenings to detect biomarkers and postulate therapies are well-served by the availability of appropriate *in vitro* model systems. Fortunately, the field of diabetic retinopathy research has several models to exploit for these purposes.

The rat R28 retinal precursor cells, immortalized with the 12S, nontumorigenic portion of the adenovirus E1A gene, provide researchers with an inexhaustible supply of cells for studying structural and functional differentiation in the retina. The R28 cells were developed in the laboratory of Dr. Gail Seigel and have been used extensively to analyze mechanisms of diabetes-induced neuronal cell death [242]. Notable among the studies are discoveries of several biomarkers for retinal neuronal cell death that may have implications for diabetic retinopathy including sphingolipids [208], p44/p42 mitogen-activated protein kinase [243], glucosamine-induced PERK activation [244], hexosamines [244], advanced glycation end products (AGE) [245], IL-1β [246], and glutamate receptor/calcium homeostasis [247] to name a few.

Useful also for studies of retinal neuronal cell death is the RGC-5 cell line. In 2001, Krishnamoorthy and colleagues published the development of the RGC-5 cell line [197]. It was reportedly derived by transforming postnatal day 1 rat retinal cells with $_2$ E1A virus [248]. Recharacterization of the RGC-5 cell line suggests that it is actually of mouse origin [249]. When originally described, the RGC-5 cells expressed Thy-1.2, Brn-3C, neuritin, NMDA-R1 and GABA receptors, which are neuronal markers characteristic of intact ganglion cells. RGC-5 cells do not express glial fibrillary acidic protein (GFAP) a marker of glial cells. In early passage, the cells were sensitive to glutamate (5 mM) and that sensitivity could be increased if the cells were treated with succinyl concanavalin A (S Con A) [248]. Interestingly, recent studies by Van Bergen and co-workers showed that RGC-5 cells no longer express the ganglion cell marker protein Thy 1.2, whether differentiated using S Con A or not [249]. Our laboratory obtained this cell line shortly after it was first available in 2001 and our stock continues to express Thy1.2, however the sensitivity to homocysteine, glutamate and oxidative stress is diminished [250, 251]. There are over 100 papers published since 2001 using the RGC-5 cells. Many studies explore alterations of genes/proteins associated with diabetic retinopathy. Indeed, in the last 12 months, there have been over 30 publications using these cells for such analyses. Many of these reports used the RGC-5 cell to determine gene/protein alterations or therapeutic intervention strategies under conditions intended to mimic diabetes. For example, Osborne et al. exposed the cells to retinal ischemia and oxidative stress to determine effects of latanoprost acid [252]. Other studies have examined the PI3-kinase/Akt pathway [253], 5-S-GAD as a neuroprotectant [254], bevacizumab toxicity [255], the ERK1/2 pathways [256], and tafluprost [257].

The RGC-5 cell lines proliferate quickly and are not neuronal in appearance. However, when incubated with the protein kinase inhibitor staurosporine, a more neuronal phenotype is achieved. RGC-5 cells treated with 316 nM staurosporine express neurites, become postmitotic and nonapoptotic, and alter their kinase phosphorylation patterns [258, 259]. In addition, they are positive for a number of neuronal markers including Thy 1.2 [258]. Staurosporine-induced differentiated RGC-5 cells are sensitive to oxidative stress induced by hydrogen peroxide, ischemia, glucose deprivation and plasminogen activators [260, 261]. Recent studies in our lab examined the susceptibility of these differentiated RGC-5 cells to homocysteine. We observed that when differentiated, the cells increase expression of active caspase-3, which is not observed in non-staurosporine-treated cells [250]. The staurosporine-treated cells were not susceptible to homocysteine until applied in exceedingly high dosages. Thus, RGC-5 cells are a useful neuronal model, but data must be interpreted cautiously acknowledging the limitations of the *in vitro* system.

A more labor-intensive method to study proteins/genes in ganglion cells that are affected by diabetic retinopathy is to purify the cells from the intact animal. Our laboratory has adapted methodology from the Barres laboratory [262] to immunopan ganglion cells from neonatal mice [263]. The cells are viable for 22 days when maintained in a supplemented neurobasal media [263]. They extend neurite processes characteristic of ganglion cells. Using these cells we have been able to demonstrate exquisite sensitivity to oxidative stress, a factor implicated in diabetic retinopathy, that leads to upregulation of the xCT protein component of the glutamate-cystine exchanger system x_c^- [263]. Owing to the role of this exchanger in generation of the antioxidant glutathione, it likely represents a mechanism to enhance ganglion cell survival under stress. The primary ganglion cell cultures are sensitive to physiologically-relevant levels of glutamate and homocysteine; they die by an apoptotic mechanism when exposed to these excitotoxins, a phenomenon that is inhibited by treatment with the σR1 ligand (+)-pentazocine [264]. We have preliminary evidence that when primary ganglion cells are treated with the oxidative stressor xanthine:xanthine oxidase, numerous pro-apoptotic genes are upregulated including Bad, Bak, Fas-L and Trail [265]. These data are noteworthy since xanthine oxidase is involved in free radical production in type 1 diabetic human patients [266]. We have used the primary ganglion cells to explore expression of a number of genes/proteins and found them to reflect the expression in intact mouse retina faithfully [267–270]. Studies are underway to explore ER stress genes whose expression is altered as a consequence of factors associated with diabetic retinopathy.

In summary, diabetic retinopathy in human patients is characterized by neuronal and vascular alterations that can ultimately lead to devastating vision loss. Oxidative stress is crucial in mediating vascular dysfunctions associated with DR via modification of retinal lipids, proteins and DNA. This effect is associated with switching on various signaling pathways that include but not limited to upregulation of angiogenic factors, activation of NF-kB and production of inflammatory mediators. Therefore, early detection of the biochemical effect of oxidative stress such as by-products of lipid peroxidation, DNA and protein oxidative damage in diabetic patients could serve as biomarkers which may be effective in early prediction and treatment of DR. Interestingly, variations in oxidative stress related genes among diabetic patients plays also a role as an additional risk factor that makes some patients more vulnerable to develop DR than others. Thus, gene polymorphism could be also used as predictive marker for increased susceptibility to develop DR in diabetic patient. There is strong evidence from analysis of human tissue that retinal ganglion cells are particularly susceptible in diabetic retinopathy. Investigators are using a variety of model systems, both *in vivo* (rat and mouse) and *in vitro* (ganglion cell lines and primary cultures) to determine key genes/proteins whose expression is changed under diabetic conditions with the aim of uncovering the most promising targets for successful intervention. Using relevant *in vitro* models and confirming the data in relevant *in vivo* models should pave the way for important biomarkers to be identified and targeted in treating this disease.

References

1. Klein R, Klein BE, Moss SE (1984) Visual impairment in diabetes. Ophthalmology 91:1–9
2. Leal EC, Manivannan A, Hosoya K, Terasaki T, Cunha-Vaz J, Ambrosio AF, Forrester JV (2007) Inducible nitric oxide synthase isoform is a key mediator of leukostasis and blood-retinal barrier breakdown in diabetic retinopathy. Invest Ophthalmol Vis Sci 48:5257–5265
3. Antonetti DA, Lieth E, Barber AJ, Gardner TW (1999) Molecular mechanisms of vascular permeability in diabetic retinopathy. Semin Ophthalmol 14:240–248

4. Adamis AP (2002) Is diabetic retinopathy an inflammatory disease? Br J Ophthalmol 86:363–365
5. Kowluru RA, Tang J, Kern TS (2001) Abnormalities of retinal metabolism in diabetes and experimental galactosemia. VII. Effect of long-term administration of antioxidants on the development of retinopathy. Diabetes 50:1938–1942
6. Brownlee M (2001) Biochemistry and molecular cell biology of diabetic complications. Nature 414:813–820
7. Kowluru RA, Abbas SN (2003) Diabetes-induced mitochondrial dysfunction in the retina. Invest Ophthalmol Vis Sci 44:5327–5334
8. Giardino I, Fard AK, Hatchell DL, Brownlee M (1998) Aminoguanidine inhibits reactive oxygen species formation, lipid peroxidation, and oxidant-induced apoptosis. Diabetes 47:1114–1120
9. Barber AJ (2003) A new view of diabetic retinopathy: a neurodegenerative disease of the eye. Prog Neuropsychopharmacol Biol Psychiatry 27:283–290
10. Antonetti DA, Barber AJ, Bronson SK, Freeman WM, Gardner TW, Jefferson LS, Kester M, Kimball SR, Krady JK, LaNoue KF, Norbury CC, Quinn PG, Sandirasegarane L, Simpson IA (2006) Diabetic retinopathy: seeing beyond glucose-induced microvascular disease. Diabetes 55:2401–2411
11. Smith SB (2007) The impact of diabetes on neuronal, glial and vascular cells of the retina: implications for the pathogenesis of diabetic retinopathy. In: Barnstable C, Tink J (eds) Retinal degenerations: biology, diagnostics, and therapeutics. Humana Press, Totowa, NJ, 165–184
12. Bresnick GH, Korth K, Groo A, Palta M (1984) Electroretinographic oscillatory potentials predict progression of diabetic retinopathy. Preliminary report. Arch Ophthalmol 102:1307–1311
13. Lopes de Faria JM, Katsumi O, Cagliero E, Nathan D, Hirose T (2001) Neurovisual abnormalities preceding the retinopathy in patients with long-term type 1 diabetes mellitus. Graefes Arch Clin Exp Ophthalmol 239:643–648
14. Mizutani M, Kern TS, Lorenzi M (1996) Accelerated death of retinal microvascular cells in human and experimental diabetic retinopathy. J Clin Invest 97:2883–2890
15. Anderson RE, Rapp LM, Wiegand RD (1984) Lipid peroxidation and retinal degeneration. Curr Eye Res 3:223–227
16. Green SP, Cairns B, Rae J, Errett-Baroncini C, Hongo JA, Erickson RW, Curnutte JT (2001) Induction of gp91-phox, a component of the phagocyte NADPH oxidase, in microglial cells during central nervous system inflammation. J Cereb Blood Flow Metab 21:374–384
17. Droge W (2002) Free radicals in the physiological control of cell function. Physiol Rev 82:47–95
18. Saran M, Michel C, Bors W (1990) Reaction of NO with O2-. implications for the action of endothelium-derived relaxing factor (EDRF). Free Radic Res Commun 10:221–226
19. Huie RE, Padmaja S (1993) The reaction of no with superoxide. Free Radic Res Commun 18:195–199
20. Mohr S, Xi X, Tang J, Kern TS (2002) Caspase activation in retinas of diabetic and galactosemic mice and diabetic patients. Diabetes 51:1172–1179
21. Kowluru RA (2003) Effect of reinstitution of good glycemic control on retinal oxidative stress and nitrative stress in diabetic rats. Diabetes 52:818–823
22. Al-Shabrawey M, Bartoli M, El-Remessy AB, Ma G, Matragoon S, Lemtalsi T, Caldwell RW, Caldwell RB (2008) Role of NADPH oxidase and Stat3 in statin-mediated protection against diabetic retinopathy. Invest Ophthalmol Vis Sci 49:3231–3238
23. Inoguchi T, Sonta T, Tsubouchi H, Etoh T, Kakimoto M, Sonoda N, Sato N, Sekiguchi N, Kobayashi K, Sumimoto H, Utsumi H, Nawata H (2003) Protein kinase C-dependent increase in reactive oxygen species (ROS) production in vascular tissues of diabetes: role of vascular NAD(P)H oxidase. J Am Soc Nephrol 14(8 Suppl 3):S227–S232
24. Sonta T, Inoguchi T, Tsubouchi H, Sekiguchi N, Kobayashi K, Matsumoto S, Utsumi H, Nawata H (2004) Evidence for contribution of vascular NAD(P)H oxidase to increased oxidative stress in animal models of diabetes and obesity. Free Radic Biol Med 37:115–123

25. Griendling KK, Sorescu D, Ushio-Fukai M (2000) NAD(P)H oxidase: role in cardiovascular biology and disease. Circ Res 86:494–501
26. Al-Shabrawey M, Rojas M, Sanders T, Behzadian A, El-Remessy A, Bartoli M, Parpia AK, Liou G, Caldwell RB (2008) Role of NADPH oxidase in retinal vascular inflammation. Invest Ophthalmol Vis Sci 49:3239–3244
27. Al-Shabrawey M, Bartoli M, El-Remessy AB, Platt DH, Matragoon S, Behzadian MA, Caldwell RW, Caldwell RB (2005) Inhibition of NAD(P)H oxidase activity blocks vascular endothelial growth factor overexpression and neovascularization during ischemic retinopathy. Am J Pathol 167:599–607
28. Adams ER, Dratz EA, Gizachew D, Deleo FR, Yu L, Volpp BD, Vlases M, Jesaitis AJ, Quinn MT (1997) Interaction of human neutrophil flavocytochrome b with cytosolic proteins: transferred-NOESY NMR studies of a gp91phox C-terminal peptide bound to p47phox. Biochem J 325(Pt 1):249–257
29. Wenger RH, Marti HH, Schuerer-Maly CC, Kvietikova I, Bauer C, Gassmann M, Maly FE (1996) Hypoxic induction of gene expression in chronic granulomatous disease-derived B-cell lines: oxygen sensing is independent of the cytochrome b558-containing nicotinamide adenine dinucleotide phosphate oxidase. Blood 87:756–761
30. Heyworth PG, Curnutte JT, Nauseef WM, Volpp BD, Pearson DW, Rosen H, Clark RA (1991) Neutrophil nicotinamide adenine dinucleotide phosphate oxidase assembly. Translocation of p47-phox and p67-phox requires interaction between p47-phox and cytochrome b558. J Clin Invest 87:352–356
31. Koga H, Terasawa H, Nunoi H, Takeshige K, Inagaki F, Sumimoto H (1999) Tetratricopeptide repeat (TPR) motifs of p67(phox) participate in interaction with the small GTPase Rac and activation of the phagocyte NADPH oxidase. J Biol Chem 274:25051–25060
32. Ishida S, Yamashiro K, Usui T, Amano S, Ogura Y, Hida T, Oguchi Y (2004) Significance of leukocytes in the regulation of retinal edema, ischemia, and angiogenesis. Nippon Ganka Gakkai Zasshi 108:193–201
33. Miyamoto K, Hiroshiba N, Tsujikawa A, Ogura Y (1998) In vivo demonstration of increased leukocyte entrapment in retinal microcirculation of diabetic rats. Invest Ophthalmol Vis Sci 39:2190–2194
34. Kinukawa Y, Shimura M, Tamai M (1999) Quantifying leukocyte dynamics and plugging in retinal microcirculation of streptozotosin-induced diabetic rats. Curr Eye Res 18:49–55
35. Nonaka A, Kiryu J, Tsujikawa A, Yamashiro K, Miyamoto K, Nishiwaki H, Honda Y, Ogura Y (2000) PKC-beta inhibitor (LY333531) attenuates leukocyte entrapment in retinal microcirculation of diabetic rats. Invest Ophthalmol Vis Sci 41:2702–2706
36. Wong RK, Pettit AI, Davies JE, Ng LL (2002) Augmentation of the neutrophil respiratory burst through the action of advanced glycation end products: a potential contributor to vascular oxidant stress. Diabetes 51:2846–2853
37. Mohanty P, Hamouda W, Garg R, Aljada A, Ghanim H, Dandona P (2000) Glucose challenge stimulates reactive oxygen species (ROS) generation by leucocytes. J Clin Endocrinol Metab 85:2970–2973
38. Brownlee M (2005) The pathobiology of diabetic complications: a unifying mechanism. Diabetes 54:1615–1625
39. Kanwar M, Chan PS, Kern TS, Kowluru RA (2007) Oxidative damage in the retinal mitochondria of diabetic mice: possible protection by superoxide dismutase. Invest Ophthalmol Vis Sci 48:3805–3811
40. Kowluru RA, Atasi L, Ho YS (2006) Role of mitochondrial superoxide dismutase in the development of diabetic retinopathy. Invest Ophthalmol Vis Sci 47(4):1594–1599
41. Kowluru RA, Kowluru V, Xiong Y, Ho YS (2006) Overexpression of mitochondrial superoxide dismutase in mice protects the retina from diabetes-induced oxidative stress. Free Radic Biol Med 41:1191–1196
42. Heinzel B, John M, Klatt P, Bohme E, Mayer B (1992) Ca2+/calmodulin-dependent formation of hydrogen peroxide by brain nitric oxide synthase. Biochem J 281(Pt 3):627–630

43. Mayer B, Heinzel B, Klatt P, John M, Schmidt K, Bohme E (1992) Nitric oxide synthase-catalyzed activation of oxygen and reduction of cytochromes: reaction mechanisms and possible physiological implications. J Cardiovasc Pharmacol 20(Suppl 12):S54–S56
44. Presta A, Liu J, Sessa WC, Stuehr DJ (1997) Substrate binding and calmodulin binding to endothelial nitric oxide synthase coregulate its enzymatic activity. Nitric Oxide 1:74–87
45. Tanner FC, Noll G, Boulanger CM, Luscher TF (1991) Oxidized low density lipoproteins inhibit relaxations of porcine coronary arteries. Role of scavenger receptor and endothelium-derived nitric oxide. Circulation 83:2012–2020
46. Jay MT, Chirico S, Siow RC, Bruckdorfer KR, Jacobs M, Leake DS, Pearson JD, Mann GE (1997) Modulation of vascular tone by low density lipoproteins: effects on L-arginine transport and nitric oxide synthesis. Exp Physiol 82:349–360
47. Taverna MJ, Elgrably F, Selmi H, Selam JL, Slama G (2005) The T-786C and C774T endothelial nitric oxide synthase gene polymorphisms independently affect the onset pattern of severe diabetic retinopathy. Nitric Oxide 13:88–92
48. Miwa K, Nakamura J, Hamada Y, Naruse K, Nakashima E, Kato K, Kasuya Y, Yasuda Y, Kamiya H, Hotta N (2003) The role of polyol pathway in glucose-induced apoptosis of cultured retinal pericytes. Diabetes Res Clin Pract 60:1–9
49. Sun W, Oates PJ, Coutcher JB, Gerhardinger C, Lorenzi M (2006) A selective aldose reductase inhibitor of a new structural class prevents or reverses early retinal abnormalities in experimental diabetic retinopathy. Diabetes 55:2757–2762
50. El-Remessy AB, Abou-Mohamed G, Caldwell RW, Caldwell RB (2003) High glucose-induced tyrosine nitration in endothelial cells: role of eNOS uncoupling and aldose reductase activation. Invest Ophthalmol Vis Sci 44:3135–3143
51. Glomb MA, Monnier VM (1995) Mechanism of protein modification by glyoxal and glycolaldehyde, reactive intermediates of the Maillard reaction. J Biol Chem 270:10017–10026
52. Stitt AW (2003) The role of advanced glycation in the pathogenesis of diabetic retinopathy. Exp Mol Pathol 75:95–108
53. Li L, Renier G (2006) Activation of nicotinamide adenine dinucleotide phosphate (reduced form) oxidase by advanced glycation end products links oxidative stress to altered retinal vascular endothelial growth factor expression. Metabolism 55:1516–1523
54. Mohamed AK, Bierhaus A, Schiekofer S, Tritschler H, Ziegler R, Nawroth PP (1999) The role of oxidative stress and NF-kappaB activation in late diabetic complications. Biofactors 10:157–167
55. Kowluru RA (2005) Effect of advanced glycation end products on accelerated apoptosis of retinal capillary cells under in vitro conditions. Life Sci 76:1051–1060
56. Moore TC, Moore JE, Kaji Y, Frizzell N, Usui T, Poulaki V, Campbell IL, Stitt AW, Gardiner TA, Archer DB, Adamis AP (2003) The role of advanced glycation end products in retinal microvascular leukostasis. Invest Ophthalmol Vis Sci 44:4457–4464
57. Tawfik A, Sanders T, Kahook K, Akeel S, Elmarakby A, Al-Shabrawey M (2009) Suppression of retinal peroxisome proliferator-activated receptor gamma in experimental diabetes and oxygen-induced retinopathy: role of NADPH oxidase. Invest Ophthalmol Vis Sci 50:878–884
58. Hammes HP, Hoerauf H, Alt A, Schleicher E, Clausen JT, Bretzel RG, Laqua H (1999) N(epsilon)(carboxymethyl)lysin and the AGE receptor RAGE colocalize in age-related macular degeneration. Invest Ophthalmol Vis Sci 40:1855–1859
59. Spencer L, Mann C, Metcalfe M, Webb M, Pollard C, Spencer D, Berry D, Steward W, Dennison A (2009) The effect of omega-3 FAs on tumour angiogenesis and their therapeutic potential. Eur J Cancer 45:2077–2086
60. Koya D, King GL (1998) Protein kinase C activation and the development of diabetic complications. Diabetes 47:859–866
61. Ishii H, Jirousek MR, Koya D, Takagi C, Xia P, Clermont A, Bursell SE, Kern TS, Ballas LM, Heath WF, Stramm LE, Feener EP, King GL (1996) Amelioration of vascular dysfunctions in diabetic rats by an oral PKC beta inhibitor. Science 272:728–731

62. Stauble B, Boscoboinik D, Tasinato A, Azzi A (1994) Modulation of activator protein-1 (AP-1) transcription factor and protein kinase C by hydrogen peroxide and D-alpha-tocopherol in vascular smooth muscle cells. Eur J Biochem 226:393–402
63. Oikawa T, Shimamura M, Ashino H, Nakamura O, Kanayasu T, Morita I, Murota S (1992) Inhibition of angiogenesis by staurosporine, a potent protein kinase inhibitor. J Antibiot (Tokyo) 45:1155–1160
64. Xia P, Aiello LP, Ishii H, Jiang ZY, Park DJ, Robinson GS, Takagi H, Newsome WP, Jirousek MR, King GL (1996) Characterization of vascular endothelial growth factor's effect on the activation of protein kinase C, its isoforms, and endothelial cell growth. J Clin Invest 98:2018–2026
65. Koya D, Jirousek MR, Lin YW, Ishii H, Kuboki K, King GL (1997) Characterization of protein kinase C beta isoform activation on the gene expression of transforming growth factor-beta, extracellular matrix components, and prostanoids in the glomeruli of diabetic rats. J Clin Invest 100:115–126
66. Kowluru RA (2001) Diabetes-induced elevations in retinal oxidative stress, protein kinase C and nitric oxide are interrelated. Acta Diabetol 38:179–185
67. Wu Y, Wu G, Qi X, Lin H, Qian H, Shen J, Lin S (2006) Protein kinase C beta inhibitor LY333531 attenuates intercellular adhesion molecule-1 and monocyte chemotactic protein-1 expression in the kidney in diabetic rats. J Pharmacol Sci 101:335–343
68. Ohshiro Y, Ma RC, Yasuda Y, Hiraoka-Yamamoto J, Clermont AC, Isshiki K, Yagi K, Arikawa E, Kern TS, King GL (2006) Reduction of diabetes-induced oxidative stress, fibrotic cytokine expression, and renal dysfunction in protein kinase Cbeta-null mice. Diabetes 55:3112–3120
69. Frank RN (2004) Diabetic retinopathy. N Engl J Med 350:48–58
70. Frank RN (1995) The galactosemic dog. A valid model for both early and late stages of diabetic retinopathy. Arch Ophthalmol 113:275–276
71. Duh E, Aiello LP (1999) Vascular endothelial growth factor and diabetes: the agonist versus antagonist paradox. Diabetes 48:1899–1906
72. Joussen AM, Poulaki V, Le ML, Koizumi K, Esser C, Janicki H, Schraermeyer U, Kociok N, Fauser S, Kirchhof B, Kern TS, Adamis AP (2004) A central role for inflammation in the pathogenesis of diabetic retinopathy. FASEB J 18:1450–1452
73. Khan ZA, Chakrabarti S (2007) Cellular signaling and potential new treatment targets in diabetic retinopathy. Exp Diabetes Res 2007:31867
74. Cheng TH, Shih NL, Chen SY, Loh SH, Cheng PY, Tsai CS, Liu SH, Wang DL, Chen JJ (2001) Reactive oxygen species mediate cyclic strain-induced endothelin-1 gene expression via Ras/Raf/extracellular signal-regulated kinase pathway in endothelial cells. J Mol Cell Cardiol 33:1805–1814
75. Chen S, Khan ZA, Barbin Y, Chakrabarti S (2004) Pro-oxidant role of heme oxygenase in mediating glucose-induced endothelial cell damage. Free Radic Res 38:1301–1310
76. Lutty GA, McLeod DS, Merges C, Diggs A, Plouet J (1996) Localization of vascular endothelial growth factor in human retina and choroid. Arch Ophthalmol 114:971–977
77. Aiello LP, Wong JS (2000) Role of vascular endothelial growth factor in diabetic vascular complications. Kidney Int Suppl 77:S113–S119
78. Lu M, Kuroki M, Amano S, Tolentino M, Keough K, Kim I, Bucala R, Adamis AP (1998) Advanced glycation end products increase retinal vascular endothelial growth factor expression. J Clin Invest 101:1219–1224
79. Caldwell RB, Bartoli M, Behzadian MA, El-Remessy AE, Al-Shabrawey M, Platt DH, Liou GI, Caldwell RW (2005) Vascular endothelial growth factor and diabetic retinopathy: role of oxidative stress. Curr Drug Targets 6:511–524
80. Tsai GY, Cui JZ, Syed H, Xia Z, Ozerdem U, McNeill JH, Matsubara JA (2009) Effect of N-acetylcysteine on the early expression of inflammatory markers in the retina and plasma of diabetic rats. Clin Experiment Ophthalmol 37:223–231
81. Zheng Z, Chen H, Zhao H, Liu K, Luo D, Chen Y, Yang X, Gu Q, Xu X (2009) PEDF inhibits JAK2/STAT3-mediated VEGF upregulation under high glucose condition through a mitochondrial ROS pathway in vitro. Invest Ophthalmol Vis Sci 51:64–71

82. Bolick DT, Srinivasan S, Kim KW, Hatley ME, Clemens JJ, Whetzel A, Ferger N, Macdonald TL, Davis MD, Tsao PS, Lynch KR, Hedrick CC (2005) Sphingosine-1-phosphate prevents tumor necrosis factor-{alpha}-mediated monocyte adhesion to aortic endothelium in mice. Arterioscler Thromb Vasc Biol 25:976–981
83. Velazquez E, Winocour PH, Kesteven P, Alberti KG, Laker MF (1991) Relation of lipid peroxides to macrovascular disease in type 2 diabetes. Diabet Med 8:752–758
84. Kedziora-Kornatowska KZ, Luciak M, Blaszczyk J, Pawlak W (1998) Lipid peroxidation and activities of antioxidant enzymes in erythrocytes of patients with non-insulin dependent diabetes with or without diabetic nephropathy. Nephrol Dial Transplant 13:2829–2832
85. Liang XC, Guo SS, Wang XD (1996) Study on relationship of lipid peroxide in coronary heart disease with and without diabetes. Zhongguo Zhong Xi Yi Jie He Za Zhi 16:29–31
86. Stringer MD, Gorog PG, Freeman A, Kakkar VV (1989) Lipid peroxides and atherosclerosis. BMJ 298:281–284
87. Griesmacher A, Kindhauser M, Andert SE, Schreiner W, Toma C, Knoebl P, Pietschmann P, Prager R, Schnack C, Schernthaner G, Mueller MM (1995) Enhanced serum levels of thiobarbituric-acid-reactive substances in diabetes mellitus. Am J Med 98:469–475
88. Losada M, Alio JL (1996) Malondialdehyde serum concentration in type 1 diabetic with and without retinopathy. Doc Ophthalmol 93:223–229
89. Gallou G, Ruelland A, Campion L, Maugendre D, Le Moullec N, Legras B, Allannic H, Cloarec L (1994) Increase in thiobarbituric acid-reactive substances and vascular complications in type 2 diabetes mellitus. Diabete Metab 20:258–264
90. Kesavulu MM, Giri R, Kameswara Rao B, Apparao C (2000) Lipid peroxidation and antioxidant enzyme levels in type 2 diabetics with microvascular complications. Diabetes Metab 26:387–392
91. Pan HZ, Zhang H, Chang D, Li H, Sui H (2008) The change of oxidative stress products in diabetes mellitus and diabetic retinopathy. Br J Ophthalmol 92:548–551
92. Nishigaki I, Hagihara M, Tsunekawa H, Maseki M, Yagi K (1981) Lipid peroxide levels of serum lipoprotein fractions of diabetic patients. Biochem Med 25:373–378
93. Dursun E, Dursun B, Suleymanlar G, Ozben T (2005) Effect of haemodialysis on the oxidative stress and antioxidants in diabetes mellitus. Acta Diabetol 42:123–128
94. Polak M, Zagorski Z (2004) Lipid peroxidation in diabetic retinopathy. Ann Univ Mariae Curie Sklodowska Med 59:434–437
95. Adaikalakoteswari A, Balasubramanyam M, Rema M, Mohan V (2006) Differential gene expression of NADPH oxidase (p22phox) and hemoxygenase-1 in patients with Type 2 diabetes and microangiopathy. Diabet Med 23:666–674
96. Armstrong D, Abdella N, Salman A, Miller N, Rahman EA, Bojancyzk M (1992) Relationship of lipid peroxides to diabetic complications. Comparison with conventional laboratory tests. J Diabetes Complications 6:116–122
97. Armstrong D, al-Awadi F (1991) Lipid peroxidation and retinopathy in streptozotocin-induced diabetes. Free Radic Biol Med 11:433–436
98. Augustin AJ, Breipohl W, Boker T, Lutz J, Spitznas M (1993) Increased lipid peroxide levels and myeloperoxidase activity in the vitreous of patients suffering from proliferative diabetic retinopathy. Graefes Arch Clin Exp Ophthalmol 231:647–650
99. Augustin AJ, Spitznas M, Koch F, Grus F, Boker T (1995) Indicators of oxidative tissue damage and inflammatory activity in epiretinal membranes of proliferative diabetic retinopathy, proliferative vitreoretinopathy and macular pucker. Ger J Ophthalmol 4:47–51
100. Diffley JM, Wu M, Sohn M, Song W, Hammad SM, Lyons TJ (2009) Apoptosis induction by oxidized glycated LDL in human retinal capillary pericytes is independent of activation of MAPK signaling pathways. Mol Vis 15:135–145
101. Wu M, Chen Y, Wilson K, Chirindel A, Ihnat MA, Yu Y, Boulton ME, Szweda LI, Ma JX, Lyons TJ (2008) Intraretinal leakage and oxidation of LDL in diabetic retinopathy. Invest Ophthalmol Vis Sci 49:2679–2685
102. Song W, Barth JL, Yu Y, Lu K, Dashti A, Huang Y, Gittinger CK, Argraves WS, Lyons TJ (2005) Effects of oxidized and glycated LDL on gene expression in human retinal capillary pericytes. Invest Ophthalmol Vis Sci 46:2974–2982

103. Fredrikson GN, Anand DV, Hopkins D, Corder R, Alm R, Bengtsson E, Shah PK, Lahiri A, Nilsson J (2009) Associations between autoantibodies against apolipoprotein B-100 peptides and vascular complications in patients with type 2 diabetes. Diabetologia 52:1426–1433
104. Morrow JD, Hill KE, Burk RF, Nammour TM, Badr KF, Roberts LJ 2nd (1990) A series of prostaglandin F2-like compounds are produced in vivo in humans by a non-cyclooxygenase, free radical-catalyzed mechanism. Proc Natl Acad Sci U S A 87:9383–9387
105. Morrow JD, Harris TM, Roberts LJ 2nd (1990) Noncyclooxygenase oxidative formation of a series of novel prostaglandins: analytical ramifications for measurement of eicosanoids. Anal Biochem 184:1–10
106. Morrow JD, Minton TA, Roberts LJ 2nd (1992) The F2-isoprostane, 8-epi-prostaglandin F2 alpha, a potent agonist of the vascular thromboxane/endoperoxide receptor, is a platelet thromboxane/endoperoxide receptor antagonist. Prostaglandins 44:155–163
107. Morrow JD, Awad JA, Boss HJ, Blair IA, Roberts LJ 2nd (1992) Non-cyclooxygenase-derived prostanoids (F2-isoprostanes) are formed in situ on phospholipids. Proc Natl Acad Sci U S A 89:10721–10725
108. Tang M, Cyrus T, Yao Y, Vocun L, Pratico D (2005) Involvement of thromboxane receptor in the proatherogenic effect of isoprostane F2alpha-III: evidence from apolipoprotein E- and LDL receptor-deficient mice. Circulation 112:2867–2874
109. Khasawneh FT, Huang JS, Mir F, Srinivasan S, Tiruppathi C, Le Breton GC (2008) Characterization of isoprostane signaling: evidence for a unique coordination profile of 8-iso-PGF(2alpha) with the thromboxane A(2) receptor, and activation of a separate cAMP-dependent inhibitory pathway in human platelets. Biochem Pharmacol 75:2301–2315
110. Joy AP, Cowley EA (2008) 8-iso-PGE2 stimulates anion efflux from airway epithelial cells via the EP4 prostanoid receptor. Am J Respir Cell Mol Biol 38:143–152
111. Nourooz-Zadeh J (2008) Key issues in F2-isoprostane analysis. Biochem Soc Trans 36(Pt 5):1060–1065
112. Nourooz-Zadeh J, Pereira P (2000) F(2) isoprostanes, potential specific markers of oxidative damage in human retina. Ophthalmic Res 32:133–137
113. Dandona P, Thusu K, Cook S, Snyder B, Makowski J, Armstrong D, Nicotera T (1996) Oxidative damage to DNA in diabetes mellitus. Lancet 347:444–445
114. Kaneto H, Fujii J, Suzuki K, Kasai H, Kawamori R, Kamada T, Taniguchi N (1994) DNA cleavage induced by glycation of Cu, Zn-superoxide dismutase. Biochem J 304(Pt 1):219–225
115. Krapfenbauer K, Birnbacher R, Vierhapper H, Herkner K, Kampel D, Lubec G (1998) Glycoxidation, and protein and DNA oxidation in patients with diabetes mellitus. Clin Sci (Lond) 95:331–337
116. Ha H, Kim C, Son Y, Chung MH, Kim KH (1994) DNA damage in the kidneys of diabetic rats exhibiting microalbuminuria. Free Radic Biol Med 16:271–274
117. Nishikawa T, Sasahara T, Kiritoshi S, Sonoda K, Senokuchi T, Matsuo T, Kukidome D, Wake N, Matsumura T, Miyamura N, Sakakida M, Kishikawa H, Araki E (2003) Evaluation of urinary 8-hydroxydeoxy-guanosine as a novel biomarker of macrovascular complications in type 2 diabetes. Diabetes Care 26:1507–1512
118. Kakimoto M, Inoguchi T, Sonta T, Yu HY, Imamura M, Etoh T, Hashimoto T, Nawata H (2002) Accumulation of 8-hydroxy-2′-deoxyguanosine and mitochondrial DNA deletion in kidney of diabetic rats. Diabetes 51:1588–1595
119. Suzuki S, Hinokio Y, Komatu K, Ohtomo M, Onoda M, Hirai S, Hirai M, Hirai A, Chiba M, Kasuga S, Akai H, Toyota T (1999) Oxidative damage to mitochondrial DNA and its relationship to diabetic complications. Diabetes Res Clin Pract 45:161–168
120. Hinokio Y, Suzuki S, Hirai M, Chiba M, Hirai A, Toyota T (1999) Oxidative DNA damage in diabetes mellitus: its association with diabetic complications. Diabetologia 42:995–998
121. Berlett BS, Stadtman ER (1997) Protein oxidation in aging, disease, and oxidative stress. J Biol Chem 272:20313–20316
122. Stadtman ER, Berlett BS (1997) Reactive oxygen-mediated protein oxidation in aging and disease. Chem Res Toxicol 10:485–494
123. Levine RL (2002) Carbonyl modified proteins in cellular regulation, aging, and disease. Free Radic Biol Med 32:790–796

124. Dalle-Donne I, Giustarini D, Colombo R, Rossi R, Milzani A (2003) Protein carbonylation in human diseases. Trends Mol Med 9:169–176
125. Nystrom T (2005) Role of oxidative carbonylation in protein quality control and senescence. EMBO J 24:1311–1317
126. Cakatay U (2005) Protein oxidation parameters in type 2 diabetic patients with good and poor glycaemic control. Diabetes Metab 31:551–557
127. Ramakrishna V, Jailkhani R (2007) Evaluation of oxidative stress in Insulin Dependent Diabetes Mellitus (IDDM) patients. Diagn Pathol 2:22
128. Margetis PI, Antonelou MH, Petropoulos IK, Margaritis LH, Papassideri IS (2009) Increased protein carbonylation of red blood cell membrane in diabetic retinopathy. Exp Mol Pathol 87:76–82
129. El-Remessy AB, Al-Shabrawey M, Khalifa Y, Tsai NT, Caldwell RB, Liou GI (2006) Neuroprotective and blood-retinal barrier-preserving effects of cannabidiol in experimental diabetes. Am J Pathol 168:235–244
130. Zhan X, Du Y, Crabb JS, Gu X, Kern TS, Crabb JW (2008) Targets of tyrosine nitration in diabetic rat retina. Mol Cell Proteomics 7:864–874
131. Squadrito GL, Pryor WA (1998) Oxidative chemistry of nitric oxide: the roles of superoxide, peroxynitrite, and carbon dioxide. Free Radic Biol Med 25:392–403
132. Beckman JS (1996) Oxidative damage and tyrosine nitration from peroxynitrite. Chem Res Toxicol 9:836–844
133. El-Remessy AB, Bartoli M, Platt DH, Fulton D, Caldwell RB (2005) Oxidative stress inactivates VEGF survival signaling in retinal endothelial cells via PI 3-kinase tyrosine nitration. J Cell Sci 118(Pt 1):243–252
134. El-Remessy AB, Khalil IE, Matragoon S, Abou-Mohamed G, Tsai NJ, Roon P, Caldwell RB, Caldwell RW, Green K, Liou GI (2003) Neuroprotective effect of (–)Delta9-tetrahydrocannabinol and cannabidiol in N-methyl-D-aspartate-induced retinal neurotoxicity: involvement of peroxynitrite. Am J Pathol 163:1997–2008
135. Ali TK, Matragoon S, Pillai BA, Liou GI, El-Remessy AB (2008) Peroxynitrite mediates retinal neurodegeneration by inhibiting nerve growth factor survival signaling in experimental and human diabetes. Diabetes 57:889–898
136. Ramakrishnan S, Rajesh M, Sulochana KN (2007) Eales' disease: oxidant stress and weak antioxidant defence. Indian J Ophthalmol 55:95–102
137. Julius U, Drel VR, Grassler J, Obrosova IG (2009) Nitrosylated proteins in monocytes as a new marker of oxidative-nitrosative stress in diabetic subjects with macroangiopathy. Exp Clin Endocrinol Diabetes 117:72–77
138. Cakatay U, Kayali R (2006) The evaluation of altered redox status in plasma and mitochondria of acute and chronic diabetic rats. Clin Biochem 39:907–912
139. Alderman CJ, Shah S, Foreman JC, Chain BM, Katz DR (2002) The role of advanced oxidation protein products in regulation of dendritic cell function. Free Radic Biol Med 32:377–385
140. Witko-Sarsat V, Friedlander M, Capeillere-Blandin C, Nguyen-Khoa T, Nguyen AT, Zingraff J, Jungers P, Descamps-Latscha B (1996) Advanced oxidation protein products as a novel marker of oxidative stress in uremia. Kidney Int 49:1304–1313
141. Baskol G, Gumus K, Oner A, Arda H, Karakucuk S (2008) The role of advanced oxidation protein products and total thiols in diabetic retinopathy. Eur J Ophthalmol 18:792–798
142. Witko-Sarsat V, Nguyen Khoa T, Jungers P, Drueke T, Descamps-Latscha B (1998) Advanced oxidation protein products: oxidative stress markers and mediators of inflammation in uremia. Adv Nephrol Necker Hosp 28:321–341
143. Witko-Sarsat V, Friedlander M, Nguyen Khoa T, Capeillere-Blandin C, Nguyen AT, Canteloup S, Dayer JM, Jungers P, Drueke T, Descamps-Latscha B (1998) Advanced oxidation protein products as novel mediators of inflammation and monocyte activation in chronic renal failure. J Immunol 161:2524–2532
144. Kalousova M, Skrha J, Zima T (2002) Advanced glycation end-products and advanced oxidation protein products in patients with diabetes mellitus. Physiol Res 51:597–604

145. Vivekanadan-Giri A, Wang JH, Byun J, Pennathur S (2008) Mass spectrometric quantification of amino acid oxidation products identifies oxidative mechanisms of diabetic end-organ damage. Rev Endocr Metab Disord 9:275–287
146. Sampathkumar R, Balasubramanyam M, Sudarslal S, Rema M, Mohan V, Balaram P (2005) Increased glutathionylated hemoglobin (HbSSG) in type 2 diabetes subjects with microangiopathy. Clin Biochem 38:892–899
147. Selvam R, Anuradha CV (1988) Lipid peroxidation and antiperoxidative enzyme changes in erythrocytes in diabetes mellitus. Indian J Biochem Biophys 25:268–272
148. Kaji H, Kurasaki M, Ito K, Saito T, Saito K, Niioka T, Kojima Y, Ohsaki Y, Ide H, Tsuji M, Kondo T, Kawakami Y (1985) Increased lipoperoxide value and glutathione peroxidase activity in blood plasma of type 2 (non-insulin-dependent) diabetic women. Klin Wochenschr 63:765–768
149. Hagglof B, Marklund SL, Holmgren G (1983) CuZn superoxide dismutase, Mn superoxide dismutase, catalase and glutathione peroxidase in lymphocytes and erythrocytes in insulin-dependent diabetic children. Acta Endocrinol (Copenh) 102:235–239
150. Hartnett ME, Stratton RD, Browne RW, Rosner BA, Lanham RJ, Armstrong D (2000) Serum markers of oxidative stress and severity of diabetic retinopathy. Diabetes Care 23:234–240
151. Siemianowicz K, Gminski J, Telega A, Wojcik A, Posielezna B, Grabowska-Bochenek R, Francuz T (2004) Blood antioxidant parameters in patients with diabetic retinopathy. Int J Mol Med 14:433–437
152. Gupta MM, Chari S (2005) Lipid peroxidation and antioxidant status in patients with diabetic retinopathy. Indian J Physiol Pharmacol 49:187–192
153. Ford ES, Mokdad AH, Giles WH, Brown DW (2003) The metabolic syndrome and antioxidant concentrations: findings from the Third National Health and Nutrition Examination Survey. Diabetes 52:2346–2352
154. Chowdhury TA, Dyer PH, Kumar S, Barnett AH, Bain SC (1999) Genetic determinants of diabetic nephropathy. Clin Sci (Lond) 96:221–230
155. Chowdhury TA, Kumar S, Barnett AH, Bain SC (1995) Nephropathy in type 1 diabetes: the role of genetic factors. Diabet Med 12:1059–1067
156. Parving HH, Tarnow L, Rossing P (1996) Genetics of diabetic nephropathy. J Am Soc Nephrol 7:2509–2517
157. Tarnow L, Rossing P, Nielsen FS, Fagerudd JA, Poirier O, Parving HH (2000) Cardiovascular morbidity and early mortality cluster in parents of type 1 diabetic patients with diabetic nephropathy. Diabetes Care 23:30–33
158. Levy A, Asleh R, Blum S, Levy N, Miller-Lotan R, Kalet-Litman S, Anbinder Y, Lache O, Nakhoul F, Asaf R, Farbstein D, Pollak M, Soloveichik Y, Strauss M, Alshiek J, Livshits A, Schwartz A, Awad H, Jad K, Goldenstein H (2009) Haptoglobin: basic and clinical aspects. Antioxid Redox Signal 12:293–304
159. Nakhoul FM, Miller-Lotan R, Awaad H, Asleh R, Levy AP (2007) Hypothesis – haptoglobin genotype and diabetic nephropathy. Nat Clin Pract Nephrol 3:339–344
160. Petrovic MG, Cilensek I, Petrovic D (2008) Manganese superoxide dismutase gene polymorphism (V16A) is associated with diabetic retinopathy in Slovene (Caucasians) type 2 diabetes patients. Dis Markers 24:59–64
161. Ye LX, Yang MP, Qiu H, Guo KQ, Yan JS (2008) Association of the polymorphism in manganese superoxide dismutase gene with diabetic retinopathy in Chinese type 2 diabetic patients. Zhonghua Yi Xue Yi Chuan Xue Za Zhi 25:452–454
162. Hovnik T, Dolzan V, Natasa Ursic B, Trebusak Podkrajsek K, Battelino T (2009) Genetic polymorphisms in genes encoding antioxidant enzymes are associated with diabetic retinopathy in type 1 diabetes. Diabetes Care 32:2258–2262
163. Katnik I, Jadach J (1996) Haptoglobin concentration in serum and other body fluids measured by comparison of its reactivity with hemoglobin and concanavalin A. Arch Immunol Ther Exp (Warsz) 44:45–50
164. Dobryszycka W (1997) Biological functions of haptoglobin – new pieces to an old puzzle. Eur J Clin Chem Clin Biochem 35:647–654

165. Nakhoul FM, Marsh S, Hochberg I, Leibu R, Miller BP, Levy AP (2000) Haptoglobin genotype as a risk factor for diabetic retinopathy. JAMA 284:1244–1245
166. Everse J, Hsia N (1997) The toxicities of native and modified hemoglobins. Free Radic Biol Med 22:1075–1099
167. Miller YI, Altamentova SM, Shaklai N (1997) Oxidation of low-density lipoprotein by hemoglobin stems from a heme-initiated globin radical: antioxidant role of haptoglobin. Biochemistry 36:12189–12198
168. Lim SK, Kim H, Bin Ali A, Lim YK, Wang Y, Chong SM, Costantini F, Baumman H (1998) Increased susceptibility in Hp knockout mice during acute hemolysis. Blood 92:1870–1877
169. Langlois MR, Delanghe JR, De Buyzere ML, Bernard DR, Ouyang J (1997) Effect of haptoglobin on the metabolism of vitamin C. Am J Clin Nutr 66:606–610
170. Delanghe JR, Langlois MR (2002) Haptoglobin polymorphism and body iron stores. Clin Chem Lab Med 40:212–216
171. Brouwers A, Langlois M, Delanghe J, Billiet J, De Buyzere M, Vercaemst R, Rietzschel E, Bernard D, Blaton V (2004) Oxidized low-density lipoprotein, iron stores, and haptoglobin polymorphism. Atherosclerosis 176:189–195
172. Chen Y, Huang H, Zhou J, Doumatey A, Lashley K, Chen G, Agyenim-Boateng K, Eghan BA, Acheampong J, Fasanmade O, Johnson T, Akinsola FB, Okafor G, Oli J, Ezepue F, Amoah A, Akafo S, Adeyemo A, Rotimi CN (2007) Polymorphism of the endothelial nitric oxide synthase gene is associated with diabetic retinopathy in a cohort of West Africans. Mol Vis 13:2142–2147
173. Ezzidi I, Mtiraoui N, Mohamed MB, Mahjoub T, Kacem M, Almawi WY (2008) Endothelial nitric oxide synthase Glu298Asp, 4b/a, and T-786C polymorphisms in type 2 diabetic retinopathy. Clin Endocrinol (Oxf) 68:542–546
174. Taverna MJ, Sola A, Guyot-Argenton C, Pacher N, Bruzzo F, Chevalier A, Slama G, Reach G, Selam JL (2002) eNOS4 polymorphism of the endothelial nitric oxide synthase predicts risk for severe diabetic retinopathy. Diabet Med 19:240–245
175. Joussen AM, Poulaki V, Mitsiades N, Kirchhof B, Koizumi K, Dohmen S, Adamis AP (2002) Nonsteroidal anti-inflammatory drugs prevent early diabetic retinopathy via TNF-alpha suppression. FASEB J 16:438–440
176. Joussen AM, Poulaki V, Qin W, Kirchhof B, Mitsiades N, Wiegand SJ, Rudge J, Yancopoulos GD, Adamis AP (2002) Retinal vascular endothelial growth factor induces intercellular adhesion molecule-1 and endothelial nitric oxide synthase expression and initiates early diabetic retinal leukocyte adhesion in vivo. Am J Pathol 160:501–509
177. Golubnitschaja O, Jaksche A, Moenkemann H, Yeghiazaryan K, Karl SE, Trog D, Schild HH, Loffler KU (2005) Molecular imaging system for possible prediction of active retinopathy in patients with Diabetes mellitus. Amino Acids 28:229–237
178. Oshitari T, Hata N, Yamamoto S (2008) Endoplasmic reticulum stress and diabetic retinopathy. Vasc Health Risk Manag 4:115–122
179. Imai H, Singh RS, Fort PE, Gardner TW (2009) Neuroprotection for diabetic retinopathy. Dev Ophthalmol 44:56–68
180. Daley ML, Watzke RC, Riddle MC (1987) Early loss of blue-sensitive color vision in patients with type I diabetes. Diabetes Care 10:777–781
181. Roy MS, Gunkel RD, Podgor MJ (1986) Color vision defects in early diabetic retinopathy. Arch Ophthalmol 10:225–228
182. Sokol S, Moskowitz A, Skarf B, Evans R, Molitch M, Senior B (1985) Contrast sensitivity in diabetics with and without background retinopathy. Arch Ophthalmol 103:51–54
183. Hirsh J, Puklin J (1982) Reduced contrast sensitivity may precede clinically observable retinopathy in type 1 diabetes. In: Acta XXIV international congress of ophthalmology, San Francisco. New York, Lippincott
184. Bresnick GH (1986) Diabetic retinopathy viewed as a neurosensory disorder. Arch Ophthalmol 104:989–990
185. Wolter JR (1961) Diabetic retinopathy. Am J Ophthalmol 51:1123–1141
186. Bloodworth JM Jr (1962) Diabetic retinopathy. Diabetes 11:1–22

187. Fortune B, Schneck ME, Adams AJ (1999) Multifocal electroretinogram delays reveal local retinal dysfunction in early diabetic retinopathy. Invest Ophthalmol Vis Sci 40:2638–2651
188. Greco AV, Di Leo MA, Caputo S, Falsini B, Porciatti V, Marietti G, Ghirlanda G (1994) Early selective neuroretinal disorder in prepubertal type 1 (insulin-dependent) diabetic children without microvascular abnormalities. Acta Diabetol 31:98–102
189. Di Leo MA, Caputo S, Falsini B, Porciatti V, Greco AV, Ghirlanda G (1994) Presence and further development of retinal dysfunction after 3-year follow up in IDDM patients without angiographically documented vasculopathy. Diabetologia 37:911–916
190. Ghirlanda G, Di Leo MA, Caputo S, Falsini B, Porciatti V, Marietti G, Greco AV (1991) Detection of inner retina dysfunction by steady-state focal electroretinogram pattern and flicker in early IDDM. Diabetes 40:1122–1127
191. Parisi V, Uccioli L (2001) Visual electrophysiological responses in persons with type 1 diabetes. Diabetes Metab Res Rev 17:12–18
192. Ghirlanda G, Di Leo MA, Caputo S, Cercone S, Greco AV (1997) From functional to microvascular abnormalities in early diabetic retinopathy. Diabetes Metab Rev 13:15–35
193. Greenstein VC, Shapiro A, Zaidi Q, Hood DC (1992) Psychophysical evidence for post-receptoral sensitivity loss in diabetics. Invest Ophthalmol Vis Sci 33:2781–2790
194. Bearse MA Jr, Han Y, Schneck ME, Barez S, Jacobsen C, Adams AJ (2004) Local multifocal oscillatory potential abnormalities in diabetes and early diabetic retinopathy. Invest Ophthalmol Vis Sci 45:3259–3265
195. Bresnick GH, Palta M (1987) Predicting progression to severe proliferative diabetic retinopathy. Arch Ophthalmol 105:810–814
196. Bengtsson B, Heijl A, Agardh E (2005) Visual fields correlate better than visual acuity to severity of diabetic retinopathy. Diabetologia 48:2494–2500
197. Barber AJ, Lieth E, Khin SA, Antonetti DA, Buchanan AG, Gardner TW (1998) Neural apoptosis in the retina during experimental and human diabetes. Early onset and effect of insulin. J Clin Invest 102:783–791
198. Abu-El-Asrar AM, Dralands L, Missotten L, Al-Jadaan IA, Geboes K (2004) Expression of apoptosis markers in the retinas of human subjects with diabetes. Invest Ophthalmol Vis Sci 45:2760–2766
199. Abu El-Asrar AM, Dralands L, Missotten L, Geboes K (2007) Expression of antiapoptotic and proapoptotic molecules in diabetic retinas. Eye (Lond) 21:238–245
200. Abu El-Asrar AM, Van den Steen PE, Al-Amro SA, Missotten L, Opdenakker G, Geboes K (2007) Expression of angiogenic and fibrogenic factors in proliferative vitreoretinal disorders. Int Ophthalmol 27:11–22
201. Oshitari T, Yamamoto S, Hata N, Roy S (2008) Mitochondria- and caspase-dependent cell death pathway involved in neuronal degeneration in diabetic retinopathy. Br J Ophthalmol 92:552–556
202. Nielsen N, Jackson C, Spurling G, Cranstoun P (2011) Nondiabetic retinal pathology – prevalence in diabetic retinopathy screening. Aust Fam Physician 40:529–532
203. Schmidt KG, Bergert H, Funk RH (2008) Neurodegenerative diseases of the retina and potential for protection and recovery. Curr Neuropharmacol 6:164–178
204. Laniado-Schwartzman M, Lavrovsky Y, Stoltz RA, Conners MS, Falck JR, Chauhan K, Abraham NG (1994) Activation of nuclear factor kappa B and oncogene expression by 12(R)-hydroxyeicosatrienoic acid, an angiogenic factor in microvessel endothelial cells. J Biol Chem 269:24321–24327
205. Puvanendran K, Devathasan G, Wong PK (1983) Visual evoked responses in diabetes. J Neurol Neurosurg Psychiatry 46:643–647
206. Oshitari T, Roy S (2005) Diabetes: a potential enhancer of retinal injury in rat retinas. Neurosci Lett 390:25–30
207. Yang LP, Sun HL, Wu LM, Guo XJ, Dou HL, Tso MO, Zhao L, Li SM (2009) Baicalein reduces inflammatory process in a rodent model of diabetic retinopathy. Invest Ophthalmol Vis Sci 50:2319–2327

208. Fox TE, Han X, Kelly S, Merrill AH 2nd, Martin RE, Anderson RE, Gardner TW, Kester M (2006) Diabetes alters sphingolipid metabolism in the retina: a potential mechanism of cell death in diabetic retinopathy. Diabetes 55:3573–3580
209. Nakazawa T, Takahashi H, Nishijima K, Shimura M, Fuse N, Tamai M, Hafezi-Moghadam A, Nishida K (2007) Pitavastatin prevents NMDA-induced retinal ganglion cell death by suppressing leukocyte recruitment. J Neurochem 100:1018–1031
210. Zhang J, Wu Y, Jin Y, Ji F, Sinclair SH, Luo Y, Xu G, Lu L, Dai W, Yanoff M, Li W, Xu GT (2008) Intravitreal injection of erythropoietin protects both retinal vascular and neuronal cells in early diabetes. Invest Ophthalmol Vis Sci 49:732–742
211. Zhang B, Ma JX (2008) SERPINA3K prevents oxidative stress induced necrotic cell death by inhibiting calcium overload. PLoS One 3:e4077
212. Reiter CE, Wu X, Sandirasegarane L, Nakamura M, Gilbert KA, Singh RS, Fort PE, Antonetti DA, Gardner TW (2006) Diabetes reduces basal retinal insulin receptor signaling: reversal with systemic and local insulin. Diabetes 55:1148–1156
213. Fort PE, Freeman WM, Losiewicz MK, Singh RS, Gardner TW (2009) The retinal proteome in experimental diabetic retinopathy: up-regulation of crystallins and reversal by systemic and periocular insulin. Mol Cell Proteomics 8:767–779
214. Schlingemann RO, Witmer AN (2009) Treatment of retinal diseases with VEGF antagonists. Prog Brain Res 175:253–267
215. Kilic U, Kilic E, Jarve A, Guo Z, Spudich A, Bieber K, Barzena U, Bassetti CL, Marti HH, Hermann DM (2006) Human vascular endothelial growth factor protects axotomized retinal ganglion cells in vivo by activating ERK-1/2 and Akt pathways. J Neurosci 26:12439–12446
216. Nishijima K, Ng YS, Zhong L, Bradley J, Schubert W, Jo N, Akita J, Samuelsson SJ, Robinson GS, Adamis AP, Shima DT (2007) Vascular endothelial growth factor-A is a survival factor for retinal neurons and a critical neuroprotectant during the adaptive response to ischemic injury. Am J Pathol 171:53–67
217. Penn JS, Madan A, Caldwell RB, Bartoli M, Caldwell RW, Hartnett ME (2008) Vascular endothelial growth factor in eye disease. Prog Retin Eye Res 27:331–371
218. Kern TS, Barber AJ (2008) Retinal ganglion cells in diabetes. J Physiol 586(Pt 18):4401–4408
219. Barber AJ, Antonetti DA, Kern TS, Reiter CE, Soans RS, Krady JK, Levison SW, Gardner TW, Bronson SK (2005) The Ins2Akita mouse as a model of early retinal complications in diabetes. Invest Ophthalmol Vis Sci 46:2210–2218
220. Gastinger MJ, Kunselman AR, Conboy EE, Bronson SK, Barber AJ (2008) Dendrite remodeling and other abnormalities in the retinal ganglion cells of Ins2 Akita diabetic mice. Invest Ophthalmol Vis Sci 49:2635–2642
221. Gastinger MJ, Singh RS, Barber AJ (2006) Loss of cholinergic and dopaminergic amacrine cells in streptozotocin-diabetic rat and Ins2Akita-diabetic mouse retinas. Invest Ophthalmol Vis Sci 47:3143–3150
222. Smith SB, Duplantier J, Dun Y, Mysona B, Roon P, Martin PM, Ganapathy V (2008) In vivo protection against retinal neurodegeneration by sigma receptor 1 ligand (+)-pentazocine. Invest Ophthalmol Vis Sci 49:4154–4161
223. Yoshioka M, Kayo T, Ikeda T, Koizumi A (1997) A novel locus, Mody4, distal to D7Mit189 on chromosome 7 determines early-onset NIDDM in nonobese C57BL/6 (Akita) mutant mice. Diabetes 46:887–894
224. Eizirik DL, Cardozo AK, Cnop M (2008) The role for endoplasmic reticulum stress in diabetes mellitus. Endocr Rev 29:42–61
225. Barile GR, Pachydaki SI, Tari SR, Lee SE, Donmoyer CM, Ma W, Rong LL, Buciarelli LG, Wendt T, Horig H, Hudson BI, Qu W, Weinberg AD, Yan SF, Schmidt AM (2005) The RAGE axis in early diabetic retinopathy. Invest Ophthalmol Vis Sci 46:2916–2924
226. Sohn EJ, Kim YS, Kim CS, Lee YM, Kim JS (2009) KIOM-79 prevents apoptotic cell death and AGEs accumulation in retinas of diabetic db/db mice. J Ethnopharmacol 121:171–174
227. Ning X, Baoyu Q, Yuzhen L, Shuli S, Reed E, Li QQ (2004) Neuro-optic cell apoptosis and microangiopathy in KKAY mouse retina. Int J Mol Med 13:87–92

228. Martin PM, Roon P, Van Ells TK, Ganapathy V, Smith SB (2004) Death of retinal neurons in streptozotocin-induced diabetic mice. Invest Ophthalmol Vis Sci 45:3330–3336
229. Sanderson TH, Kumar R, Murariu-Dobrin AC, Page AB, Krause GS, Sullivan JM (2009) Insulin activates the PI3K-Akt survival pathway in vulnerable neurons following global brain ischemia. Neurol Res 31:947–958
230. De Felice FG, Vieira MN, Bomfim TR, Decker H, Velasco PT, Lambert MP, Viola KL, Zhao WQ, Ferreira ST, Klein WL (2009) Protection of synapses against Alzheimer's-linked toxins: insulin signaling prevents the pathogenic binding of Abeta oligomers. Proc Natl Acad Sci U S A 106:1971–1976
231. Chaturvedi N, Porta M, Klein R, Orchard T, Fuller J, Parving HH, Bilous R, Sjolie AK (2008) Effect of candesartan on prevention (DIRECT-Prevent 1) and progression (DIRECT-Protect 1) of retinopathy in type 1 diabetes: randomised, placebo-controlled trials. Lancet 372:1394–1402
232. Sjolie AK, Klein R, Porta M, Orchard T, Fuller J, Parving HH, Bilous R, Chaturvedi N (2008) Effect of candesartan on progression and regression of retinopathy in type 2 diabetes (DIRECT-Protect 2): a randomised placebo-controlled trial. Lancet 372:1385–1393
233. Cooper ME (2010) DIRECT study: a commentary. Diab Vasc Dis Res 7:319–320
234. Watanabe M, Osada J, Aratani Y, Kluckman K, Reddick R, Malinow MR, Maeda N (1995) Mice deficient in cystathionine beta-synthase: animal models for mild and severe homocyst(e) inemia. Proc Natl Acad Sci U S A 92:1585–1589
235. Aydemir O, Turkcuoglu P, Guler M, Celiker U, Ustundag B, Yilmaz T, Metin K (2008) Plasma and vitreous homocysteine concentrations in patients with proliferative diabetic retinopathy. Retina 28:741–743
236. Wright AD, Martin N, Dodson PM (2008) Homocysteine, folates, and the eye. Eye (Lond) 22:989–993
237. Brazionis L, Rowley K Sr, Itsiopoulos C, Harper CA, O'Dea K (2008) Homocysteine and diabetic retinopathy. Diabetes Care 31:50–56
238. Roedl JB, Bleich S, Reulbach U, Rejdak R, Naumann GO, Kruse FE, Schlotzer-Schrehardt U, Kornhuber J, Junemann AG (2007) Vitamin deficiency and hyperhomocysteinemia in pseudoexfoliation glaucoma. J Neural Transm 114:571–575
239. Roedl JB, Bleich S, Reulbach U, Rejdak R, Kornhuber J, Kruse FE, Schlotzer-Schrehardt U, Junemann AG (2007) Homocysteine in tear fluid of patients with pseudoexfoliation glaucoma. J Glaucoma 16:234–239
240. Ganapathy PS, Moister B, Roon P, Mysona BA, Duplantier J, Dun Y, Moister TK, Farley MJ, Prasad PD, Liu K, Smith SB (2009) Endogenous elevation of homocysteine induces retinal neuron death in the cystathionine-beta-synthase mutant mouse. Invest Ophthalmol Vis Sci 50:4460–4470
241. Ganapathy PS, Roon P, Moister TK, Mysona B, Smith SB (2009) Diabetes accelerates retinal neuronal cell death in a mouse model of endogenous hyperhomocysteinemia. Ophthalmol Eye Dis 1:3–11
242. Seigel GM (1996) Establishment of an E1A-immortalized retinal cell culture. In Vitro Cell Dev Biol Anim 32:66–68
243. Nakanishi Y, Nakamura M, Mukuno H, Kanamori A, Seigel GM, Negi A (2006) Latanoprost rescues retinal neuro-glial cells from apoptosis by inhibiting caspase-3, which is mediated by p44/p42 mitogen-activated protein kinase. Exp Eye Res 83:1108–1117
244. Kline CL, Schrufer TL, Jefferson LS, Kimball SR (2006) Glucosamine-induced phosphorylation of the alpha-subunit of eukaryotic initiation factor 2 is mediated by the protein kinase R-like endoplasmic-reticulum associated kinase. Int J Biochem Cell Biol 38:1004–1014
245. Shen J, Wu Y, Xu JY, Zhang J, Sinclair SH, Yanoff M, Xu G, Li W, Xu GT (2009) ERK- and Akt-dependent neuroprotection by Erythropoietin (Epo) against glyoxal-AGEs via modulation of Bcl-xL, Bax and BAD. Invest Ophthalmol Vis Sci 51:35–46
246. Abcouwer SF, Shanmugam S, Gomez PF, Shushanov S, Barber AJ, Lanoue KF, Quinn PG, Kester M, Gardner TW (2008) Effect of IL-1beta on survival and energy metabolism of R28 and RGC-5 retinal neurons. Invest Ophthalmol Vis Sci 49:5581–5592

247. Santiago AR, Rosa SC, Santos PF, Cristovao AJ, Barber AJ, Ambrosio AF (2006) Elevated glucose changes the expression of ionotropic glutamate receptor subunits and impairs calcium homeostasis in retinal neural cells. Invest Ophthalmol Vis Sci 47:4130–4137
248. Krishnamoorthy RR, Agarwal P, Prasanna G, Vopat K, Lambert W, Sheedlo HJ, Pang IH, Shade D, Wordinger RJ, Yorio T, Clark AF, Agarwal N (2001) Characterization of a transformed rat retinal ganglion cell line. Brain Res Mol Brain Res 86:1–12
249. Van Bergen NJ, Wood JP, Chidlow G, Trounce IA, Casson RJ, Ju WK, Weinreb RN, Crowston JG (2009) Recharacterization of the RGC-5 retinal ganglion cell line. Invest Ophthalmol Vis Sci 50:4267–4272
250. Ganapathy PS, Dun Y, Ha Y, Duplantier J, Allen JB, Farooq A, Bozard BR, Smith SB (2010) Sensitivity of staurosporine-induced differentiated RGC-5 cells to homocysteine. Curr Eye Res 35:80–90
251. Ganapathy PS, White RE, Ha Y, Bozard BR, McNeil PL, Caldwell RW, Kumar S, Black SM, Smith SB (2011) The role of N-methyl-D-aspartate receptor activation in homocysteine-induced death of retinal ganglion cells. Invest Ophthalmol Vis Sci 52:5515–5524
252. Osborne NN, Ji D, Majid A, Fawcett R, Sparatore A, Soldato P (2009) ACS67, a hydrogen sulphide-releasing derivative of latanoprost acid, attenuates retinal ischemia and oxidative stress to RGC-5 cells in culture. Invest Ophthalmol Vis Sci 51:284–294
253. Biswas SK, Zhao Y, Sandirasegarane L (2009) Imatinib induces apoptosis by inhibiting PDGF- but not insulin-induced PI 3-kinase/Akt survival signaling in RGC-5 retinal ganglion cells. Mol Vis 15:1599–1610
254. Koriyama Y, Ohno M, Kimura T, Kato S (2009) Neuroprotective effects of 5-S-GAD against oxidative stress-induced apoptosis in RGC-5 cells. Brain Res 1296:187–195
255. Sharma RK, Chalam KV (2009) In vitro evaluation of bevacizumab toxicity on a retinal ganglion cell line. Acta Ophthalmol 87:618–622
256. Schallenberg M, Charalambous P, Thanos S (2009) GM-CSF regulates the ERK1/2 pathways and protects injured retinal ganglion cells from induced death. Exp Eye Res 89:665–677
257. Kanamori A, Naka M, Fukuda M, Nakamura M, Negi A (2009) Tafluprost protects rat retinal ganglion cells from apoptosis in vitro and in vivo. Graefes Arch Clin Exp Ophthalmol 247:1353–1360
258. Frassetto LJ, Schlieve CR, Lieven CJ, Utter AA, Jones MV, Agarwal N, Levin LA (2006) Kinase-dependent differentiation of a retinal ganglion cell precursor. Invest Ophthalmol Vis Sci 47:427–438
259. Lieven CJ, Millet LE, Hoegger MJ, Levin LA (2007) Induction of axon and dendrite formation during early RGC-5 cell differentiation. Exp Eye Res 85:678–683
260. Iizuka Y, Hong S, Kim CY, Kim SK, Seong GJ (2008) Agmatine pretreatment protects retinal ganglion cells (RGC-5 cell line) from oxidative stress in vitro. Biocell 32:245–250
261. Harvey R, Chintala SK (2007) Inhibition of plasminogen activators attenuates the death of differentiated retinal ganglion cells and stabilizes their neurite network in vitro. Invest Ophthalmol Vis Sci 48:1884–1891
262. Barres BA, Silverstein BE, Corey DP, Chun LL (1988) Immunological, morphological, and electrophysiological variation among retinal ganglion cells purified by panning. Neuron 1:791–803
263. Dun Y, Mysona B, Van Ells T, Amarnath L, Ola MS, Ganapathy V, Smith SB (2006) Expression of the cystine-glutamate exchanger (xc-) in retinal ganglion cells and regulation by nitric oxide and oxidative stress. Cell Tissue Res 324:189–202
264. Dun Y, Thangaraju M, Prasad P, Ganapathy V, Smith SB (2007) Prevention of excitotoxicity in primary retinal ganglion cells by (+)-pentazocine, a sigma receptor-1 specific ligand. Invest Ophthalmol Vis Sci 48:4785–4794
265. Ha Y, Dun Y, Thangaraju M, Duplantier J, Dong Z, Liu K, Ganapathy V, Smith SB (2011) Sigma receptor 1 modulates endoplasmic reticulum stress in retinal neurons. Invest Ophthalmol Vis Sci 52:527–540
266. Desco MC, Asensi M, Marquez R, Martinez-Valls J, Vento M, Pallardo FV, Sastre J, Vina J (2002) Xanthine oxidase is involved in free radical production in type 1 diabetes: protection by allopurinol. Diabetes 51:1118–1124

267. Kutty RK, Samuel W, Chen S, Vijayasarathy C, Dun Y, Mysona B, Wiggert B, Smith SB (2006) Immunofluorescence analysis of the expression of Norpeg (Rai14) in retinal Muller and ganglion cells. Neurosci Lett 404:294–298
268. Umapathy NS, Gnana-Prakasam JP, Martin PM, Mysona B, Dun Y, Smith SB, Ganapathy V, Prasad PD (2007) Cloning and functional characterization of the proton-coupled electrogenic folate transporter and analysis of its expression in retinal cell types. Invest Ophthalmol Vis Sci 48:5299–5305
269. Dun Y, Duplantier J, Roon P, Martin PM, Ganapathy V, Smith SB (2008) Serine racemase expression and D-serine content are developmentally regulated in neuronal ganglion cells of the retina. J Neurochem 104:970–978
270. Umapathy NS, Dun Y, Martin PM, Duplantier JN, Roon P, Prasad P, Smith SB, Ganapathy V (2008) Expression and function of system N glutamine transporters (SN1/SN2 or SNAT3/SNAT5) in retinal ganglion cells. Invest Ophthalmol Vis Sci 49:5151–5160

Diabetic Retinopathy: The Need for Predictive, Preventive and Personalized Approach to Management

Tatjana Josifova and Paul Bernhard Henrich

Abstract For over six decades, there has been continuous improvement in our understanding of fundamental processes underlying the ocular complications of diabetes. Recent studies provide us with a very precious and detailed understanding of molecular mechanisms involved in the evolution of diabetic retinopathy (DR). The novel therapeutic approaches include intravitreal steroid and Anti-Vascular Endothelial Growth Factor (VEGF) therapy modalities. The latest clinical investigations include randomized multicenter clinical trials. The results seem to be encouraging with respect to short-term control of diabetic macular edema, but recurrence can occur after initiation of treatment. Laser photocoagulation and surgical methods, such as pars plana vitrectomy, have undergone some changes in the last years, but in general, the principle of their action remain the same. Combination of ocular pharmacotherapy and laser photocoagulation is a subject of ongoing investigations.

In order to achieve better results, basic scientific investigations should take into consideration clinical implication of personalized medicine. The individual variations reveal the need for different interventional modalities. Thus, the scientific progress should lead to "tailor made" treatment modalities in medicine in general and with respect to diabetic retinopathy in particular. In this context, it is to remember that about 20 years before the discovery of insulin, Dr. Eliot P. Joslin advised the diabetic patients to "Live, so that you may profit from some new discovery".

Keywords Diabetic retinopathy • Diabetic macular edema • Blood-retina barrier • Vascular endothelial growth factor • Ischemic changes • Pharmacotherapy • Diabetic retinopathy prevention

T. Josifova (✉) • P.B. Henrich
Eye Clinic, University Hospital, Mittlere Strasse 91, CH-9031 Basel, Switzerland
e-mail: tjosifova@uhbs.ch

M.S. Mozaffari (ed.), *New Strategies to Advance Pre/Diabetes Care: Integrative Approach by PPPM*, Advances in Predictive, Preventive and Personalised Medicine 3,
DOI 10.1007/978-94-007-5971-8_8,

1 Diabetes Mellitus

The number of individuals with diabetes mellitus, worldwide, is estimated to be about 346 million. In 2004, an estimated 3.4 million people died from consequences of high blood sugar. More than 80% of diabetes deaths occur in low-and middle-income countries. World Health Organization (WHO) predicts that diabetes-related deaths will double between 2005 and 2030. The overall risk of death among people with diabetes is at least double the risk of their peers without diabetes [1]. Healthy diet, regular physical activity, maintaining a normal body weight and avoiding tobacco use can prevent or delay type 2 diabetes [2, 3].

Type 1 diabetes (previously known as insulin-dependent, juvenile or childhood-onset) is characterized by deficient insulin production and requires daily administration of insulin. The cause of type 1 diabetes is not known and it is not preventable according to current knowledge. Symptoms include excessive excretion of urine (polyuria), thirst (polydipsia), constant hunger, weight loss, vision changes and fatigue. These symptoms may occur suddenly. On the other hand, type 2 diabetes (formerly called non-insulin-dependent or adult-onset) results from the body's ineffective use of insulin.

Type 2 diabetes comprises 90% of people with diabetes around the world, and is largely the result of excessive body weight and physical inactivity [4]. Symptoms may be similar to those of type 1 diabetes, but are often less marked. As a result, the disease may be diagnosed several years after its onset, once complications have already occurred. Until recently, this type of diabetes was seen only in adults but it is now also occurring in children due to the increasing prevalence of obesity and overweight.

Gestational diabetes is hyperglycemia with onset or first recognition during pregnancy. Symptoms of gestational diabetes are similar to type 2 diabetes. Gestational diabetes is most often diagnosed through prenatal screening, rather than reported symptoms.

Impaired glucose tolerance (IGT) and impaired fasting glycemia (IFG) are intermediate conditions in the transition between normoglycemia and diabetes. People with IGT or IFG are at high risk of progressing to type 2 diabetes, although this is not inevitable [5].

Diabetic complications are among the most severe involving the heart, blood vessels, eyes, kidneys and nerves. Approximately 50% of people with diabetes die of cardiovascular disease (primarily heart disease and stroke). Combined with reduced blood flow, neuropathy in the lower extremity, it increases the chance of foot ulcers and eventual limb amputation. Diabetes is among the leading causes of kidney failure; about 10–20% of people with diabetes die of kidney failure [6–8].

Diabetic neuropathy affects up to 50% of people with diabetes. Although many different problems can occur as a result of diabetic neuropathy, common symptoms are tingling, pain, numbness, or weakness in the feet and hands [9].

2 Diabetic Retinopathy

One of the most severe complications of diabetes is DR which in both forms of non-proliferative and proliferative DR (NPDR and PDR, respectively) can lead to vision loss. One of the most common reasons for loss of vision in DR is diabetic maculopathy, presented as a diabetic macular edema (DME) or ischemic maculopathy [10, 11]. The second most common reason for visual loss in diabetic patients is PDR, which, in fact, represents the leading cause for blindness. The ischemic changes in PDR that can induce production of neovascularization (NV) and fibrotic tissue may provoke recurrent retinal and vitreal bleeding. This can lead to tractional retinal detachment and serious eye complications as rubeotic glaucoma.

Diabetic retinopathy is a leading cause of blindness in middle-aged and older Americans. The single greatest source of vision loss in diabetes is macular edema, a pathology that is a direct consequence of diabetic blood–retinal barrier breakdown [12, 13].

Although the management of risk factors, including hyperglycemia, hyperlipidemia, and hypertension, has been shown to ameliorate diabetes-induced vision loss, the exact pathophysiological mechanisms that are involved in this process remain to be elucidated. Whereas the role of high blood glucose has been suggested to be the primary catalyst for the biomolecular and cellular changes seen in the retina, less is known regarding the intraocular biochemical changes associated with the mechanisms that potentially contribute to the pathogenesis of PDR [8, 9].

One of the risk factors for the DR progression is hypertension. In order to prove this statement several studies have simultaneously measured the vitreous cytokine concentrations, in a group of patients with hypertension, using flow cytometry. Both monocyte chemotactic protein 1 and interleukin 8 levels were elevated in the vitreous of patients with diabetic retinopathy and concurrent hypertension. These findings may help to explain the epidemiologic and clinical evidence that systemic hypertension exacerbates diabetic retinopathy [14, 15].

2.1 Predictive Factors for Ischemic Changes in Diabetic Retinopathy

Proliferative diabetic retinopathy is characterized by pathological retinal neovascularization (NV). Accumulating evidence has indicated that high levels of circulating endothelial progenitor cells (EPCs) are an important risk factor for DR [16].

Paradoxically, the reduction and dysfunction of circulating EPCs has been extensively reported in diabetic patients. EPCs are differentially altered in the various vascular complications of diabetes mellitus, exhibiting distinct behaviors in terms of angiogenic response to ischemia and growth factors. They play a potent role in motivating vascular precursors to induce pathological NV. These results suggest that neurotrophic factors may induce neoangiogenesis through EPC activation, leading to pathological retinal NV. It has been proposed that NV in the ischemic retina might be regulated by overexpression of neurotrophic factors.

VEGF, a potent vasopermeability factor, may be operative in the pathogenesis of diabetic blood–retinal barrier breakdown. Retinal VEGF levels are up-regulated in diabetes and coincide with blood–retinal barrier breakdown in rodents and humans. The major high-affinity VEGF receptors, are similarly up-regulated in the diabetic retina and localize to the retinal vasculature [17, 18]. Although VEGF is required for experimental retinal NV, its definitive role in diabetic blood–retinal barrier breakdown is unknown [19]. VEGF concentration is elevated in the vitreous of patients with PDR as compared with those of patients with non-proliferative disease (NPDR). Increased levels of VEGF are also present in patients with PDR compared to quiescent proliferative disease or from non-diabetic patients without NV. Intraviteral VEGF concentration is correlated both with ischemic retinal changes and with DME.

Several findings support the conclusion that VEGF can induce intraocular NV resembling that of PDR. Growth of retinal microvascular endothelial cells in culture is increased by VEGF presence. Repetitive intravitreal injections of recombinant human VEGF are sufficient to produce iris NV in a nonhuman primate leading to ectropion uveae and neovascular glaucoma. The new vessels originate from the retinal VEGF-producing vasculature and grow toward the photoreceptor layer. Although increased permeability can occur in the absence of NV as is often observed with DME, a universal characteristic of retinal proliferation is a corresponding increase in vascular permeability. VEGF is a very effective inducer of permeability, being 50,000 times more potent in the dermal microvasculature than histamine in this regard. Data suggest that VEGF may exert its effects on retinal vascular permeability by altering tight junction proteins such as occluding and adherence junction proteins (e.g. VE-cadherin). Although the role of VEGF in NPDR is less firmly established than it is in proliferative disease, recent studies suggest it may be an important factor in the development of earlier stages of DR. VEGF production is not only present in the early stage of DR, but also in the vitrectomized eyes, with vitreal concentration of ten times more than serum concentration. Persistent secretion of VEGF in the vitreal cavity can explain the high incidence of neovascular glaucoma in the previously vitrectomized eyes [19].

Experimental findings show that intravitreal injections of physiological concentration of VEGF in rats alter retinal blood flow and venous calibre. The changes in animal models are in the same manner as observed in diabetic patients with increasing severity of DR. Moreover, diabetes accentuates the retina's response to VEGF. Such expression and response could theoretically result in positive feedback loop that might eventually induce enough retinal ischemia and VEGF expression to stimulate intraocular neovascularization. This hypothesis raises the idea that inhibitors of VEGF action might prove beneficial not only for neovascular and permeability complications of diabetes, but also in prevention of retinopathy progression in non-proliferative stages.

Studies have demonstrated that leptin is an angiogenic factor, and an increase in intravitreous leptin concentrations in diabetic patients with PDR has also been described. The study that has investigated the role of leptin concentration intravitrealy, has shown that leptin concentrations were lower in the vitreous fluid than in serum samples from both, diabetic patients with PDR ($P<0.001$) and controls ($P<0.001$). The intraocular production of leptin is not critically involved in the etiopathogenesis of PDR. The results suggested that serum diffusion is a relevant source of leptin in vitreous fluid [20].

Hyperglycemia itself induces numerous metabolic changes including increases in oxidative stress, polyol pathway flux, advanced glytation end products and diacylglycerol production. One of the most important enzymes in this pathway is protein kinase C (PKC). Different isoforms predominate in different body tissues and respond differently to various cytokines. In diabetes, PKC activation is observed in tissues in which complications are most prevalent, including the retina, peripheral nerves, kidneys and the heart. In the early stage of diabetes, PKC-β is activated in the retina by hyperglycemia-induced de novo synthesis of diacylglycerol. PKC activation is a step in the hypoxic and hyperglycemic stimulation of VEGF expression and in VEGF enhancement of endothelial cell survival [21]. PKC-β activation is required for VEGF to induce its permeability effects and to some degree its proliferative effects as well. Both physiologic concentration of VEGF and direct activation of PKC result in a rapid three to five-fold increase in retinal vascular permeability in rats. In animal models, the intraviterally and orally administrated PKC-β inhibitor dramatically suppressed the VEGF-induced permeability.

Atrial natriuretic peptide (ANP) has been recently described as an endogenous inhibitor of the synthesis and angiogenic action of vascular endothelial growth factor [17]. ANP was determined by radioimmunoassay in plasma and vitreous humor samples collected from diabetic patients with and without PDR and from non-diabetic subjects [22]. ANP was also immunoehistochemically localized in the epiretinal membranes of patients with PDR.

Diabetic patients with active neovascularization have significantly higher levels of ANP in the vitreous humor than those without active PDR. Diabetic patients without PDR were also found to have significantly higher vitreous ANP levels than non-diabetic patients.

In the fibrovascular epiretinal tissue of these patients, ANP is localized in vascular, glial, fibroblast-like and retinal pigment epithelium cells.

Recent studies have investigated reactive oxygen species (ROS) levels, which are among the potential selective factors for PDR. Also, ROS levels were investigated among different groups of PDR patients. Accordingly, it was shown that ROS levels were significantly elevated in the vitreous fluid of PDR patients, and patients with a more advanced clinical PDR appearance had higher ROS levels. These findings suggest an association between ROS and the pathogenesis of PDR. Further research is necessary to prove the hypothesis that ROS might be correlated with PDR severity [23].

3 VEGF-Initiated Blood–Retinal Barrier Breakdown in Early Diabetes

On a cellular level, VEGF has been implicated in many different mechanisms, which lead to macular edema (ME). It has been proved that VEGF decreases the occludin protein responsible for the tightness of the intracellular junctions [24]. VEGF also induces rapid phosphorylation of the tight junction protein occludin and zonula occludens1 resulting in the break-down of the blood-retina barrier. VEGF-induced blood-retina barrier breakdown appears to be affected via nitric oxide pathway. VEGF also increases paracellular transport without altering the solvent drag reflection coefficient. Furthermore VEGF activation of protein kinase C stimulates occludin phosphorylation and contributes to endothelial permeability, a mechanism that could be seen in different pathological conditions followed by ME [25, 26].

3.1 Inflammatory Factors in DR

Recently, there has been growing evidence indicating that inflammation is an important factor in the pathogenesis of PDR. A number of pro-inflammatory cells have been isolated from the vitreal body which along with ischemic factors, play a very important role in the development of the PDR [27]. Inflammatory cytokines include chemotactic cytokines which are referred to as chemokines. Interleukin-8 (IL-8) is the prototype of C-X-C chemokine which has been recognized as a potent chemoattractant and activator of neutrophils and T lymphocytes but not monocytes. Monocyte chemoattractant protein-1 (MCP-1) is the best-characterized C-C chemokine and exhibits chemoattractant potential for monocytes and lymphocytes but not for neutrophils. Therefore, IL-8 and MCP-1 are involved in the recruitment of inflammatory cells. The presence of these inflammatory cells is associated with local secretion of angiogenic factors which are crucial for the development of PDR. These findings provide the rationale for steroid therapy in the treatment of DME [28].

Some studies have investigated specific pro-inflammatory cells and have shown that the pro-inflammatory cytokines IL-8 and MCP-1 are increased in the vitreous fluid of PDR patients without an increase in the anti-inflammatory cytokine, IL-10. In addition, both sIL-8 and MCP-1 intravitreous levels are correlated with PDR activity, thus suggesting that these cytokines may be pathogenically important in PDR. The imbalance between pro-inflammatory and anti-inflammatory cytokines, rather than pro-inflammatory cytokines alone, is a primary factor in determining the development of PDR [29–32].

Among the inflammatory factors that play a considerable role in detecting the risk factors of the development of severe changes in the DR is the level of sIL-6R in the vitreous fluid. One of the newest studies has shown the difference of IL-6R concentration between diabetic and non-diabetic patients, which could be considered as a potential biomarker of PDR. Recent study results indicate increased levels of sIL-6R in vitreous fluid of patients with PDR and NPDR compared to the non-diabetic group [612.7 +/− 233.8 and 746.3 +/− 523.1 vs. 215.4 +/− 98.3 pg/ml, (mean +/− SD), respectively]. Vitreous levels of sIL-6R in the PDR and pre-proliferative diabetic retinopathy (PPDR) group were significantly higher than those in the non-diabetic group (PDR group, $P<0.0001$; NPDR group, $P<0.01$). Further, serum levels of sIL-6R were 39.38 +/− 9.43 ng/ml in the PDR group and 22.84 +/− 5.32 ng/ml in the control group. Statistical analysis of the samples indicated that serum sIL-6R levels in the PDR group were significantly higher than those of the control group ($P<0.01$). A partial correlation analysis showed a significant correlation between the levels of sIL-6R and VEGF in the vitreous in the PDR group ($r=0.34$, $P<0.05$) [33, 34].

3.2 *Cysteine-Rich 61 (Cyr61) Is One of the Angiogenic Factors Involved in Proliferative Diabetic Retinopathy (PDR)*

One recent study measured and compared the vitreous levels of Cyr61 and VEGF in patients with PDR. Previous studies have shown that Cyr61 expression is associated with proliferative epiretinal membranes [35]. Vitreous Cyr61 levels were significantly higher in active PDR patients, and diabetic macular edema patients compared with non-diabetic control patients ($P<0.01$). Vitreous Cyr61 levels were related to different states of PDR and correlated with VEGF levels in PDR patients. Pretreatment with bevacizumab did not inhibit vitreous Cyr61 levels in active PDR patients. Cysteine-rich 61 might mediate angiogenesis and post-angiogenic fibrosis in PDR. The fact that bevacizumab does not inhibit the Cyr61 production means that anti-VEGF therapy likely cannot inhibit one of the cascade mechanisms in PDR. Furthermore investigations should be performed not only on the isoforms of VEGF, but also on the predictive factors that may stimulate the production of VEGF isoforms [35].

3.3 *Predictive Factors for PDR*

The diabetes-induced alterations in the vitreous protein composition in diabetic patients with PDR may contribute to the manifestation of the disease. The vitreous humor can be utilized in a variety of studies aimed toward the discovery of new targets for the treatment or prevention of PDR and the identification of novel disease mechanisms. In the last few years, proteomics has been applied to explore proteins which were differentially expressed in patients with proliferative diabetic retinopathy and in non-diabetic patients. Previous studies have identified some proteins that are differentially expressed in the vitreous of PDR patients. Many of the proteins are involved in such processes as angiogenesis, cellular proliferation, acute phase response, vascular permeability changes, and oxygen-induced vessel loss.

Endothelial progenitor cells (EPCs) isolated from diabetic patients have a profound inability to migrate in vitro. Investigations on animal models have shown that normal EPCs are better able to repopulate degenerative (acellular) retinal capillaries in chronic diabetes and acute ischemia/reperfusion (I/R) injury in neonatal oxygen-induced retinopathy animal models. These studies demonstrate that healthy huCD34 (+) cells can effectively repair injured retina and that there is defective repair of vasculature in patients with diabetes. Defective EPCs may be amenable to pharmacological manipulation and restoration of the cells' natural robust reparative function [17].

In order to investigate the probable pathophysiological diabetic retinopathy mechanisms, diabetic post mortem eyes ($n=8$) were compared with eyes (n=8) from non-diabetic donors matched by age. Messenger ribonucleic acid (mRNA) of ApoA1 (quantitative reverse-transcriptase polymerase chain reaction) was measured separately in the neuroretina and retinal pigment epithelium (RPE). ApoA1 was assessed by immunofluorescence (confocal laser microscopy) and Western blot analysis. The presence of early diabetic retinal damage was evaluated by measuring the rate of apoptosis and glial activation. In the vitreous fluid of diabetic patients with PDR the levels of apo A-I and apo H are elevated. It has been shown that a higher expression of apo A-I and apo H mRNAs exists in the diabetic retina [36]. The study which has investigated the ischemic meditators, has shown the role of apolipoprotein A-I and apolipoprotein H levels, mediated from retinal pigmented epithelium which was the main contributor of the differences between the retinal samples from diabetic and non-diabetic post mortem donors. ApoA1 overexpression is an early event in the retina of diabetic patients and can be involved in the pathophysiology of diabetic retinopathy. In addition, RPE is the main source of ApoA1 within the retina. These findings may be relevant to aiming new treatment strategies toward reducing the development of diabetic retinopathy.

4 Investigation Techniques

Up to date investigation techniques include the biomicroscopy, ocular coherence tomography (OCT), fluorescein angiography (FA) and microperimetry [37–39]. The investigation techniques provide information regarding persistent changes, but not the risk for future morphological and functional changes (Figs. 1 and 2). Furthermore, the prognosis for additional functional development cannot be predicted in patients in early stage of changes.

OCT was first proposed by Hee as an important diagnostic tool in diabetic patients. Hee and associates concluded in 1995 that cross-sectional imaging of the macula can be useful for objectively monitoring retinal thickness with high resolution in patients with ME [40]. Nowadays, OCT is widely used to detect and monitor DME, mostly based on the scan profile and macular mapping modes. These now allow remarkably accurate definition of the different characteristics of DME, such as diffuse swelling, cystoid cavities and hard exudates. The disadvantages of the OCT technique include lack of conclusive information regarding choroidal changes and

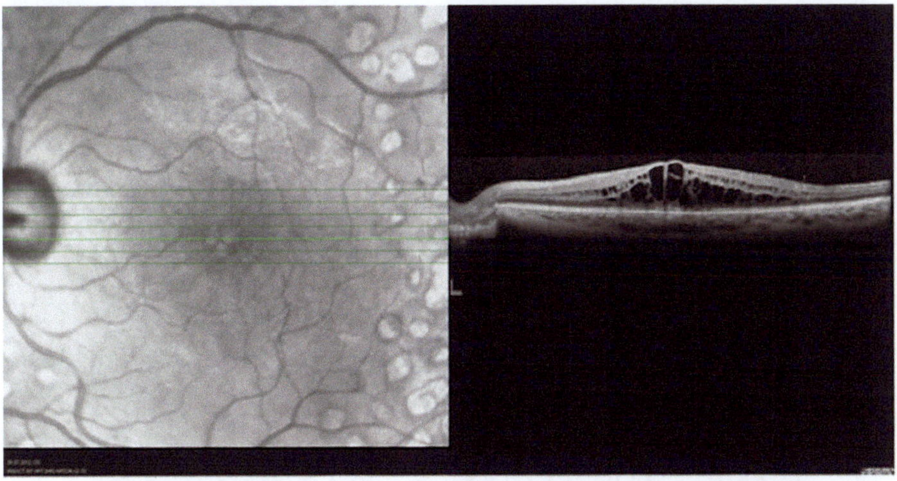

Fig. 1 Simultaneous fundus autofluorescence and spectral domain OCT of cystoid DME

Fig. 2 OCT image: retinal thickening in DME

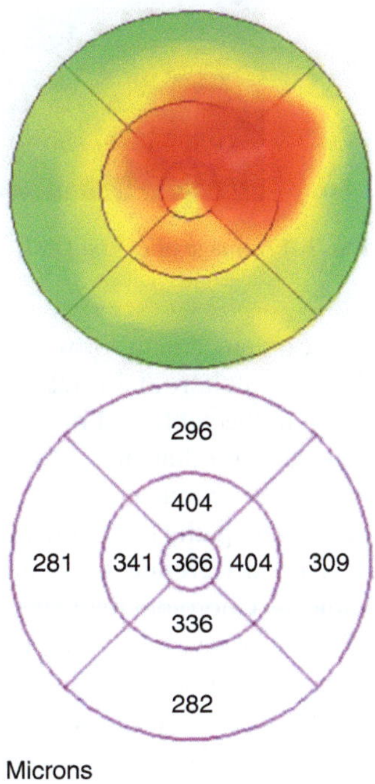

blood-retina break down. In addition, there is no appropriate information about photoreceptor function.

One of the factors that might influence the prognosis of DR is choroidal thickness. Investigation of choroidal thickness in DR patients has been performed with high-definition raster scanning using spectral-domain optical coherence tomography with

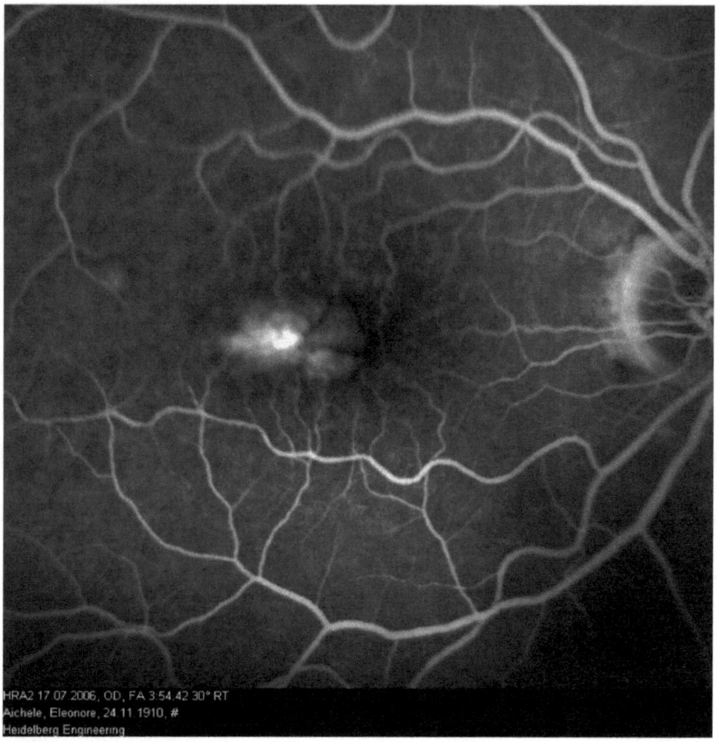

Fig. 3 FA image of cystod DME

frame enhancement software. The study included normal as well as diabetic patients. The mean choroidal thickness showed a pattern of thinnest choroid nasally, thickening in the subfoveal region, and thinning again temporally in normal subjects and patients with diabetes. Mean subfoveal choroidal thickness was thinner in patients with diabetic macular edema (63.3 μm, 27.2%, $P<0.05$) or treated proliferative diabetic retinopathy (69.6 μm, 30.0%, $P<0.01$), compared with normal subjects. There was no difference between non-proliferative diabetic retinopathy and normal subjects. Choroidal thickness is altered in diabetes and may be related to the severity of retinopathy. Presence of DME is associated with a significant decrease in the choroidal thickness [41].

One of the hypotheses could be that the choroidal vasculature that is in charge of oxygen supply, in phase of atrophic changes, cannot respond well and thus leads either to persistent macular changes or retinal ischemic changes that can underlie retinal neovascularization [41].

Fluorescein angiography (FA) is one of the oldest diagnostic methods that can show the diabetic maculopathy changes and ischemic retinal changes (Fig. 3).

The advantage of FA compared to other diagnostic methods is that not only the macula changes can be detected, but the monitoring of the retinal periphery can give us very useful information about the initial ischemic retinal changes. Standard FA investigations include a seven-field of investigation to detect the peripheral ischemic changes and retinal neovascularization. Ischemic changes in the anterior part of the retina could not be always detected with standard FA investigation and cannot afford monitoring of the hidden areas of non-perfused retina which can lead to severe ocular complications such as rubeotic glaucoma. Ultra-wide field FA is one of the newest diagnostic methods for clinical use, but unfortunately it is not in routine use and not yet readily available for detecting the peripheral retinal changes in diabetic patients. Clinical results show that ultra-wide-field fluorescein angiogram imaging, on average, demonstrated 3.2 times more total retinal surface area than the standard (seven field technique) 7SF. When compared with 7SF, ultra-wide-field fluorescein angiography showed 3.9 times more non-perfusion areas ($P<0.001$), 1.9 times more neovascularization ($P=0.036$), and 3.8 times more peripheral laser spots from the previous panretinal photocoagulation ($P<0.001$), which is a very important tool for further treatment decisions [39].

5 Ocular Pharmacotherapy

Laser photocoagulation is the "gold standard" of care for DME. Still a substantial group of patients do not respond to laser therapy. On the other hand, latest reports consider a lot of retinal damage as a result of the laser photocoagulation. The aim of the laser photocoagulation in most of the diabetic patients is considered to prevent the serious diabetic complications. Previously, the laser treatment for DME was the more commonly used therapy. It has been reported that 3 years after initial laser treatment, visual acuity improved in 14.5% of patients with DME, did not change in 60.9% and decreased in 24.6% [42, 43]. Unfortunately efficient treatment for ischemic macula changes still does not exist. The patients who develop ischemic changes achieve only clinical control without any efficient treatment of the ischemic diabetic maculopathy.

Most recent studies have focused on pharmacotherapy and its target components. Intravitreal injections have replaced the laser photocoagulation at the initial phase of treatment, as well as in recurrent DME changes.

6 Steroids

Steroids play an important role in inhibiting many factors that can lead to DME. The anti-inflammatory, angiostatic and antipermeability properties of these compounds can give us the rationale for steroid therapy in chronic DME. Corticosteroids have been shown to interfere with many regulatory components of gene expression inhibiting the expression of VEGF and key pro-inflammatory genes (tumor necrosis factor α and other inflammatory cytokines). The anti-inflammatory activity of steroids is also related to the inhibition of the phospholipase A2 pathway, to lower the generation of prostaglandins, and to reduce leukocyte chemotaxis. Several steroid-releasing implants have been designed to provide long-term drug delivery to the macula. These include non-biodegradable and biodegradable dexamethasone, flucinolone acetonide, and triamcinolon implants. Most of the studies report that the steroid implant effects on the DME are between 3 and 4 months. The recurrence of macular edema after this period of time is present in most patients. Side effects include increased intraocular pressure and cataract formation [44–46].

A number of diabetic conditions can be successfully treated in the early phase of morphological changes. Unfortunately with the current diagnostic methods, at the time when the changes are detected, the treatment with the recent medications cannot provide enough satisfactory results. An up-to date study has evaluated the effectiveness of Intravitreal Triamcinolone Acetonide (IVTA) on the reduction of diffuse DME with different optical coherence tomographic patterns. Each preoperative macular OCT image was classified according to its appearance as follows: sponge-like diffuse retinal thickening, cystoid macular edema (CME), and serous retinal detachment (SRD). The effectiveness of IVTA on diffuse diabetic macular edema was dependent on the optical coherence tomographic pattern, and IVTA was found to be more effective in patients with CME, while IVTA was less effective in those with SRD [47–49]. The results show that advanced stages of macular changes cannot be satisfactorily treated with steroid therapy [50].

Leukocyte adhesion plays an important role in DME. The endothelial damage resulting from the leukocyte adherence to vessel walls is mediated by nitric oxide adhesion molecules and other inflammatory mediators. Subtenon administration of Triamcinolone inhibits leukocyte-endothelium interaction in the retina and down regulates adhesion molecules of retinal vascular endothelium and thus decreases the retinal thickness [51].

"Up-to date" treatment modalities include different medications that can be combined for intravitreal administration.

A recent study evaluated the effect of intravitreal ranibizumab or triamcinolone in eyes which receiving additional focal/grid laser for DME and panretinal photocoagulation. Patients with DME received focal/grid laser, while diabetic retinopathy patients received prompt panretinal photocoagulation. Patients were randomly assigned to sham (n = 123), 0.5-mg ranibizumab (n = 113) at baseline and 4 weeks, and 4-mg triamcinolone at baseline and sham at 4 weeks (n = 109). Treatment was at investigator discretion from 14 weeks to 56 weeks. The results showed that the addition of one intravitreal triamcinolone injection or two intravitreal ranibizumab injections in eyes receiving focal/grid laser for diabetic macular edema and panretinal photocoagulation is associated with better visual acuity and decreased macular edema at 14 weeks. Whether continued long-term intravitreal treatment is beneficial cannot be determined from this study [51]. From our daily clinical practice we also know that combined laser and steroid, or anti-VEGF therapy can bring us better long-term results in the treatment of DME, compared to the standard monotherapy–pharmacotherapy or laser.

Among the trials that have examined the effectiveness of steroid therapy is a study which evaluated a Dexamethason implant 0.7 mg (Ozurdex) intravitreal, implanted in eyes with following vitrectomy for long lasting ME. Treatment with the dexamethasone intravitreal implant led to statistically and clinically significant improvements in both vision and vascular leakage from DME in difficult-to-treat vitrectomized eyes, but numerous adverse events such as conjunctival hemorrhage, conjunctival hyperemia, eye pain and increased intraocular pressure were observed [52].

Another recent study analyzed the efficacy and safety of triple therapy (subtenon triamcinolone, intravitreal bevacizumab [1.25 mg], and focal/grid laser) for refractory DME. In the triple therapy group, mean central foveal thickness significantly regressed from a baseline value of 441 to 298 μm at month 12 ($P<0.001$), but there was no significant change in best corrected visual acuity (BCVA). In the control group, there were no sustained significant changes of central foveal thickness or BCVA. A subgroup analysis of seven eyes in the triple therapy group with baseline BCVA of ≤20/100 showed significant BCVA improvements from 4 weeks to 9 months. The maximum improvement was achieved at 6 months, when the mean BCVA improved by 9.5 letters. Sustained reduction of central foveal thickness was achieved with triple therapy over the 1-year study period. Significant visual improvement was seen only in patients with worse baseline BCVA, but not in the triple therapy group as a whole. Significant side effects of intraocular pressure rise, ptosis, and cataractogenesis were encountered in the triple therapy group [51].

None of the current studies that have evaluated the effectiveness of steroid therapy on DME showed the results for more than 1 year. In our daily clinical practice we are aware that many patients treated with intravitreal drugs achieve an improvement of macular edema on OCT. Unfortunately, however, structural improvements are not always followed by functional improvements. The recurrence of macula changes could be seen in many patients, as soon as the concentration and the duration of the steroid therapy decreased.

7 VEGF Inhibitors

7.1 Pegaptanib (Macugen)

Pegaptanib sodium (Macugen, Eytech Pfizer) is an anti VEGF aptamer, a small piece of RNA that self-folds into a shape that binds to and blocks the effects of VEGF165 isoform. The Macugen Diabetic Retinopathy Study in a phase 2 Study compared the use of three different doses of Pegaptanib (0.3, 1, or 3 mg) with sham injection. Patients assigned to the Pegaptanib group had better outcomes in terms of visual acuity. Patients receiving 0.3 and 0.5 mg Pegaptanib gained ten or more letters of vision (34% vs. 10%, $P=0.0003$) and 15 or more letters (18% vs 7%, $P=0.12$) compared to the control group. Mean central thickness decreased by 68 μm compared to the sham group ($P=0.02$) [53].

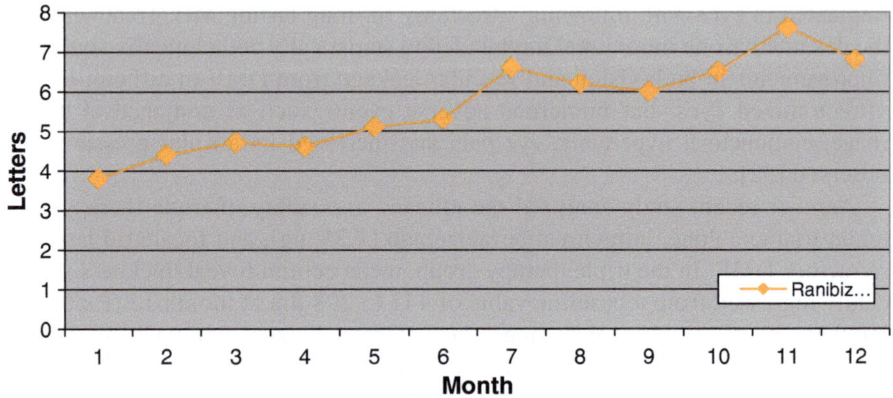

Fig. 4 Monotherapy with ranimizumab (RESTORE Study Results)

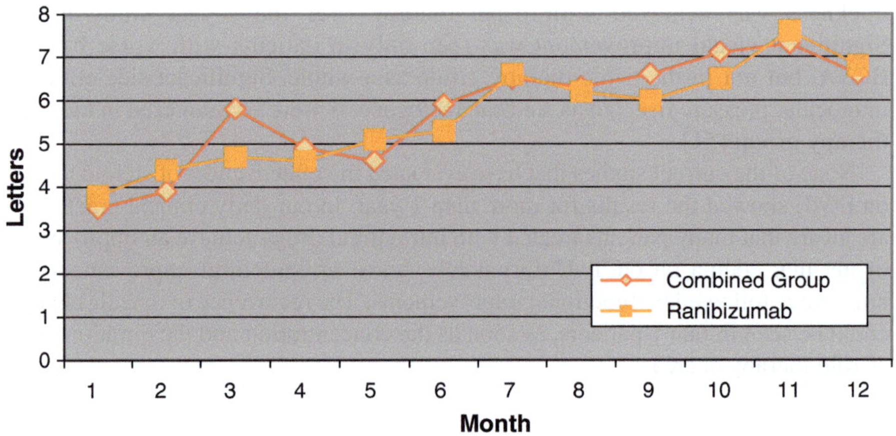

Fig. 5 Ranimizumab and laser therapy for DME (RESTORE Study Results)

7.2 Ranimizumab (Lucentis)

Ranimizumab (Lucentis, Genetech) is a humanized antigen-binding fragment directed against VEGF. It consists of two parts: a non-binding human sequence (humanized), making it less antigenic in humans, and a high affinity binding epitope (Fab fragment) derived from the mouse, which serves to bind the antigen. A phase 3 study on Efficacy and Safety of Ranimizumab in patients with visual impairment due to diabetic macular edema (RESTORE) showed that Ranibzumab mono or adjuvant therapy to laser has better results (Figs. 4 and 5) than laser therapy alone [54–56].

7.3 *Bevacizumab (Avastin)*

Bevazizumab (Avastin, Genentech) is full length humanized antibody against VEGF. It is used "off-label" intravitreally not only for treatment of DME but also in cases with retinal and iris NV. After the first report of the off-label use of intravitreal Bevacizumab in treating choroidal neovascularization, its use has rapidly gained popularity among ophthalmologist and is now widespread. Injection of Bevacizumab may promote faster resorption of vitreous hemorrhage, by stopping active leakage while administrating intravitreally. In one prospective study, 126 patients with chronic diffuse DME were treated with repeated intravitreal injections of Bevacizumab (1.25 mg). Patients were observed in intervals of 4–12 weeks for a period of 6–12 months. Within this period 48% received at least three injections of Bevacizumab. Visual acuity changes among the patients were significant with 5.1 ETDRS letters improvement from baseline at month 12 [26, 57, 58].

The loading dosage for the anti-VEGF drugs is three injections, repeated monthly. After the last drug administration the OCT findings [59] and the visual acuity are the main criteria for the renewal of the treatment. The disadvantage of the intravitreal therapy is that in most patients it should be repeated several times. Among the other complications (e.g., corneal defects, increased intraocular pressure and bleeding), the most severe complication (i.e., endophthalmitis) can occur with higher incidence in diabetic compared to non-diabetic patients.

Among drugs which still do not have clinical application in ophthalmology is Infliximab, an engineered antibody against tumour necrosis factor α. It is used intravenously in the treatment of systemic inflammatory diseases. Direct intravitreal administration has been investigated in a rabbit model and has been proven safe at dose of up to 1.7 mg [60].

Pigment epithelium-derived factor (PEDF) is an endogenous inhibitor of angiogenesis and its levels correlate directly with oxygen concentrations in animal models. In a murine model of angiogenic retinopathy, administration of PEDF showed complete inhibition of retinal vascular abnormalities. These results suggest that PEDF could be used in the treatment of DR.

Matrix metaloproteinases (MMPs) are endogenous peptidases that degrade extracellular matrix elements involved in endothelial migration. Intraocular MMP inhibition has been investigated in animal studies [60].

VEGF-Trap is a fusion protein of portions of VEGF receptors 1 and 2 and the Fc region of a human Ig G. It binds all VEGF isoforms and has been studied in a phase 1 study, which showed a good safety profile and encouraging efficacy results [25].

Sirolimus is an immunosuppressant drug approved for systemic use in renal transplantation. It inhibits the mammalian target of rapamycin. It can be injected either subconjuctivally or intravitreally. The phase 1 study data with sirolimus are encouraging in patients with DME [60].

Bevasiranib is a short interfering RNA designed to silence the VEGF gene. A phase 2 randomized trial study investigated the results of application of three doses of Bevasiranib and showed decrease in macular thickness between weeks 8–12 [60].

8 Current Status of Novel Therapeutic Agents in Clinical Trial

Significant improvement in diagnostic tools has greatly changed the management of DME. The clinical use of OCT has been important in determining a correct interpretation of pathogenic mechanisms involved in the appearance of DME. OCT is a precious tool when determining the efficacy of different treatment modalities [61–64]. Significant mechanistic discoveries, continued clinical investigation with new data from controlled clinical trials are now forthcoming each year. Many large pharmaceutical and biotechnology companies currently have active programs devoted to the identification, development and testing of novel approaches applicable to DME or DR.

A variety of therapeutic approaches in early stages of development of DME or DR include the gene therapy, anti-inflammatory agents, receptor tyrosine kinase inhibitors, integrin blockers, advanced glytation end products inhibitors, and vitreoretinal interface disruptors such as microplasmine.

The "Gold Standard" for over 40 years in the therapy of DR has been and remains laser photocoagulation [65, 66]. More recently, the new pattern scan laser has been introduced for clinical practice. Thermal damage induced by photocoagulation is influenced by laser wavelength, spot size, power intensity and exposure. By changing these parameters, resultant retinal collateral damage can be minimized.

The pattern scan laser photocoagulation (Opti Medica Corp. Santa Clara, California, USA) is a semi-automated photocoagulator system that allows delivery of multiple laser burns per activation of foot pedal, in a pen configured pattern array produced by scanner [67, 68].

It employs short duration pulses (10–30 ms) compared to 50–300 ms in conventional lasers. The result is significantly less thermal damage to the surrounding retinal tissue compared to the conventional photocoagulation [69].

Indications for laser treatment in diabetic patients include:

- Neovascularization at the disc (NVD) equal to or greater than 1/4-1/3 disc area
- Vitreous hemorrhage (VH) or pre retinal hemorrhage with NVD or neovascularization elsewhere (NVE), greater than ½ disc area.
- Neovascularization of iris (NVI) with NVD/NVE with or without VH.

Possible indications for laser treatment are the following: severe NPDR 4-2-1 Rule.

If two of the following meet the criteria then severe non-proliferative diabetic retinopathy (NPDR) laser photocoagulation can be initiated:

- Venous dilatation equal or greater than that shown in standard photograph 6A in 2 or more quadrants;
- Intraretinal microvascular abnormalities (IRMAs) equal to or greater than that shown in standard photograph 8A in at least one quadrant;
- Microaneurysms and hemorrhages equal to or greater than standard photograph 2A in all four retinal quadrants.

In cases of severe PDR with vitreal hemorrhage the only option is performing the pars plana vitrectomy (ppv) alone or in combination with cataract – phako surgery. In 1992 were published the surgical results after vitrectomy in DME. Nowadays not only ppv is recommended in persistent DME, but also ILM peeling. During postoperative follow-up, study results have shown no recurrent edema up to 31 months. The advantage of avoiding recurrences could be offset by potential damage of Müller cells whose basal membrane forms the ILM [70].

Several studies suggest that patients with diabetes may have an increased risk of developing postoperative endophthalmitis. Conditions such as cataract operation or ppv can lead to severe infections more frequently in diabetic patients compared to control patients. In a case control study of endophthalmitis following secondary lens implantation, 50% of patients had a history of diabetes compared with 5.9% of control patients. In a report of 162 patients treated for acute postoperative endophthlamitis, 21% had diabetes. Vitrectomy for complications of PDR often requires longer surgical time and postoperative recovery. The Endophthalmitis Vitrectomy Study reported that only 39% of diabetic patients compared with 55% of non-diabetic patients achieved a final visual acuity of 0.50 or better [70].

9 What Is the Economic Impact of Diabetes?

Diabetes and its complications have a significant economic impact on individuals, families, health systems and countries. For example, WHO estimates that in the period 2006–2015, China will lose $558 billion in national income due to heart disease, stroke and diabetes alone [71–73].

10 How Can the Burden of Diabetes Be Reduced?

10.1 Prevention

Simple lifestyle measures have been shown to be effective in preventing or delaying the onset of type 2 diabetes. To help prevent type 2 diabetes and its complications, recommended changes in lifestyle include [74, 75]:

- achieve and maintain healthy body weight;
- remain physically active – at least 30 min of regular, moderate-intensity activity on most days (more activity is required for weight control);
- eat a healthy diet of between three and five servings of fruit and vegetables a day and reduce sugar and saturated fats intake;
- avoid tobacco use – smoking increases the risk of cardiovascular diseases.

10.2 Measures Taken by WHO to Prevent and Control Diabetes

WHO aims to stimulate and support the adoption of effective measures for the surveillance, prevention and control of diabetes and its complications, particularly in low and middle-income countries.

To reach this goal, the WHO:

- provides scientific guidelines for diabetes prevention;
- develops norms and standards for diabetes care;
- builds awareness on the global epidemic of diabetes, including partnership with the International Diabetes Federation in the celebration of World Diabetes Day;
- conducts surveillance of diabetes and its risk factors.

11 Conclusion

What will be the current and future strategies in the treatment of diabetes and its complications?

As a result of insufficient standard therapy protocols, and despite increasing numbers of studies in the field, we still lack effective measures for early detection of diabetic changes. Furthermore, a standard therapy in every case of DME cannot produce satisfactory results. Yet, the ophthalmologist has to address the issue of which therapeutic modality would be indicated in early stage of the disease. Recent studies have shown that laser photocoagulation, steroid or anti-VEGF therapy can be used as treatment modalities in DR. The question remains as to which of these treatments can be effective in different patients?

Another problem in studying the development DR changes is the deficiency of information on animal models. All diabetic rodent models examined do not develop typical NV identical to that in PDR patients. The streptozotocon-induced diabetic rat model is a commonly used type 1 diabetes model. This model develops increased pericytes loss, increased vascular permeability, but no NV, even in cases of long hyperglycemia duration.

These limitations emphasize the need for a "tailor made" strategy in order to be effective in management of each patient. The "initial point of disease detection and treatment" should be recognized much earlier if we want to be efficient and reduce burden of diabetes and its complications. In this context, the goal of personalized medicine is to analyze each disease in a multidisciplinary way so that the profile of one patient could be established after considering multifunctional risk factors. This approach and establishing the idea of Predictive Preventive and Personalized Medicine may be of particular value to those many patients we see in our daily practice who do not respond sufficiently to standard treatment.

References

1. Diabetes Control and Complications Trial Research Group (1993) The effect of intensive treatment of diabetes on the development and progression of long-term complications in insulin-dependent diabetes mellitus. N Engl J Med 329:977–986
2. Ackerman SJ (1992) Benefits of preventive programs in eye care are visible on the bottom line: a new nationwide effort to improve eye care for people with diabetes gets backing from a study on the cost-effectiveness of screening for retinopathy. Diabetes Care 15:580–581
3. Williams R, Airey M, Baxter H, Forrester J, Kennedy-Martin T, Girach A (2004) Epidemiology of diabetic retinopathy and macular edema: a systematic review. Eye 18:963–983
4. Varma R, Torres M, Pena F, Klein R, Azen S, Los Angeles Latino Eye Study Group (2004) Prevalence of diabetic retinopathy in adult Latinos: the Los Angeles Latino eye study. Ophthalmology 111:1298–1306
5. Brinchmann-Hansen O, Dahl-Jorgensen K, Sandvik I, Hanssen KF (1992) Blood glucose concentrations and progression of diabetic retinopathy: the seven year results of the Oslo study. BMJ 304:19–22
6. Klein R (1994) Eye-care delivery for people with diabetes: an unmet need. Diabetes Care 17:614–615
7. Klein R, Sharrett AR, Klein BE, Moss SE, Folsom AR, Wong TY, Brancati FL, Hubbard LD, Couper D, ARIC Group (2002) The association of arteriosclerosis, vascular risk factors, and retinopathy in adults with diabetes: the arteriosclerosis risk in community study. Ophthalmology 109:1225–1234
8. Weinberger D, Fink-Cohen S, Gaton DD, Priel E, Yassur Y (1995) Non-retino-vascular leakage in diabetic maculopathy. Br J Ophthalmol 79:728–731
9. United Kingdom Prospective Diabetes Study Group (1998) Effect of intensive blood-glucose control with metformin on complications in overweight patients with type 2 diabetes (UKPDS 34). Lancet 352:854–865
10. Chew EY, Klein ML, Ferris FL 3rd, Remaley NA, Murphy RP, Chantry K, Hoogwerf BJ, Miller D (1996) Association of elevated serum lipid levels with retinal hard exudates in diabetic retinopathy. ETDRS Report Number 22. Arch Ophthalmol 114:1079–1084
11. Girach A, Lund-Andersen H (2007) Diabetic macular edema: a clinical overview. Int J Clin Pract 61:88–97
12. Gardner TW, Antonetti DA, Barber AJ, LaNoue KF, Levison SW (2002) Diabetic retinopathy: more than meets the eye. Surv Ophthalmol 47:253–262
13. Wilkinson CP, Ferris FL, Klein RE, Lee PP, Agardh CD, Davis M, Dills D, Kampik A, Pararajasegaran R, Verdaguer JT, Global Diabetic Retinopathy Project Group (2003) Proposed international clinical diabetic retinopathy and diabetic macular edema disease severity scales. Ophthalmology 110:1677–1682
14. Matthews DR, Stratton IM, Aldington SJ, Holman RR, Kohner EM, UK Prospective Diabetes Study Group (2004) Risks of progression of retinopathy and vision loss related to tight blood pressure control in type 2 diabetes mellitus UKPDS 69. Arch Ophthalmol 122:1631–1640
15. The United Kingdom Prospective Diabetes Study Group (1998) Tight blood pressure control and risk of macrovascular and microvascular complications in type 2 diabetes. UKPDS 38. Br Med J 317:703–713
16. Xia P, Aiello LP, Ishii H, Jiang ZY, Park DJ, Robinson GS, Takagi H, Newsome WP, Jirousek MR, King GL (1996) Characterization of vascular endothelial growth factor's effect on the activation of protein kinase C, its isoforms, and endothelial cell growth. J Clin Invest 98:2018–2026
17. Donahue ML, Phelps DL, Watkins RH, LoMonaco MB, Horowiz S (1996) Retinal vascular endothelial growth factor (VEGF) mRNA expression altered in relation to neovascularization in oxygen induced retinopathy. Curr Eye Res 15:175–184
18. Miller JW, Adamis AP, Shima DT, D'Amore PA, Moulton RS, O'Reilly MS, Folkman J, Dvorak HF, Brown LF, Berse B, Yeo TK, Yeo KT (1994) Vascular endothelial growth factor/vascular

permeability factor is temporally and spatially correlated with ocular angiogenesis in a primate model. Am J Pathol 145:574–584
19. Derevjanik NL, Vinores SA, Xiao W-H, Mori K, Turon T, Hudish T, Dong S, Campochiaro PA (2002) Quantitative assessment of the integrity of blood-retinal barrier in mice. Invest Ophthalmol Vis Sci 43:2462–2467
20. Park HY, Kwon HM, Lim HJ, Hong BK, Lee JY, Jang Y, Cho SY, Kim HS (2001) Potential role of leptin angiogenesis:leptin induces endothelial cell proliferation and expression of matrix metalloproteinases in vivo and in vitro. Exp Mol Med 33:95–102
21. Dorey CK, Aouididi S, Reynaud X, Dvorak HF, Brown LF (1996) Correlation of vascular permeability factor/vascular endothelial growth factor with extaretinal neovascularization in the rat. Arch Ophthalmol 114:1210–1217
22. Eyetech Study Group (2002) Preclinical and phase 1A clinical evaluation of an Anti-VEGF pegylated aptamer (EYE001) for the treatment of exudative age-related macular degeneration. Retina 22:143–152
23. Xeh PT, Yang CM, Huamg JS, Chien CT, Yang CH, Chiang YH, Shih YF (2008) Vitreous levels of reactive oxygen species in proliferative diabetic retinopathy. Ophthalmology 115:734–737
24. Antonetti DA, Lieth E, Barber AJ, Gardner TW (1999) Molecular mechanisms of vascular permeability in diabetic retinopathy. Semin Ophthalmol 14:240–248
25. Aiello LP, Northrup JM, Keyt BA, Takagi H, Iwamoto MA (1995) Hypoxic regulation of vascular endothelial growth factor in retinal cells. Arch Ophthalmol 113:1538–1544
26. Pierce EA, Foley ED, Smith LE (1997) Regulation of vascular endothelial growth factor by oxygen in a model of retinopathy of prematurity. Arch Ophthalmol 114:1219–1228
27. Growth Hormone Antagonist for Proliferative Diabetic Retinopathy Study Group (2000) The effect of a growth hormone receptor antagonist drug on proliferative diabetic retinopathy. Ophthalmology 108:2266–2272
28. Ishibashi T, Miki K, Sorgente N, Patterson R, Ryan SJ (1985) Effects of intravitreal administration of steroids on experimental subretinal neovascularization in the subhuman primate. Arch Ophthalmol 103:708–711
29. BenEzra D, Maftzir G (1996) Antibodies to IL-1 and TNF-α but not to bFGF or VEGF inhibit angiogenesis. Invest Ophthalmol Vis Sci 37:4664–4670
30. Hernández C, Segura RM, Fonollosa A, Carrasco E, Francisco G, Simó R (2005) Interleukin-8, monocyte chemoattractant protein-1 and IL-10 in the vitreous fluid of patients with proliferative diabetic retinopathy. Diabet Med 22:719–722
31. Joussen AM, Murata T, Tsujikawa A, Kirchhof B, Bursell SE, Adamis AP (2001) Leukocyte-mediated endothelial cell injury and death in the diabetic retina. Am J Pathol 158:147–152
32. Mohr S (2004) Potential new strategies to prevent the development of diabetic retinopathy. Expert Opin Investig Drugs 13:189–198
33. Burgansky-Eliash Z, Barak A, Barash H, Nelson DA, Pupko O, Lowenstein A, Grinvald A, Rubinstein A (2012) Increased retinal blood flow velocity in patients with early diabetes mellitus. Retina 32:112–119
34. Nauck M, Roth M, Tamm M, Eickelberg O, Wieland H, Stulz P, Perruchoud AP (1997) Induction of vascular endothelial growth factor by platelet-activating factor and platelet-derived growth factor is down regulated by corticosteroids. Am J Respir Cell Mol Biol 16:398–406
35. You JJ, Yang CM, Cheng MS, Yang CH (2012) Elevation of angiogenic factor Cysteine-rich 61 levels in vitreous of patients with proliferative diabetic retinopathy. Retina 32:103–111
36. Simo R, Higuera M, García-Ramírez M, Canals F, García-Arumí J, Hernández C (2008) Elevation of apolipoprotein A-I and apolipoprotein H levels in the vitreous fluid and over expression in the retina of diabetic patients. Arch Ophthalmol 126:1076–1081
37. Bolz M, Schmidt-Erfruth U, Deak G, Mylonas G, Kriechbaum K, Scholda C, Diabetic Retinopathy Research Group Vienna (2009) Optical coherence tomographic hyperreflective foci: a morphologic sign of lipid extravasation in diabetic macular edema. Ophthalmology 116:914–920

38. Brown JC, Solomin SD, Bressler SB, Schachat AP, DiBernardo C, Bressler NM (2004) Detection of diabetic fovel edema: contact lens biomicroscopy compared with optical coherence tomography. Arch Ophthalmol 122:330–335
39. Vujosevic S, Midena E (2007) Perimetry and the fundus. Diabet Retinopathy 5:177–187
40. Hee MR, Izat JA, Swarson EA (1995) Optical coherence tomography of the human retina. Arch Ophthalmol 113:325–332
41. Regatieri CV, Branchini L, Carmody J, Fujimoto JG, Duker JS (2012) Choroidal thickness in patients with diabetic retinopathy analyzed by spectral-domain optical coherence tomography. Retina 32:563–568
42. Early Treatment Diabetic Retinopathy Study Research Group (1991) Early treatment diabetic retinopathy study design and baseline patient characteristics. ETDRS report number 7. Ophthalmology 98:741–756
43. Early Treatment Diabetic Retinopathy Study Research Group (1985) Photocoagulation for diabetic macular edema. Early Treatment Diabetic Retinopathy Study report number 1. Arch Ophthalmol 103:1796–1806
44. Bakri SJ, Kaiser PK (2005) Posterior subtenon triamcinolone acetonide for refractory diabetic macular edema. Am J Ophthalmol 139:290–294
45. Chieh JJ, Roth DB, Liu M, Belmont J, Nelson M, Regillo C, Martidis A (2005) Intravitreal triamcinolone acetonide for diabetic macular edema. Retina 25:828–834
46. Jonas JB, Hayler JK, Sofker A, Panda-Jonas S (2001) Intravitreal injection of crystalline cortisone as adjunctive treatment of proliferative diabetic retinopathy. Am J Ophthalmol 131:468–471
47. Jonas JB, Kamppeter BA, Harder B, Vossmerbaeumer U, Sauder G, Spandau UH (2006) Intravitreal triamcinolone acetonide for diabetic macular edema: a prospective randomized study. J Ocul Pharmacol Ther 22:200–207
48. Jonas JB, Sofker A (2001) Intraocular injection of crystalline cortisone as adjunctive treatment of diabetic macular edema. Am J Ophthalmol 132:425–427
49. Ozdemir H, Karacorlu M, Karacorlu S (2005) Serous macular detachment in diabetic cystoid macular edema. Acta Ophthalmol Scand 83:63–66
50. Massin P, Audren F, Haouchine B, Erginay A, Bergmann JF, Benosman R, Caulin C, Gaudric A (2004) Intravitreal triamcinolone acetonide for diabetic diffuse macular edema: preliminary results of prospective controlled trial. Ophthalmology 111:218–224
51. Chan CK, Lai TY, Mohamed S, Lee VY, Liu DT, Li CL, Lam DS (2012) Combined high-dose sub-tenon Triamcinolone, intravitreal Bevacizumab, and laser photocoagulation for refractory diabetic macular edema: A Pilot Study. Retina 32:672–678
52. Boyer DS, Faber D, Gupta S, Patel SS, Tabandeh H, Li XY, Liu CC, Lou J, Whitcup SM, Ozurdex CHAMPLAIN Study Group (2011) Dexamethasone intravitreal implant for treatment of diabetic macular edema in vitrectomized patients. Retina 31:915–923
53. Adamis AP, Altaweel M, Bressler NM, Cunningham ET Jr, Davis MD, Goldbaum M, Gonzales C, Guyer DR, Barrett K, Patel M, Macugen Diabetic Retinopathy Study Group (2006) Changes in retinal neovascularization after pegaptanib (Macugen) therapy in diabetic individuals. Ophthalmology 113:23–28
54. Chun DW, Heier JS, Topping TM, Duker JS, Bankert JM (2006) A pilot study of multiple intravitreal injections of ranibizumab in patients with center-involving clinically significant diabetic macular edema. Ophthalmology 113:1706–1712
55. Diabetic Retinopathy Clinical Research Network, Scott IU, Edwards AR, Beck RW, Bressler NM, Chan CK, Elman MJ, Friedman SM, Greven CM, Maturi RK, Pieramici DJ, Shami M, Singerman LJ, Stockdale CR (2007) A phase 2 randomized clinical trial of intravitreal bevacizumab for diabetic macular edema. Ophthalmology 114:1860–1867
56. Gillies MC (2006) What we don't know about Avastin might hurt us. Arch Ophthalmol 124:1478–1479
57. Kook D, Wolf A, Kreutzer T, Neubauer A, Strauss R, Ulbig M, Kampik A, Haritoglou C (2008) Long-term effect of intravitreal bevacizumab (avastin) in patients with chronic diffuse macular edema. Retina 28:1053–1060

58. Pierce EA, Avery RL, Foley ED, Aiello LP, Smith LE (1995) Vascular endothelial growth factor/vascular permeability factor expression in a mouse model of retinal neovascularization. Proc Natl Acad Sci U S A 92:905–909
59. Catier A, Tadayoni R, Paques M, Erginay A, Haouchine B, Gaudric A, Massin P (2005) Characterization of macular edema according to various etiologies by optical coherence tomography. Am J Ophthalmol 140:200–206
60. Bandello F, Parodi M, Lanzetta P, Loewenstein A, Massin P, Menchini F, Veritti D (2010) Diabetic macular edema. Dev Ophthalmol 47:73–105
61. Kang SW, Park CY, Ham DI (2004) The correlation between fluorescein angiographic an optical coherence tomographic features in clinically significant diabetic macular edema. Am J Ophthalmol 137:313–322
62. Massin P, Girach A, Erginay A, Gaudric A (2006) Optical coherence tomography: a key to the future management of patients with diabetic macular oedema. Acta Ophthalmol Scand 84:466–474
63. Otani T, Kishi S, Maruyama Y (1999) Patterns of diabetic macular edema with optical coherence tomography. Am J Ophthalmol 127:688–693
64. Diabetic Retinopathy Study Group (1981) Photocoagulation treatment of proliferative diabetic retinopathy: relationship of adverse treatment effects to retinopathy severity. Diabetic retinopathy study report no. 5. Dev Ophthalmol 2:248–261
65. Fong DS, Girach A, Boney A (2007) Visual side effects of successful scatter laser photocoagulation surgery for proliferative diabetic retinopathy: a literature review. Retina 27:816–824
66. Blumenkranz MS, Yellachich D, Andersen DE, Wiltberger MW, Mordaunt D, Marcellino GR, Palanker D (2006) Semi automated patterned scanning laser for retinal photocoagulation. Retina 26:370–376
67. Mathur R (2012) Viteoretinal surgery. Laser Diabet Retinopathy 20:215–223
68. Sanghavi C, McLauchlan R, Delgado C, Young L, Charles SJ, Marcellino G, Stanga PE (2008) Initail experience with the pascal photocoagulator: a pilot study of 75 procedures. Br J Ophthalmol 92:1061–1064
69. Harbour JW, Smiddy WE, Flynn HWJ, Rubsamen PE (1996) Vitrectomy for diabetic macular edema associated with a thickened and taut posterior hyaloid membrane. Am J Ophthalmol 121:405–413
70. Smiddy WE, Feuer W, Irvine WD, Flynn HW Jr, Blankenship GW (1995) Vitrectomy for complications of proliferative diabetic retinopathy. Functional outcomes. Ophthalmology 102:1688–1695
71. Javitt JC, Canner JK, Sommer A (1989) Cost effectiveness of current approaches to the control of retinopathy in type 1 diabetic. Ophthalmology 96:255–264
72. Javitt JC, Aiello LP (1996) Cost-effectiveness of detecting and treating diabetic retinopathy. Ann Intern Med 124:164–169
73. King H, Aubert RE, Hermann WH (1998) Global burden of diabetes, 1995–2025: prevalence, numerical estimates, and projections. Diabetes Care 21:1414–1431
74. Kupfer C (1989) The challenge of transferring research results into patient care. Ophthalmology 96:737–738
75. Diabetes Control and Complications Trial Research Group (1996) Lifetime benefits and costs of intensive therapy as practiced in diabetes control and complications trial. J Am Med Assoc 276:1409–1415

The Role of Inflammatory Cytokines in Diabetic Nephropathy: Potential Use as Predictive Markers for Early Detection and Progression

Ahmed A. Elmarakby, Babak Baban, Jun Yao Liu, and Mahmood S. Mozaffari

Abstract Nephropathy is a major complication of diabetes mellitus and the leading cause of end-stage renal disease. The pathogenic roles of hyperglycemia and hypertension (systemic and/or intraglomerular) in diabetic nephropathy are well-established. Importantly, increasing evidence indicates that activated innate immunity and inflammation are major players in pathogenesis of diabetic nephropathy. This notion is supported by the recent literature indicating that both cytokines-chemokines and pro-fibrotic growth factors are pivotal players in the progression of diabetic nephropathy, as they effectively exacerbate inflammatory and fibrotic processes culminating in end-stage renal disease. Accordingly, we will focus on several predominant cytokines-chemokines as potential predictive markers for diabetic renal disease. These cytokines-chemokines may be helpful not only as biomarkers to monitor the progression of the disease but also to evaluate the impact of interventional modalities aimed at halting eventual manifestation of end-stage renal disease in diabetic patients.

Keywords Renal disease • Major complication of diabetes • Cytokines/chemokines • Advanced predictive diagnostics • Disease monitoring • Personalized patient treatment

---

A.A. Elmarakby (✉)
Department of Oral Biology, College of Dental Medicine, CL 2126,
Georgia Regents University, Augusta, GA 30912, USA
e-mail: aelmarakby@gru.edu

B. Baban • J.Y. Liu • M.S. Mozaffari
Department of Oral Biology, College of Dental Medicine, Georgia Regents University, Augusta, GA 30912, USA
e-mail: mmozaffa@gru.edu

M.S. Mozaffari (ed.), *New Strategies to Advance Pre/Diabetes Care: Integrative Approach by PPPM*, Advances in Predictive, Preventive and Personalised Medicine 3,
DOI 10.1007/978-94-007-5971-8_9, © Springer Science+Business Media Dordrecht 2013

1 Introduction

Nephropathy is a microvascular complication of both type 1 and type 2 diabetes mellitus [1]. It is the leading cause of end-stage renal disease (ESRD), and associated morbidity and mortality, worldwide [2–4]. The course of diabetic nephropathy remains unpredictable and the pathogenesis of progression is not completely understood. A number of risk factors have been suggested in the pathogenesis of diabetic nephropathy including low birth weight, endothelial dysfunction, smoking, renin-angiotensin system stimulation, obesity, hyperglycemia and hypertension (Fig. 1) [5–7]. The development of diabetic nephropathy is a gradual process which mainly starts with progression from normal albuminuria to microalbuminuria [urinary albumin excretion (UAE) 30–300 mg/24 h] for 5–10 years [1, 8, 9]. Thus, microalbuminuria could be a predictive marker for development of overt diabetic nephropathy in the uncontrolled diabetic patient. Nonetheless, with progression of the disease, blood pressure increases associated with marked macroalbuminuria (UAE >300 mg/day) and a relentless decline in glomerular filtration rate (GFR) [8, 9].

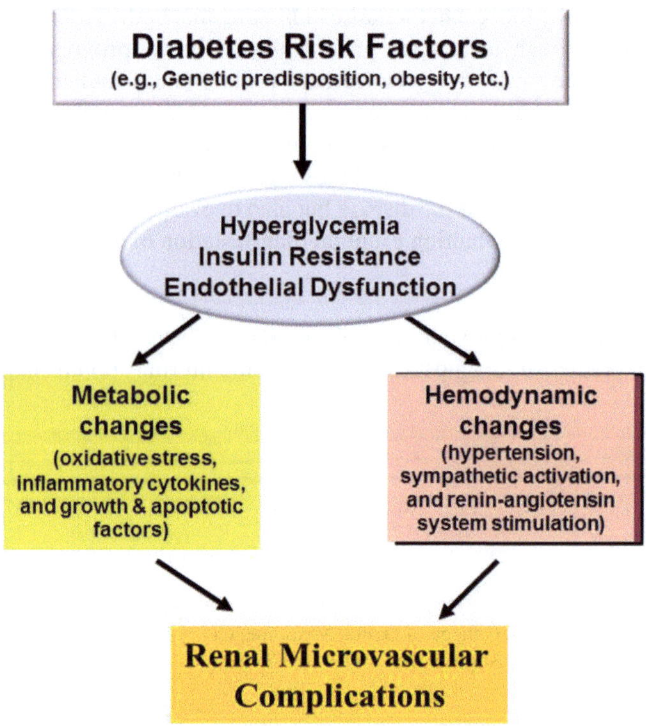

Fig. 1 A number of risk factors contribute to eventual manifestation of insulin resistance/hyperglycemia. In turn, hyperglycemia results in a myriad of metabolic and hemodynamic abnormalities that are intimately associated with microvascular complication of diabetes including nephropathy

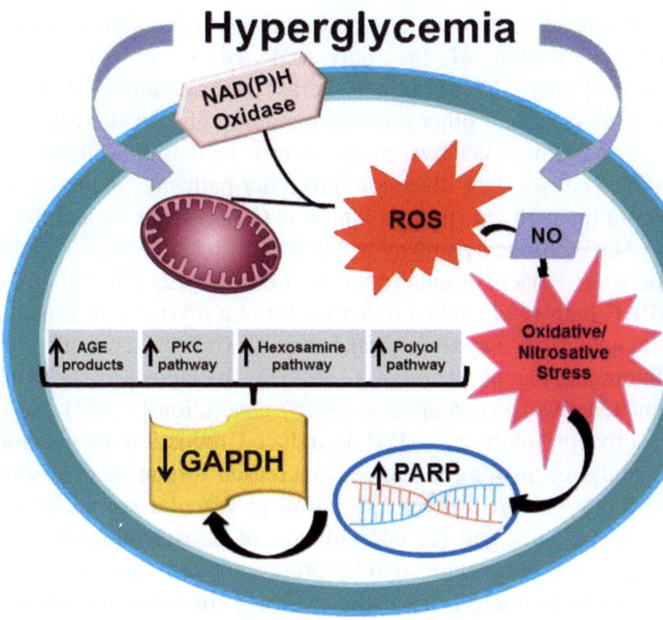

Fig. 2 Proposed mechanisms of intracellular effects of hyperglycemia. Hyperglycemia causes overproduction of ROS such as superoxide (e.g., from mitochondria and NAD(P)H oxidase) which, in turn, can react with NO to form peroxynitrite thereby culminating in increased oxidative/nitrosative stress. This results in DNA injury and overactivation of PARP. In turn, increased PARP activity causes addition of ADP-ribose polymer onto the GAPDH resulting in its inhibition. Inhibition of GAPDH will promote build-up of glycolytic intermediates and their subsequent diversion into and augmentation of the sorbitol/polyol, hexosamine, PKC and glycosylation pathways. *NADPH* reduced nicotinamide adenine dinucleotide phosphate, *eNOS* endothelial nitric oxide synthase, *PKC* protein kinase C, *ROS* reactive oxygen species, *NO* nitric oxide, *PARP* poly(ADP-ribose) polymerase, *GAPDH* glyceraldehydes-3-phosphate dehydrogenase, *AGE* advanced glycation end (products)

Hyperglycemia is a well-recognized causal factor for development of diabetic microvascular complications including nephropathy [10–12]. Under normal physiological conditions, there is a balance in the generation of oxygen free radicals and the antioxidant defense mechanisms that ameliorate free radical toxicity [13, 14]. Impairment in the oxidant/antioxidant equilibrium results in an elevation in oxidative stress in numerous pathological conditions including diabetes mellitus leading to cellular damage [14]. Accordingly, it is suggested that hyperglycemia results in increased generation of superoxide from a number of sources including the mitochondria, the NADPH oxidase and the uncoupled nitric oxide synthase (NOS) (Fig. 2) [15–21]. Superoxide exerts a diverse array of effects that are considered detrimental to the cell. For example, superoxide could scavenge nitric oxide leading to the generation of peroxynitrite [14, 22]. This will have the dual effects of removing/attenuating the protective effects of nitric oxide on the vasculature and generation of peroxynitrite which could damage proteins [18, 22]. However, as shown in Fig. 2, increased generation of superoxide and peroxynitrite also results in DNA damage and subsequent activation of the poly-ADP ribose polymerase [23, 24]. In turn, activation of poly-ADP ribose polymerase exerts a number of effects including poly-ADP ribosylation of other proteins including glycerladehyde-3-phosphate dehydrogenase, rendering it inactive. Subsequently, increased availability of glycolytic intermediates causes their diversion into other pathways including the polyol pathway, hexosamine pathway, protein kinase C (PKC) pathway and the advanced glycation end (AGE) products pathway which are ultimately linked to manifestation of diabetic micro- and macrovascular complications (Fig. 2) [25–29]. For example, activation of PKC pathway is linked to activation of a myriad of downstream pathways. These include (a) increased nuclear factor-κB (NF-κB) activation and subsequent pro-inflammatory gene expression [30, 31], (b) increased NADPH oxidase and subsequent reactive oxygen species (ROS) generation [32–34], (c) increased plasminogen activator inhibitor-1 (PAI-1)-induced reduction in fibrinolysis and consequent vascular occlusion [35, 36], (d) increased formation of vascular endothelial growth factor (VEGF) thereby resulting in vascular permeability changes and angiogenesis [37, 38] and (e) upregulation of transforming growth factor-β (TGF-β), collagen and fibronectin thereby promoting capillary occlusion and the increase in extracellular matrix deposition [39–41]. In summary, enhanced ROS production in experimental and clinical diabetes has been linked to increased vasoconstriction, vascular smooth muscle cell growth and migration, endothelial dysfunction, modification of extracellular matrix proteins, among other effects.

Careful control of glycemic status minimizes the symptoms of diabetic complications indicating that hyperglycemia is the main driving force behind the development of diabetic complications; however, strict glycemic control is difficult to be maintained. Current treatment strategies for diabetic nephropathy include glycemic and blood pressure control, low protein-diet, lipid lowering drugs, and interference with the renin-angiotensin system. Although these therapeutic interventions low the progression of diabetic nephropathy, the burden and mortality rate of the disease remain very high and the majority of patients with diabetic nephropathy continue to progress to end-stage renal disease. Therefore, more detailed understanding of molecular mechanisms for disease progression is needed. Both oxidative stress and inflammation are intimately linked with the development of diabetic nephropathy. Increases in oxidative stress can increase the production of inflammatory cytokines and likewise, an increase in inflammatory cytokines can stimulate the production of free radicals.

Clearly, the pathogenic mechanisms of hyperglycemia-induced diabetic complications are complex. However, as alluded to earlier, one consequence of hyperglycemia-induced ROS generation is activation of pro-inflammatory cascades which, in turn, increases transcription of genes encoding cytokines-chemokines, growth factors and extracellular matrix proteins (Fig. 3) [42, 43]. Thus, this chapter will focus on the emerging role of inflammatory cytokines-chemokines in relation to diabetic nephropathy. Given the devastating consequences of diabetic renal disease, early detection through identification of predictive biomarkers could potentially lead to prevention of disease progression, reduced morbidity and mortality and ultimately reduced heath care costs.

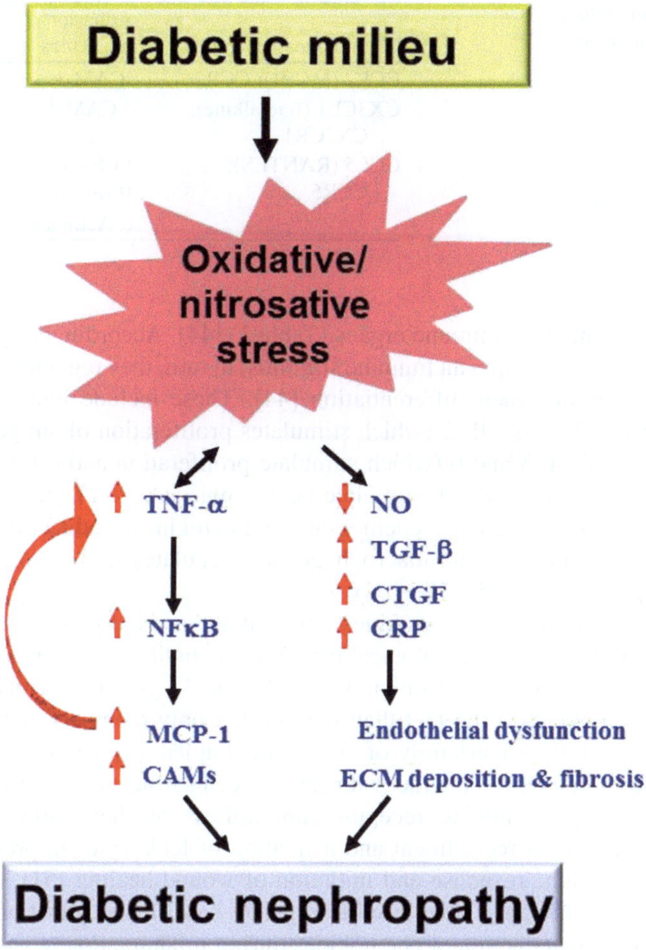

Fig. 3 Schematic diagram showing the proposed relationship between hyperglycemia, oxidative stress and inflammatory cytokines production in the pathogenesis and progression of diabetic nephropathy. *ROS* reactive oxygen species, *TNF*-α tumor necrosis factor-α, *NFκB* nuclear transcription factor-κ, *MCP-1* monocyte chemoattractant protein-1, *CAMs* cellular adhesion molecules, *NO* nitric oxide, *TGF*-β transforming growth factor-β, *CTGF* connective tissue growth factor, *CRP* C-reactive protein

2 Inflammatory Cytokines-Chemokines and Diabetic Nephropathy

Cytokines are redundant secreted polypeptides that exert a myriad of effects and regulate diverse physiological responses through autocrine, paracrine and juxtacrine actions. They exert important pleiotropic effects including regulation and determination of the nature of immune responses, immune cell trafficking and

Table 1 Inflammatory indicators in diabetic nephropathy

Chemokines; receptors	Adhesion molecules	Cytokines
CCL2 (MCP1); CCR2	ICAM-1	IL-1
CX3CL1 (fractalkine); CX3CR1	VCAM-1	IL-6
CCL5 (RANTES); CCR5	ECSAM	IL-18
	E-selectin	TNF-α
	α-Actinin 4	

cellular arrangements in immune organs (Table 1) [44]. Accordingly, cytokines are produced de novo following an immune stimulus; in turn, they regulate proliferation of immune cells and their differentiation [44]. These include interleukin (IL)-1 (which activates T cells), IL-2 (which stimulates proliferation of antigen-activated T and B cells), IL-4, 5 and 6 (which stimulate proliferation and differentiation of B-cells), IL-3 and IL-7 (which stimulate hematopoiesis), IL-17 (e.g., T-helper 17 cells and neutrophils; regulates chemokines and cytokines production), and IL-18 (e.g., activates monocytes and macrophages and regulates IL-13 production by T and natural killer cells) [45–47].

Chemokines (chemotactic cytokines), also small secreted proteins (7–10 kDa), play crucial roles in many pathological processes including infection, allergic reaction, autoimmune disease and inflammation (Table 1) [48, 49]. A distinguishing feature of chemokines is the fact that they are the only members of the cytokine family that act on the superfamily of G-protein coupled serpentine receptors [48]. Once released, chemokines interact with specific cell surface receptors (e.g., CC, C, CXC and CX3C chemokine receptor subfamilies) of their target cells [50]. Consequently, there is recruitment and activation of leukocytes in preparation for mounting an immune response and initiation of wound healing [51]. Constitutive production of chemokines is believed to regulate basal leukocytes trafficking. On the other hand, the induction of chemokines during inflammation (e.g., in response to early-response cytokines such as IL-1 and tumor necrosis factor) is crucial for recruitment of leukocytes to loci of immune reaction, a process regulated by mitogen-activated protein kinases [52, 53]. A noted feature of the process is the observation that the same stimulus can elicit production of multiple chemokines form the same cell thereby augmenting the inflammatory response. On the other hand, hemostatic chemokines are involved in adaptive immune responses such as lymphocyte trafficking, antigen sampling in secondary lymphoid tissues and immune surveillance [54].

Increasing evidence points to critical roles of pro-inflammatory cytokines in pathogenesis of diabetic nephropathy. For example, IL-1 is believed to increase vascular permeability and proliferation of mesangial cells and matrix deposition [4, 55]. On the other hand, IL-6 reportedly upregulates mesangial cell proliferation, increases fibronectin expression and affects extracellular matrix dynamics of mesangial cells and podocytes along with increased expression of adhesion molecules on endothelial cells and vascular smooth muscle cells [4, 56, 57]. Still other cytokines, such as tumor necrosis factor-α (TNF-α), impair balance among vasodilator and vasoconstriction mediators, upregulate production of ROS thereby contributing to alterations in glomerular capillary permeability barrier [4, 58]. Importantly, tubular epithelial cells express IL-18, an immunoregulatory cytokine belonging to the IL-1 superfamily; IL-18 induces the releases of interferon γ thereby leading to production of other inflammatory indicators (e.g., IL-1, TNF, and adhesion molecules) [59]. Collectively, these effects are believed to contribute to the functional alterations associated with diabetic nephropathy such as albuminuria and dysregulation of sodium homeostasis. For example, IL-18 is strongly associated with the progression of diabetic nephropathy leading to the suggestion that it may be a useful indicator for early detection of renal dysfunction in type 2 diabetes mellitus [59, 60]. Importantly, chemokine-induced inflammatory cell recruitment into renal tissue is a critical feature of various forms of renal disease including diabetic nephropathy. For example, during early phase of diabetic nephropathy, monocyte chemoattractant protein-1 (MCP-1) induces macrophage recruitment and accumulation; this has been reported both for experimental models of diabetes and diabetic patients [61–63]. For example, the db/db mouse is a widely used animal model of type 2 diabetic nephropathy as this mouse model lacks functional leptin receptor and confers marked degree of obesity and type 2 diabetes [64]. Knocking out of MCP-1 in db/db mice targeting macrophage recruitment and activation results in reduction of renal inflammation, albuminuria, glomerular and tubular injury, renal fibrosis and improves glomerular filtration rate [65]. Chemokines also increase expression of intercellular and vascular cellular adhesion molecules (ICAM-1, VCAM-1) and E-selectin leading to recruitment of more monocytes and macrophages to site of inflammation [66]. Furthermore, chemokines-induced proinflammatory gene activation in diabetes leads to further production of cytokines such as TNF-α, interleukins and interferon-γ thereby establishing a self-perpetuating cascade [67, 68]. The contributing factors that stimulate expression of these genes are components of the diabetic milieu including ROS, oxidized lipids, reduced nitric oxide (NO), increased angiotensin II, free fatty acids (FFA) and AGE products [4, 69]. Both endothelial cells and macrophages contribute to altered vascular reactivity and coagulation through increased expression of plasminogen activator inhibitor-1 (PAI-1) as well as through platelet activation and increased generation of coagulation factors such as fibrinogen and factor VIII [70]. Importantly, a frequent accompanying feature of patients with type 2 diabetes is overweight/obesity and adipose tissue is known to contribute importantly to the inflammatory process in these individuals in both vascular and nonvascular tissues [70]. Pro-inflammatory and procoagulant mediators released by adipose cells in obese subjects exert both local and systemic effects on vascular metabolism and function [71]. The adipose tissue of obese subjects contains activated macrophages that together with adipocytes produce inflammatory mediators such as MCP-1, macrophage inhibitory factor (MIF), TNFα, and IL-6 as well as vasoactive substances such as angiotensinogen and endothelin [71–73] and all together exacerbate inflammatory cycle.

The following describes the most relevant inflammatory markers that are currently used or are under investigation as predictive/diagnostic markers for diabetic nephropathy.

2.1 C-Reactive Protein (CRP)

CRP is an acute phase protein produced by liver cells in response to various inflammatory stimuli; it is found in the blood [4]. CRP is a member of the pentraxin family of oligomeric proteins which is believed to play a fundamental role in natural host defense and innate immunity [4, 74]. As a member of the class of acute-phase reactants, plasma CRP level rises dramatically during acute inflammatory processes [75, 76]. This is primarily due to the rise in the serum concentration of macrophage-derived IL-6, the most important stimulator of CRP production. Thus, CRP is a very sensitive predictor of inflammation and it has been associated with various inflammatory diseases such as atherosclerosis, diabetes and myocardial infarction [4, 77]. Immunoregulatory functions of CRP include enhancement of leukocyte reactivity, complement fixation, modulation of platelet activation and clearance of cellular debris from sites of active inflammation. It is well known that serum CRP levels could serve as a sensitive circulating marker of inflammation as CRP is independently associated with an increased risk of cardiovascular diseases and diabetes mellitus [78, 79]. In addition, experimental and clinical studies suggest that CRP and IL-6 are sensitive physiological markers of subclinical systemic inflammation which are associated with insulin resistance and hyperglycemia [80]. Further, CRP level increases at early stages of diabetic nephropathy and is independently associated with biomarkers of glomerular and tubulointerestitial damage (e.g., urinary albumin excretion and urinary N-acetyl-β-glucosaminidase) [81–83]. Collectively, these observations suggest that assessment of serum CRP may serve as a useful predictive marker for detection and progression of diabetic nephropathy (Table 2).

2.2 Tumor Necrosis Factor-Alpha (TNF-α)

Monocytes and macrophages are key inflammatory cells and components of the innate immunity [84]. Infiltration of monocytes into renal tissue increases the release of inflammatory mediators, including TNF-α which is a pleiotropic cytokine, produced mainly in macrophages and is involved in systemic inflammation [84]. It is well established that TNF-α plays a significant pathophysiological role in different experimental models of renal diseases such as lupus nephritis, crescentic glomerulonephritis, and the remnant kidney model of nephropathy [85–87]. In diabetic nephropathy, renal expression of TNF-α is increased compared to kidneys of non-diabetic animals [88]. Macrophages from obese db/db mice display enhanced expression of TNF-α. Glomerular and tubulointerestitial TNFα gene expression is

Table 2 Several inflammatory molecules in diabetic nephropathy: potential predictive roles

Molecule	Role in diabetic nephropathy	Possible prognostic value
CRP	An acute phase reactant that plays a fundamental role in natural host defense and innate immunity during renal injury	Plasma CRP could be used as an indicator of early stages of diabetic nephropathy
TNF-α	Is involved in systemic inflammation via increasing the formation of lymphocyte chemoattractant factors and cell surface expression of intercellular adhesion molecule-1 which has been implicated in the development of diabetic nephropathy	Urinary TNF-α levels could be used as a marker for the development and progression of diabetic nephropathy
MCP-1	Mediates macrophage infiltration and accumulation in diabetic nephropathy	Urinary MCP-1 levels is an important marker to detect the progression of later stages of diabetic nephropathy
ICAM-1	Promotes leukocyte adhesion and is involved in renal infiltration of macrophages in diabetic nephropathy	Urinary ICAM-1 could be used as an early marker for monitoring the progression of diabetic nephropathy
TGF-β	TGF-β plays an important role in the development of glomerulosclerosis and interstitial fibrosis in diabetes via enhancing glomerular extracellular matrix formation	Urinary TGF-β could be used as an early marker for the progression of diabetic nephropathy
CCN2	CCN2 is the key factor in stimulating connective tissue cell proliferation, extracellular matrix production, and other pro-fibrotic properties of TGF-β during diabetes	1. Urinary CCN2 levels: may be a marker of early stage of diabetic nephropathy 2. Plasma CCN2 levels: may be a marker for the late stage of diabetic nephropathy
IL-18	It is strongly associated with diabetic nephropathy	Early marker of renal dysfunction in type 2 diabetes

also increased in diabetic rats [83, 89]. Exposure of tubular epithelial cells to TNF-α increased the synthesis and secretion of lymphocyte chemoattractant factors as well as the cell surface expression of intercellular adhesion molecule-1 which has been implicated in the development of renal injury in diabetes [90]. It is important to note that TNF-α may not only be produced in the diabetic kidney by infiltrating macrophages but also intrinsically by renal cells such as endothelial, mesangial, glomerular and tubular epithelial cells [82, 83, 89]. Additionally, the cytotoxic effects of TNF-α can directly induce damage to glomerular, mesangial and epithelial cells [58, 83]. TNF-α also promotes the local generation of superoxide, which affects the barrier function of the glomerular capillary wall resulting in enhanced albumin permeability, independently of diabetic hemodynamic effects and inflammatory cytokines activation [58]. TNF-α has stimulatory effects on sodium uptake by proximal tubule cells contributing to sodium retention and renal hypertrophy [91]. In experimental models of diabetic nephropathy, renal TNF-α gene expression was also increased in diabetic rats, with a significant correlation between renal TNF-α expression and urinary TNF-α level [89, 92]. The increase in renal TNF-α production in diabetes appears to be related to hyperglycemia and formation of AGE products [84]. Diabetic patients had approximately three-fold higher serum TNF-α than non-diabetic individuals; however, serum TNF-α levels were increased only in diabetic patients with micro- or macroalbuminuria as well as in subjects with overt nephropathy and renal insufficiency suggesting a significant relationship between serum TNF-α and urinary albumin excretion [83]. Diabetic patients with micro- or macroalbuminuria also had elevated TNF-α expression and urinary concentrations of TNF-α [89, 92]. Urinary concentration of TNF-α may represent the local production of this cytokine within the kidney as experimental studies have demonstrated a direct and independent association between renal TNF-α expression and urinary TNF-α excretion with the severity of renal injury [89]. The elevation in urinary TNF-α concentration in diabetic patients with increased urinary albumin excretion was further exacerbated as diabetic nephropathy progressed [83]. Importantly, urinary TNF-α is significant and independently related to clinical markers of both glomerular and tubulointerestitial injury [83]. Thus, assessment of urinary TNF-α levels could be used as a predictive marker for the development and progression of renal injury in diabetic nephropathy (Table 2).

2.3 Intercellular Adhesion Molecules-1 (ICAM-1)

ICAM-1 is a cell-surface protein with five immunoglobulin-like domains [3]. Its production is induced by inflammatory cytokines such as TNF-α, interleukin-1, and interferon-γ [93]. Activation of PKC and shear stress in diabetes also induce ICAM-1 [94, 95]. ICAM-1 expression on the vascular endothelium and its binding to β_2 leukocyte integrins promotes leukocyte adhesion to the endothelium [3]. Clinically, ICAM-1 expression is up-regulated in renal diseases such as human glomerulonephritis and in experimental animal models of renal diseases such as glomerulonephritis, renal ablation and ischemia/reperfusion injury [96–99]. Renal ICAM-1 expression is increased during diabetic nephropathy [100, 101]. For example, studies demonstrated that macrophage infiltration and expression of adhesion molecules including ICAM-1 and selectins increase in the kidneys of patients with diabetic nephropathy, streptozotocin (STZ)-induced diabetic rats, and in type 2 diabetes [101–103]. Blocking ICAM-1 signaling, with anti–ICAM-1 antibody, prevented leukocyte influx into the glomeruli and reduced renal injury in diabetes [104]. Glomerular infiltration of macrophages and renal and glomerular hypertrophy increased in *ICAM-1*$^{+/+}$ diabetic mice and these changes were accompanied with overproduction of TGF-β, accumulation of glomerular type IV collagen, and albuminuria [3]. *ICAM-1*$^{-/-}$ diabetic mice showed a decrease in glomerular macrophage infiltration and hypertrophy and reduced glomerular TGF-β and type IV collagen when compared with *ICAM-1*$^{+/+}$ diabetic mice [3]. Infiltration of CD4$^+$ cells was decreased in the glomeruli of diabetic ICAM-1-deficient *db/db* obese mice compared with ICAM-1-intact *db/db* mice suggesting that ICAM-1 is critically involved in renal infiltration of macrophages in diabetic nephropathy [105]. Clinically, it has been shown that patients with both type 1 and type 2 diabetes mellitus and diabetic nephropathy have elevated concentrations of plasma ICAM-1 compared with subjects without renal injury indicating that assessment of plasma ICAM-1 levels can be used as a predictor of the development of renal damage [106, 107]. Recent studies also showed that renal ICAM-1 levels increased in diabetic patients and in experimental models of diabetic nephropathy [108, 109]. These findings suggest that early assessment of plasma and urinary ICAM-1 levels could be a useful marker for diagnosis of diabetic nephropathy and preventing disease progression (Table 2). Importantly, modulation of ICAM-1 activity may be a new venue for treating diabetic nephropathy especially with the current trials developing ICAM-1 antagonists [110].

2.4 Monocyte Chemoattractant Protein-1 (MCP-1)

Chemokine (C-C motif) ligand 2 (CCL2) is a small cytokine belonging to the CC chemokine family that is also known as MCP-1 [111]. MCP-1 mediates macrophage migration and recruits monocytes, memory T cells, and dendritic cells to sites of tissue injury and infection [111, 112]. It is a monomeric polypeptide, with a molecular weight of approximately 13 kDa [113]. As with many other CC chemokines, CCL2 is located on chromosome 17 in humans [113]. The cell surface receptors that bind CCL2 are chemokine receptor 2 and 4 (CCR2 and CCR4) [114].

The role of MCP-1 in the pathogenesis of diabetic renal complications is clearly established. For example, high concentrations of glucose and AGE products stimulate MCP-1 production from cultured mesangial cells, podocytes, and renal tubular epithelial cells *in vitro* [115–117]. Increased glomerular infiltration of macrophages was detected before deposition of extracellular matrix component [118] and MCP-1 was upregulated and has been shown to mediate macrophage infiltration and accumulation during the development of early diabetic nephropathy [7, 119]. In addition, increased amounts of MCP-1 were detected in renal biopsies and urine of patients with diabetic nephropathy [62, 120, 121]. MCP-1 knockout mice have recently been shown to be protected from streptozotocin (STZ) induced diabetic nephropathy and glomerular and tubulointerestitial injury were abrogated in these mice [119]. Blockade of the MCP-1/CCR2 also reduced glomerulosclerosis in STZ-induced diabetic mice [122]. These data suggest that MCP-1 plays a crucial role in the progression of diabetic nephropathy and inhibition of MCP-1 signaling could be an important therapeutic goal in the treatment of diabetic nephropathy.

Urinary MCP-1 was elevated in patients with diabetic nephropathy, but with different patterns and implications for progression of renal disease [7]. Elevated urinary MCP-1 levels in diabetic patients with macroalbuminuria were prognostic for deterioration in kidney function [7]. It was found that urinary MCP-1 levels were better correlated with the rate of deterioration of glomerular filtration rate (GFR) than urinary protein/creatinine ratio suggesting that urinary MCP-1 levels could

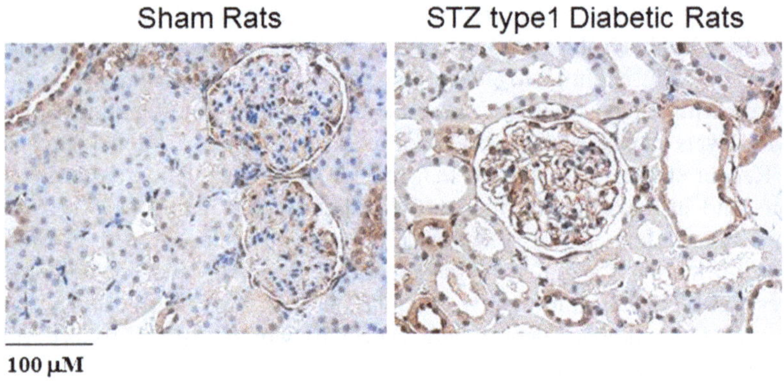

Fig. 4 Shows immunohistochemical assessment of collagen IV deposition in the kidney section from streptozotocin-induced type 1 diabetic rats and sham control rat. Collagen IV deposition (*dark brown*) increased in the kidney section from diabetic vs. sham control rat (images are shown at 200×)

provide a better prognostic tool in the latter stage of diabetic nephropathy where patients have symptoms of macroalbuminuria and worsening of GFR (Table 2) [7]. As indicated earlier, hypertension and poor glycemic control are established factors in the progression of diabetic nephropathy. However, neither HbA1c nor blood pressure assessment showed a significant correlation with urinary MCP-1 levels in patients with diabetic nephropathy [7]. Collectively, these observations raise the possibility that measurement of urinary MCP-1 levels could provide more accurate prognostic information in relation to the progression of diabetic nephropathy than blood pressure and glycemic control assessment.

2.5 *Transforming Growth Factor-β (TGF-β)*

Glomerular hypertrophy is a characteristic feature of diabetic nephropathy [123]. Recent evidence suggests that pro-fibrotic growth factors are involved in the pathogenesis of diabetic glomerular hypertrophy; these include increased mesangial matrix that is mainly composed of type IV collagen and fibronectin (Fig. 4) [123, 124]. This may relate to the fact that diabetes mellitus upregulates TGF-β expression and signaling, leading to increased collagen and fibronectin formation [125]. TGF-β, a prosclerotic cytokine, is a member of the transforming growth factor superfamily that controls proliferation and cellular differentiation; three isozymes have been identified and designated as TGF-β1, TGF-β2 and TGF-β3. Several experimental and clinical studies demonstrate a major role for TGF-β in development of glomerulosclerosis and interstitial fibrosis in diabetes [126] as TGF-β plays a central role in the enhancement of glomerular extracellular matrix production in diabetic nephropathy [125–127]. This is evident from studies indicating that TGF-β expression and activity increase in glomerular mesangial cells and proximal tubule cells during

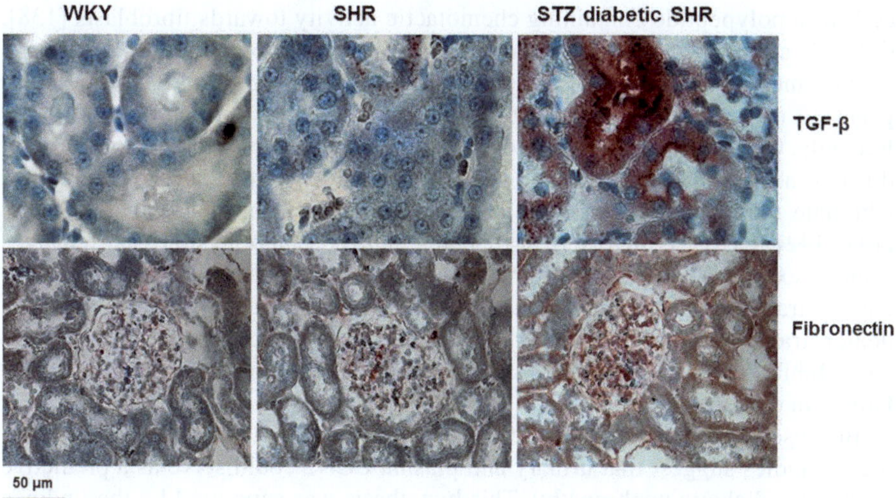

Fig. 5 Shows immunohistochemical assessment of TGF-β and fibronectin expression in kidney sections from normotensive Wistar Kyoto rats (WKY), spontaneously hypertensive rats (SHR) and streptozotocin-induced type 1 diabetic SHR. TGF-β and fibronectin expressions (*dark brown*) increased in the kidneys of diabetic vs. control WKY rat or SHR (images are shown at 200×)

diabetes [128–130]. TGF-β mRNA also increased in mesangial cells, podocytes and tubular epithelial cells in diabetic nephropathy and was further increased with the disease progression especially in the glomeruli [126]. TGF-β is also involved in interstitial fibrosis, another important histopathological change that correlates with diabetic renal dysfunction [43, 131]. We have recently shown that renal TGF-β expression increased in spontaneously hypertensive rats compared to normotensive rats and this increase was exacerbated upon induction of diabetes with streptozotocin in spontaneously hypertensive rats (SHR). The increase in TGF-β expression was also associated with increased in fibronectin expression in diabetic SHR compared to normotensive or control SHR (Fig. 5) [132]. Thus, as would be expected, inhibition of TGF-β prevented glomerular enlargement and reduced fibrosis in experimental models of diabetic nephropathy [133, 134]. The observation that urinary TGF-β excretion is elevated in diabetic patients with micro- and macroalbuminuria suggests a role for TGF-β in the development of diabetic nephropathy [135, 136]. Thus assessment of urinary TGF-β levels could serve as a predictive marker for early detection and monitoring of the progression of diabetic nephropathy (Table 2).

2.6 *Connective Tissue Growth Factor (CTGF)*

CCN2, formerly named connective tissue growth factor (CTGF), is now known to be a major downstream effector of TGF-β signaling [137, 138]. CCN2 is a 36- to 38-kDa protein that was first identified in conditioned media of endothelial cells as a polypeptide containing chemotactic activity towards fibroblasts [138]. CCN2 is a crucial factor in extracellular matrix production and other profibrotic activity mediated by TGF-β [138, 139]. It also plays an important role in angiogenesis, cell adhesion, migration, proliferation and differentiation [140]. Recently, CCN2 has been shown to play an important role in the pathogenesis of diabetic nephropathy [138]. High concentrations of glucose and AGE products stimulate the production of both TGF-β and CCN2 in mesangial cell cultures [141, 142]. CCN2 is involved in diabetes-induced pathophysiological changes such as extracellular matrix synthesis, cell migration, and epithelial-to-mesenchymal transition [143, 144]. Furthermore, upregulation of CCN2 has also been demonstrated in human and experimental models of diabetic nephropathy [145–147]. Inhibition of CCN2 signaling preserved the structure and function of the kidney in diabetic mice [148].

Because CCN2 is a secreted protein that can be detected in biological fluids, recent studies suggest that urinary and plasma CCN2 could serve as a predictive marker for diabetic nephropathy. This hypothesis was supported by the findings that both urinary CCN2 excretion and plasma CCN2 levels are elevated in patients with diabetic nephropathy [149, 150]. Although healthy individuals excrete low levels of urinary CCN2, patients with diabetic nephropathy and experimental animal models of diabetic nephropathy exhibit significant CCN2 excretion [149, 151, 152]. Interestingly, urinary CCN2 levels were highest at the early stage of diabetic nephropathy when patients were microalbuminuric and prospective follow up showed that elevation in urinary CCN2 predicted worsening of microalbuminuria [7]. In the streptozotocin-induced type 1 diabetes, urinary CCN2 excretion was elevated as early as 2 weeks following the development of diabetes, peaked during the early progression of diabetic nephropathy, and then decreased as animals became proteinuric [151]. The early increase in the urinary CCN2 in diabetic nephropathy supports the idea that renal fibrosis starts early in the pathogenesis of diabetic nephropathy, and that CCN2 is an important triggering factor in this process [7].

Plasma CCN2 was also evaluated as a possible marker for diabetic nephropathy and was found to be higher in patients with diabetic nephropathy than in patients with normoalbuminuria [138]. Plasma CCN2 level correlated with rate of decline in GFR and was an independent predictor of both end-stage renal disease and mortality in patients with type 1 diabetic nephropathy [138]. These data suggest that elevated plasma CCN2 is associated with decline in renal function in type 1 diabetic patients with severe proteinuria and assessment of plasma CCN2 has unique potential as a prognostic biomarker of renal function decline. Thus, assessment of plasma CCN2 levels could be used to improve prediction of end-stage renal disease and mortality in patients with type 1 diabetic nephropathy. The unique predictive value of plasma CCN2 in the progression of diabetic nephropathy, particularly in patients with severe proteinuria, suggests that CCN2 could also be used as a biomarker not only in diabetic patients who are likely to develop clinical nephropathy, but also in those who will exhibit rapid disease progression despite receiving appropriate treatment (Table 2).

3 Interleukins (ILs)

3.1 IL-1

IL-1 plays an important role in the progression of cardiovascular diseases as it increases chemotactic factors and adhesion molecules expression, enhances vascular endothelial permeability, and stimulates the proliferation of mesangial cells and matrix synthesis [4, 46]. IL-1 was first implicated in the development of diabetic nephropathy when glomerular basement proteins isolated from streptozotocin-induced diabetic male rats had significantly greater macrophage, TNF-α, and IL-1 production compared to control rats [153]. Renal IL-1 levels are also elevated in streptozotocin-induced type 1 diabetes in male Sprague–Dawley rats compared with to control rats [154] where renal expression of IL-1 is significantly correlated with urinary albumin excretion [155] suggesting the importance of assessment of renal IL-1 expression levels in the pathogenesis of diabetic nephropathy.

3.2 IL-6

Renal IL-6 expression is positively related to mesangial proliferation, tubular atrophy in diverse models of renal disease, supporting the role of IL-6 in the progression of renal disease [4]. Serum IL-6 levels are significantly higher in patients with type 2 diabetic nephropathy compared to levels observed in diabetic patients without nephropathy suggesting a role for IL-6 in the pathogenesis of diabetic nephropathy [156–158]. In fact, serum IL-6 levels are similar in type 2 diabetic patients with normal albumin excretion and microalbuminuria, but significantly increased in patients with severe diabetic nephropathy where macroalbuminuria is quite evident [159]. Serum IL-6 levels are also significantly greater in diabetic patients with overt proteinuria compared to normoalbuminuric and microalbuminuric patients [57, 160]. Using high resolution *in situ* hybridization in kidney biopsies of Japanese patients with diabetic nephropathy, it was found that interstitial expression of IL-6 mRNA correlated significantly with the degree of interstitial injury [161]. Clinical reports are in agreement with studies in experimental animals where renal expression of IL-6 mRNA is increased in streptozotocin diabetic rats compared to controls and levels are significantly associated with urinary albumin excretion [155]. These data support a role for IL-6 in the progression of diabetic nephropathy in the latter stages of the disease as assessment of serum IL-6 could provide a diagnostic tool for the late stage of diabetic nephropathy.

3.3 IL-18

IL-18 is a potent inflammatory cytokine secreted from activated monocytes/macrophages and it is known to induce interferon-γ (IFN-γ) which increases functional chemokine receptor expression in human mesangial cells [54]. IL-18 also stimulates the production of other inflammatory cytokines including IL-1, TNF-α, and IL-6, upregulates ICAM-1 expression, and induces endothelial cell apoptosis [158]. This raises the possibility that an increase in IL-18 is diabetic nephropathy may precede the observed increase in IL-6. IL-18 is constitutively expressed in renal tubular epithelia, and infiltrating monocytes, macrophages, and T cells and proximal tubular cells have all been all identified as potential sources of IL-18 production [46, 162]. Expression of IL-18 is increased in renal biopsies from patients with diabetic nephropathy in proximal and epithelial tubular cells [163] and patients with type 2 diabetes have significantly higher serum and urinary levels of IL-18 compared to healthy controls [158, 159, 164–166]. Moreover, there is a positive correlation between IL-18 levels in diabetic patients and the development of urinary albumin excretion, with the highest IL-18 levels found in patients with microalbuminuria and clinical albuminuria [158, 159, 164, 165] suggesting the potential diagnostic value of urinary and serum IL-18 in diabetic nephropathy.

3.4 Oxidative Stress Stimulates ILs Production

Increasing evidence supports a role for ILs in diabetic nephropathy; however, the stimulus for increased ILs production during diabetes remains elusive. A role for oxidative stress is implicated in increased ILs production during diabetes as hyperglycemia-induced oxidative stress and AGE products have been shown to increase inflammatory cytokines production. For example, acute hyperglycemia increases plasma IL-6 and IL-18 concentrations in healthy volunteers with either normal or impaired glucose tolerance profile and this effect was blocked by infusion of the antioxidant glutathione prior to the glucose infusion, suggesting that oxidative stress mediates the increase in ILs production [167]. In diabetic patients with poor glycemic control, TBARS, a marker of oxidative stress, and IL-6 levels are independently correlated with C-reactive protein, suggesting that oxidative stress promotes low-grade systemic inflammation in patients with type II diabetes [168]. Hyperglycemia increased oxidative stress and AGE products and their receptors (RAGE). In turn, RAGE stimulation activates NFκB which may lead to an increase in ILs production [169, 170]. This effect is aside from the direct ability of oxidative stress to activate transcription factors such as NFκB and activator protein-1 (AP-1) [171–174] leading to the transcription of ILs (either directly or through the induction of other cytokines).

ILs production could also increase in response to increased oxidative stress via the stimulation of alternative pathways in diabetes such as renin-angiotensin system activation. For example, angiotensin receptor blockers suppress oxidative stress and inflammation and provide protection against the progression of diabetic nephropathy in diabetic patients [175]. Treatment of streptozotocin-induced diabetic male Sprague–Dawley rats with the angiotensin converting enzyme inhibitor enalapril decreases the elevation in IL-6 expression, leading to a decrease in urinary IL-6 excretion and a decrease in albuminuria [155]. Alternatively, treatment of human renal proximal tubular epithelial cells with TGF-β increases IL-18 mRNA expression

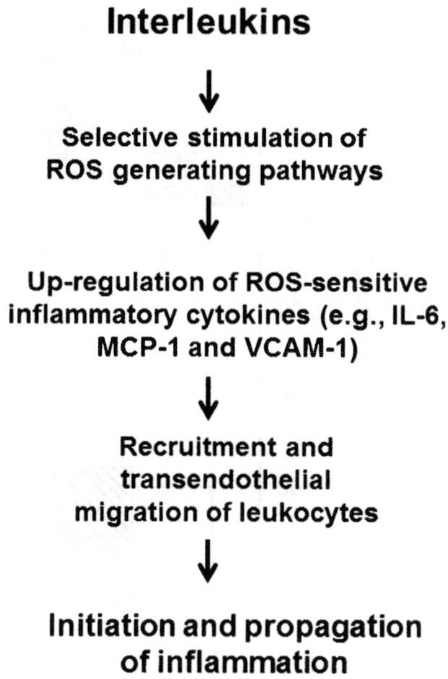

Fig. 6 Schematic diagram showing the proposed role of interleukins (ILs) and reactive oxygen species (ROS) in relation to initiation and propagation of inflammation in cardiovascular diseases

[163], and TGF-β is known to be increased by oxidative stress. Collectively, these observations suggest that elevated oxidative stress during diabetes is a potential mediator for increased ILs production and increased ILs levels could further increase oxidative stress thereby exacerbating the inflammatory cycle to trigger and propagate inflammation (Fig. 6).

4 Inflammatory Cytokines – Lessons from a Diabetic Animal Model

The obese Zucker rat (OZR) has an autosomal recessive mutation of the *fa* gene encoding the leptin receptor [176, 177]. As a result, it manifests marked obesity which becomes increasingly prominent, starting at about 6 weeks of age, compared to their lean counterparts. As shown in Fig. 7, both fasting plasma glucose and insulin levels are (significantly) higher in (6-month-old) OZR than lean Zucker rats (LZR) suggestive of marked insulin resistance of OZR; this is reflected from the calculated homeostatic model assessment (HOMA) index of insulin resistance, based on fasting plasma glucose and insulin levels, which is (significantly) higher for the OZR than LZR [177].

The marked compensatory hyperinsulinemia of OZR is associated with prominent structural changes of the endocrine pancreas. Microscopic examination of the pancreas from OZR reveals markedly enlarged hyperplastic islets of Langerhans exhibiting benign cellular proliferation which results in markedly irregular and jagged peripheral outline compared to those of the LZR which show sharply demarcated,

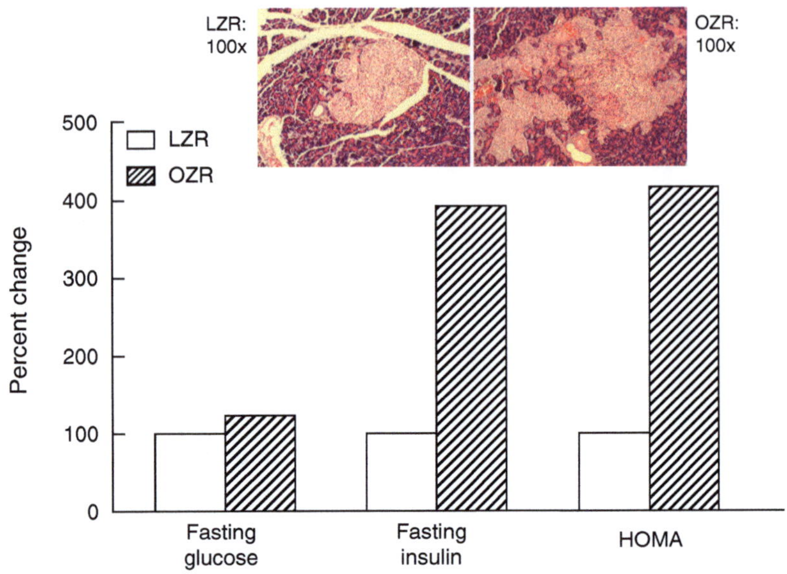

Fig. 7 Bar graphs show fasting plasma glucose and insulin levels as well as the homeostatic model assessment (HOMA) insulin resistance index of (6-month-old) obese Zucker rats (OZR), expressed as percent of values of lean Zucker rats (LZR). Also shown are hematoxylin-eosin stained sections of pancreas from LZR and OZR rats (insets, 100×)

small islets of Langerhans (Fig. 7). We have shown that pancreatic islets of OZR display prominent intercellular deposits of collagen fibers [177]. Collectively, these histological findings are consistent with the interpretation of islet hyperplasia with interstitial fibrosis of pancreatic islets of OZR.

A noted physical feature of the OZR is marked accumulation of fat within the abdominal cavity and consequent effective encasement of the kidney with fat deposits. Interestingly, local release of autocrine and paracrine factors is believed to contribute importantly to the pathogenesis of cardiac abnormalities in animals with dietary fat-induced obesity [178]. Nonetheless, the relevance of this mechanism for encapsulated organs, such as the kidney, remains to be established. More importantly, however, is the observation that OZR kidney displays numerous lipid droplets compared to LZR kidney (Fig. 8, panels a and b for LZR and OZR kidneys, respectively). Ectopic lipid accumulation is suggested to contribute importantly to organ dysfunction in obesity/type diabetes (i.e., lipotoxicity) as adipocytes are sources of a myriad factors including those promoting inflammatory changes [179]. Indeed, renal tissue of OZR shows foci of interstitial CD68 positive cells; whereas, examination of renal tissue from LZR does not show CD68 positive cells (Fig. 8, panels c and d for LZR and OZR kidneys, respectively). Consistent with the notion of upregulation of inflammatory processes associated with type 2 diabetes/obesity, the OZR group showed (significant) elevation in urinary excretion of MCP-1 compared to LZR (Fig. 9a) [177]. Interestingly, however, plasma CRP was similar between OZR and LZR [0.62±0.04 (n=8) vs. 0.58±0.07 (n=5) ng/ml, respectively). As described

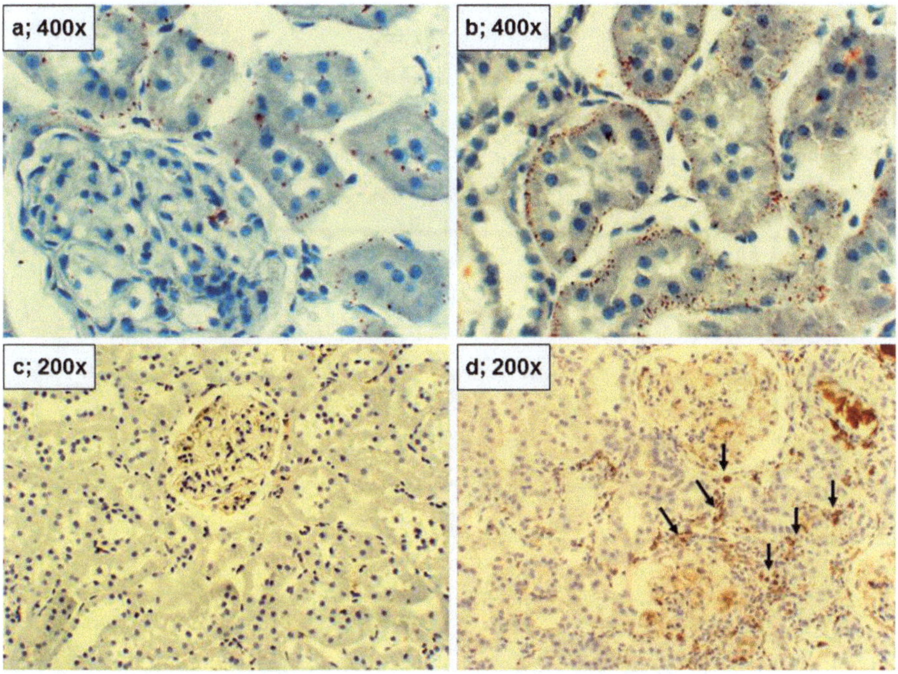

Fig. 8 Oil-Red-O stained kidney tissue from (6-month-old) OZR show more numerous lipid droplets compared to age-matched LZR (panels **a** and **b**; 400×). Also shown are immunostaining for CD 68 positive cells (*arrows*) in kidney tissue from experimental groups (panels **c** and **d**; 200×)

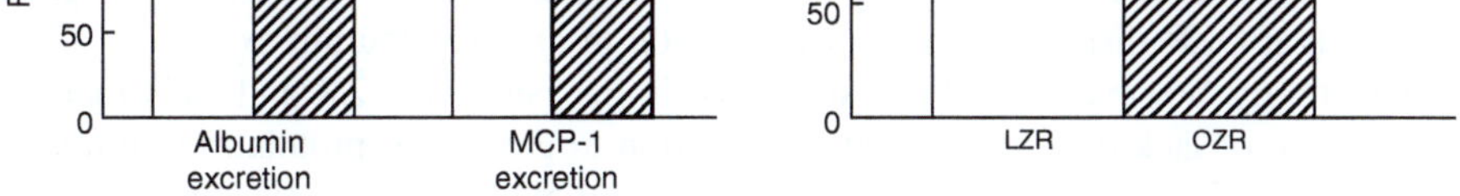

Fig. 9 Bar graphs show urinary excretions of albumin and monocyte chemoattractant protein-1 (MCP-1) expressed as the percent of the LZR group (panel **a**). Panel **b** shows that renal tissue from (6-month-old) OZR displayed higher level of intercellular adhesion molecule-1 (ICAM-1) compared to age-matched LZR. Also shown are representative blots for ICAM-1 and β-actin (i.e., loading control). Data are means ± SEM of 7–9 animals/group. $^{*}p < 0.05$ compared to LZR

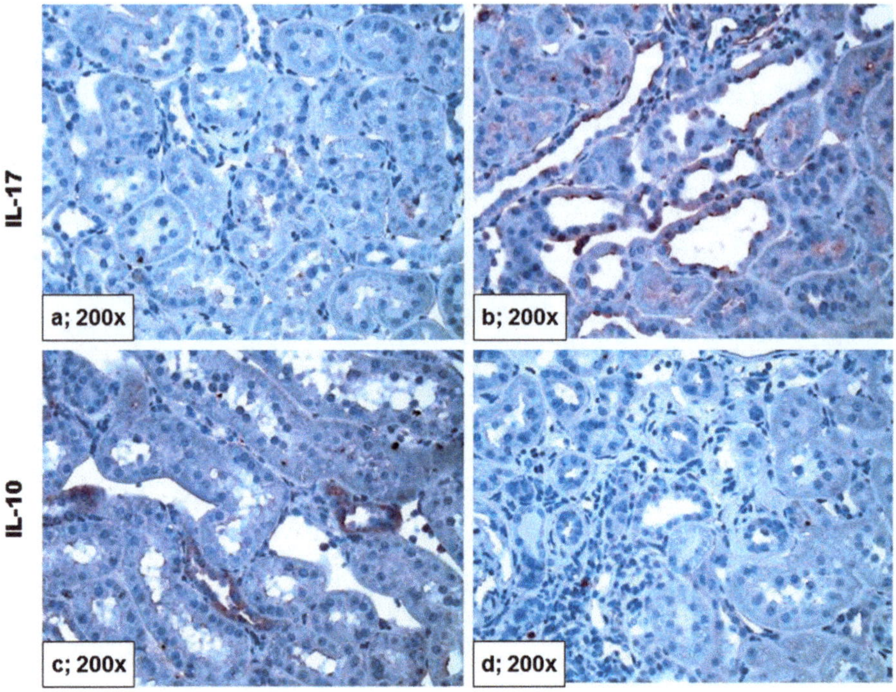

Fig. 10 Panels show interleukin (IL)-17 (**a**, **b**) and IL-10 (**c**, **d**) immunostaining in renal tissue of lean (**a**, **c**) and obese (**b**, **d**) Zucker rats

earlier, plasma CRP is an index of early changes in inflammatory status providing a likely explanation for lack of a significant differential between LZR and OZR at 6 months of age. Taken together, these observations reinforce the notion that while plasma CRP may serve as a useful predictive index at early stages of obesity/type 2 diabetes, urinary excretion of MCP-1 may be helpful in monitoring the progression of renal disease and the efficacy of interventional modalities. It is noteworthy that the kidney of OZR displays increased levels of ICAM-1 compared to that of LZR (Fig. 9b). This finding is consistent with our recent demonstration that other indices of inflammation and oxidative stress (i.e., tissue cyclooxygenase-2 and nitrotyrosine) are increased in kidneys of OZR than LZR [177]. As a functional correlate, Fig. 9a shows that OZR display (significant) elevation in urinary albumin excretion [177]. Interestingly, however, blood pressure was similar between OZR and LZR thereby suggestive of lack of a differential contribution to genesis of proteinuria (although greater sensitivity of OZR kidney to the same level of blood pressure cannot be ruled out) [177]. Nonetheless, it is important to note that multiple mechanisms contribute to genesis of proteinuria including oxidative stress and inflammatory chemokines and cytokines some of which were significantly elevated in OZR than LZR as described above [177]. More recently, we have shown that kidneys of OZR display increased immunostaining for IL-17 compared to their age-matched lean controls (Fig. 10a, b). IL-17 is novel pro-inflammatory cytokine which was originally discovered to be produced by a subset of T helper cells (i.e., Th17); indeed, its discovery has added to the previous repertoire of T helper cells (i.e., Th-1 and Th-2). Nonetheless, aside from modulation of T function, IL-17 is expressed in other cells including those of the lung, liver, spleen and kidney [47]. On the other hand, renal tubules of LZR show immunostaining for the anti-inflammatory cytokine, IL-10, compared to those of OZR (Fig. 10c, d). These novel observations indicate that an imbalance in pro- and anti-inflammatory cytokines in favor of the former contributes importantly to the pathogenesis of nephropathy in this animal model of prediabetes/insulin resistance. Thus, elucidation of the temporal and spatial relationships of cytokine expressions in renal tissue of animal models of diabetes and their comparison/correlation with those in plasma and/or urine should ultimately lead to identification of novel inflammatory molecules as predictive markers for early detection of renal disease in diabetes mellitus.

5 Conclusion

Over the past decade, a close association between inflammation and diabetes has become increasingly clear. It is also well-established that diabetic nephropathy is a devastating complication of diabetes that often leads to end-stage renal disease. The relationship between inflammatory cytokines-chemokines levels and the development and progression of diabetic nephropathy is very complex and requires further elucidation. However, it is increasingly becoming clear that many inflammatory cytokines-chemokines play significant roles in this scenario, including adipokines, chemoattractant cytokines, adhesion molecules, and pro-fibrotic cytokines. Because current treatment of diabetic nephropathy is still suboptimal, early assessment of levels of some inflammatory cytokines could potentially help with early diagnosis and halting disease progression to end-stage renal disease. Also, a better understanding of the role of inflammatory cytokines-chemokines in the progression of diabetic nephropathy should facilitate the development of novel treatment(s) and improvement of current therapeutic strategies. Nonetheless, given the complex web of mechanisms that contribute to the genesis of diabetic nephropathy, a pressing challenge is to identify predictive markers for early detection of the disease. However, it is unlikely that this important task can be relegated to a single marker. Rather, in order to increase the power of predictive strategies, multiple markers will be required to increase the likelihood of early detection of the disease process. Another important consideration is the need for to identify minimally- or non-invasive sources for obtaining these biomarkers such as those from the blood (or its components) and/or urine. Clearly, temporal changes in the levels of these biomarkers from such sources should reflect changes occurring in target organs such as the kidney. To that end, the OZR and db/db mice can serve as valuable investigational tools to compare and contrast the levels of molecular biomarkers in the organ of interest to those of blood (or its components) and/or urine.

References

1. Batlle D (2003) Clinical and cellular markers of diabetic nephropathy. Kidney Int 63:2319–2330
2. Ohga S, Shikata K, Yozai K, Okada S, Ogawa D, Usui H, Wada J, Shikata Y, Makino H (2007) Thiazolidinedione ameliorates renal injury in experimental diabetic rats through anti-inflammatory effects mediated by inhibition of NF-kappaB activation. Am J Physiol Renal Physiol 292:F1141–F1150
3. Okada S, Shikata K, Matsuda M, Ogawa D, Usui H, Kido Y, Nagase R, Wada J, Shikata Y, Makino H (2003) Intercellular adhesion molecule-1-deficient mice are resistant against renal injury after induction of diabetes. Diabetes 52:2586–2593
4. Rivero A, Mora C, Muros M, Garcia J, Herrera H, Navarro-Gonzalez JF (2009) Pathogenic perspectives for the role of inflammation in diabetic nephropathy. Clin Sci (Lond) 116:479–492
5. Rossing P (2006) Prediction, progression and prevention of diabetic nephropathy. The Minkowski Lecture 2005. Diabetologia 49:11–19
6. Rossing P (2006) Diabetic nephropathy: worldwide epidemic and effects of current treatment on natural history. Curr Diab Rep 6:479–483
7. Tam FW, Riser BL, Meeran K, Rambow J, Pusey CD, Frankel AH (2009) Urinary monocyte chemoattractant protein-1 (MCP-1) and connective tissue growth factor (CCN2) as prognostic markers for progression of diabetic nephropathy. Cytokine 47:37–42
8. Hovind P, Rossing P, Tarnow L, Smidt UM, Parving HH (2001) Progression of diabetic nephropathy. Kidney Int 59:702–709
9. Parving HH (2001) Diabetic nephropathy: prevention and treatment. Kidney Int 60:2041–2055
10. Brown WV (2008) Microvascular complications of diabetes mellitus: renal protection accompanies cardiovascular protection. Am J Cardiol 102:10L–13L
11. Steinke JM (2009) The natural progression of kidney injury in young type 1 diabetic patients. Curr Diab Rep 9:473–479
12. Lewko B, Stepinski J (2009) Hyperglycemia and mechanical stress: targeting the renal podocyte. J Cell Physiol 221:288–295
13. Anzalone R, La Rocca G, Di Stefano A, Magno F, Corrao S, Carbone M, Loria T, Lo Iacono M, Eleuteri E, Colombo M, Cappello F, Farina F, Zummo G, Giannuzzi P (2009) Role of endothelial cell stress in the pathogenesis of chronic heart failure. Front Biosci 14:2238–2247
14. Xiao H, Li Y, Qi J, Wang H, Liu K (2009) Peroxynitrite plays a key role in glomerular lesions in diabetic rats. J Nephrol 22:800–808
15. Munusamy S, MacMillan-Crow LA (2009) Mitochondrial superoxide plays a crucial role in the development of mitochondrial dysfunction during high glucose exposure in rat renal proximal tubular cells. Free Radic Biol Med 46:1149–1157
16. Serpillon S, Floyd BC, Gupte RS, George S, Kozicky M, Neito V, Recchia F, Stanley W, Wolin MS, Gupte SA (2009) Superoxide production by NAD(P)H oxidase and mitochondria is increased in genetically obese and hyperglycemic rat heart and aorta before the development of cardiac dysfunction. The role of glucose-6-phosphate dehydrogenase-derived NADPH. Am J Physiol Heart Circ Physiol 297:H153–H162
17. Weidig P, McMaster D, Bayraktutan U (2004) High glucose mediates pro-oxidant and antioxidant enzyme activities in coronary endothelial cells. Diabetes Obes Metab 6:432–441
18. Zheng L, Kern TS (2009) Role of nitric oxide, superoxide, peroxynitrite and PARP in diabetic retinopathy. Front Biosci 14:3974–3987
19. Thum T, Fraccarollo D, Schultheiss M, Froese S, Galuppo P, Widder JD, Tsikas D, Ertl G, Bauersachs J (2007) Endothelial nitric oxide synthase uncoupling impairs endothelial progenitor cell mobilization and function in diabetes. Diabetes 56:666–674
20. San Martin A, Foncea R, Laurindo FR, Ebensperger R, Griendling KK, Leighton F (2007) Nox1-based NADPH oxidase-derived superoxide is required for VSMC activation by advanced glycation end-products. Free Radic Biol Med 42:1671–1679

21. San Martin A, Du P, Dikalova A, Lassegue B, Aleman M, Gongora MC, Brown K, Joseph G, Harrison DG, Taylor WR, Jo H, Griendling KK (2007) Reactive oxygen species-selective regulation of aortic inflammatory gene expression in Type 2 diabetes. Am J Physiol Heart Circ Physiol 292:H2073–H2082
22. Liang JH, Li YN, Qi JS, Jia XX (2010) Peroxynitrite-induced protein nitration is responsible for renal mitochondrial damage in diabetic rat. J Endocrinol Invest 33:140–146
23. Narasimhan P, Fujimura M, Noshita N, Chan PH (2003) Role of superoxide in poly(ADP-ribose) polymerase upregulation after transient cerebral ischemia. Brain Res Mol Brain Res 113:28–36
24. Ahmad R, Rasheed Z, Ahsan H (2009) Biochemical and cellular toxicology of peroxynitrite: implications in cell death and autoimmune phenomenon. Immunopharmacol Immunotoxicol 31:388–396
25. Szabo C (2009) Role of nitrosative stress in the pathogenesis of diabetic vascular dysfunction. Br J Pharmacol 156:713–727
26. Hammes HP (2003) Pathophysiological mechanisms of diabetic angiopathy. J Diabetes Complications 17:16–19
27. Ceriello A (2003) New insights on oxidative stress and diabetic complications may lead to a "causal" antioxidant therapy. Diabetes Care 26:1589–1596
28. Forbes JM, Fukami K, Cooper ME (2007) Diabetic nephropathy: where hemodynamics meets metabolism. Exp Clin Endocrinol Diabetes 115:69–84
29. Brownlee M (2005) The pathobiology of diabetic complications: a unifying mechanism. Diabetes 54:1615–1625
30. Kumar A, Hawkins KS, Hannan MA, Ganz MB (2001) Activation of PKC-beta(I) in glomerular mesangial cells is associated with specific NF-kappaB subunit translocation. Am J Physiol Renal Physiol 281:F613–F619
31. Park CW, Kim JH, Lee JH, Kim YS, Ahn HJ, Shin YS, Kim SY, Choi EJ, Chang YS, Bang BK (2000) High glucose-induced intercellular adhesion molecule-1 (ICAM-1) expression through an osmotic effect in rat mesangial cells is PKC-NF-kappa B-dependent. Diabetologia 43:1544–1553
32. Wei XF, Zhou QG, Hou FF, Liu BY, Liang M (2009) Advanced oxidation protein products induce mesangial cell perturbation through PKC-dependent activation of NADPH oxidase. Am J Physiol Renal Physiol 296:F427–F437
33. Thallas-Bonke V, Thorpe SR, Coughlan MT, Fukami K, Yap FY, Sourris KC, Penfold SA, Bach LA, Cooper ME, Forbes JM (2008) Inhibition of NADPH oxidase prevents advanced glycation end product-mediated damage in diabetic nephropathy through a protein kinase C-alpha-dependent pathway. Diabetes 57:460–469
34. Tojo A, Asaba K, Onozato ML (2007) Suppressing renal NADPH oxidase to treat diabetic nephropathy. Expert Opin Ther Targets 11:1011–1018
35. Baccora MH, Cortes P, Hassett C, Taube DW, Yee J (2007) Effects of long-term elevated glucose on collagen formation by mesangial cells. Kidney Int 72:1216–1225
36. Ha H, Lee HB (2005) Reactive oxygen species amplify glucose signalling in renal cells cultured under high glucose and in diabetic kidney. Nephrology (Carlton) 10(Suppl):S7–S10
37. Xia L, Wang H, Munk S, Frecker H, Goldberg HJ, Fantus IG, Whiteside CI (2007) Reactive oxygen species, PKC-beta1, and PKC-zeta mediate high-glucose-induced vascular endothelial growth factor expression in mesangial cells. Am J Physiol Endocrinol Metab 293:E1280–E1288
38. Frank RN (2009) Treating diabetic retinopathy by inhibiting growth factor pathways. Curr Opin Investig Drugs 10:327–335
39. Wu D, Peng F, Zhang B, Ingram AJ, Kelly DJ, Gilbert RE, Gao B, Krepinsky JC (2009) PKC-beta1 mediates glucose-induced Akt activation and TGF-beta1 upregulation in mesangial cells. J Am Soc Nephrol 20:554–566
40. Meier M, Menne J, Park JK, Holtz M, Gueler F, Kirsch T, Schiffer M, Mengel M, Lindschau C, Leitges M, Haller H (2007) Deletion of protein kinase C-epsilon signaling pathway induces glomerulosclerosis and tubulointerstitial fibrosis in vivo. J Am Soc Nephrol 18:1190–1198

41. Meier M, Park JK, Overheu D, Kirsch T, Lindschau C, Gueler F, Leitges M, Menne J, Haller H (2007) Deletion of protein kinase C-beta isoform in vivo reduces renal hypertrophy but not albuminuria in the streptozotocin-induced diabetic mouse model. Diabetes 56:346–354
42. Wolf G (2004) New insights into the pathophysiology of diabetic nephropathy: from haemodynamics to molecular pathology. Eur J Clin Invest 34:785–796
43. Kanwar YS, Wada J, Sun L, Xie P, Wallner EI, Chen S, Chugh S, Danesh FR (2008) Diabetic nephropathy: mechanisms of renal disease progression. Exp Biol Med (Maywood) 233:4–11
44. Goldberg RB (2009) Cytokine and cytokine-like inflammation markers, endothelial dysfunction, and imbalanced coagulation in development of diabetes and its complications. J Clin Endocrinol Metab 94:3171–3182
45. Sarafidis PA, Ruilope LM (2006) Insulin resistance, hyperinsulinemia, and renal injury: mechanisms and implications. Am J Nephrol 26:232–244
46. Navarro-Gonzalez JF, Mora-Fernandez C (2008) The role of inflammatory cytokines in diabetic nephropathy. J Am Soc Nephrol 19:433–442
47. D'Acquisto F, Maione F, Pederzoli-Ribeil M (2010) From IL-15 to IL-33: the never-ending list of new players in inflammation. Is it time to forget the humble aspirin and move ahead? Biochem Pharmacol 79:525–534
48. Puneet P, Moochhala S, Bhatia M (2005) Chemokines in acute respiratory distress syndrome. Am J Physiol Lung Cell Mol Physiol 288:L3–L15
49. Ransohoff RM (2009) Chemokines and chemokine receptors: standing at the crossroads of immunobiology and neurobiology. Immunity 31:711–721
50. Moser B, Loetscher P (2001) Lymphocyte traffic control by chemokines. Nat Immunol 2:123–128
51. Bhatia M, Moochhala S (2004) Role of inflammatory mediators in the pathophysiology of acute respiratory distress syndrome. J Pathol 202:145–156
52. Keane MP, Strieter RM (2000) Chemokine signaling in inflammation. Crit Care Med 28:N13–N26
53. Lacotte S, Brun S, Muller S, Dumortier H (2009) CXCR3, inflammation, and autoimmune diseases. Ann N Y Acad Sci 1173:310–317
54. Ward SG, Marelli-Berg FM (2009) Mechanisms of chemokine and antigen-dependent T-lymphocyte navigation. Biochem J 418:13–27
55. Sedor JR, Konieczkowski M, Huang S, Gronich JH, Nakazato Y, Gordon G, King CH (1993) Cytokines, mesangial cell activation and glomerular injury. Kidney Int Suppl 39:S65–S70
56. Royall JA, Berkow RL, Beckman JS, Cunningham MK, Matalon S, Freeman BA (1989) Tumor necrosis factor and interleukin 1 alpha increase vascular endothelial permeability. Am J Physiol 257:L399–L410
57. Dalla Vestra M, Mussap M, Gallina P, Bruseghin M, Cernigoi AM, Saller A, Plebani M, Fioretto P (2005) Acute-phase markers of inflammation and glomerular structure in patients with type 2 diabetes. J Am Soc Nephrol 16(Suppl 1):S78–S82
58. McCarthy ET, Sharma R, Sharma M, Li JZ, Ge XL, Dileepan KN, Savin VJ (1998) TNF-alpha increases albumin permeability of isolated rat glomeruli through the generation of superoxide. J Am Soc Nephrol 9:433–438
59. Fujita T, Shimizu C, Fuke Y, Satomura A, Abe M, Kaizu K, Matsumoto K, Soma M (2011) Serum interleukin-18 binding protein increases with behavior different from IL-18 in patients with diabetic nephropathy. Diabetes Res Clin Pract 92:e66–e69
60. Uzu T, Yokoyama H, Itoh H, Koya D, Nakagawa A, Nishizawa M, Maegawa H, Yokomaku Y, Araki S, Abiko A, Haneda M (2011) Elevated serum levels of interleukin-18 in patients with overt diabetic nephropathy: effects of miglitol. Clin Exp Nephrol 15:58–63
61. Chiarelli F, Cipollone F, Mohn A, Marini M, Iezzi A, Fazia M, Tumini S, De Cesare D, Pomilio M, Pierdomenico SD, Di Gioacchino M, Cuccurullo F, Mezzetti A (2002) Circulating monocyte chemoattractant protein-1 and early development of nephropathy in type 1 diabetes. Diabetes Care 25:1829–1834

62. Morii T, Fujita H, Narita T, Shimotomai T, Fujishima H, Yoshioka N, Imai H, Kakei M, Ito S (2003) Association of monocyte chemoattractant protein-1 with renal tubular damage in diabetic nephropathy. J Diabetes Complications 17:11–15
63. Takebayashi K, Matsumoto S, Aso Y, Inukai T (2006) Association between circulating monocyte chemoattractant protein-1 and urinary albumin excretion in nonobese Type 2 diabetic patients. J Diabetes Complications 20:98–104
64. Nam DH, Lee MH, Kim JE, Song HK, Kang YS, Lee JE, Kim HW, Cha JJ, Hyun YY, Kim SH, Han SY, Han KH, Han JY, Cha DR (2012) Blockade of cannabinoid receptor 1 improves insulin resistance, lipid metabolism, and diabetic nephropathy in db/db mice. Endocrinology 153:1387–1396
65. Chow FY, Nikolic-Paterson DJ, Ma FY, Ozols E, Rollins BJ, Tesch GH (2007) Monocyte chemoattractant protein-1-induced tissue inflammation is critical for the development of renal injury but not type 2 diabetes in obese db/db mice. Diabetologia 50:471–480
66. Pace TW, Miller AH (2009) Cytokines and glucocorticoid receptor signaling. Relevance to major depression. Ann N Y Acad Sci 1179:86–105
67. Hohmeier HE, Tran VV, Chen G, Gasa R, Newgard CB (2003) Inflammatory mechanisms in diabetes: lessons from the beta-cell. Int J Obes Relat Metab Disord 27(Suppl 3):S12–S16
68. Kim KA, Lee MS (2009) Recent progress in research on beta-cell apoptosis by cytokines. Front Biosci 14:657–664
69. Rudijanto A (2007) The expression and down stream effect of lectin like-oxidized low density lipoprotein 1 (LOX-1) in hyperglycemic state. Acta Med Indones 39:36–43
70. Fonseca VA (2007) Rationale for the use of insulin sensitizers to prevent cardiovascular events in type 2 diabetes mellitus. Am J Med 120:S18–S25
71. Ritchie SA, Connell JM (2007) The link between abdominal obesity, metabolic syndrome and cardiovascular disease. Nutr Metab Cardiovasc Dis 17:319–326
72. Karalis KP, Giannogonas P, Kodela E, Koutmani Y, Zoumakis M, Teli T (2009) Mechanisms of obesity and related pathology: linking immune responses to metabolic stress. FEBS J 276:5747–5754
73. Vernochet C, Peres SB, Farmer SR (2009) Mechanisms of obesity and related pathologies: transcriptional control of adipose tissue development. FEBS J 276:5729–5737
74. Ridker PM, Paynter NP, Rifai N, Gaziano JM, Cook NR (2008) C-reactive protein and parental history improve global cardiovascular risk prediction: the Reynolds Risk Score for men. Circulation 118:2243–2251
75. Correale M, Brunetti ND, De Gennaro L, Di Biase M (2008) Acute phase proteins in atherosclerosis (acute coronary syndrome). Cardiovasc Hematol Agents Med Chem 6:272–277
76. Steptoe A, Hamer M, Chida Y (2007) The effects of acute psychological stress on circulating inflammatory factors in humans: a review and meta-analysis. Brain Behav Immun 21:901–912
77. Blake GJ, Ridker PM (2002) Tumour necrosis factor-alpha, inflammatory biomarkers, and atherogenesis. Eur Heart J 23:345–347
78. Buckley DI, Fu R, Freeman M, Rogers K, Helfand M (2009) C-reactive protein as a risk factor for coronary heart disease: a systematic review and meta-analyses for the U.S. Preventive Services Task Force. Ann Intern Med 151:483–495
79. Greenfield JR, Campbell LV (2006) Relationship between inflammation, insulin resistance and type 2 diabetes: 'cause or effect'? Curr Diabetes Rev 2:195–211
80. Pickup JC, Mattock MB, Chusney GD, Burt D (1997) NIDDM as a disease of the innate immune system: association of acute-phase reactants and interleukin-6 with metabolic syndrome X. Diabetologia 40:1286–1292
81. Saraheimo M, Teppo AM, Forsblom C, Fagerudd J, Groop PH (2003) Diabetic nephropathy is associated with low-grade inflammation in Type 1 diabetic patients. Diabetologia 46:1402–1407
82. Navarro JF, Mora C, Maca M, Garca J (2003) Inflammatory parameters are independently associated with urinary albumin in type 2 diabetes mellitus. Am J Kidney Dis 42:53–61

83. Navarro JF, Mora C, Muros M, Garcia J (2006) Urinary tumour necrosis factor-alpha excretion independently correlates with clinical markers of glomerular and tubulointerstitial injury in type 2 diabetic patients. Nephrol Dial Transplant 21:3428–3434
84. Sugimoto H, Shikata K, Wada J, Horiuchi S, Makino H (1999) Advanced glycation end products-cytokine-nitric oxide sequence pathway in the development of diabetic nephropathy: aminoguanidine ameliorates the overexpression of tumour necrosis factor-alpha and inducible nitric oxide synthase in diabetic rat glomeruli. Diabetologia 42:878–886
85. Javaid B, Quigg RJ (2005) Treatment of glomerulonephritis: will we ever have options other than steroids and cytotoxics? Kidney Int 67:1692–1703
86. Aringer M, Smolen JS (2008) The role of tumor necrosis factor-alpha in systemic lupus erythematosus. Arthritis Res Ther 10:202
87. Feldmann M, Pusey CD (2006) Is there a role for TNF-alpha in anti-neutrophil cytoplasmic antibody-associated vasculitis? Lessons from other chronic inflammatory diseases. J Am Soc Nephrol 17:1243–1252
88. Fornoni A, Ijaz A, Tejada T, Lenz O (2008) Role of inflammation in diabetic nephropathy. Curr Diabetes Rev 4:10–17
89. Navarro JF, Milena FJ, Mora C, Leon C, Claverie F, Flores C, Garcia J (2005) Tumor necrosis factor-alpha gene expression in diabetic nephropathy: relationship with urinary albumin excretion and effect of angiotensin-converting enzyme inhibition. Kidney Int Suppl 68:S98–S102
90. Ishikura H, Takahashi C, Kanagawa K, Hirata H, Imai K, Yoshiki T (1991) Cytokine regulation of ICAM-1 expression on human renal tubular epithelial cells in vitro. Transplantation 51:1272–1275
91. DiPetrillo K, Coutermarsh B, Gesek FA (2003) Urinary tumor necrosis factor contributes to sodium retention and renal hypertrophy during diabetes. Am J Physiol Renal Physiol 284:F113–F121
92. Kalantarinia K, Awad AS, Siragy HM (2003) Urinary and renal interstitial concentrations of TNF-alpha increase prior to the rise in albuminuria in diabetic rats. Kidney Int 64:1208–1213
93. Wertheimer SJ, Myers CL, Wallace RW, Parks TP (1992) Intercellular adhesion molecule-1 gene expression in human endothelial cells. Differential regulation by tumor necrosis factor-alpha and phorbol myristate acetate. J Biol Chem 267:12030–12035
94. Lane TA, Lamkin GE, Wancewicz E (1989) Modulation of endothelial cell expression of intercellular adhesion molecule 1 by protein kinase C activation. Biochem Biophys Res Commun 161:945–952
95. Ballermann BJ, Dardik A, Eng E, Liu A (1998) Shear stress and the endothelium. Kidney Int Suppl 67:S100–S108
96. Kawasaki K, Yaoita E, Yamamoto T, Tamatani T, Miyasaka M, Kihara I (1993) Antibodies against intercellular adhesion molecule-1 and lymphocyte function-associated antigen-1 prevent glomerular injury in rat experimental crescentic glomerulonephritis. J Immunol 150:1074–1083
97. Wada J, Shikata K, Makino H, Morioka S, Hirata K, Ota K, Tamatani T, Miyasaka M, Horiuchi T, Noji S, Nishikawa K, Myokai F, Taniguchi S, Kanwar Y, Ota Z (1996) The critical role of intercellular adhesion molecule-1 in Masugi nephritis in rats. Nephron 73:264–272
98. Miyatake N, Shikata K, Sugimoto H, Kushiro M, Shikata Y, Ogawa S, Hayashi Y, Miyasaka M, Makino H (1998) Intercellular adhesion molecule 1 mediates mononuclear cell infiltration into rat glomeruli after renal ablation. Nephron 79:91–98
99. Kelly KJ, Williams WW Jr, Colvin RB, Meehan SM, Springer TA, Gutierrez-Ramos JC, Bonventre JV (1996) Intercellular adhesion molecule-1-deficient mice are protected against ischemic renal injury. J Clin Invest 97:1056–1063
100. Matsui H, Suzuki M, Tsukuda R, Iida K, Miyasaka M, Ikeda H (1996) Expression of ICAM-1 on glomeruli is associated with progression of diabetic nephropathy in a genetically obese diabetic rat, Wistar fatty. Diabetes Res Clin Pract 32:1–9

101. Coimbra TM, Janssen U, Grone HJ, Ostendorf T, Kunter U, Schmidt H, Brabant G, Floege J (2000) Early events leading to renal injury in obese Zucker (fatty) rats with type II diabetes. Kidney Int 57:167–182
102. Lavaud S, Michel O, Sassy-Prigent C, Heudes D, Bazin R, Bariety J, Chevalier J (1996) Early influx of glomerular macrophages precedes glomerulosclerosis in the obese Zucker rat model. J Am Soc Nephrol 7:2604–2615
103. Hirata K, Shikata K, Matsuda M, Akiyama K, Sugimoto H, Kushiro M, Makino H (1998) Increased expression of selectins in kidneys of patients with diabetic nephropathy. Diabetologia 41:185–192
104. Sugimoto H, Shikata K, Hirata K, Akiyama K, Matsuda M, Kushiro M, Shikata Y, Miyatake N, Miyasaka M, Makino H (1997) Increased expression of intercellular adhesion molecule-1 (ICAM-1) in diabetic rat glomeruli: glomerular hyperfiltration is a potential mechanism of ICAM-1 upregulation. Diabetes 46:2075–2081
105. Chow FY, Nikolic-Paterson DJ, Ozols E, Atkins RC, Tesch GH (2005) Intercellular adhesion molecule-1 deficiency is protective against nephropathy in type 2 diabetic db/db mice. J Am Soc Nephrol 16:1711–1722
106. Clausen P, Jacobsen P, Rossing K, Jensen JS, Parving HH, Feldt-Rasmussen B (2000) Plasma concentrations of VCAM-1 and ICAM-1 are elevated in patients with Type 1 diabetes mellitus with microalbuminuria and overt nephropathy. Diabet Med 17:644–649
107. Guler S, Cakir B, Demirbas B, Yonem A, Odabasi E, Onde U, Aykut O, Gursoy G (2002) Plasma soluble intercellular adhesion molecule 1 levels are increased in type 2 diabetic patients with nephropathy. Horm Res 58:67–70
108. Xiang G, Schinzel R, Simm A, Sebekova K, Heidland A (2001) Advanced glycation end products impair protein turnover in LLC-PK1: amelioration by trypsin. Kidney Int Suppl 78:S53–S57
109. Galkina E, Ley K (2006) Leukocyte recruitment and vascular injury in diabetic nephropathy. J Am Soc Nephrol 17:368–377
110. Anderson ME, Siahaan TJ (2003) Targeting ICAM-1/LFA-1 interaction for controlling autoimmune diseases: designing peptide and small molecule inhibitors. Peptides 24:487–501
111. Carr MW, Roth SJ, Luther E, Rose SS, Springer TA (1994) Monocyte chemoattractant protein 1 acts as a T-lymphocyte chemoattractant. Proc Natl Acad Sci USA 91:3652–3656
112. Xu LL, Warren MK, Rose WL, Gong W, Wang JM (1996) Human recombinant monocyte chemotactic protein and other C-C chemokines bind and induce directional migration of dendritic cells in vitro. J Leukoc Biol 60:365–371
113. Mehrabian M, Sparkes RS, Mohandas T, Fogelman AM, Lusis AJ (1991) Localization of monocyte chemotactic protein-1 gene (SCYA2) to human chromosome 17q11.2-q21.1. Genomics 9:200–203
114. Craig MJ, Loberg RD (2006) CCL2 (Monocyte Chemoattractant Protein-1) in cancer bone metastases. Cancer Metastasis Rev 25:611–619
115. Ha H, Yu MR, Choi YJ, Kitamura M, Lee HB (2002) Role of high glucose-induced nuclear factor-kappaB activation in monocyte chemoattractant protein-1 expression by mesangial cells. J Am Soc Nephrol 13:894–902
116. Yamagishi S, Inagaki Y, Okamoto T, Amano S, Koga K, Takeuchi M, Makita Z (2002) Advanced glycation end product-induced apoptosis and overexpression of vascular endothelial growth factor and monocyte chemoattractant protein-1 in human-cultured mesangial cells. J Biol Chem 277:20309–20315
117. Hall-Stoodley L, Hu FZ, Gieseke A, Nistico L, Nguyen D, Hayes J, Forbes M, Greenberg DP, Dice B, Burrows A, Wackym PA, Stoodley P, Post JC, Ehrlich GD, Kerschner JE (2006) Direct detection of bacterial biofilms on the middle-ear mucosa of children with chronic otitis media. JAMA 296:202–211
118. Young BA, Johnson RJ, Alpers CE, Eng E, Gordon K, Floege J, Couser WG, Seidel K (1995) Cellular events in the evolution of experimental diabetic nephropathy. Kidney Int 47:935–944

119. Chow FY, Nikolic-Paterson DJ, Ozols E, Atkins RC, Rollin BJ, Tesch GH (2006) Monocyte chemoattractant protein-1 promotes the development of diabetic renal injury in streptozotocin-treated mice. Kidney Int 69:73–80
120. Mezzano S, Aros C, Droguett A, Burgos ME, Ardiles L, Flores C, Schneider H, Ruiz-Ortega M, Egido J (2004) NF-kappaB activation and overexpression of regulated genes in human diabetic nephropathy. Nephrol Dial Transplant 19:2505–2512
121. Banba N, Nakamura T, Matsumura M, Kuroda H, Hattori Y, Kasai K (2000) Possible relationship of monocyte chemoattractant protein-1 with diabetic nephropathy. Kidney Int 58:684–690
122. Tesch GH (2008) MCP-1/CCL2: a new diagnostic marker and therapeutic target for progressive renal injury in diabetic nephropathy. Am J Physiol Renal Physiol 294:F697–F701
123. Chen S, Jim B, Ziyadeh FN (2003) Diabetic nephropathy and transforming growth factor-beta: transforming our view of glomerulosclerosis and fibrosis build-up. Semin Nephrol 23:532–543
124. Chiarelli F, Gaspari S, Marcovecchio ML (2009) Role of growth factors in diabetic kidney disease. Horm Metab Res 41:585–593
125. Yokoyama H, Deckert T (1996) Central role of TGF-beta in the pathogenesis of diabetic nephropathy and macrovascular complications: a hypothesis. Diabet Med 13:313–320
126. Ziyadeh FN, Wolf G (2008) Pathogenesis of the podocytopathy and proteinuria in diabetic glomerulopathy. Curr Diabetes Rev 4:39–45
127. Nicholas SB (2003) Advances in pathogenetic mechanisms of diabetic nephropathy. Cell Mol Biol (Noisy-le-Grand) 49:1319–1325
128. Sharma K, Ziyadeh FN (1995) Hyperglycemia and diabetic kidney disease. The case for transforming growth factor-beta as a key mediator. Diabetes 44:1139–1146
129. Leehey DJ, Singh AK, Alavi N, Singh R (2000) Role of angiotensin II in diabetic nephropathy. Kidney Int Suppl 77:S93–S98
130. Phillips AO (1998) Diabetic nephropathy: the modulating influence of glucose on transforming factor beta production. Histol Histopathol 13:565–574
131. Gilbert RE, Cooper ME (1999) The tubulointerstitium in progressive diabetic kidney disease: more than an aftermath of glomerular injury? Kidney Int 56:1627–1637
132. Elmarakby AA, Faulkner J, Baban B, Saleh MA, Sullivan JC (2012) Induction of hemeoxygenase-1 reduces glomerular injury and apoptosis in diabetic spontaneously hypertensive rats. Am J Physiol Renal Physiol 302:F791–F800
133. Benigni A, Zoja C, Campana M, Corna D, Sangalli F, Rottoli D, Gagliardini E, Conti S, Ledbetter S, Remuzzi G (2006) Beneficial effect of TGFbeta antagonism in treating diabetic nephropathy depends on when treatment is started. Nephron Exp Nephrol 104:e158–e168
134. Gagliardini E, Benigni A (2006) Role of anti-TGF-beta antibodies in the treatment of renal injury. Cytokine Growth Factor Rev 17:89–96
135. Rivarola EW, Moyses-Neto M, Dantas M, Da-Silva CG, Volpini R, Coimbra TM (1999) Transforming growth factor beta activity in urine of patients with type 2 diabetes and diabetic nephropathy. Braz J Med Biol Res 32:1525–1528
136. Sharma K, McGowan TA (2000) TGF-beta in diabetic kidney disease: role of novel signaling pathways. Cytokine Growth Factor Rev 11:115–123
137. Leask A, Abraham DJ (2004) TGF-beta signaling and the fibrotic response. FASEB J 18:816–827
138. Nguyen TQ, Tarnow L, Jorsal A, Oliver N, Roestenberg P, Ito Y, Parving HH, Rossing P, van Nieuwenhoven FA, Goldschmeding R (2008) Plasma connective tissue growth factor is an independent predictor of end-stage renal disease and mortality in type 1 diabetic nephropathy. Diabetes Care 31:1177–1182
139. Locatelli F, Pozzoni P, Del Vecchio L (2004) Renal replacement therapy in patients with diabetes and end-stage renal disease. J Am Soc Nephrol 15(Suppl 1):S25–S29
140. Perbal B (2004) CCN proteins: multifunctional signalling regulators. Lancet 363:62–64
141. Riser BL, Denichilo M, Cortes P, Baker C, Grondin JM, Yee J, Narins RG (2000) Regulation of connective tissue growth factor activity in cultured rat mesangial cells and its expression in experimental diabetic glomerulosclerosis. J Am Soc Nephrol 11:25–38

142. Blom IE, van Dijk AJ, de Weger RA, Tilanus MG, Goldschmeding R (2001) Identification of human ccn2 (connective tissue growth factor) promoter polymorphisms. Mol Pathol 54:192–196
143. Blom IE, van Dijk AJ, Wieten L, Duran K, Ito Y, Kleij L, deNichilo M, Rabelink TJ, Weening JJ, Aten J, Goldschmeding R (2001) In vitro evidence for differential involvement of CTGF, TGFbeta, and PDGF-BB in mesangial response to injury. Nephrol Dial Transplant 16:1139–1148
144. Burns WC, Twigg SM, Forbes JM, Pete J, Tikellis C, Thallas-Bonke V, Thomas MC, Cooper ME, Kantharidis P (2006) Connective tissue growth factor plays an important role in advanced glycation end product-induced tubular epithelial-to-mesenchymal transition: implications for diabetic renal disease. J Am Soc Nephrol 17:2484–2494
145. Adler SG, Kang SW, Feld S, Cha DR, Barba L, Striker L, Striker G, Riser BL, LaPage J, Nast CC (2001) Glomerular mRNAs in human type 1 diabetes: biochemical evidence for microalbuminuria as a manifestation of diabetic nephropathy. Kidney Int 60:2330–2336
146. Ito Y, Aten J, Bende RJ, Oemar BS, Rabelink TJ, Weening JJ, Goldschmeding R (1998) Expression of connective tissue growth factor in human renal fibrosis. Kidney Int 53:853–861
147. Roestenberg P, van Nieuwenhoven FA, Joles JA, Trischberger C, Martens PP, Oliver N, Aten J, Hoppener JW, Goldschmeding R (2006) Temporal expression profile and distribution pattern indicate a role of connective tissue growth factor (CTGF/CCN-2) in diabetic nephropathy in mice. Am J Physiol Renal Physiol 290:F1344–F1354
148. Guha M, Xu ZG, Tung D, Lanting L, Natarajan R (2007) Specific down-regulation of connective tissue growth factor attenuates progression of nephropathy in mouse models of type 1 and type 2 diabetes. FASEB J 21:3355–3368
149. Gilbert RE, Akdeniz A, Weitz S, Usinger WR, Molineaux C, Jones SE, Langham RG, Jerums G (2003) Urinary connective tissue growth factor excretion in patients with type 1 diabetes and nephropathy. Diabetes Care 26:2632–2636
150. Roestenberg P, van Nieuwenhoven FA, Wieten L, Boer P, Diekman T, Tiller AM, Wiersinga WM, Oliver N, Usinger W, Weitz S, Schlingemann RO, Goldschmeding R (2004) Connective tissue growth factor is increased in plasma of type 1 diabetic patients with nephropathy. Diabetes Care 27:1164–1170
151. Riser BL, Cortes P, DeNichilo M, Deshmukh PV, Chahal PS, Mohammed AK, Yee J, Kahkonen D (2003) Urinary CCN2 (CTGF) as a possible predictor of diabetic nephropathy: preliminary report. Kidney Int 64:451–458
152. Skarbinski J, James EM, Causer LM, Barber AM, Mali S, Nguyen-Dinh P, Roberts JM, Parise ME, Slutsker L, Newman RD (2006) Malaria surveillance – United States, 2004. MMWR Surveill Summ 55:23–37
153. Hasegawa G, Nakano K, Sawada M, Uno K, Shibayama Y, Ienaga K, Kondo M (1991) Possible role of tumor necrosis factor and interleukin-1 in the development of diabetic nephropathy. Kidney Int 40:1007–1012
154. Sassy-Prigent C, Heudes D, Mandet C, Belair MF, Michel O, Perdereau B, Bariety J, Bruneval P (2000) Early glomerular macrophage recruitment in streptozotocin-induced diabetic rats. Diabetes 49:466–475
155. Navarro JF, Milena FJ, Mora C, Leon C, Garcia J (2006) Renal pro-inflammatory cytokine gene expression in diabetic nephropathy: effect of angiotensin-converting enzyme inhibition and pentoxifylline administration. Am J Nephrol 26:562–570
156. Sekizuka K, Tomino Y, Sei C, Kurusu A, Tashiro K, Yamaguchi Y, Kodera S, Hishiki T, Shirato I, Koide H (1994) Detection of serum IL-6 in patients with diabetic nephropathy. Nephron 68:284–285
157. Shikano M, Sobajima H, Yoshikawa H, Toba T, Kushimoto H, Katsumata H, Tomita M, Kawashima S (2000) Usefulness of a highly sensitive urinary and serum IL-6 assay in patients with diabetic nephropathy. Nephron 85:81–85
158. Wong CK, Ho AW, Tong PC, Yeung CY, Kong AP, Lun SW, Chan JC, Lam CW (2007) Aberrant activation profile of cytokines and mitogen-activated protein kinases in type 2 diabetic patients with nephropathy. Clin Exp Immunol 149:123–131

159. Moriwaki Y, Yamamoto T, Shibutani Y, Aoki E, Tsutsumi Z, Takahashi S, Okamura H, Koga M, Fukuchi M, Hada T (2003) Elevated levels of interleukin-18 and tumor necrosis factor-alpha in serum of patients with type 2 diabetes mellitus: relationship with diabetic nephropathy. Metabolism 52:605–608
160. Pickup JC, Chusney GD, Thomas SM, Burt D (2000) Plasma interleukin-6, tumour necrosis factor alpha and blood cytokine production in type 2 diabetes. Life Sci 67:291–300
161. Suzuki D, Miyazaki M, Naka R, Koji T, Yagame M, Jinde K, Endoh M, Nomoto Y, Sakai H (1995) In situ hybridization of interleukin 6 in diabetic nephropathy. Diabetes 44:1233–1238
162. Melnikov VY, Ecder T, Fantuzzi G, Siegmund B, Lucia MS, Dinarello CA, Schrier RW, Edelstein CL (2001) Impaired IL-18 processing protects caspase-1-deficient mice from ischemic acute renal failure. J Clin Invest 107:1145–1152
163. Miyauchi K, Takiyama Y, Honjyo J, Tateno M, Haneda M (2009) Upregulated IL-18 expression in type 2 diabetic subjects with nephropathy: TGF-beta1 enhanced IL-18 expression in human renal proximal tubular epithelial cells. Diabetes Res Clin Pract 83:190–199
164. Altinova AE, Yetkin I, Akbay E, Bukan N, Arslan M (2008) Serum IL-18 levels in patients with type 1 diabetes: relations to metabolic control and microvascular complications. Cytokine 42:217–221
165. Nakamura A, Shikata K, Hiramatsu M, Nakatou T, Kitamura T, Wada J, Itoshima T, Makino H (2005) Serum interleukin-18 levels are associated with nephropathy and atherosclerosis in Japanese patients with type 2 diabetes. Diabetes Care 28:2890–2895
166. Araki S, Haneda M, Koya D, Sugimoto T, Isshiki K, Chin-Kanasaki M, Uzu T, Kashiwagi A (2007) Predictive impact of elevated serum level of IL-18 for early renal dysfunction in type 2 diabetes: an observational follow-up study. Diabetologia 50:867–873
167. Esposito K, Nappo F, Marfella R, Giugliano G, Giugliano F, Ciotola M, Quagliaro L, Ceriello A, Giugliano D (2002) Inflammatory cytokine concentrations are acutely increased by hyperglycemia in humans: role of oxidative stress. Circulation 106:2067–2072
168. Arnalich F, Hernanz A, Lopez-Maderuelo D, Pena JM, Camacho J, Madero R, Vazquez JJ, Montiel C (2000) Enhanced acute-phase response and oxidative stress in older adults with type II diabetes. Horm Metab Res 32:407–412
169. Schmidt AM, Yan SD, Wautier JL, Stern D (1999) Activation of receptor for advanced glycation end products: a mechanism for chronic vascular dysfunction in diabetic vasculopathy and atherosclerosis. Circ Res 84:489–497
170. Miyata T, Dan T (2008) Inhibition of advanced glycation end products (AGEs): an implicit goal in clinical medicine for the treatment of diabetic nephropathy? Diabetes Res Clin Pract 82(Suppl 1):S25–S29
171. Meyer M, Schreck R, Baeuerle PA (1993) H2O2 and antioxidants have opposite effects on activation of NF-kappa B and AP-1 in intact cells: AP-1 as secondary antioxidant-responsive factor. EMBO J 12:2005–2015
172. Schreck R, Rieber P, Baeuerle PA (1991) Reactive oxygen intermediates as apparently widely used messengers in the activation of the NF-kappa B transcription factor and HIV-1. EMBO J 10:2247–2258
173. Kim SJ, Angel P, Lafyatis R, Hattori K, Kim KY, Sporn MB, Karin M, Roberts AB (1990) Autoinduction of transforming growth factor beta 1 is mediated by the AP-1 complex. Mol Cell Biol 10:1492–1497
174. Nath KA, Grande J, Croatt A, Haugen J, Kim Y, Rosenberg ME (1998) Redox regulation of renal DNA synthesis, transforming growth factor-beta1 and collagen gene expression. Kidney Int 53:367–381
175. Ogawa S, Mori T, Nako K, Kato T, Takeuchi K, Ito S (2006) Angiotensin II type 1 receptor blockers reduce urinary oxidative stress markers in hypertensive diabetic nephropathy. Hypertension 47:699–705
176. Chander PN, Gealekman O, Brodsky SV, Elitok S, Tojo A, Crabtree M, Gross SS, Goligorsky MS (2004) Nephropathy in Zucker diabetic fat rat is associated with oxidative and nitrosative

stress: prevention by chronic therapy with a peroxynitrite scavenger ebselen. J Am Soc Nephrol 15:2391–2403

177. Mozaffari MS, Abdelsayed R, Liu JY, Wimborne H, El-Remessy A, El-Marakby A (2009) Effects of chromium picolinate on glycemic control and kidney of the obese Zucker rat. Nutr Metab (Lond) 6:51
178. Swifka J, Weiss J, Addicks K, Eckel J, Rosen P (2008) Epicardial fat from guinea pig: a model to study the paracrine network of interactions between epicardial fat and myocardium? Cardiovasc Drugs Ther 22:107–114
179. Szendroedi J, Roden M (2009) Ectopic lipids and organ function. Curr Opin Lipidol 20:50–56

Preventing Hyperglycemia and Tissue Injury in Diabetes: The Dynamic Role of 2,3 Indoleamine Dioxygenase (IDO) in Diabetes and Its Complications

Manuela Bartoli, Folami Lamoke, and Babak Baban

Abstract The rapid rise in the number of people affected by diabetes mellitus urges the scientific community with the development of appropriate tools to prevent the occurrence of hyperglycemia and its deleterious effects in patients at risk. A major obstacle in achieving this important goal of the preventive medicine is represented by the heterogeneity of the disease as well as of the targeted organs.

Recent clinical and experimental studies, however, have demonstrated that chronic inflammatory processes are involved in the pathogenesis of diabetes and its complications. Thus, the monitoring of these inflammatory events has important therapeutic and diagnostic value. In particular, the identification within the immune system of critical "pathogenic hubs" governing diabetes-related inflammatory processes will enable clinicians and scientist to develop appropriate preventive strategies.

The 2,3 indoleamine dioxygenase (IDO) may represent a potential biomarker of diabetes-related inflammatory disease. This critical catabolic enzyme has been recently shown to be a central regulator of acute and chronic inflammatory responses. While little is known about its specific contribution to the pathogenesis of diabetes and its complications, a number of important preliminary evidence, here discussed, would suggest its potential involvement in these complex multifactorial diseases.

Keywords 2,3 indoleamine dioxygenase • Diabetes prevention • Biomarker • Preventive medicine • Diabetes • Diabetic complications

M. Bartoli
Department of Ophthalmology, Georgia Regents University, Augusta, GA, USA

Department of Pharmacology and Toxicology, Medical College of Georgia, Augusta, GA, USA

F. Lamoke
Department of Pharmacology and Toxicology, Medical College of Georgia, Augusta, GA, USA

B. Baban (✉)
Department of Oral Biology (CL2122), College of Dental Medicine, Georgia Regents University, 15th Street, Augusta, GA 30912, USA
e-mail: bbaban@gru.edu

M.S. Mozaffari (ed.), *New Strategies to Advance Pre/Diabetes Care: Integrative Approach by PPPM*, Advances in Predictive, Preventive and Personalised Medicine 3,
DOI 10.1007/978-94-007-5971-8_10,

1 Introduction

Diabetes mellitus (DM) is now recognized as a major health emergency in the Western world and in many developing countries. It is estimated that, by year 2013, 330 million people will be affected by diabetes with estimated costs to the Public Health Systems exceeding $174 billion/year in United States and 90€ billion/year in Europe (*from: Organization for Economic Co-operation and Development, Center for Disease Control and World Health Organization*). These alarming data urge the scientific community for the identification of new strategies to prevent and/or combat the development of diabetes and its complications.

Disease prevention is a major goal to halt the occurrence of diabetes or its complications, however, the heterogeneity of these pathologies and the multiplicity of targeted organs, make it very difficult to identify "common" pathogenic pathways, thus, limiting the development of therapies and/or diagnostic tools. Nevertheless, new clinical and experimental studies have determined that chronic inflammatory events are tightly associated with the development of diabetes and its complications, thus, suggesting that a clear understanding of these mechanisms hold tremendous diagnostic and therapeutic value.

The interplay of systemic and tissue specific inflammatory events is coordinated by a plethora of immune receptors and soluble cytokines in a self-sustainable and often redundant process. The identification of this "diabetes-associated immune response" is of significant clinical relevance. In the last two decades the 2,3 indoleamine dioxygenase (IDO) has emerged as a critical regulator of innate immunity, because of its ability to modulate survival and differentiation of T cells and antigen presenting cells (APCs). A large number of studies are beginning to clarify that IDO is implicated in human pathologies in which inflammation and immune tolerance are involved, these include, but are not limited to, cancer and autoimmune diseases.

Altered IDO expression and activity can be found in these pathological conditions and may dynamically vary during disease progression, thus suggesting that its monitoring can be used as a reliable and informative diagnostic tool. Although IDO has been implicated in the etiology of type 1 diabetes, its specific contribution to the pathogenesis of type 2 diabetes or diabetes-related complications has not been clarified yet. In the next few paragraphs we will analyze data that, directly and/or indirectly, implicate IDO function in the induction and progression of diabetes and its complications and will put forward the rationale for the realization of future and more specific studies.

2 IDO Structure and Function

Indoleamine-pyrrole 2,3-dioxygenase (IDO) [1] is a cytosolic heme-containing enzyme catalyzing the first and rate-limiting step of the oxidative tryptophan catabolism leading to tryptophan starvation and production of tryptophan metabolites known as kynurenines (Fig. 1) [2].

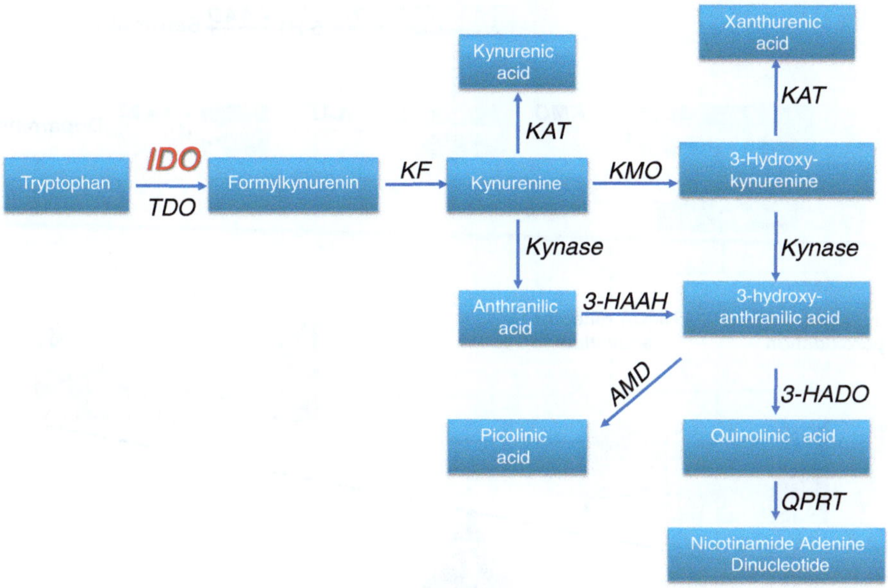

Fig. 1 Tryptophan metabolism IDO is the rate-limiting enzyme involved in the oxidative catabolism of the tryptophan. As a result of IDO-mediated catalytic activity is the production of biologically active metabolic intermediates including kynurenine and quinolinic acid. Monitoring of IDO activity can be achieved by High Liquid Pressure Chromatographic analysis of kynurenine and assessment of tryptophan/kynurenine ratio. Abbreviations: *IDO* Indoleamine 2,3-dioxygenase, *TDO* Tryptophan 2,3-dioxygenase, *KF* Kynurenine formylase, *KAT* Kynurenine aminotransferase, *KMO* Kynurenine mono-oxygenase, *Kynase* Kynureninase, *QPRT* Quinolinic acid phosphoribosylotransferase, *3-HADO* 3-Hydroxyanthranilic acid 3,4 -dioxygenase, *3-HAAH* 3- Hydroxyanthranilic acid hydroxylase, *AMD* Aminocarboxy muconate semialdehyde decarboxylase

IDO proteins (IDO1 and its paralogue IDO2) are encoded by two homologous genes (*ido1* and *ido2*) mapped on chromosome 8 [3–5]. Conserved interferon responsive elements (ISRE, GAS) are located in the promoter region of these IDO genes, thus, explaining the ability of interferons (IFN) to regulate IDO gene expression in several cell types [6–9]. IDO genes are most prominently expressed in immune cells, however, a number of other cell types express it (such as, epithelial and endothelial cells and trophoblasts) [10], thus, suggesting a potential role for this enzyme outside the immune system [1].

IDO enzymatic activity is required for the oxidative metabolism of the essential amino acid tryptophan (Fig. 1) [11]. This catabolic pathway involves the formation of highly biologically active intermediate metabolites, such as kynurenines and quinolinic acid, of which expression levels are monitored to assess IDO activity [12, 13]. As tryptophan metabolism is part of the serotonin and melatonin metabolism, it has been shown that levels of IDO and its metabolites may be directly implicated in aging and in the pathogenesis of metabolic and neurological disorders [14] (Fig. 2). Importantly, during this metabolic process IDO consumes superoxide radicals, thus entailing that IDO possess antioxidant abilities [15]. Hyperactive IDO, however,

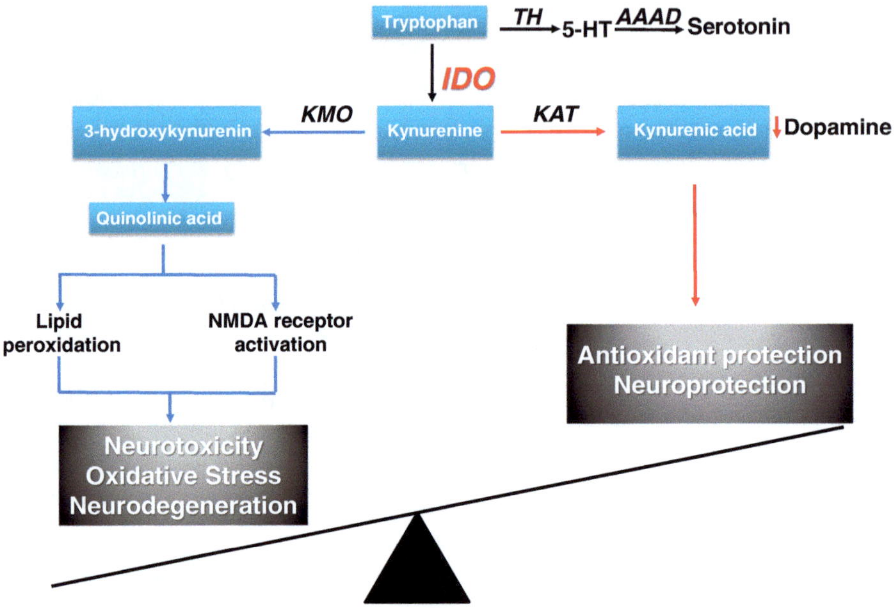

Fig. 2 Tryptophan-Kynurenine pathway. The metabolism of the essential amino acid tryptophan is here described. Altered tryptophan metabolism may have profound metabolic consequences. IDO by controlling tryptophan intracellular availability may switch cell from an antioxidant protected state to a pro-oxidant and toxic state. This is partially due to the effects of IDO in depleting tryptophan availability for the synthesis of serotonin leading to senescence and neurotoxicity. Furthermore enhanced IDO activity leads to increased levels of quinolic acid quinolic acid which promotes oxidative stress and neurotoxicity. Abbreviations: *IDO* Indoleamine 2,3-dioxygenase, *KAT* Kynurenine aminotransferase, *KMO* Kynurenine mono-oxygenase, *TH* tryptophan hydroxylase, *5-HT* 5-Hydroxy tryptamine, *AAAD* aromatic L-amino acid decarboxylase

can lead to generation of metabolites, which, on the contrary, possess pro-oxidant effects [15]. IDO enzymatic activity is regulated by heme and substrate bioavailability, redox potentials, nitric oxide and arginase [12, 13]. Protein post-translational modifications have also been identified and their modifications confer IDO signaling ability in dendritic cells [16]. Much of IDO biological effects have been tightly linked to the tryptophan/kynurenines pathway and, in particular, to the regulation of tryptophan bioavailability in immunocompetent cells and of kynurenines levels in the microenvironment [1].

2.1 IDO and Immunity

Although IDO is present in cells outside of the immune system, its immunomodulatory activity has been studied more extensively [17, 18]. IDO was initially involved in host-defense mechanisms [17, 18], but in 1998 Munn and Colleagues by showing that IDO inhibition, by 1-methyl-DL-tryptophan (1-MT), in pregnancy would cause rejection of semi allogeneic, but not syngeneic, fetuses in mice [10], disclosed a new critical function for this catalytic enzyme as regulator of immune responses and immune tolerance [18]. A large number of studies have subsequently confirmed the dynamic role of IDO in the immune system by dissecting and characterizing its stimulatory and regulatory properties [18].

To explain IDO dual biological activity, it has been postulated that IDO initially responds promoting stimulatory effects to which follows a number of immunosuppressive events to limit and/or control exaggerated inflammatory responses. This dual dynamic role of IDO in the immune response requires both enzymatic and non-enzymatic properties [1].

Regulation of tryptophan bioavailability is an important determinant of T cell function and activity and the protein kinase general control nonderepressing-2 (GCN2) serves as molecular sensor for the regulation of intracellular tryptophan levels during stress responses [19]. IDO by means of its enzymatic activity has been shown to mediate a bidirectional flow of information between CTLA-4-expressing T cells and accessory cells of the immune system. This represents an important effector pathway for T regulatory cells (Treg), to promote and sustain peripheral tolerance [20, 21]. In addition, IDO-mediated kynurenines formation leads to induction of Foxp3+ Treg cells, thus, promoting immunosuppression and tolerogenicity [22].

Recent studies have disclosed signaling function of IDO in dendritic cells (DCs) which appear to be independent of its enzymatic activity and rather associated with IDO protein post-translational modifications including phosphorylation and de-phosphorylation [18]. This non-enzymatic function of IDO in DCs results in a phenotypic switch toward a regulatory activity for the maintenance of long-term immune homeostasis and immune-related functions. Altered IDO-mediated signaling activity in DCs has been implicated in a number of pathological conditions including allergic, autoimmune, chronic inflammatory diseases and also have been shown to play a role in promoting immune tolerance in transplantation and immune-evasion mechanisms in cancer (Fig. 3) [21, 23–25].

The dynamic function of IDO in innate and adaptive immunity is tightly regulated by a number of inflammatory cytokines and this may be very relevant in the hyperglycemic milieu, as enhanced levels of these inflammatory mediators are involved in the pathogenesis of diabetes and its complications (see chapter "The Role of Inflammatory Cytokines in Diabetic Nephropathy: Potential Use as Predictive Markers for Early Detection and Progression"). Indeed, IDO stimulatory and/or regulatory functions are coordinated and dictated by a number of soluble factors primarily including: interferons (IFN), transforming growth factor beta (TGFβ) and interleukin 6 (IL-6). In response to inflammatory stimuli IFN-γ is induced and promotes IDO expression and activity in competent cells. As a result of this stimulatory event, IDO-mediated tryptophan starvation and increased kynurenines levels promote Treg maturation. Subsequently, Treg, through TGFβ production, mediates IDO signaling function promoting its regulatory effects. These mechanisms are halted by IL-6 which stimulates IDO proteasomal degradation by promoting its interaction with the suppressor of cytokines signaling 3 (SOCS3) [26, 27]. Systemic and local changes in cytokines levels, may, therefore, profoundly affect IDO activity and result in differential modulation of immune responses.

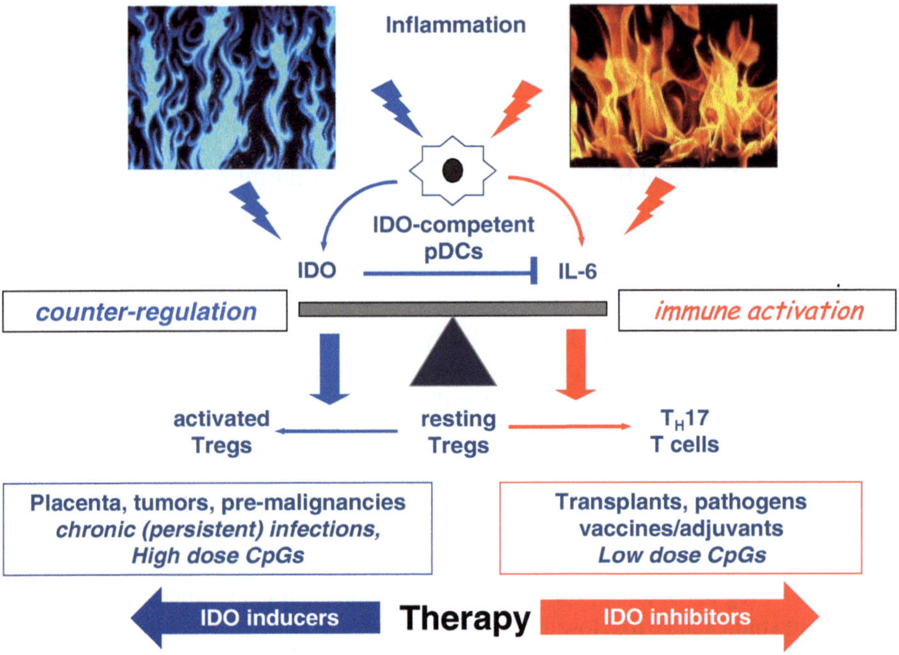

Fig. 3 The dual role of IDO in the immune system. IDO contributes to the regulation of the immune response by means of stimulatory and regulatory activities. Altered IDO activity is involved in human pathologies and strategies to halt and/or promote IDO activity in these disease conditions have been suggested and/or developed. Therapeutic modulation of IDO activity can be achieved by treatment with CpGs compounds which have demonstrated to be potential modulators of immune responses. Abbreviations: *IDO* Indoleamine 2,3-dioxygenase, *DCs* Dendritic cells, *pDCs* plasmacytoid dendritic cells, *Tregs* T lymphocytes regulatory cells, *TH17* T inflammatory

3 IDO and Human Pathologies

Because of IDO's critical involvement in the regulation of immune response, it is not surprising to know that altered IDO activity has been implicated in a number of human pathologies involving dysregulated immune surveillance. Here, we will only briefly overview IDO's implications in human pathologies and remind the reader to refer to more specific publications [17, 18].

Many lessons on IDO functions have been learned from studies defining its role in a number of bacterial and parasitic infections [18, 28] and in determining the benefits of manipulating its activity during infection diseases [28]. However, IDO's role in autoimmunity and immune tolerance finds its most important applications in human pathology.

In line with IDO's function in dampening exaggerated inflammatory responses, increasing evidence supports the notion that IDO may delay the onset and progression of autoimmune diseases. In a murine model of, experimental autoimmune encephalomyelitis (EAE), evidence has been provided suggesting that IDO suppresses T cell activation. Indeed, IDO inhibition, by 1-MT, in EAE resulted in enhanced inflammation, tissue damage and rapid disease progression [29].

Our group has also obtained similar results using a murine model of experimental uveitis [unpublished data]. We have found that lipopolysaccharide (LPS)-induced retinal inflammation is associated with significant decrease of IDO expression. Furthermore, genetic loss of IDO, in IDO-knockout mice, exacerbated LPS-induced retinal inflammation, enhanced production of inflammatory cytokines and, ultimately, resulted in rapid and more pronounced tissue damage [unpublished data].

In rheumatoid arthritis (RA), IDO behaves differently, as enhanced IDO expression is found in dendritic cells in synovial fluids of affected patients [30, 31]. However, IDO effects in RA appear to be due to B cells recruitment and function in areas of inflammation, with a mechanism which has not been completely understood [32].

IDO's ability to promote immune tolerance has provided a rationale to investigate the hypothesis that IDO plays an essential role in preventing graft rejection. Several studies have, indeed, demonstrated that enhancement of IDO activity ameliorates grafted tissue survival and prevents rejection [33, 34]. The specific cellular and molecular mechanisms explaining IDO effects in promoting tolerance and suppressing Graft-Versus-Host Disease (GVHD) are still under investigation. However, several hypotheses have been put forward and these involve: kynurenine-mediated T cell anergy, suppression of activated myeloid cells and functional interaction with Toll-like receptor 7 and 8 (TLR7/8) [35, 36].

One of the most elegant but also devastating consequences of dysregulated immune response is the evasion of tumor cells from immune surveillance. Not surprisingly, IDO has been shown to play an important role in protecting cancer cells from immune rejection [37–40]. Not only IDO expression can be found increased in tumor cells, but also in tumor draining lymph nodes where tumor antigens first drain and tumor-derived DCs first migrate [38]. A large number of clinical and experimental studies are strongly supporting the notion that IDO plays a key role in suppressing immune rejection of tumor cells and that suppression of IDO hold important therapeutic applications in combating human cancer [16]. Although several models of IDO function in supporting escape of tumor cells from immune control have been drawn, an interesting venue is the possibility that enhanced IDO expression in cancer is linked to suppression of the gene Bin1 of which expression levels are, indeed, attenuated in several human malignancies [39]. Decreased Bin1 expression and enhanced IDO expression and activity, may be predictive of malignancy and metastatic potential, thus may hold important diagnostic value [39]. However, data from another study demonstrated no relation between IDO and Bin1 [40] and that nor IDO or Bin1 expression levels had any correlation with survival rate [40]. These contrasting results suggest that more studies will be necessary to better understand IDO implication in cancer.

4 IDO and Diabetes

Genetic and environmental factors contribute to the pathogenesis of type 1 and type 2 diabetes: the two main forms of diabetes mellitus. Although very different in nature and in etiologies, these two pathological conditions share the common feature of uncontrolled increases in circulating levels of glucose and its by-products. Hyperglycemia, in turn, has devastating toxic effects which result in the development of diabetic complications, macro- and micro-angiopathies (see chapter "Diabetes Mellitus: New Challenges and Innovative Therapies"). There is a great heterogeneity of responses and of contributing factors to the pathogenesis of diabetes and its complications and this makes it very difficult to develop effective therapies. Disease prevention remains the most logical strategy to halt the occurrence of diabetes and its complications. To date, however, there is not a definite picture about specific risk factors which can be considered as highly predictors of disease development and progression. Because diabetes incidence is rising with epidemic proportion, the identification of these predictive factors represents a major research emphasis area in the scientific community.

Diabetic patients, once diagnosed, endure a life-long co-existence with the disease which often translates to a life-long commitment with a large list of drugs with potential side effects. Ideally, the identification of common pathogenic events may facilitate the development of common therapies, thus helping in restricting the number of therapeutic options to fewer, but more effective.

Recent clinical and experimental studies have determined that chronic inflammatory events are strongly implicated in the pathogenesis of both forms of diabetes and, also, of its complications [41–43]. As inflammatory responses are self-sustainable and redundant processes, involving a large number of cells, antigens and soluble factors, it may be very difficult to identify key components to be targeted pharmacologically to prevent and/or halt hyperglycemia-induced inflammatory processes. However, there is little doubt that a better understanding of the inflammatory processes involved in diabetes and its complications is key to the identification of these "pathogenic hubs" that once identified, can be monitored as predictor of disease development and progression.

The immunological processes associated with this hyperglycemia-induced inflammatory response are likely to offer important pathogenic clues and to provide a diagnostic tool to assess the development and progression of diabetes and its complications. As IDO has been shown to play a critical role in modulating inflammatory responses in a number of pathological conditions characterized by dysregulated immune response, it is logical to speculate its involvement in diabetes-related immunity and inflammatory processes.

To date, there are no definite proofs directly implicating hyperactive or dampened IDO activity in the pathogenesis of any form of diabetes or of its complications. However a large number of circumstantial evidence point to IDO involvement in these pathogenic processes and certainly, warrant more specific and thorough studies to determine its contribution to these pathological processes.

4.1 IDO and Type 1 Diabetes

The origin of type 1 diabetes is not fully understood, however, loss of tolerance to self-determinants, activation of autoreactive lymphocytes and subsequent damage of target organs are implicated in the disease [44]. Type 1 diabetes is an autoimmune disease that results from the T-cell mediated destruction of the insulin-producing β-cells of the pancreatic islets of Langerhans [19, 44]. Much work has been focused on determining ways to inhibit T-cell proliferation to control immune response leading to β-cell destruction. As described in the previous paragraphs, tryptophan catabolism has been shown to play a role in the immune response by inhibiting T-cell proliferation [19, 45, 46]. In addition, localized IDO expression has been correlated with increased tryptophan depletion and decreased T-cell reactivity in the surrounding microenvironment [19].

IDO expression and activity is decreased in the dendritic cell population of NOD mice, a genetic model of type 1 diabetes, suggesting that poor IDO expression in dendritic cell may play a role in the development of diabetes [44]. However in an experimental model of autoimmune diabetes IDO expression differs from that reported in NOD mice. In this experimental model of streptozotocin-induced type 1 diabetes, the disease in wild-type animals was accompanied by up-regulation of IDO in the pancreatic lymph nodes [22]. However, the disease was surprisingly exacerbated by in vivo administration of an IDO inhibitor and IDO was identified as the critical downstream effector of Toll-like receptor 9 (TLR9) in regulating inflammation/autoimmunity in these conditions [22].

IDO also has an immune tolerogenic role in pregnancy. Cells at the maternal fetal interface also express IDO which consumes local tryptophan in order to exert immune tolerance of the fetus [23, 24]. Administration of a key pregnancy hormone has been shown to inhibit diabetogenic T-cells and the progression of type 1 diabetes by upregulating the expression of IDO [23]. These factors have made IDO induction a therapeutic target for mediating immune tolerance in autoimmune diseases such as type 1 diabetes.

Available treatment options for type 1 diabetic patients are based on repeated daily injections of insulin; therefore, IDO has been utilized experimentally to create better treatment options. Despite intensive insulin replacement therapy, patients are unable to attain adequate blood glucose regulation to completely prevent the development and progression of long-term complications of insulin-dependent diabetes mellitus [19, 45]. Transplantation of insulin-producing donor islet cells has been an alternative strategy which aims to replace the efficient insulin regulation that is lost [19, 46]. Successful islet transplantations have included rigorous immunosuppressive regimen to avoid both recurring autoimmunity and allo-rejection [19, 46]. However, successful islet transplantation has not been achieved without the consequence of patients needing life-long administration of systemic immunosuppressive drugs. Despite its beneficial outcomes, the use of these long-term immunosuppressive drugs is limited to patients that are non-compliant with the management of exogenous insulin delivery [19, 46]. In cases of young, compliant patients, the risks of life-long insulin therapy are preferred to the risks of drug toxicity from a continuous immunosuppressive regimen [19, 46]. Genetic modification by overexpression of IDO in donor islets ex vivo before transplantation has been explored as a means to improve islet replacement therapy. This gene therapy approach has been successful in protecting islet grafts from an autoimmune attack using NOD (donor) and immunodeficient NOD*scid* (transplant recipients) mice [19, 46]. In a study by Jalili et al., local expression of IDO in bystander syngeneic fibroblasts could prevent islet allogeneic immune response in vitro, thus further supporting a potential therapeutic role for IDO in islet transplant in type 1 diabetic patients [47]. IDO is just one of an identified group of the therapeutic genes utilized in this strategy due to its known immunomodulatory and anti-inflammatory actions which result in the inhibition of inflammatory T-cell proliferation.

Other strategies aim at blunting autoreactivity without compromising immunity. One study determined that autoimmune diabetes results in the conversion of naïve low-avidity autoreactive CD8+ T-cells into autoregulatory cells [24]. In this study, treatments with nanoparticles coated with human diabetes-relevant pHLA complexes (pMHC-NP) expanded cognate autoregulatory T cells while restoring normoglycemia in NOD-mice by IFNγ- and IDO-mediated suppression of local autoantigen presentation [24].

4.2 IDO and Type 2 Diabetes

Much less defined and studied is the role of IDO in the pathogenesis of insulin resistance, metabolic syndrome and type 2 diabetes.

In two different review papers, Oxenkrug has described a potential role for IDO in the pathogenesis of age-related disease including: hypertension, obesity, dyslipidemia, type 2 diabetes, menopause, late onset depression, vascular cognitive impairment, impairment of immune defense, and some forms of cancer, e.g., breast and prostate [48, 49]. He has put forward the hypothesis that chronic inflammation might trigger and/or maintain the development of these metabolic and neuroendocrine disorders through IDO transcriptional induction by pro-inflammatory cytokines. In turn, activation of IDO will result in shifting the tryptophan metabolism from serotonin synthesis to formation of kynurenines (Fig. 2). Lower serotonin production is associated with mental depression while increased formation of kynurenines might contribute to the development of metabolic syndromes and also to the induction of apoptotic, neurotoxic, pro-oxidative and pro-inflammatory effects [48, 49].

These data would imply that enhanced IDO activity is a potential risk factor for type 2 diabetes. However, one can argue that in patients affected by insulin resistance and/or overt type 2 diabetes there is a significant increase in circulating levels of IL-6 [50]. Increased IL-6 production is also associated with development and progression of diabetic complications [51, 52]. As described before, IL-6 is a suppressor of IDO function by promoting its proteasomal degradation through SOCS3 [53]. Abnormal IL-6 expression and function in diabetes, however, may differentially affect different cell types including T-cells, DCs and endothelial cell at different stages during disease development and progression [54].

The lack of systematic studies to specifically determine IDO contributing role to these processes makes it difficult to specifically define its role in the pathogenesis of type 2 diabetes, but certainly warrant further studies in human subjects and in experimental models. Very informative studies could involve the assessment of T-cells subpopulation and DCs in diabetic patients to extrapolate IDO function *in vivo*. Accurate measurements of the kynurenine pathways may also be a good approach. Finally, studies can be performed in animal models of diabetes treated with inhibitors of IDO (e.g. 1-MT or genetic manipulation using knockout mice). Until these experiments are conducted no definite conclusions on IDO contribution to metabolic syndrome and/or type 2 diabetes should be drawn.

5 IDO and Diabetic Complications

Much of diabetes morbidity and mortality is due to its macrovascular and microvascular complications. Hyperglycemia and its by-products (Amadori compounds and advanced glycation-end products) are central culprit for the development of diabetic complications.

Two large clinical studies have clarified that a strict glycemic control can significantly prevent the development and progression of diabetic microangiopathies [55–57]. However, poor patient compliance, epigenetic mechanisms leading to diabetic memory [55–57] and the deleterious effects of hypoglycemia [57] demonstrate that along with glycemic control, therapies are necessary to halt hyperglycemia-induced tissue injury.

Several clinical and experimental studies show that the diabetic milieu promotes the production of pro-inflammatory factors at both tissue and systemic levels, thus supporting the notion that low-grade chronic inflammation play a key pathogenic role in diabetic complications [41, 42, 52, 58]. Halting the production of these various inflammatory factors is a critical therapeutic goal for the prevention and cure of diabetic complications. The redundancy of this process, however, makes it difficult to select which one or which class of inflammatory factors should be more effectively blocked and the use of general anti-inflammatory drugs (i.e. NSAID and steroids) may result in side effects or in not efficient inhibition. Despite these challenges, there is little doubt that the characterization of the inflammatory events leading to diabetic complications is of primary interest for their prevention in patients at risk.

The question that we are addressing here is whether IDO may find a role in the "immunity" involved in diabetes and its complications and whether the monitoring of its activity may hold predictive value. To date, no information is available on IDO's direct influence in the inflammatory processes involved in the development of diabetic complications. However, a number of studies have emphasized a potential contributing role of IDO in atherosclerosis, retinopathy and nephropathy.

5.1 IDO and Diabetic Macroangiopathy

Atherosclerosis is greatly accelerated in diabetic patients [59] and this effect is implicated in the development of both macro- and microvascular complications.

Despite the original observation implicating enhanced IDO activity in the development of age-related and metabolic disease including atherosclerosis, several clinical studies have determined that there is no clear association between enhanced kynurenine metabolism, due to hyperactive IDO, and atherosclerosis or cardiovascular disease [60, 61].

On the other hand, the ability of IDO to modulate DCs suppressive and/or stimulatory activity has, recently gained interest in the field of cardiovascular medicine and studies show that plasmacytoid DCs possess protective role in atherosclerosis, possibly by dampening T-cell proliferation and activity in peripheral lymphoid tissue [62].

These preliminary observations, however, are far from giving us a clear landscape on IDO function in diabetic complications. In the next paragraphs we will analyze the evidence implicating IDO in diabetic microvascular complications, particularly diabetic retinopathy and nephropathy.

5.2 IDO and Diabetic Retinopathy

Diabetic retinopathy (DR) is the leading cause of blindness in adults worldwide. Eighty percent of the diabetic patients will develop some form of visual impairment within the first 10 years of disease, about 25% of this population of patients will be legally blind within 20 years of diagnosis (see chapters "Correction of Diabetes-Induced Endothelial Progenitor Dysfunction to Promote Retinal Vascular Repair" and "Diabetic Retinopathy: The Need for Predictive, Preventive and Personalized Approach to Management"). Inflammation is a key causative factor for DR, thus suggesting the involvement of immunological event in its pathogenesis.

IDO is expressed in eye and retina [16, 63–65]. Altered IDO expression is found in several ocular pathologies, particularly affecting the anterior chamber of the eye [65]. IDO expression is down-regulated in experimental uveitis and in our studies IDO knockout mice show much higher sensitivity to LPS-induced retinal tissue damage [unpublished data]. These data would suggest that inhibition of IDO in retina promotes inflammation, therefore, supporting the hypothesis of a protective effect of IDO in inflammatory retinopathy.

However, plasma levels of kynurenine and IDO mRNA are found elevated in patients affected by diabetic retinopathy [66]. On the contrary, we have found a marked decrease in IDO expression levels in retina of streptozotocin-induced diabetic mice after 8 and 12 weeks of hyperglycemia and also in human post-mortem retinas of diabetic donors [unpublished data]. These contrasting results while confirming a role for IDO in the immunity associated with the induction and progression of diabetic retinopathy, fail to define its specific role in this process. Further studies using both experimental and clinical approaches are justified and needed to better characterize IDO's role in the pathogenesis of diabetic retinopathy and to explore its monitoring as a predictive tool of disease induction and progression.

5.3 *IDO and Diabetic Nephropathy*

Diabetic nephropathy (DN) is a leading cause of end stage renal disease: a major morbidity factor in diabetic patients [67]. A large number of clinical and experimental studies have shown that, as in the case of other diabetic complications, diabetic nephropathy is associated with enhanced inflammatory processes (see chapter "The Role of Inflammatory Cytokines in Diabetic Nephropathy: Potential Use as Predictive Markers for Early Detection and Progression"). Whether or not IDO plays a role in this devastating and potentially deadly pathology still remains to be elucidated.

As for the case of diabetic retinopathy, circumstantial evidence would suggest a potential role of IDO in mediating hyperglycemia-induced renal disease and failure. The lack of specific studies leave us speculating a potential role of IDO in the pathogenesis of diabetic nephropathy, largely based on evidence involving tryptophan metabolism with renal disease and failure. Indeed, in patients undergoing hemodialysis a significant decrease in tryptophan level and an increase in kynurenine have been observed, thus demonstrating enhanced IDO activity [68, 69]. We have also conducted preliminary studies and determined that IDO expression is significantly decreased in kidney of diabetic mice (streptozotocin-induced) at 12 and 16 weeks of hyperglycemia (unpublished data). Therefore, as in the case of diabetic retinopathy, the specific function of IDO in the diabetic kidney remains controversial and needs to be elucidated.

Other studies have focused attention on the potential role of IDO as immunosuppressor to prevent rejection in transplanted kidneys [70], further emphasizing the importance of examining in detail the contribution of IDO to pathologies leading to end stage renal disease, such as diabetic nephropathy.

6 Conclusions and Recommendations

The rise of diabetes incidence to epidemic proportion urges the development of highly specific diagnostic and therapeutic tools to identify and cure patients at risk.

The heterogeneity of the disease and of the targeted organs makes it very difficult to identify common pathogenic events. Nevertheless, several clinical and experimental studies have determined that inflammation represents a central common pathogenic factor for the induction and progression of diabetes and its complications. The understanding of diabetes associated "immunity" linking hyperglycemia to inflammation and tissue damage is, therefore, of primary interest for the prevention of diabetes and diabetes-induced micro- and macrovascular complications.

IDO has been shown to play a critical role as inducer and regulator of innate immunity affecting T-cells and APCs biology. In this chapter, we have discussed a number of circumstantial evidence supporting a potential role for IDO in the pathogenesis of both type1 and type 2 diabetes, as well as of diabetic complications, such as atherosclerosis, nephropathy and retinopathy. In addition, evidence has been provided of potential applications for IDO in preventing rejection of pancreatic beta islets transplant, a critical therapeutic approach for type 1 diabetic patients.

The summary of the data presented demonstrates that the information available is insufficient to precisely clarify the specific role of IDO in the process of induction and progression of diabetes and hyperglycemia-related diseases. Some controversial data may be attributable to time-, tissue- and cell type-dependent activity of IDO, further underlying its "dynamic role" role as regulator of the immune system and of inflammatory responses. More studies are necessary, using both experimental (i.e. IDO knockout mice) and clinical tools (large populations of diabetic patients), to resolve these controversies and to help on drawing a clear picture of IDO as contributing factor to the etiology of diabetes and its complications.

There is very little doubt, however, that a better knowledge of IDO function in diabetes hold tremendous potential in preventative medicine. Monitoring of IDO activity in diabetic patients suffering end stage renal failure could also be very useful to determine potential kidney rejection in transplanted patients. At the same time, the role of IDO in cancer and in stem cells survival and homing could be very important in diabetic patients as predictor of malignancy and wound healing ability, respectively. IDO monitoring can be also useful in gestational diabetes, as this may affect fetus survival.

Finally, more studies are also necessary to determine the effects of hyperglycemia on dendritic cells and on IDO ability to modulate their biological functions in the hyperglycemic milieu. Dendritic cells have gained increasing interest for their involvement in cardiovascular disease and a number of very interesting studies are now at the forefront of the research aimed at developing preventive tools in cardiovascular disease.

References

1. Kaul K, Hodgkinson A, Tarr JM, Kohner EM, Chibber R (2010) Is inflammation a common retinal-renal-nerve pathogenic link in diabetes? Curr Diabetes Rev 6:294–303
2. Sugimoto H, Oda S, Otsuki T, Hino T, Yoshida T, Shiro Y (2006) Crystal structure of human indoleamine 2,3-dioxygenase: catalytic mechanism of O2 incorporation by a heme-containing dioxygenase. Proc Natl Acad Sci USA 103:2611–2616
3. Ball HJ, Sanchez-Perez A, Weiser S, Austin CJ, Astelbauer F, Miu J, McQuillan JA, Stocker R, Jermiin LS, Hunt NH (2007) Characterization of an indoleamine 2,3-dioxygenase-like protein found in humans and mice. Gene 396:203–213
4. Ball HJ, Yuasa HJ, Austin CJ, Weiser S, Hunt NH (2009) Indoleamine 2,3-dioxygenase-2; a new enzyme in the kynurenine pathway. Int J Biochem Cell Biol 41:467–471

5. Metz R, Duhadaway JB, Kamasani U, Laury-Kleintop L, Muller AJ, Prendergast GC (2007) Novel tryptophan catabolic enzyme IDO2 is the preferred biochemical target of the antitumor indoleamine 2,3-dioxygenase inhibitory compound D-1-methyl-tryptophan. Cancer Res 67:7082–7087
6. Hayashi T, Beck L, Rossetto C, Gong X, Takikawa O, Takabayashi K, Broide DH, Carson DA, Raz E (2004) Inhibition of experimental asthma by indoleamine 2,3-dioxygenase. J Clin Invest 114:270–279
7. Beutelspacher SC, Tan PH, McClure MO, Larkin DF, Lechler RI, George AJ (2006) Expression of indoleamine 2,3-dioxygenase (IDO) by endothelial cells: implications for the control of alloresponses. Am J Transplant 6:1320–1330
8. Mohib K, Wang S, Guan Q, Mellor AL, Sun H, Du C, Jevnikar AM (2008) Indoleamine 2,3-dioxygenase expression promotes renal ischemia-reperfusion injury. Am J Physiol Renal Physiol 295:F226–F234
9. Baban B, Chandler P, McCool D, Marshall B, Munn DH, Mellor AL (2004) Indoleamine 2,3-dioxygenase expression is restricted to fetal trophoblast giant cells during murine gestation and is maternal genome specific. J Reprod Immunol 61:67–77
10. Munn DH, Zhou M, Attwood JT, Bondarev I, Conway SJ, Marshall B, Brown C, Mellor AL (1998) Prevention of allogeneic fetal rejection by tryptophan catabolism. Science 281:1191–1193
11. Chen Y, Guillemin GJ (2009) Kynurenine pathway metabolites in humans: disease and healthy states. Int J Tryptophan Res 2:1–19
12. Littlejohn TK, Takikawa O, Truscott RJ, Walker MJ (2003) Asp274 and his346 are essential for heme binding and catalytic function of human indoleamine 2,3-dioxygenase. J Biol Chem 278:29525–29531
13. Johnson BA 3rd, Baban B, Mellor AL (2009) Targeting the immunoregulatory indoleamine 2,3 dioxygenase pathway in immunotherapy. Immunotherapy 1:645–661
14. Fuvesi J, Rajda C, Bencsik K, Toldi J, Vecsei L (2012) The role of kynurenines in the pathomechanism of amyotrophic lateral sclerosis and multiple sclerosis: therapeutic implications. J Neural Transm 119:225–234
15. Thomas SR, Stocker R (1999) Redox reactions related to indoleamine 2,3-dioxygenase and tryptophan metabolism along the kynurenine pathway. Redox Rep 4:199–220
16. Heitger A (2011) Regulation of expression and function of IDO in human dendritic cells. Curr Med Chem 18:2222–2233
17. Baban B, Penberthy WT, Mozaffari MS (2010) The potential role of indoleamine 2,3 dioxygenase (IDO) as a predictive and therapeutic target for diabetes treatment: a mythical truth. EPMA J 1:46–55
18. Fallarino F, Grohmann U, Puccetti P (2012) Indoleamine 2,3-dioxygenase: from catalyst to signaling function. Eur J Immunol 42:1932–1937
19. Alexander AM, Crawford M, Bertera S, Rudert WA, Takikawa O, Robbins PD, Trucco M (2002) Indoleamine 2,3-dioxygenase expression in transplanted NOD Islets prolongs graft survival after adoptive transfer of diabetogenic splenocytes. Diabetes 51:356–365
20. Hayaishi O, Hirata F, Ohnishi T, Henry JP, Rosenthal I, Katoh A (1977) Indoleamine 2,3-dioxygenase: incorporation of 18O2– and 18O2 into the reaction products. J Biol Chem 252:3548–3550
21. Christen S, Peterhans E, Stocker R (1990) Antioxidant activities of some tryptophan metabolites: possible implication for inflammatory diseases. Proc Natl Acad Sci USA 87:2506–2510
22. Fallarino F, Volpi C, Zelante T, Vacca C, Calvitti M, Fioretti MC, Puccetti P, Romani L, Grohmann U (2009) IDO mediates TLR9-driven protection from experimental autoimmune diabetes. J Immunol 183:6303–6312
23. Ueno A, Cho S, Cheng L, Wang J, Hou S, Nakano H, Santamaria P, Yang Y (2007) Transient upregulation of indoleamine 2,3-dioxygenase in dendritic cells by human chorionic gonadotropin downregulates autoimmune diabetes. Diabetes 56:1686–1693
24. Tsai S, Shameli A, Yamanouchi J, Clemente-Casares X, Wang J, Serra P, Yang Y, Medarova Z, Moore A, Santamaria P (2010) Reversal of autoimmunity by boosting memory-like autoregulatory T cells. Immunity 32:568–580

25. Penberthy WT (2007) Pharmacological targeting of IDO-mediated tolerance for treating autoimmune disease. Curr Drug Metab 8:245–266
26. Orabona C, Pallotta MT, Volpi C, Fallarino F, Vacca C, Bianchi R, Belladonna ML, Fioretti MC, Grohmann U, Puccetti P (2008) SOCS3 drives proteasomal degradation of indoleamine 2,3-dioxygenase (IDO) and antagonizes IDO-dependent tolerogenesis. Proc Natl Acad Sci USA 105:20828–20833
27. Orabona C, Belladonna ML, Vacca C, Bianchi R, Fallarino F, Volpi C, Gizzi S, Fioretti MC, Grohmann U, Puccetti P (2005) Cutting edge: silencing suppressor of cytokine signaling 3 expression in dendritic cells turns CD28-Ig from immune adjuvant to suppressant. J Immunol 174:6582–6586
28. Zelante T, Fallarino F, Bistoni F, Puccetti P, Romani L (2009) Indoleamine 2,3-dioxygenase in infection: the paradox of an evasive strategy that benefits the host. Microbes Infect 11:133–141
29. Platten M, Ho PP, Youssef S, Fontoura P, Garren H, Hur EM, Gupta R, Lee LY, Kidd BA, Robinson WH, Sobel RA, Selley ML, Steinman L (2005) Treatment of autoimmune neuroinflammation with a synthetic tryptophan metabolite. Science 310:850–855
30. Zhu L, Ji F, Wang Y, Zhang Y, Liu Q, Zhang JZ, Matsushima K, Cao Q, Zhang Y (2006) Synovial autoreactive T cells in rheumatoid arthritis resist IDO-mediated inhibition. J Immunol 177:8226–8233
31. Takakubo Y, Takagi M, Maeda K, Tamaki Y, Sasaki A, Asano T, Fukushima S, Kiyoshige Y, Orui H, Ogino T, Yamakawa M (2008) Distribution of myeloid dendritic cells and plasmacytoid dendritic cells in the synovial tissues of rheumatoid arthritis. J Rheumatol 35:1919–1931
32. Scott GN, DuHadaway J, Pigott E, Ridge N, Prendergast GC, Muller AJ, Mandik-Nayak L (2009) The immunoregulatory enzyme IDO paradoxically drives B cell-mediated autoimmunity. J Immunol 182:7509–7517
33. Jasperson LK, Bucher C, Panoskaltsis-Mortari A, Mellor AL, Munn DH, Blazar BR (2009) Inducing the tryptophan catabolic pathway, indoleamine 2,3-dioxygenase (IDO), for suppression of graft-versus-host disease (GVHD) lethality. Blood 114:5062–5070
34. Ingelsten M, Gustafsson K, Oltean M, Karlsson-Parra A, Olausson M, Haraldsson B, Nyström J (2009) Is indoleamine 2,3-dioxygenase important for graft acceptance in highly sensitized patients after combined auxiliary liver-kidney transplantation? Transplantation 88:911–919
35. Furset G, Floisand Y, Sioud M (2008) Impaired expression of indoleamine 2, 3-dioxygenase in monocyte-derived dendritic cells in response to toll-like receptor-7/8 ligands. Immunology 123:263–271
36. Ratajczak P, Janin A, Peffault de Larour R, Koch L, Roche B, Munn D, Blazar BR, Socié G (2012) IDO in human gut graft-versus-host disease. Biol Blood Marrow Transplant 18:150–155
37. Uyttenhove C, Pilotte L, Theate I, Stroobant V, Colau D, Parmentier N, Boon T, Van den Eynde BJ (2003) Evidence for a tumoral immune resistance mechanism based on tryptophan degradation by indoleamine 2,3-dioxygenase. Nat Med 9:1269–1274
38. Munn DH, Sharma MD, Hou D, Baban B, Lee JR, Antonia SJ, Messina JL, Chandler P, Koni PA, Mellor AL (2004) Expression of indoleamine 2,3-dioxygenase by plasmacytoid dendritic cells in tumor-draining lymph nodes. J Clin Invest 114:280–290
39. Muller AJ, DuHadaway JB, Donover PS, Sutanto-Ward E, Prendergast GC (2005) Inhibition of indoleamine 2,3-dioxygenase, an immunoregulatory target of the cancer suppression gene Bin1, potentiates cancer chemotherapy. Nat Med 11:312–319
40. Gao YF, Peng RQ, Li J, Ding Y, Zhang X, Wu XJ, Pan ZZ, Wan DS, Zeng YX, Zhang XS (2009) The paradoxical patterns of expression of indoleamine 2,3-dioxygenase in colon cancer. J Transl Med 7:71
41. Caldwell RB, Bartoli M, Behzadian MA, El-Remessy AE, Al-Shabrawey M, Platt DH, Liou GI, Caldwell RW (2005) Vascular endothelial growth factor and diabetic retinopathy: role of oxidative stress. Curr Drug Targets 6:511–524
42. Al-Shabrawey M, Rojas M, Sanders T, Behzadian A, El-Remessy A, Bartoli M, Parpia AK, Liou G, Caldwell RB (2008) Role of NADPH oxidase in retinal vascular inflammation. Invest Ophthalmol Vis Sci 49:3239–3244

43. Elmarakby AA, Faulkner J, Baban B, Sullivan JC (2012) Induction of hemeoxygenase-1 reduces renal oxidative stress and inflammation in diabetic spontaneously hypertensive rats. Int J Hypertens 2012:957235
44. Grohmann U, Fallarino F, Bianchi R, Vacca C, Orabona C, Belladonna ML, Fioretti MC, Puccetti P (2003) Tryptophan catabolism in nonobese diabetic mice. Adv Exp Med Biol 527:47–54
45. Grohmann U, Fallarino F, Bianchi R, Orabona C, Vacca C, Fioretti MC, Puccetti P (2003) A defect in tryptophan catabolism impairs tolerance in nonobese diabetic mice. J Exp Med 198:153–160
46. Bertera S, Alexander AM, Crawford ML, Papworth G, Watkins SC, Robbins PD, Trucco M (2004) Gene combination transfer to block autoimmune damage in transplanted islets of Langerhans. Exp Diabesity Res 5:201–210
47. Jalili RB, Rayat GR, Rajotte RV, Ghahary A (2007) Suppression of islet allogeneic immune response by indoleamine 2,3 dioxygenase-expressing fibroblasts. J Cell Physiol 213:137–143
48. Oxenkrug GF (2011) Interferon-gamma-inducible kynurenines/pteridines inflammation cascade: implications for aging and aging-associated psychiatric and medical disorders. J Neural Transm 118:75–85
49. Oxenkrug GF (2010) Metabolic syndrome, age-associated neuroendocrine disorders, and dysregulation of tryptophan-kynurenine metabolism. Ann N Y Acad Sci 1199:1–14
50. Badawi A, Klip A, Haddad P, Cole DE, Bailo BG, El-Sohemy A, Karmali M (2010) Type 2 diabetes mellitus and inflammation: prospects for biomarkers of risk and nutritional intervention. Diabetes Metab Syndr Obes 3:173–186
51. Koleva-Georgieva DN, Sivkova NP, Terzieva D (2011) Serum inflammatory cytokines IL-1beta, IL-6, TNF-alpha and VEGF have influence on the development of diabetic retinopathy. Folia Med (Plovdiv) 53:44–50
52. Rojas M, Zhang W, Lee DL, Romero MJ, Nguyen DT, Al-Shabrawey M, Tsai NT, Liou GI, Brands MW, Caldwell RW, Caldwell RB (2010) Role of IL-6 in angiotensin II-induced retinal vascular inflammation. Invest Ophthalmol Vis Sci 51:1709–1718
53. Pallotta MT, Orabona C, Volpi C, Grohmann U, Puccetti P, Fallarino F (2010) Proteasomal degradation of indoleamine 2,3-dioxygenase in CD8 dendritic cells is mediated by suppressor of cytokine signaling 3 (SOCS3). Int J Tryptophan Res 3:91–97
54. Cheng S, Rhee EP, Larson MG, Lewis GD, McCabe EL, Shen D, Palma MJ, Roberts LD, Dejam A, Souza AL, Deik AA, Magnusson M, Fox CS, O'Donnell CJ, Vasan RS, Melander O, Clish CB, Gerszten RE, Wang TJ (2012) Metabolite profiling identifies pathways associated with metabolic risk in humans. Circulation 125:2222–2231
55. Fong DS, Aiello LP, Ferris FL 3rd, Klein R (2004) Diabetic retinopathy. Diabetes Care 27:2540–2553
56. Reichard P, Nilsson BY, Rosenqvist U (1993) The effect of long-term intensified insulin treatment on the development of microvascular complications of diabetes mellitus. N Engl J Med 329:304–309
57. Stratton IM, Kohner EM, Aldington SJ, Turner RC, Holman RR, Manley SE, Matthews DR (2001) UKPDS 50: risk factors for incidence and progression of retinopathy in type II diabetes over 6 years from diagnosis. Diabetologia 44:156–163
58. Kern TS (2007) Contributions of inflammatory processes to the development of the early stages of diabetic retinopathy. Exp Diabetes Res 2007:95103
59. Puri R, Kataoka Y, Uno K, Nicholls SJ (2012) The distinctive nature of atherosclerotic vascular disease in diabetes: pathophysiological and morphological insights. Curr Diab Rep 12:280–285
60. Elovainio M, Hurme M, Jokela M, Pulkki-Raback L, Kivimaki M, Hintsanen M, Hintsa T, Lehtimäki T, Viikari J, Raitakari OT, Keltikangas-Järvinen L (2011) Moderating effect of indoleamine 2,3-dioxygenase (IDO) activation in the association between depressive symptoms and carotid atherosclerosis: evidence from the young finns study. J Affect Disord 133:611–614
61. Niinisalo P, Raitala A, Pertovaara M, Oja SS, Lehtimäki T, Kähönen M, Reunanen A, Jula A, Moilanen L, Kesäniemi YA, Nieminen MS, Hurme M (2008) Indoleamine 2,3-dioxygenase

activity associates with cardiovascular risk factors: the health 2000 study. Scand J Clin Lab Invest 68:767–770
62. Daissormont IT, Christ A, Temmerman L, Sampedro Millares S, Seijkens T, Manca M, Rousch M, Poggi M, Boon L, van der Loos C, Daemen M, Lutgens E, Halvorsen B, Aukrust P, Janssen E, Biessen EA (2011) Plasmacytoid dendritic cells protect against atherosclerosis by tuning T-cell proliferation and activity. Circ Res 109:1387–1395
63. Malina HZ, Martin XD (1996) Indoleamine 2,3-dioxygenase: antioxidant enzyme in the human eye. Graefes Arch Clin Exp Ophthalmol 234:457–462
64. Park CY, Yang SH, Chuck RS, Gehlbach PL, Park CG (2010) The role of indoleamine 2,3-dioxygenase in retinal pigment epithelial cell-mediated immune modulation. Ocul Immunol Inflamm 18:24–31
65. Chen X, Liu L, Yang P, Wu C, Jin H, Xing L, Li B, Zhou H, Huang X, Zhu L (2006) Indoleamine 2,3-dioxygenase (IDO) is involved in promoting the development of anterior chamber-associated immune deviation. Immunol Lett 107:140–147
66. Munipally PK, Agraharm SG, Valavala VK, Gundae S, Turlapati NR (2011) Evaluation of indoleamine 2,3-dioxygenase expression and kynurenine pathway metabolites levels in serum samples of diabetic retinopathy patients. Arch Physiol Biochem 117:254–258
67. Jawa A, Kcomt J, Fonseca VA (2004) Diabetic nephropathy and retinopathy. Med Clin North Am 88:1001–1036, xi
68. Schefold JC, Zeden JP, Fotopoulou C, von Haehling S, Pschowski R, Hasper D, Volk HD, Schuett C, Reinke P (2009) Increased indoleamine 2,3-dioxygenase (IDO) activity and elevated serum levels of tryptophan catabolites in patients with chronic kidney disease: a possible link between chronic inflammation and uraemic symptoms. Nephrol Dial Transplant 24:1901–1908
69. Koenig P, Nagl C, Neurauter G, Schennach H, Brandacher G, Fuchs D (2010) Enhanced degradation of tryptophan in patients on hemodialysis. Clin Nephrol 74:465–470
70. De Miguel MP, Fuentes-Julian S, Blazquez-Martinez A, Pascual CY, Aller MA, Arias J, Arnalich-Montiel F (2012) Immunosuppressive properties of mesenchymal stem cells: advances and applications. Curr Mol Med 12:574–591

An Integrative Approach to Chronic Wounds in Patients with Diabetes: PPPM in Action

Jack C. Yu, Robert Dinsmore, Mohamad Masoumy, Jacqueline Sequoia, and Babak Baban

Abstract This chapter is a synopsis of diabetic wounds. We begin with an overview of the scope of the problem and examine tissue equilibrium in quantitative details, incorporating concepts of basic biomechanics and endurance limits. The key points are the interdependence of cell and the extracellular matrix (ECM): with the cells producing ECM and ECM affecting cell fate and behavior. This is followed by biochemical and cellular basis of pathogenesis of diabetic wounds. There is clearly a dysregulation of energy metabolism with sustained nutrient and caloric excess. With persistent intake of carbohydrates and fat exceeding the nominal capacity to process, predictable detrimental effects are inevitable. It is at this level, prevention should be seriously contemplated and implemented. It is much better to not be obese and diabetic to begin with. However, once overweight and hyperglycemic, the local regional

J.C. Yu (✉)
Section of Plastic Surgery, Medical College of Georgia at Georgia Regents University, Augusta, 30912, Georgia

Plastic and Reconstructive Surgery, Georgia Regents University, 1467 Harper Street, HB 5040, Augusta 30912-4080, Georgia
e-mail: jyu@georgiahealth.edu

R. Dinsmore
Section of Plastic Surgery, Medical College of Georgia at Georgia Regents University, Augusta 30912, Georgia

M. Masoumy
Department of General Surgery, Medical College of Georgia at Georgia Regents University, Augusta 30912, Georgia

J. Sequoia
Department of Family Medicine, Medical College of Georgia at Georgia Regents University, Augusta 30912, Georgia

B. Baban
Department of Oral Biology, College of Dental Medicine, Georgia Regents University, Augusta 30912, Georgia

M.S. Mozaffari (ed.), *New Strategies to Advance Pre/Diabetes Care: Integrative Approach by PPPM*, Advances in Predictive, Preventive and Personalised Medicine 3, DOI 10.1007/978-94-007-5971-8_11,

and the systemic effects combine to cause cellular swelling, apoptosis, necrosis, ischemia, hypoxia, ischemia-reperfusion, protein glycation, neuropathy, microangiopathy, persistent inflammation, and eventually tissue breakdown and chronic wounds. When the normal sequence of events is disrupted, a diabetic wound can remain in a pro-inflammatory stage and resist innate and therapeutic attempts to regain the normal cadence. We then inventoried the current, standard, clinical protocols and available commercial products, emphasizing those few that are evidence-based. Of particular importance are the roles of oxygen as free radicals are highly detrimental but paradoxically some evidences suggest a beneficial role of hyperbaric oxygen. The personalized approach will be realized with a proposed proteomic assessment of the wound, to precisely determine what are in excess and what are insufficient to allow targeted and individualized therapy. The advances in stem cell biotechnology and tissue derivatives are translating into the clinical arena; and together with surgical managements, constitute the subjects of the last two sections.

Keywords Diabetes • Chronic Wounds • Inflammation • Biomarkers • Wound management treatment

Objectives

This chapter reviews the current understanding of pathogenesis of chronic wounds in patients with diabetes and their management. Specifically, the objectives are for the reader to be able to:

1. Understand the scope of the problem and the magnitude of the burden imposed by diabetic wounds;
2. Explain why patients with diabetes are more prone to wound complications;
3. Compare and contrast the cellular and molecular events of diabetic wound healing and normal wound healing;
4. Formulate an appropriate, evidence-based treatment strategy to manage chronic wounds in patients with diabetes;
5. Understand the complex interplay at the organ and tissue levels involved in diabetic wounds;
6. Explain the current role of biologically derived materials useful in diabetic wound management;
7. Develop the correct affective perspective towards taking care of diabetic patients with chronic wounds.

1 Introduction

Chronic wounds pose a problem of increasing magnitude both to society and to patients in all corners of the world, both in developed and developing countries, and both rural and urban areas. A chronic wound occurs when, after injury, the healing process does not proceed through an orderly and timely progression from inflammation through proliferation to remodeling. Complete healing can be interrupted or delayed by intrinsic or extrinsic factors or combinations of both [1]. Chronic wounds represent a very heavy burden to patients, health care professionals, and the health care systems, affecting 29 million patients worldwide and growing at compound annual growth rates from 6 to 10% as of 2008 [2]. Diabetic foot ulcer is the leading cause for hospital admissions and frequently leads to below the knee amputation. There is increasing awareness of acute and chronic wounds and the need to manage them in an aging population [3]. The rise in the prevalence of diabetes and obesity contribute greatly to the development and difficulty in the healing of these wounds, and thus foster the increase in the incidence of persistent chronic wounds [4].

One difficulty in discussing chronic wounds is the lack of a precise, universally accepted definition. For some centers, a wound is termed chronic if it has not healed in 3 months, others use 8 weeks. Another definition of chronic wound focuses on the rate of healing as the key determinant. But even here there is disagreement. Some authors define a chronic wound as one which does not decrease in size by 30% per month, others use 50% [5], while others propose that neither size, complexity, nor time to healing is sufficient [6]. Rather, a chronic wound is one that does not progress through the normal stages of wound healing, inflammation, proliferation, and remodeling, in an orderly or timely manner. Unfortunately, this definition is tautological: a wound that does not heal normally is abnormal. True, but useless. As a result, numerous more quantifiable definitions have been used many with methodological weakness from linearizing the non-linear nature of wound healing. If the rate of healing is expressed in terms of absolute area healed per day, the result tends to exaggerate the rate of healing for large wounds, because of their larger surface area. When expressed as a percentage of initial area healed per day, the healing rate tends to exaggerate the rate of healing for small wounds. The only measurement found not to be influenced by wound size is the advancement of the wound margin towards the wound center per day [7]. Thus the most commonly and clinically validated expectation for a normal rate of healing comes from the pressure sore literature. Pressure sores that ultimately go on to heal, typically progress at a rate of just under 50% reduction in area in 4 weeks [8].

The current standard of treatment typically includes reduction in bioburden by repeated debridement and frequent dressing changes. Once systemic and local conditions have been optimized, the definitive coverage can be complete healing by secondary intention or, more often, by skin grafts or surgical flaps [9]. The time needed and costs are very high, with annual expenditure estimated to be more than $150 billion for pressure ulcer related chronic wounds alone [10].

Chronic wounds typically fall into one of three types: venous stasis ulcers, pressure ulcers, and diabetic ulcers. Venous stasis ulcers account for the majority of chronic wounds, and generally affect the elderly. They are caused by improper function of micro valves in the veins that normally prevent retrograde flow in the venous system. The dysfunction of these valves impedes the normal circulation of blood in the legs, causing increase in venous pressure and edema, leading to tissue damage and impaired wound healing. Although these problems may occur in the upper extremities, these effects are magnified in the legs due to their distance from the heart

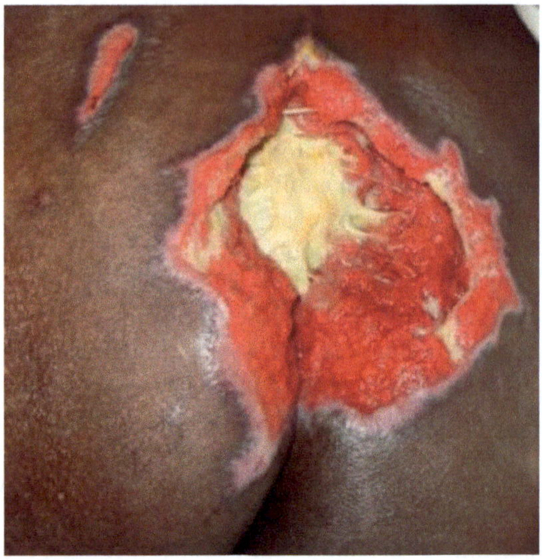

Fig. 1 Pressure ulcer. A large deep chronic sacral pressure ulcer with central wet necrotic tissue called slough surrounded by health red granulations tissue. At the junction of normal skin and the wound is newly formed epithelium

and dependent position, and the majority of venous stasis ulcers occur in the medial aspect of the lower leg [11].

Pressure ulcers comprise the second major type of chronic wounds. These typically occur in people who are bedridden or whose mobility is severely limited, including those with diabetes. When pressure on the tissue is greater than the capillary perfusion pressure, necessary for nutritive blood flow, typically around 23 mmHg or 2.6 kPa, ischemia and hypoxia are inevitable. Histological observation of rat soft tissue exposed to 11.5 kPa for 6 h showed minimum changes compared to controls [12]. If the external pressure is high enough and the time period is long enough to produce cellular and tissue injury, then when pressure is relieved and blood flow resumes, ischemia-reperfusion injury can occur (Fig. 1). If the damage incurred is greater than the body's ability to repair before the next episode of ischemia, tissue breakdown results in ulceration. Some areas of the body that are particularly susceptible to pressure ulcers include the heels, metatarsal heads, shoulder blades, ischium, sacrum, the greater trochanter, and the triangular area at the base of the spine forming the posterior part of the pelvis [13]. The sensory impairment caused by diabetic neuropathy makes diabetic patients particularly susceptible to this process [14]. Indeed, the third category of chronic wounds, diabetic foot ulcers is thought to have dual proximate causations.

Diabetic ulcers are chronic wounds which occur in patients with long standing diabetes, most frequently affecting the feet. Diabetic patients are particularly susceptible to developing ulcers because of the long-term consequences of hyperglycemia. These ulcers are often difficult to heal, and they are the focus of this chapter. In addition to the endocrine and energy metabolism perturbations, diabetes mellitus also impairs the immune, peripheral nervous, and the circulatory systems at both the macro- and microvascular levels [15]. Repeated injury due to sensory deficits, compounded by defective reparative capacity and impaired immune systems, can cause even the smallest cut or bruise to become dangerously infected. Individuals with diabetes are also two to four times more likely than non-diabetic individuals to have peripheral arterial disease (PAD), a term that covers an array of medical problems caused by obstruction of the large arteries in the arms or legs. A more severe form of PAD is critical limb ischemia (CLI), a leading cause of lower limb amputations in the civilian population. Diabetic patients are five times more likely to have amputations, with higher mortality and death at a younger age [16].

Even though there have been great advances in both basic and applied biomedical sciences over the past four to five decades, little substantive improvement has been made in preventing morbidity and disability from chronic wounds. The best available evidence for chronic wound care in a Cochrane review of 342 unique publications showed only 18 studies met the required standards [17]. At 4 weeks, there is a <50% healing rate. This improves to approximately 80% at 12 weeks and can be predicted if the percent surface reduction exceeds 53% at 4 weeks [18]. Among the many factors contributing to nonhealing or poor healing wounds, local microbial biota coupled with impairment in the production of cytokines by resident inflammatory cells and fibroblasts, as well as reduced angiogenesis are crucial [19]. Recent evidence suggests that stem cells, exogenous growth factors, and biologically derived tissue substitutes may have the potential to improve the results of management of complex chronic diabetic wounds. In order to effectively treat diabetic patients, efforts must be multidisciplinary with educated professionals implementing standards of care and maximizing preventive measures. The ultimate goal is to standardize chronic wound treatments and best define predictive markers and preventive measures for the diabetic patient [20, 21]. Realizing this goal begins with a comprehensive understanding of the disease process and current treatment modalities available. This chapter will provide a detailed and comprehensive assessment in the current state of the art regarding diabetic wound healing, from bench to bedside, and from molecules to the patients and society.

2 Fundamental Laws of Tissue Dynamics

Both normal adult tissue and chronic wounds exist in a state of homeostasis, with the production of tissue mass precisely balanced by the amount of degradation, given sufficient period of time to allow averaging [22]. Thermodynamically, these are open systems with inflow of material and energy and output of products, waste, and heat. The difference between normal tissue and chronic wounds is that the chronic wound starts at a lower overall level and without normal integumentary integrity. The initial loss in integrity is the essence of injury. Tissue disruptions occur when an external force, F, measured in kg m/s^2, known as a Newton or N, produces sufficient stress, σ, which is force per unit area, F/A, to exceed what the ultimate stress is for that tissue as a material. The unit for stress is therefore N/m^2, which is also known as a Pa. The amount of tissue deformation resulting from the

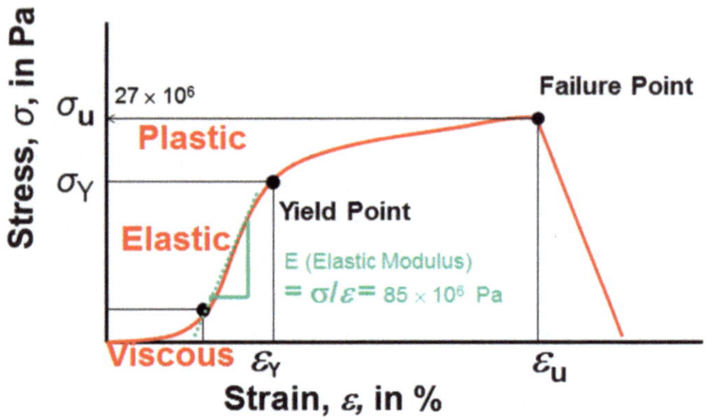

Fig. 2 Theoretical stress-strain curve of human skin. This curve shows the viscoelastic behavior with initial "toe-in" (viscous) phase followed by the linear (elastic phase) leading up to permanent (plastic) deformation and eventually to breaking (failure)

stress as a fraction of the original tissue dimension, expressed by $\Delta L/L_0$, is known as strain, represented by ε. The strain at which tissue ruptures is the ultimate strain and the maximum sustainable stress is called the ultimate stress [23]. For human skin, the ultimate strain, measured in tension at a strain rate of 0.012/s and in parallel to the Langer Lines, is 40%, and the ultimate tensile stress is 27×10^6 Pa or 27 MPa. The stress-strain curve, from which ratio of stress to strain, σ/ε, can be calculated is very important as it describes how materials behave under stress. σ/ε is the definition of elastic modulus or Young's modulus (Fig. 2). Rather than pure elastic deformation, almost all biological tissues undergo visco-elastic deformation, as such the strain rate is important, and at the strain of 0.012/s, the human skin elastic modulus parallel to the Langer's Lines is 85 MPa [24].

There are two basic types of injury mechanisms which can produce integumentary interruption, or an open wound, based on strain rate. High strain rates result from sudden impact, such as when a diabetic patient strikes his or her great toe against the leg of a table while walking, causing immediate tissue disruption due to the resulting strain exceeding ultimate strain. The second mechanism is the same as that which occurs in a pressure ulcer. When a steady stress, even below the ultimate stress but at a level high enough to impede the nutrient blood flow, is applied for sufficient time can cause tissue breakdown. This is the most basic proximate cause of pressure ulcer formation. For all living tissues, there is an inverse relationship between the amount of pressure and time required to form an ulcer [25], approximated by the equation below:

$$k = Pt \tag{1}$$

k is a constant and P is the pressure or compressive stress and t is time. Expressing in logarithms,

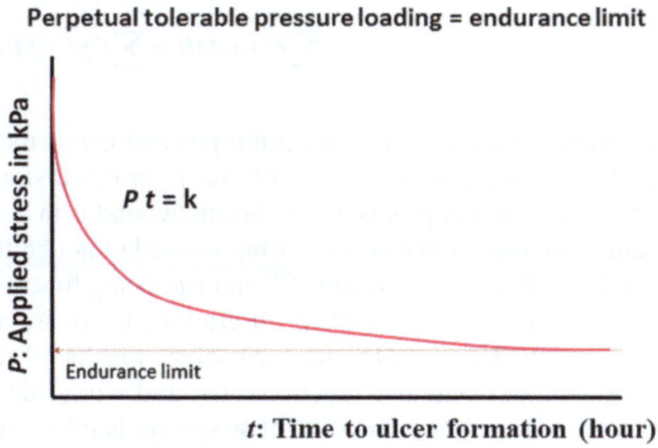

Fig. 3 Pressure (*P*) – time (*t*) curve. The theoretical curve showing *P t*=k: the higher the pressure the shorter the time to ulcer formation. All points above the *P-t* curve are non-sustainable and thus "ulcerogenic"

$$\log P + \log t = \log k \quad \text{or} \ \log P + \log t = \text{constant} \tag{2}$$

This relationship, the sum of several variables equal to a constant, is the essence of zero-sum game and it describes many relationships in nature, whenever an overall constraint or limit exits [26]. Below certain pressure level, like the rheobase of stimulus-duration curve, it can be tolerated indefinitely, that pressure is the endurance limit and is 23 mmHg or 3,000 Pa, 3 kPa, for human skin and corresponds to the capillary closing pressure. Different tissues have different architecture and ischemia tolerance; therefore possess different characteristic *P-t* curves, with the sole of the foot being the most resistant. The normal tissue equilibrium cannot exist without blood flow (Fig. 3).

In patients with normal sensation, pressure induces ischemia and ischemia produces pain and discomfort as cellular hypoxia develops. Such sensory input alerts the patient to change posture and relief the pressure before more tissue damage can occur, thus keeping below the *P-t* curve. The pressure which produces pain, tested at 0.5 cm^2 using a constant application rate of 0.68 N/s, is called the pressure-pain threshold or PPT [27]. Diabetic neuropathy interferes with this important sensory feedback, causing the pressure time product to exceed the limit and resulting in injury to tissues. This mechanism contributes greatly to the high incidence of diabetic foot ulcers.

The goal in the treatment of chronic wounds is to temporarily increase tissue production rate above tissue degradation rate so that there is a net gain of tissue. During the normal healing of an acute wound, there is a transient net gain in tissue mass with production exceeding degradation, until eventually when the wound is healed and antagonistic balance is restored. Since the mass of any tissue is the sum total of the mass of all cells and the mass of all non-cellular components, the extracellular matrices (ECM), in that tissue, the following equation describes this tissue dynamics:

$$\sum_{i}^{n} Cli + Mli = \sum_{i}^{n} Cpi + Mpi \tag{3}$$

Cli is the cell loss of various cell types and *Cpi* is the various cells produced, *Mli* is the different types of extracellular matrices lost and *Mpi* is the different type of tissue matrices produced. A chronic wound is in a steady state if the wound size remains constant that is not getting worse but not better either. The terms of Eq. 3, can be further expanded with *Cli* and *Cpi* being functions of cell size, and cell number which are affected by proliferation, necrosis, apoptosis, immigration, and emigration. *Mli* and *Mpi* are themselves results of cellular activities and they in turn alter cell fate and functions, *Cli* and *Cpi*. This circular, recursive iterative, system require judicious control as system is inherently at risk for "run-away", or feed-forward, type of explosive transition driving the system out of equilibrium. Perhaps chronic wound is the body's attempt to suppress the more detrimental potential of run-away cellular growth, the development of cancer. In addition to affecting cellular function, ECM also determines the oncotic pressure, πi in the interstitium. Together with three other variables, it directs the movement of fluids in and out of a defined local tissue space as transudates or exudates, dictated by the Starling's equation:

$$J = K[(Pc - Pi) - \alpha(\pi c - \pi i)] \tag{4}$$

J is the net movement of the fluid volume from capillary into the interstitium; *K* is the filtration coefficient which is itself a function of membrane and cell junctions, as is α which denotes the reflection coefficient of the capillary membrane. *Pc* and *Pi* are hydrostatic pressure inside the capillary and the interstitium. πi and πc are the corresponding values for oncotic pressure. Thus, *Mli* and *Mpi* are under both cellular and non-cellular controls and can be expressed:

$$M = f(J),(C) \tag{5}$$

Combining all five equations, it is amply obvious that cells are the key variables and events that cause cells to undergo proliferation, differentiation, migration, and apoptosis are complex, coordinated, highly regulated, and non-linear [28]. Importantly, the relationship between cells and matrix is recursive, cells produce matrix and matrix affect cells, making cause and effect iterative and rendering chronic wounds complex adaptive systems, CAS. It has been proposed that cell viability is the reason of the limit in Eq. 2 and cellular failure underpins the inability of chronic wounds to progress. Recent efforts at treating recalcitrant, complex, chronic wounds transfer fresh, robust, high-Hayflick constant stem cells to change the stalled wound into a transient state of imbalance; with $\sum_{i}^{n} Cli + Mli < \sum_{i}^{n} Cpi + Mpi.$ This temporary increase in the anabolic rate must be followed by a gradual decrease, ending with a new, restored equilibrium state, after having repaired the wound. Failure to achieve this will make malignancy an ever present danger.

In patients with diabetes, at least 100 variables negatively affect normal tissue dynamics and wound healing with perturbations at multiple levels, affecting both the cellular and matrix compartments and will be the focus of the following sections.

3 The Key Cellular and Biochemical Events Related to Energy Metabolism

A diabetic wound is a complex adaptive system, an evolving "ecology", with many agents playing many roles at multiple points in time and space over many magnitudes of scales from microseconds to decades and from 10^{-9} to 10^{0} m with many events overlapping and intermingling [29]. Unfortunately, describing these events and agents can only occur one sentence at a time, making the task difficult, if not impossible. However, the root cause of the problem can be summarized as sustained excess of dietary glucose and fatty acids. There is a decoupling of caloric intake and energy expenditure in the many developed countries. As the old saying goes in pharmacology 101: what separates a poison and a medicine is the amount. Excessive amount of anything can make it a poison. Food is no exception.

Persistent high fat and high carbohydrate intake together with a progressively more sedentary life style results in food intake substantially exceeding the metabolic requirements and marked increase in serum glucose and non-esterified fatty acids. Through the highly efficient transmembrane transporters, excess glucose floods the cells in the body during postprandial hyperglycemic spike, increasing the delivery of pyruvate to mitochondria, overcoming their maximum capacity to control the electrons generated from oxidative phosphorylation, and results in increased electron leakage above the nominal level, producing ever higher levels of reactive chemicals with unpaired electrons called free radicals [30, 31]. These radicals, in the form of reactive nitrogen species (RNS), and reactive oxygen species (ROS), can be extremely destructive to the cells by oxidizing important elements within and outside the nucleus. Lipids in the membranes, proteins in the cytosol and nucleus, and DNA are all targets of oxidation [30, 31]. Histone protein H2A in the serine-phosphorylated form has been quantitatively measured to show evidence of DNA stress, as shown in Fig. 4 [32, 33]. In addition, ROS activates an important pro-inflammatory transcription factor NF-κB through phosphorylation of IκB, which in the non-phosphorylated form binds and keep NF-κB from nuclear translocation. There is also increase in protein kinase c, PKC, activity from increase in glycolysis and PKC can also activate NF-κB. Nitric oxide production and concentration increase [30, 31, 34]. The end result of this mayhem is perturbation of the normal balance and function in innate immunity, in particular, inflammation.

The back-up system to glycolysis for metabolizing intracellular glucose is the sorbitol pathway, through the action of aldose reductase. The sorbitol produced, if not in excess, can be converted through a series of enzymatic actions and eventually by hexokinase back to be managed by glycolysis, serving as a shunt to temporarily "park" the excess glucose. However, if this pathway is also overwhelmed, the sorbitol

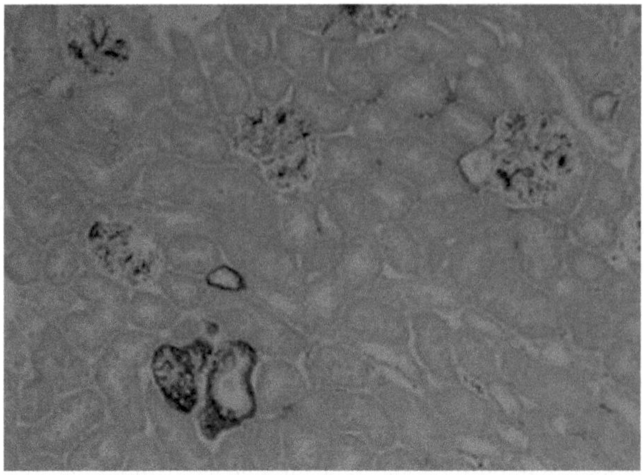

Fig. 4 Evidence of nuclear stress in the kidney of diabetic mice. This image shows the presence of γ H2AX, a measure of double-stranded DNA breaks

level elevates, increasing the osmotic pressure within the cell, causing cellular swelling [35]. The portal vein collects the absorbed glucose, amino acids, and free fatty acids after each meal and delivers them to the liver. Typically, an increase in serum glucose causes pancreatic beta islets to release insulin and insulin suppresses further hepatic glucose production. But this insulin-induced reduction in hepatic glucose release is blunted by high non-esterified free fatty acids [34]. Persistent hyperglycemia, the hallmark of both type 1 and type 2 diabetes, can also cause glycosylation of proteins such albumin and collagen, among others, which alter their shape and thus function. The end results of all these negative effects are micro-angiopathy and neuropathy [30, 35–37].

The longer the peripheral nerve, the more susceptible and symptomatic it is, since nerve conduction is a function of the length of the nerve. That is why clinically, the patient presents with glove-stocking type of dysesthesia, paresthesia, and eventually, anesthesia. Contributing to and compounding the problem is the compression-prone anatomic "choke" points: carpal and tarsal tunnels. The sorbitol-induced increase in osmotic pressure causes nerve swelling and in the confined spaces where the median nerve and the tibial nerve pass under the touch, unyielding ligaments such as transverse carpal ligament of the wrist and the flexor retinaculum of the ankle, hour-glass shaped nerve compression can produce profound increase in nerve conduction velocity and sensorimotor loss [35]. Without normal sensory protection, excessively long periods of pressure applied to the bony prominences such as the heel and the first meta-tarsal head cause prolonged tissue ischemia and hypoxia, followed intermittently with ineffective relief, setting up repeated ischemia and reperfusion injuries, leading finally to tissue breakdown and ulcer formation [38].

Oxygen is another, perhaps more illustrative, example of poison-medicine duality, playing a very important double role in that it is absolutely required but at a very precise delivery rate, and it is also absolutely dangerous if it exceeds that rate. The only organelle capable of processing oxygen is the mitochondrion, through its electron transport chain, which must be maintained by oxidative phosphorylation. During ischemia, glycolytic enzymes are up-regulated; pyruvates, no longer entering the

TCA cycle, are converted to lactate and mitochondria shut down. But like a nuclear power station, the proper powering-down does not always occur smoothly, and when it does not, spillage of dangerous material to the immediate environment is inevitable. Intra-mitochondrial protein species when released into the cytosol are potent initiators of apoptosis. When oxygenated blood returns during re-perfusion, the available functioning mitochondria in viable cells may be insufficient to handle the oxygen and more free radicals are formed causing cell death and tissue necrosis. Eighty-five percent of all below the knee amputation are associated with foot ulcers.

4 Inflammation and Diabetic Wound: Cellular and Molecular Aspects

Wound healing is a natural and dynamic response to tissue injury. It is a restorative and complex process that involves a cascade of molecular and cellular events. Wound healing is an interactive course of successive events, involving inflammatory and non-inflammatory cell types, cytokines, chemokines, enzymes and growth factors. The healing process employs the whole structural, chemical, and signaling components of cellular system to achieve tissue repair.

Normal healing is a systematic process consisting of four overlapping classic phases of hemostasis, inflammation, proliferation, and maturation. As the very first phase of the healing process, hemostasis begins with platelets activation and aggregation. Activation of platelets results in production and secretion of chemotactic and growth factors, such as platelet-derived growth factor (PDGF), proteases, and vasoactive agents (e.g., serotonin, histamine) [29, 39, 40]. Vasoactive mediators cause vasoconstriction which protects again excessive blood loss. Platelet aggregation is caused by exposed sub-endothelium collagen and results in clot formation. Platelet-induced chemokines start the inflammation, one of the early phases of a wound healing response, with the infiltration of immune cells to the wound site. The major inflammatory cells involved in the healing process are neutrophils, macrophages, and lymphocytes. Although neutrophils are the most abundant cell type for the first 48 h after injury, they are not pivotal to wound healing and their absence does not prevent the healing. Neutrophils cleanse the wound site of bacteria and necrotic matter, and release inflammatory mediators and bactericidal oxygen free radicals. Macrophages, on the other hand, are essential to wound healing and are the most important cells in the early phase. Macrophages orchestrate the complex processes of cellular proliferation and functional tissue regeneration within wounds. The phagocytic function of macrophages purges debris and bacteria. Macrophages also secrete collagenases and elastases to break down injured tissue and release cytokines. In addition, macrophages release PDGF from their α granules which can stimulate the chemotaxis and proliferation of fibroblasts and smooth muscle cells. Finally, macrophages play a crucial role in promoting angiogenesis. Macrophage-derived growth factors are the principal agents in new tissue formation, as evidenced by the fact that new tissue formation is impaired and wound repair is defective in macrophage-depleted animals. In studies in which experimental wounds are rendered monocytopenic, subsequent stages of fibroplasia and granulation tissue formation are impaired and the overall rate of wound healing is delayed. T lymphocytes are the other cell types recruited into the site of wound during the inflammatory phase, approximately 72 h following injury [29, 40, 41]. T lymphocytes are mobilized and recruited to the wound by the cellular release of interleukin 1 (IL-1), a component of inflammasome which also plays a protective role in the host response to severe injury [29, 40, 42]. Lymphocytes also participate in cellular immunity and antibody production. The proliferative phase has multiple stages, including angiogenesis, collagen deposition, granulation tissue formation, epithelialization, and wound contraction. It occurs 3–5 days following injury and overlaps with the preceding inflammatory phase. Granulation tissue is the central part of proliferative phase and includes inflammatory cells, fibroblasts, neovasculature in a matrix of fibronectin, collagen, glycosaminoglycans, and proteoglycans. Collagen deposition is also important because it increases the strength of the wound. One of the major functions of fibroblasts is the production of collagen [40, 43]. At the end of the granulation sub-phase, fibroblasts begin to commit apoptosis, converting granulation tissue from an environment rich in cells to one that consists mainly of collagen [29, 39, 40, 42, 43]. The maturation phase of tissue repair begins when collagen production equals to collagen degradation. During maturation, type III collagen, which is prevalent during proliferation, is replaced by type I collagen. The onset of the maturation phase may vary extensively, depending on the size of the wound and whether it was initially closed or left open, ranging from approximately 3 days to 3 weeks. The maturation phase can last for a year or longer, similarly depending on wound type [29, 40] (Fig. 5).

Normal wound healing progresses in an organized and predictable rhythm. If such orchestration gets disrupted, the temporal sequence is lost, and eventually stall happens if attempts to resume the original sequence fail, resulting in a chronic wound. Diabetes is one of the major factors in affecting the wound healing adversely, altering the normal healing pathway to an impaired process which leads to the formation of chronic ulcers.

Mounting evidence indicates that diabetes causes many cellular and molecular changes, which together impair wound healing, produce local ischemia, cause occlusion of the blood vessels within the edge of the wound, and form chronic ulcers. At the cellular level, diabetic patients show enervated recruitment of inflammatory cells at the early stages following injury, as well as impaired leucocyte activity such as bactericidal capacity which increase the risk of wound infection. At later stages, once the leukocytes migration is established, several animal studies on diabetic murine models (db/db) have demonstrated a significant elevated presence of inflammatory cells such as neutrophils and macrophages, leading to prolonged up-regulation of proinflammatory cytokines IL-17, IL-6, tumor necrosis factor-alpha (TNF-α) and inflammasome component, interleukin-1β (IL-1β) [29, 39, 40, 44]. As a consequence of uncontrolled and sustained recruitment of inflammatory cells, higher expression of chemokines such as macrophage inflammatory protein 2 (MIP-2/CXCL8) and monocyte chemotactic protein 1 (MCP-1/CCL2) could be detected [40, 44].

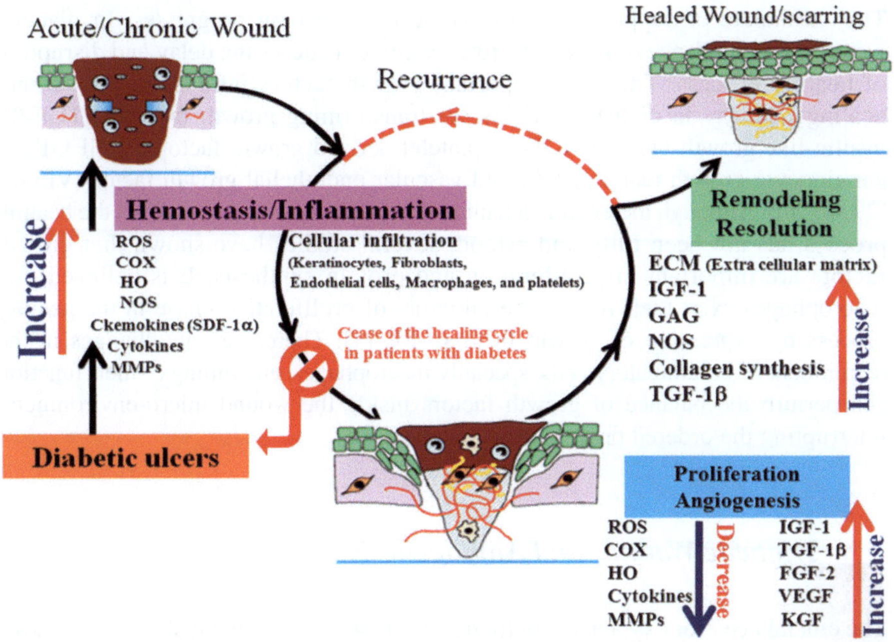

Fig. 5 Wound healing cycle in healthy versus diabetic patients. The cellular and molecular mechanisms of wound healing start with the inflammatory phase. Inflammation involves multiple signals, cell infiltration, cytokine and chemokine production, signaling molecules (e.g. NO), and endopeptidases including MMPs. In healthy individuals, the healing cycle enters the phases of angiogenesis and proliferation while in the patients with diabetes, this transfer from an inflammatory phase to angiogenesis/proliferative phase may be blocked. As a result, tissue formation and remodeling is delayed and healing cycle remains in the inflammatory stage chronically, with no wound resolution and incomplete healing cycle

Therefore, despite the primary impairments in leukocyte migration and recruitment at the early stages of injury in diabetic patients, once established, sustained and unregulated inflammatory cells recruitment into the wound causes an inappropriately prolonged inflammatory response, delaying and disrupting the normal sequencing of wound healing. It is very likely that such alterations may result in closing of "windows of functionality" and once lost, such it may never return.

4.1 *Diabetic Wounds and Growth Factors*

Growth factors are naturally occurring substances capable of stimulating cellular growth. They influence cellular proliferation, differentiation, and function. Growth factors regulate a variety of cellular processes, acting as signaling molecules between cells and inducing migration of inflammatory cells into the wound bed.

They stimulate or suppress protein synthesis as healing progresses. In diabetic wounds, an imbalance of growth factors expression causes the delay and disruption of healing process. The most important growth factors influencing the wound healing pathway in diabetic patients are transforming growth factor β (TGF-β), insulin-like growth factor 1 (IGF-1), platelet derived growth factor (PDGF), IL-6, keratinocyte growth factor (KGF) and vascular endothelial growth factor (VEGF) [29, 40, 44]. Although the exact mechanism by which growth factors affect the healing process has not been fully understood, several studies have shown that growth factors are important in proliferation and protein synthesis. It is believed that macrophages exert their role as the initiators of proliferative phase in the healing process by expression of growth factors [40, 44]. Therefore, any changes in the recruitment of inflammatory cells, specially macrophages, and timing of their function can perturb the balance of growth factors inside the wound micro-environment, interrupting the ordered process of wound healing.

4.2 Diabetic Wounds and Angiogenesis

The crucial need for oxygen and nutrients during wound healing makes angiogenesis imperative. Essentially, angiogenesis is the formation of new blood vessels from pre-existing capillaries that are then able to penetrate the wound site. Diabetic patients show irregular and abnormal angiogenesis occurring in different organs as well as in chronic and non-healing wounds. Impaired development of new blood vessels and capillary network acutely at the very early stages of injury and in non-healing diabetic wounds, limits the recruitment and migration of inflammatory cells into the wound site. As the result of this reduction of inflammatory cells, the total amount of chemokines and growth factors necessary for angiogenesis will be decreased. The oxygen supply as well as nitric oxide (NO), a potent inducer of angiogenesis and vasodilator, will decrease. Later on, in established diabetic chronic wounds, as the results of hyper-inflammatory responses, the level of NO will exceed the limit required for the normal healing, which results in blockade of angiogenesis [40, 44, 45]. Any imbalance in the level of NO, insufficient or excessive, along with irregularity in production and presence of the other angiogenic factors (such as VEGF or FGF-2), can adversely affect the healing process. Several animal studies have shown that restoration of these factors in a balanced fashion will modulate the angiogenic process, resulting in optimization of wound healing.

4.3 Diabetic Wounds and Extracellular Matrix

Extracellular matrix (ECM) is a structural framework produced by the cells that cells use to properly attach and function. Importantly, ECM provides the cues for proper cellular behavior. ECM maintain its integrity through a balanced interaction with cells in a reciprocal fashion through production and utilization of extracellular proteins such as collagen and glycosaminoglycans (GAG) [29, 40, 44, 46]. Chronic diabetic wound formation is a cycle of abnormal cellular activities causing abnormal ECM which in turn causes abnormal cell activities and the cycle repeats. Several studies have shown a reduction in the level of collagen and GAG in the diabetic wounds which results in the lower wound strength, making them more vulnerable to further trauma.

Decreased productions of collagen and GAG have reportedly attributed to the lower presence and activity of fibroblasts as well as decreased level of NO synthesis. Therefore, lower level of NO, decreased fibroblasts count and activity, low TGF-β, altered biochemical micro-environment, and higher metalloproteinases (MMP) may lead to an insufficient scaffold structure for cells to attach and function appropriately inside the wound, resulting in the delayed healing [40, 44, 46].

5 Prevention and Prediction – the Solution at the Population Level

In addition to providing a deep understanding, mathematical modeling by equations, or sets of equations, allow us to predict how a system will behave under different settings, if the system is truly deterministic. Once we can predict, we can more effectively intervene and, ultimately prevent. Chronic wounds in patients with diabetes are best modeled as complex adaptive systems, or CAS, because like all such chaotic systems, there are many interacting elements that are woven into a network of smaller networks, each affecting and being affected by others. The hallmarks of CAS are emergence and self-organization. At the simplest level, emergence means the whole is greater than the sum all individual parts. Though deterministic, the system, as a whole, behaves in a way that cannot be predicted based on the behaviors of the individual components, no matter how detailed, accurate, and precise the information is. CAS abounds in biology and requires investigation across temporal and spatial scales. Healing, or non-healing are emergent properties, they represent attractors: the states in which the system tends to settle into. Below the tissue level, the cells and extracellular matrices are the two principle interacting agents; one level below that, and the organelles, signaling and structural molecules are the key agents. Above the individual wounds and patients are the community and societal levels. One way to start the understanding of chronic wounds at the level of an individual diabetic patent is to have a complete inventory at the molecular level by protein array approach. At the present there has been no such systematic assessment of proteomic data, but it can certainly be done using samples gathered during tissue debridement (Fig. 6).

Biomarkers have long been used to discern a normal state from an abnormal one, as well as to assess the extent of a clinical disease and how patients are responding to therapeutic treatments. Classically, hemoglobin A1c has been used as the standard measure of glycemic control and its trends followed to help with overall

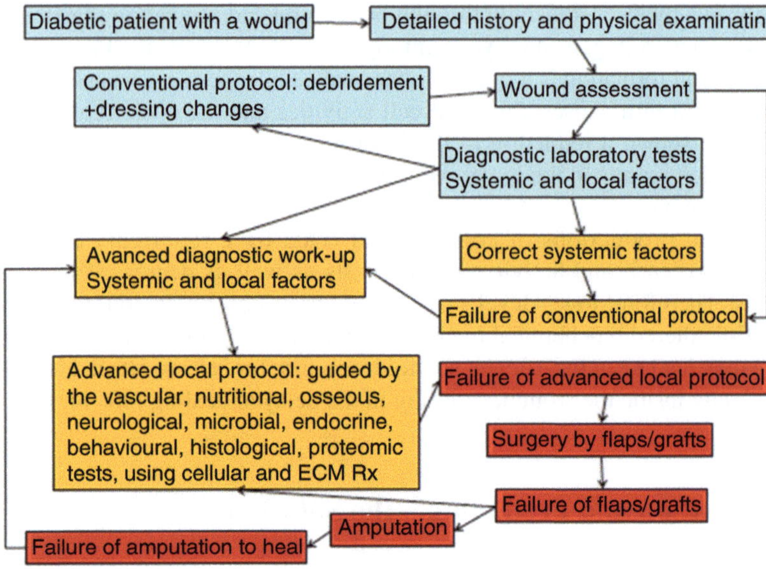

Fig. 6 Tiered management protocol for wounds in a diabetic patient. *Blue* is tier 1 protocol and should be implemented in all initial encounters. Failing that, tier 2 protocol (*gold*) is instituted. Finally, if the patient continues to not respond, then tier 3 (*red*) is necessary

management and treatment decisions. More recently, other potential biomarkers for diabetic patients have been investigated. For instance, advanced glycation end-products (AGE), resulting from carbonyl stress, have been implicated in the pathogenesis of vascular complications seen in diabetes and the soluble receptors of AGE (RAGE) has been suggested as a predictive biomarker of coronary artery disease in type 2 diabetes [47]. In addition to coronary heart disease, the pathogenesis of other complications related to diabetes, including retinopathy, neuropathy, nephropathy, vascular disease, diabetic foot, and others continue to be the focus of intense research as their mechanisms may shed light on process specific biomarkers which can assist in treatment and prevention [47–52].

Specific to patients with chronic wounds, the potential for biomarker candidates is wide-ranging and each could play a critical role in helping assess one, or all phases of wound healing and the overall healing potential of a wound. The interplay of the components involved in wound healing, as described previously, are complex, adaptive, and overlapping. The greatest limitation currently is our incomplete understanding of these systems and how they communicate with and affect one another. Some have suggested looking into the histology, cell morphology and genetic profile of a chronic wound to help dictate the extent of wound debridement [53]. Taking this idea a step further, there is great potential and utility in being able to better assess an individual's wound profile to help provide more personalized treatments. Ultimately, such measures will help in providing comprehensive, cost-effective care with better end-results in a more timely fashion. Aside from the commonly described cytokines, growth factors and pathways involved in wound healing,

Table 1 Novel biomarkers and therapeutic targets for chronic wounds in diabetes

Marker/target	Significance to wound healing	Reference
Extracellular matrix (ECM) modulators		
Matrix Metalloproteinases (MMPs)	A family of endoproteinases integral in wound healing The very strong degradative capacity of MMPs necessitates an equally strong counterpart. A family of tissue inhibitors of MMPs (TIMP 1–4) as well as other inhibitors including α 2-macroglobulin and tissue factor pathway inhibitor 2 work to counteract the degrading capacity of MMPs. Dysregulation of MMPs has deleterious effects on ECM degradation and turnover. This disturbance in balance between ECM formation and degradation has been seen in many diseases and associated with complications of diabetes including nephropathy, retinopathy, poor wound healing, and microangiopathy Up-regulated by cytokines including IL-1β, TNF-α Down-regulated by TGF- β, IL-4, corticoid hormones and insulin-like growth factors	[19, 51, 54–56]
MMP-9: TIMP-1 ratio	Suggested predictor of wound healing: the higher the ratio the more poorly the wound is predicted to heal	[56]
CD9	Cell-surface protein implicated in wound healing. Specifically it has been shown to down-regulate JNK signaling which helps attenuate MMP-9/2 transcription.	[57]
P38 MAP Kinase	P38 MAPK acts as a growth kinase (via ERK and PKB/AKT) during normal wound healing *but* can act as a stress kinase in the presence of 3-deoxyglucosone (3DG), an advanced glycation end product precursor. In the presence of 3DG-modified collagen, fibroblasts show reduced migration, proliferation and ECM production while showing increased apoptotic activity.	[58]
DNA damage and oxidative stress		
γH2Ax	Formed by phosphorylation of H2Ax (minor histone variant of H2A) in response to double-strand breaks (DSB) in DNA as part of the DNA damage-response (DDR) pathway. Helps with initiation of DSB repair and modulates checkpoint responses. An excellent marker of DNA damage and repair which is seen in chronic inflammation secondary to formation of reactive oxygen (ROS) and nitrogen species.	[49]
8 hydroxy-2-deoxyguano sine (8-OH-DG)	Another marker of oxidative stress in DNA.	[59]

(continued)

Table 1 (continued)

Marker/target	Significance to wound healing	Reference
Inflammasomes and inflammatory mediators		
Caspase 1/IL-1β, IL-18	Inflammasomes are intracellular multi-protein complexes that activate capsases which regulate the innate immune system and are involved in apoptosis, inflammation, cell proliferation and cell differentiation. This process is initiated by pathogen-associated molecular patterns (PAMPs: e.g. LPS and flagellin) and damage-associated molecular patterns (DAMPs or alarmins: e.g. high-mobility group box 1, uric acid and heat-shock proteins) The end result includes maturation and secretion of pro-inflammatory cytokines IL-1β and IL-18.	[49, 60, 61]
Plasminogen (Plg)	Plasminogen shows pronounced pro-inflammatory effects through cytokines like IL-6, which potentiates the early inflammatory response. Diabetic patients have been found to have minimal Plg enhancement and subsequently lower levels of IL-6.	[52]
Hypoxia, angiogenesis and proliferation		
Hypoxia-induced factors (HIF)	HIF1α and HIF2α regulate the expression of over 150 genes involved in metabolism, cell survival, tumor metastasis, erythropoiesis, and vascular remodeling. Some of these genes include vascular endothelial growth factor (VEGF), erythropoietin, CXCR4 and MMPs **VEGF**: Endothelial cells are its main target and it is paramount in wound healing as it is involved in vasodilation, basement membrane degradation, endothelial cell migration and proliferation helping lead to angiogenesis and maintaining endothelial integrity. Degradation of HIF1α is rapid involving a von Hippel-Lindau tumor suppressor containing protein E3 ubiquitin ligase while HIF2α degradation involves an Int6/proteasome pathway.	[50, 62, 63]
Int6/eIFe	INT6/eIF3e is a regulator of HIF2α which can be modulated to help promote angiogenesis and wound healing in normoxic conditions through enhancement of angiogenic factors.	[50]
Stromal cell-derived factor-1 α (SDF-1α)	Stromal cell-derived factor-1α assists with recruitment of endothelial progenitor cells (EPCs) to ischemic tissue and both SDF-1α and EPCs are found to be diminished in diabetic wounds. EPCs are essential to neovascularization and wound healing. Bone marrow EPCs are mobilized through activation by eNOS and localize to the injured tissue with the help of SDF-1α signaling.	[19, 62]

(continued)

Table 1 (continued)

Marker/target	Significance to wound healing	Reference
CXCR4	Stromal cell-derived factor-1α binds to CXCR4 which helps regulate inflammation and cell migration. Antagonism of CXCR4 in a diabetic model has shown to promote neovascularization and wound healing though mobilization of EPCs and enhancing their recruitment to sites of ischemia.	[19, 64]
Hoxa3 (transcription factor)	Hox genes help regulate self-renewal and differentiation in hematopoietic stem cells. BMD Gr-1 CD11b cells, a subset of myeloid cells, function as immune suppressor cells and contribute to tumor-induced and injury induced neovascularization. In diabetes, the functional profile of these cells is altered affecting their ability to stimulate neovascularization. They also show abnormal proliferation, chemotactic, adhesion and differentiation potential. Expression of HoxA3 in diabetic derived BMD-Gr 1 CD11 b cells reverses the phenotype changes seen in diabetes.	[65]

many new molecules and mechanisms which modulate inflammation, angiogenesis and ECM dynamics are being investigated. Table 1 shows a few potential novel biomarkers and therapeutic targets in diabetic patients with chronic wounds that are currently being studied.

6 Current Management Protocols – an Evidence-Based Approach

Personalization of medical care requires a panoramic view of the problem and a systematic approach. It is commonly the General Practitioner (GP), also known as Primary Care Physician (PCP), who serves as the central coordinator of care for diabetic patients. After the diagnosis of diabetes has been made, the GP will work with the patient to optimize systemic glucose control, monitor markers of disease progression, manage comorbid conditions, and provide personalized support for lifestyle modifications that can decrease diabetic complications, including chronic diabetic wounds.

In addition to blood glucose control, primary prevention of diabetic wounds involves management of co-morbidities including cardiovascular disease (CVD), anemia, nutritional deficiencies, and both functional and cognitive impairments.

Current guidelines for prevention of cardiovascular disease related diabetic complications are as follows. Blood pressure should be measured at every visit with a target of <130/80 mmHg for diabetics. Pharmacologic therapy should include an ACE (angiotensin converting enzyme) inhibitor or ARB (angiotension receptor blocker) unless contraindicated. Dyslipidemia should be monitored at least annually with a target LDL <100 mg/dL. Any diabetic patient with overt CVD or over the age of 40 with at least one other CVD risk factor should be on statin therapy, regardless of baseline lipid levels unless contraindicated. Aspirin (75–162 mg/day) therapy should be considered as a primary prevention strategy for diabetic patients with 10 year CVD risk >10%. For secondary prevention in diabetic patients with a history of CVD, aspirin therapy, or clopidogrel (75 mg/day) in the case of aspirin allergy, should be used. All diabetic patients should be advised not to smoke and evidence-based smoking cessation counseling should be provided [66].

Comorbid conditions that can undermine healing of chronic wounds in a diabetic include infection, peripheral arterial disease (PAD), anemia, malnutrition, and vitamin deficiencies. Diabetic patients over 50 years of age should be screened for peripheral arterial disease (PAD) utilizing ankle-brachial index (ABI) since abnormal ABI is an early marker of generalized atherosclerotic vessel disease and a predictor of future diabetic wounds [67, 68]. Testing for and treatment of anemia is important to both prevention and treatment of diabetic wounds. Effective oxygen delivery to tissues is impaired in anemic individuals. Insufficient caloric intake and/or protein intake is commonly overlooked and can be addressed in consultation with a nutritionist or dietician. Additionally, there are numerous micronutrient deficiencies including Vitamin A, Vitamin K, Thiamine, Riboflavin, Niacin, Vitamin D, Vitamin B6, Folate, Vitamin B12, Vitamin C, Magnesium, Iron, Zinc, Copper, Iodine, and Chromium can impair wound healing [69]. Many of these can be recognized in standard laboratory blood tests while others require astute clinical examination skills for identification of systemic signs and dermatologic lesions such as angular cheilitis presenting as fissures in the corners of the mouth (Vitamin B6) to changes in gums and observation of pinpoint petechiae (Vitamin C). Nutritional deficiencies coupled with poor glycemic control result in diminished neutrophil function and subsequent non-healing wounds [67, 69, 70].

Peripheral neuropathy will complicate the course of 30–50% of diabetic patients. Motor neuropathy can lead to foot deformities that increase the risk for skin ulceration [70]. Diabetic autonomic neuropathy (DAN) may result in shunting of arteriovenous flow and thereby reduce skin perfusion [71]. Periodic exams with sensory monitoring for development of distal symmetric polyneuropathy (DPN), a major risk factor for chronic diabetic wounds, should begin at initial diagnosis of type 2 diabetes and beginning 5 years after diagnosis of type 1 using tests for pinprick sensation, vibration perception (128 Hz tuning fork), 10-g Semmes-Weinstein monofilament test for pressure sensation at the distal plantar aspect of the great toes and metatarsal joints, and assessment of ankle reflexes [66]. Diabetic patients aged 18 years and older should have a neurological exam of their lower extremities that includes an assessment of vascular sufficiency including distal pulses, presence of foot deformity, dry or fissured skin, and appropriateness of footwear during one or more office visits performed at least every 12 months based on risk category as shown in Table 2 [72, 73].

As part of a comprehensive strategy to optimize each patient, the GP will also make referrals for appropriate laboratory and imaging studies, diabetic education, dietician or nutritional counseling, specialist care such as podiatry and vascular surgery,

Table 2 Diabetic foot wound risk categorization system

Category	Risk profile	Evaluation frequency
0	Normal	Annual
1	Peripheral, neuropathy (LOPS)	Semi-annual
2	Neuropathy, deformity, and/or PAD	Quarterly
3	Previous ulcer or amputation	Monthly to quarterly

From diabetic foot and ankle care physician performance measurement set

Table 3 Quality indicators in care for diabetic patients

Diabetic care quality indicators summary
– Hbg A1c <7.0
– Blood Pressure <130/80 mmHg
– LDL <100 mg/dL (<2.6 mmol/L)
– Statin
– ACE Inhibitor or ARB
– Aspirin
– Yearly monofilament exam
– Yearly urine micro albumin
– Smoking cessation counseling
– Immunizations
– Tetanus-Diphtheria-Pertusis (Tdap)
– Hepatitis B
– Pneumococcal
– Influenza

and weight loss surgery. There is evidence that in appropriately selected obese patients, surgical intervention can reduce or eliminate the burden of diabetes [74].

Optimal personalized care of the individual diabetic patient is enhanced by systemic quality review of diabetic patient panels to track performance on indicators such as HgbA1c, lipid monitoring, and blood pressure control among others (Table 3).

Individualized interval targets in each of the major areas of diabetic care quality should be set for each patient. These targets need to be reassessed at regular intervals to ensure gradual progress toward systemic optimization. Attaining these targets is even more important in the patient who already has a diabetic wound. Often patients with diabetic wounds have fallen short of multiple targets for optimal care and many have reached advanced co-morbid disease states. The diabetic wound can often be one of the first signs that the patient's body is weakening.

The GP must develop a secondary prevention strategy to ensure consistent early recognition and appropriate management of diabetic wounds. These strategies may vary from including "non-healing wounds" on the standard review of systems intake form to standardized protocols for regular in office skin exams. The primary care office will commonly be the first contact a patient has for care of their chronic wound. Initial assessment and documentation should not be delayed. Wound age, location, dimension (length, width, and depth), color, presence or absence of drainage, exudates, odor, location, pain, itching, bleeding, swelling, or signs of infection such as induration and erythema are characteristics to include in any documentation. Initial wound documentation should also include the presence of arterial ischemia, venous insufficiency, presence of protective sensation, and biomechanical problems [75].

Probing of wounds for visible and palpable bone is essential to early diagnosis of underlying osteomyelitis [70]. Notation of initial mechanism of injury, if known is also critical as puncture wounds can be misleading and lead to deep tissue infections that can go unrecognized until serious infection presents emergently weeks to months later. In diabetic wounds, infection is a clinical diagnosis based on signs of inflammation (warmth, swelling, redness, tenderness, or pain) and purulent discharge or odor [75]. If infection is suspected, wound should be classified on severity which will guide decision to hospitalize and antibiotic administration [75].

Debridement, cleansing, and deep tissue culture by biopsy or curettage is recommended for wounds with strong suspicion for infection [75]. Swabs, sinus tract, and inadequately debrided specimens should not be sent for culture as they often yield contaminants [76]. Since diagnosis of diabetic wound infection is clinical, positive culture is not necessary. Empiric antibiotics for infected wounds should be initiated with coverage for Staphylococcus as the main microbial organism contributing to diabetic wound infection [75, 77]. In a patient with a personal history of Methicillin-resistant *Staphylococcus aureus* (MRSA), severe infection, or where local prevalence of MRSA is high, antibiotics should be tailored appropriately.

In addition to maximizing the nutrition, blood glucose, lipids, and blood pressure control of patients presenting with their first non-healing diabetic wound, the GP needs to obtain timely assessment of the wound etiology. Depending on wound location, presence and pressure of blood supply to the wound tissue should be assessed with tools including Doppler ultrasonography, ankle-brachial index (ABI), angiograms, MRI, and transcutaneous oximetry of transcutaneous oxygen pressure (TcPO2). Timely referral for treatment of peripheral arterial disease is essential in early wound management [67, 73]. Knowledge and appropriate treatment of precipitating factors leading up to development of a wound is the key to personalizing the wound healing care plan and for optimizing prevention of future wounds in the individual patient. Training of nursing staff to properly assess and document wounds in a standard fashion will help the busy GP maintain highly personalized care while operating with efficiency.

Standardized protocols are also useful for in office management of existing wounds. Physician-directed surgical preparation of the affected lower extremity wound and a physician-applied sterile dressing at each outpatient clinic appointment has shown a reduced incidence of infection, use of systemic antibiotics, and frequency of required follow-up office visits [78]. Debridement is essential prior to collection of tissue for culture and timely debridement of early wounds has been show to decrease total healing time for diabetic wounds [79]. Adequate debridement should remove all necrotic tissue, debris, eschar (dry slough or scab covering the wound), and callus revealing a healthy bleeding edge [75].

Tertiary prevention is achieved by providing the patient timely referral to specialists who can help prevent complications from the initial diabetic wound.

These referrals may include plastic surgeons, vascular surgeons, podiatrists, endocrinologists, orthopedic surgeons among others. Following consultations and referrals the GP must synthesize the information and management provided by the specialists into a comprehensive care plan for the individual diabetic patient. It is particularly important for the GP to continue monitoring progress on issues referred to specialist care. Maintaining regular interval follow up with diabetic patients to reassess progress and risk factors is essential. Many providers would agree that diabetic patients with complications including a history of diabetic wounds should be seen at intervals no longer than every 3 months, with many needing even more frequent visits.

A commonly held belief in medicine is that the physician, who diagnosed the wound cares for the wound and follows it until it closes. It is actually not uncommon for the GP to assume the wound is being dealt with by the surgeon and not address the wound for fear of stepping on the toes of another physician. When a diabetic patient has a non-healing wound, surgical or otherwise, it is important that the GP achieve early communication with specialists regarding wound plan of care and to assess outstanding needs. There are many interventions that the GP can contribute that will improve the healing process such as early debridement, nutritional optimization, and procurement of bacterial cultures. As we strive for patient centered, personalized care, the whole medical team will be required to participate in the healing of the chronic wound. There must be open communication and collaborative care among the entire multidisciplinary team and the GP has the central role as facilitator of that those interactions [75].

7 The Role of Oxygen: Friend or Foe? ROS and HBOT

Oxygen is the first element in Group 6A (16) with an atomic number of 8. It is highly unstable due to the two unpaired electrons in its outer shell, each in a separate p* anti-bonding orbital but with similar spins, constantly seeking an electron to reach a more stable state. The reaction is called redox reaction and the in the process electron donors are said to be oxidized, and the electron receivers are reduced. Oxygen belongs to the group called chalcogens, located next to the most reactive gases, the halogens of Group 7A (17). Because of this high reactivity, oxygen is extremely dangerous if it is not managed well. The oxygen concentration in earth's atmosphere has undergone variations over the past 4.5 billion years. Reaching 1% by 2.5 billion years ago, due to the activities of photosynthetic cyanobacteria, it took the next 1.5 billion years to climb to 3% [80]. There it stayed until the Marinoan glaciation 650 million years ago when the oxygen concentration started to climb more rapidly, exceeding 10% for the first time ever. This rise was associated with the most amazing display of novel animal body forms and complexity, witnessing the appearance of ostracoderm, the first vertebrate, ushering the Cambrian explosion. The oxygen level plummeted rapidly after reaching a record high of 35%, a level at which spontaneous combustion of organic matter occur at room temperature, even when wet. Each "rapid" (in geological scale) change in oxygen concentration within the atmosphere is linked to drastic changes in the earth's biota, massive extinction followed by appearance of new species. The Permian-Triassic, or the PT Boundary, followed the first rapid decrease and the Cretaceous-Tertiary, or the KT Boundary after the second decline [81]. Oxygen, a life-sustaining poison, is without doubt a key player in our phylogeny and ontogeny. The only organelles in our cells capable of handling oxygen are the mitochondria, internalized purple bacteria which evolved 2.5 billion years ago, when atmospheric oxygen concentration was 1%. And that is the concentration range mitochondria operate today: 0.7–2 mmHg, with reported P_{50}=0.022–0.005 kPa, 1% of intracellular O_2 level [82]. It is important to point out that, like all bacterial DNA, mitochondrial DNA differs from nuclear DNA in that it is "naked", without the protection of nuclear membrane, histone proteins, or NER/BER (nucleotide and base excision repair) mechanisms [83]. The auto-oxidation rate of mitochondrial DNA varies greatly between different mammals and correlates tightly, inversely, to the maximum life expectancy of a species. It is entirely conceivable that with an open wound, the cells are exposed to atmospheric oxygen at 150 mmHg, or 20 kPa, 1,000–4,000× the P_{50} for mitochondrial oxygen processing capacity. This may well account for the senescent cells often found populating a chronic open wound. With such high O_2 tension, ROS and RNS production may exceed even the nuclear DNA protection mechanism. Recent findings of uncoordinated gene expression with proliferation of epithelium not matched by migration, thus producing the frequently seen hyperkeratotic rim surrounding a chronic open wound, support this conjecture. Like radiation therapy, chronic open wound is accelerated aging at the cellular level – for those cells exposed to elevated O_2.

The opposite of this is the lack of oxygen. Tissue hypoxia results often from ischemia, and ischemia happens when blood cannot reach target tissues. Or, alternatively, blood reaching the tissue carries insufficient oxygen, a condition called hypoxemia.

7.1 Hyperbaric Oxygen Therapy: The Science Behind the Idea

The role of hyperbaric oxygen (HBO) therapy in treatment of chronic wounds continues to be a growing area of research and controversy. Introduced in the mid-twentieth century for the treatment of gas gangrene caused by anaerobic microbes, the role of HBO therapy has expanded to treat acute and chronic pathologies including necrotizing soft tissue infections, chronically infected wounds, acute trauma, and limb or surgical flap ischemia, to name a few. The extent of its use has grown immensely over the last 20 years with the number of centers performing HBO therapy growing by nearly five times in the US [84]. Hyperbaric oxygen therapy uses a pressurized chamber to deliver 100% oxygen at pressures above 1 atm (1 atmospheric pressure). The percent oxygen in the air at sea level is ~21%. The amount of oxygen within one's blood can be calculated using the oxygen content equation:

$$1.34 \times Hgb \times (SaO_2 / 100)] + 0.003 \times PaO_2 \quad (6)$$

In the setting of room air, the patient's hemoglobin (Hgb) level and the oxygen saturation (SaO_2) are the main factors affecting the total oxygen being carried within the blood. Under standard conditions of room air at 1 atm, only a very small percentage of the total oxygen being delivered comes from the oxygen dissolved in the plasma (PaO_2: the partial pressure of oxygen diffused in arterial blood, 11–13 kPa or 75–100 mmHg). However, when patients are placed in a pressurized chamber with 100% oxygen, oxygen delivery can be improved significantly. For example, using Eq. 6, the volume of oxygen diffused within the blood with a PaO_2 of 100 mmHg is 0.3 mL/100 mL of blood (100 × 0.003). If the patient is placed in a HBOT chamber with 100% oxygen at 1 atm, this number can be increased to approximately 2 mL/100 mL of blood. At pressures of 2 or 3 atm, the calculated amount of oxygen being transported by oxygen diffused in the plasma can be doubled and tripled respectively. At these levels, the oxygen that can be delivered through simple diffusion becomes substantial, and significantly elevates tissues and vital organs oxygen level. There are standard physiologic mechanisms which sense this increased oxygen delivery to the tissues causing vasoconstriction, however the net effect is still positive.

There are many potential benefits to hyperbaric oxygen therapy in addition to improving oxygen delivery to tissues. However, hypoxia activates specific cellular mechanisms and genes which help improve oxygen delivery and promote wound healing. Oxygen delivery is improved through proliferation of fibroblasts and expression of vascular endothelial growth factor (VEGF), platelet derived growth factor (PDGF) and transforming growth factor beta (TGF-β) which is crucial in angiogenesis. Additionally, inducible nitric oxide synthase is expressed leading to formation of nitric oxide and vasodilation. Systemically, factors including erythropoietin (EPO) are expressed leading to hematopoiesis [84, 85]. Collectively, these compensatory mechanisms induced by hypoxia and tissue injury along with expression of other important factors alter the biochemical and physiologic milieu toward one more favorable in promoting wound healing. Though initiation of these processes and gene expression are dependent on the changes seen with the initial insult (hypoxia), HBO therapy may play a crucial role in the long term success of wound healing in the appropriately chosen patient in whom the normal healing has stalled. The following is a list of some of the potential benefits of HBO therapy in addition to increasing oxygen delivery as described above:

1. Edema can be decreased within injured tissue by causing vasoconstriction within the wound bed as a response to the increased oxygen delivery.
2. Promotion of fibroblasts proliferation is seen which can lead to increased extracellular matrix (ECM) synthesis.
3. Angiogenesis and epithelialization can potentially be improved.
4. As described in one of its first intended uses in the treatment of gas gangrene, HBOT is directly toxic to anaerobic pathogens and can help fight infections [84–86].

HBO therapy, obviously lethal to the anaerobic pathogens, may alter the defective signaling pathways in the diabetic patients, such as low endothelial progenitor cell (EPC) homing due to reduced stromal cell-derived factor 1 alpha (SDF-1 α) [86]. The fact that diabetic patients have altered physiology which leads to their chronic wounds is clear. These alterations affect cells crucial to wound healing in addition to EPC, including fibroblasts and macrophages, and their response to cellular signals [19, 85, 86]. HBO therapy appears to help intervene in the processes that are directly affected by and resulting from poor wound healing in diabetic patients. It is important to point out that there is insufficient understanding and empirical clinical evidence to recommend wide use of HBO therapy in chronic diabetic wounds except when specific, limited indications are met as detailed below [87].

7.2 Hyperbaric Oxygen Therapy Treatment Protocols

Even as HBOT has become more accepted in the treatment of a subtype of chronic diabetic wounds, the protocols have not been standardized and debates regarding treatment efficacy continue. A recent Cochrane review showed that centers have implemented use of HBOT from 45 min to 2 h at pressures between 2 and 3 atm [87]. Importantly, there has been more of a focus on patient selection criteria. Generally, the diabetic ulcers treated with HBOT have been present for greater than 4 weeks and are Wagner grade 2 (deep, full thickness ulcers with potential bone, tendon, joint involvement) lesions or higher. Prior to initiation of HBOT, a $TcPO_2$ (Transcutaneous oxymetry) measurement is required to show the presence of tissue hypoxia (<50 mmHg) with evidence that the blood supply is sufficient to respond to oxygen challenge by having the patient breath 100% O_2. HBOT will be helpful if $TcPO_2$ measured >200 mmHg during the oxygen challenge [88]. Several studies cited show that HBOT in chronic diabetic wounds can lead to increased wound healing, lower overall costs, and less need for major amputations [87, 89]. However, many aspects of HBOT remain controversial and unstandardized. Each institution should develop its own specifics, including treatment protocols, when they are to be initiated, how long they are to be used, and, importantly patient selection criteria. To help with these questions and help better evaluate the efficacy of HBO therapy in diabetic patients with chronic wounds, several prospective, double-blind, randomized, control clinical trial studies are currently underway [87, 89].

8 The Biological Therapeutics: Stem Cells and Tissue Derivatives

Diabetes produces a chronic inflammatory state, disrupting the sequence of events which are required for normal wound healing [19, 85, 86, 90, 91]. Cells which are central in the healing process are affected in addition to the cross talk and signaling pathways required during this highly coordinated series of events [19, 92]. Stem cells and growth factors have been investigated as possible adjuncts to wound healing therapies because of their theoretical potential to intervene and restore the appropriate balance of cells, cell function, and signaling pathways which could lead toward a state of more normal wound healing.

The currently available therapeutics derived from allogeneic or xenogeneic sources are numerous. Collectively, these are called "biologics" or biologically derived therapeutics. Most of them are marketed as acellular or de-cellularized tissue substitutes [93]. These are not grafts if no living cells are present. Some de-cellularized human dermis has sufficient tensile modulus to be used in repair of defects in fascia. Some are too weak to allow such. Both bovine and porcine tissues are frequent source of such material. A majority of the products available as skin substitutes attempt to mimic the anatomic functionality of the skin by having bilayer: a superficial layer that may act as the epidermal barrier to the wound and a deeper layer whose characteristics are designed to promote wound healing [93]. Many of these products required staged procedures after initial application of the product to a chronic wound. Prior to usage of these products, one must consider many factors including the characteristics of the wound, time frame required between stages and cost. The only commercially available tissue engineered product containing live human keratinocytes and fibroblasts have only had sporadic success and the effect size index calculated from many reports indicate a modest improvement.

Stem cells, allogeneic or autologous, represent the next major frontier. If mitochondrial oxidative DNA damage, due to the inability to repair and the inevitable nature of the insult from at or near a chronic open wound, is indeed the reason for the limit or constraint demonstrated in Eq. 2, then stem cell transfer will overcome this cellular aging by replacing senescent cells with fresh, robust, high Hayflick constant cells equipped with pristine mitochondria. Recent findings of the immune privileged status of allogeneic stem cell greatly enhance their potential clinical applications. In addition to providing a fresh source of young cells, these stem cells may modulate the innate immunity such as inflammatory response in general.

Currently, there are two main focuses on how stem cells potentially improve wound healing. First, stem cells may have a regenerative effect by differentiating into the tissues where tissue damage has occurred. In a sense, stem cells could differentiate into specific cell types to "fill the gap" of the tissue and cell types which have been damaged or lost. Second, stem cells interact with the surrounding environment through paracrine signaling mechanisms leading to biological responses (angiogenesis, fibroproliferation, chemokine release, etc.) in favor of wound healing [92]. This effect is seen through an amplification of signals via mRNA production and appears to be the more prevalent of the two, especially early on upon its application to the wound.

One of the main challenges with stem cell therapy has been developing a delivery system by which stem cells are applied using a vehicle that will minimally alter the functionality of the cells and ultimately help optimize their engraftment and total affect. Specifically, products available which have been used to this point include fibrin sealants, synthetic products and ECM-like materials [91]. Though no product currently available provides the optimal solution, ideally, a delivery system will be developed that can intimately conform to the contours of the wound and provide the right balance of extracellular components to help stem cell engraftment and function without negatively affecting the other cells required in wound healing.

Large scale, prospective, randomized controlled trials are needed before biologics, in the form of graft or implants, can achieve their maximum potential. One great impediment to this, paradoxically, is the aggressive marketing and clinical penetration, making such detailed, high level of evidence approach more difficult. It is likely that some xenogeneic implants such as de-cellularized porcine urinary bladder submucosa will "breakthrough" and become the accepted, effective choice.

Because of the perturbed signaling in diabetic chronic wounds, efforts at using topical application of exogenous growth factors have met some success. The high concentration of proteolytic enzymes within a chronic wound coupled with impaired response account for the abnormal signaling. Reports using db/db mice (a strain with leptin receptor deficiency and good model for type 2 diabetes) of improved healing by PDGF and FGF began to appear in the 1990s. VEGF and EGF have also been shown to stimulate diabetic wound healing. But to date, the only growth factor that enjoys some clinical application is rhPDGF [85]. Our own experience with this material, however, is less impressive. Because of the complex nature of the wound environment, it is unlikely that one single factor can "fix" the signaling defects. Thus, combinatorial approaches seem more logical. What is difficult is to work out what combinations, how much, what sequence, and for how long. SDF-1α, VEGF, TGF-β, FGF-2 are some reasonable candidates. These, at this time, are still to be considered highly experimental and await the further development of delivering devices, better understanding of the effects and interactions, and a new paradigm-one that uses repeated iterative functions and can allow us to model quantitatively and qualitatively the complex adaptive nature of wound healing. We are far from being able to achieve this at present. As a result, we still depend, for the most part on the medical and surgical managements.

9 The Surgical Managements: Debridement, Grafts, and Flaps

The treatment of all wounds is oriented towards one purpose: closure of the wound. For wounds that are progressing normally, little more is needed than to foster that normal progression. Indeed, the normal wound goes on to heal almost in spite of what is done to it rather than as a result of our treatment. It is only when the wound fails to progress that it is necessary to intervene in such a way as to restore that progression. In order to treat a symptom, it is necessary to first understand the cause of that symptom. The causes of delayed wound healing are numerous and often multifactorial. Although the variables are many, and continuously increasing as our understanding increases, what is known can conveniently be grouped in to one of two types: systemic factors or wound factors. Systemic factors are factors affecting the patient's general physiology as a whole, including such things as the patient's nutrition, any systemic diseases such as diabetes. The diabetic patient suffers from a multitude of factors which negatively affect wound healing as discussed in above sections, including changes in cell wall stiffness, with a resultant decrease in the flexibility of RBCs and decrease red blood cell permeability and blood flow, microangiopathy, neuropathy and impaired immune response [94]. Thus, it is not surprising that poor control of elevated hemoglobin A1c (HbA1c) levels have been shown to correspond with slower healing [95], and higher complication rates [96]. However, diabetic patients are also subject to the same wound factors of typical chronic wounds occurring in non-diabetic patients.

Wound factors are those variables associated with the wound itself, regardless of the patient's physiologic status. These factors can be manipulated directly, and are the focus of surgical intervention. It cannot be overstated that neither systemic factors nor wound factors can be ignored, by treating one in favor of the other. The optimal management of a wound while ignoring systemic factors in a physiologically impaired patient will meet with failure just as readily as a patient who is physiologically optimized, but whose wound is ignored. For best results, the systemic factors should be optimized, followed by appropriate intervention on the wound. As with the systemic factors affecting a diabetic patient, there are multiple wound factors which can impair healing. The bony structures can increase pressure and result in ischemia; bio-burden can overwhelm host defenses and result in infection; changes over time can decrease the healing response, etc. But here too, while there are many variables, they can be simplified so as to fall into one of three categories. (1) tissue perfusion, (2) bio-burden, (3) wound environment. Tissue perfusion is the most fundamental requirement for wound healing. Without adequate perfusion, even healthy tissues die. It was been demonstrated over 40 years ago that oxygen tensions of ~30 mmHg are required for wound healing and infection control [97]. Up to 1/3 of leg ulcer patients have some component of arterial insufficiency [98]. Ankle Brachial Index, or ABI, values of between 0.6 and 0.8 is associated with Peripheral Artery Disease (PAD) and should prompt a vascular work up for possible intervention to improve blood flow. An ABI less than 0.5 is associated with critical ischemia, and an ABI of 1.2 is associated with calcification and non-compressibility of arterial walls. In these patients arm-toe pressures index (ATI) may be more accurate than ankle pressures for assessing PAD [99]. Ischemia may also be caused by pressure and pressure off loading is a significant factor in facilitating healing. Pressure sores occurring on the pad of the food overlying the metatarsal head, and associated with clawing of the toes, may benefit from joint manipulation in the form of range of motion exercises for early disease. More severe clawing may be treated with percutaneous tenotomies to lengthen the extensors and allow the joint to be manipulated back to a neutral position [100]. If the wound extends to the bone, it may be necessary to excise the bony prominence to prevent pressure points. Pressure relief can also be achieved through casting, but if the underlying bony abnormality is not corrected, recidivism rates are high [101].

Edema can also contribute to tissue ischemia, even when the blood inflow is adequate, by increasing the distance for oxygen to diffuse. Decreasing edema through compressive therapy has been shown to have a positive effect on healing [9]. Negative Pressure Therapy has also been shown to greatly improve healing, very likely in some degree through a reduction in edema [102]. Another factor contributing to tissue ischemia is the fibrotic rind of tissue that develops around the chronic wound. This chronic inflammatory reaction, or "fibrin cuff", also obstructs diffusion and inhibits healing [103]. The second category of variables that inhibit wound healing is bioburden, which comes in one of three forms: soft tissue infection, osteomyelitis with or without osteonecrosis, and biofilms. Wounds with a bacterial load of greater than 10^5 organisms/g of tissue have been shown not to heal normally, while wounds with a bacterial load less than this generally go on to heal uneventfully [104]. Likewise, infected bone beneath an otherwise healthy wound will manifest through the wound, causing breakdown. These factors have been known for many years. More recently, we have recognized that bacteria colonizing the surface of wounds exist not in free-living planktonic states, but in communities called biofilms. In this setting, the bacteria live within a complex polysaccharide substrate. Bacteria in these conditions have different phenotypes, virulence, and demonstrate a dramatic increased resistance to antibiotic therapies [105].

The third category of variables affecting chronic wounds is the wound environment itself. Chronic wounds are associated with increased levels of metalloproteases, inflammatory cytokines, overabundance of neutrophils, and depleted growth factor levels [9]. Repeated cycles of division change the ability of the cells in immediate wound environment to heal. Chronic wounds may also stimulate neoplastic changes, the presence of which will prevent healing. All of these factors are related to the wound environment itself, and need to be addressed to effect healing. One of the most practical methods of addressing the wound factors is through direct debridement. The excision of the inflammatory rind, removal of all necrotic or infected tissues, and resection of bony pressure points, effectively re-sets the wound to a fresh wound status. For small wounds primary or partial closure ± the application of Negative Pressure Dressings to address edema may be all that is required to allow wound healing [100, 101]. If systemic factors have been optimized, once appropriate debridement has been accomplished, surgical closure affords the fastest resolution and healing of the wound. There are many options at the surgeons' disposal. Primary closure is always the preferred method, if possible. Often the debridement and pressure point excision can allow closure without tension. For example, in plantar wounds, ray amputation of the metatarsal pressure point may allow the wound to be easily closed longitudinally. But any such alteration in bony architecture is necessarily met with re-distribution of weight and thus new stress pattern, often increasing some at the expense of the other. Some wounds are too large to close primarily. When considering the technique for closure, it is helpful to consider the "reconstructive ladder" regarding the surgical options. If primary closure is not possible, every other option costs some other area of the body injury when implemented. Simple interventions like skin grafting or local flaps cost less, and are low on the ladder (Fig. 7). Fasciocutaneous and myocutaneous flaps transfer fascia and skin and muscle and skin respectively, are more powerful tools, but cost more in terms of patient recovery time, also in terms of operative time, effort and operative skill required. At the top of the ladder is microvascular free tissue transfer. This is a very powerful tool, but one that has been shown to have a higher complication rate of flap failure when used for chronic wounds than when used for other purposes.

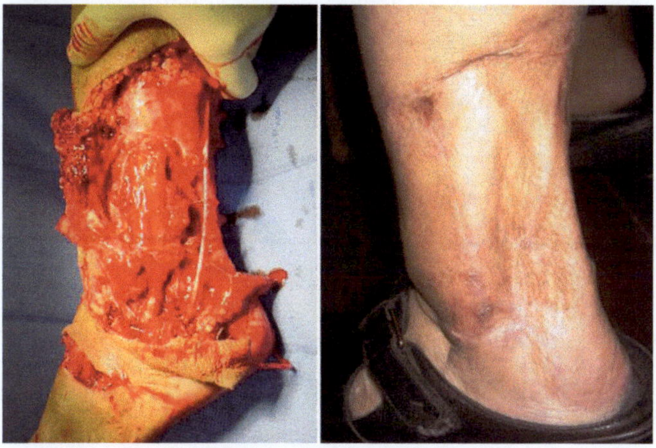

Fig. 7 Surgical management of a traumatic lower extremity wound. This large traumatic open wound of the left distal leg (*left*) was too large to close by primary approximation. The wound was debrided sequentially and covered with split-thickness skin graf

Skin grafting works well when the underlying tissue to be covered is well-vascularized and not particularly mobile. Skin grafting is limited in that it works poorly, if at all, over tendon, bone, and joints. In these settings, a local flap is more effective. The simplest form of local flap is the random pattern skin flap. These flaps are so named because they are not based on a named blood supply, but on the random blood supply of the dermal plexus. Local flaps seek to mobilize tissue from adjacent areas of relative skin laxity, into the defect. Surgeons mobilize tissue along a straight line, as an advancement flap, or V-Y advancement flap. Tissue can also be mobilized around a pivot point, as with a rotational flap. Transposition flaps lift adjacent tissues over an isthmus of intervening tissue, as in rhomboid or Dufourmentel flaps. Interpolation flaps move tissue from away from the wound under or over a bridge of intervening tissues, as with an island flap (Fig. 8).

In many cases, in order to close the primary defect, a secondary defect must be created where the donor tissue was moved from. Ideally, the tissues surrounding the secondary defect are mobile enough to be closed primarily, but occasionally such is not possible. In that case, if the secondary defect can be positioned in a favorable recipient site, it can be closed with a skin graft. For example, a flap of well vascularized skin overlying muscle may be mobilized to cover a bony prominence or exposed tendons, and a skin graft applied to the area the flap is mobilized from. Random pattern skin flaps typically work best for small defects (5 cm × 5 cm), located more proximally, where tissues are more abundant, and blood supply is more robust. When designing these flaps, the rule of thumb is that the base of the flap should not be less than 1/3 the length. However, while this works well for small flaps, it does not always hold true for larger flaps (>15 cm in length). For flaps needing to be mobilized a longer distance, it is safer to include a blood vessel (axial pattern skin flaps) or other source of blood supply, such as the underlying fascia or muscle (Fig. 9).

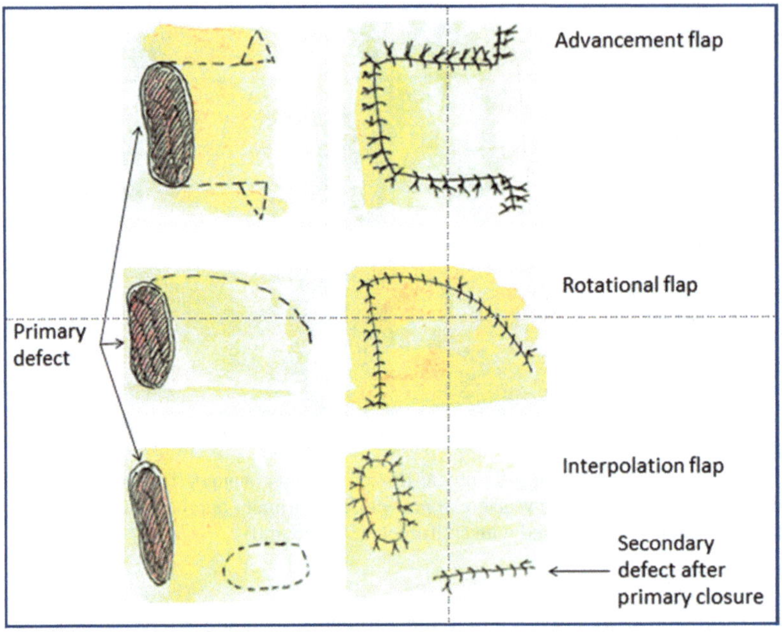

Fig. 8 The three major types of local flaps. The above depicts the three major types of local flaps. Advancement flap, rotational flap and interpolation flap (from *top* to *bottom*)

10 Recommendations

In this final part of the synopsis, we provide three lists. The first one is how physicians currently treat chronic wounds in patients with diabetes. The second is steps that are needed to alter future directions at regional, national, and global levels. The third list is how bench research can help the first two.

1. **What we do now:**
   (a) Establish that there is adequate blood flow to the region with ankle brachial index (ABI) above 0.8 and transcutaneous oximetry (TCOM) at or above 50 mmHg.
   (b) Maintain good medical management keeping serum glucose below 100 mg/dL, HbA1c below 7%, blood pressure below 130/80 mmHg, and LDL less than 2.6 mmol/L (or 100 mg/dL)
   (c) Daily dressing changes and weekly to biweekly debridement of the wound with accurate documentation of the wound size, keeping bacterial count below 10^5/g tissue. Surgical referral if less than 50% reduction in size for two consecutive months
   (d) A coordinated team approach using an axial and spoke model, with the PCP at the center working closely with endocrinologist, vascular surgeons, plastic surgeons, wound care nurses, nutritionist, podiatrist, orthopedist, and the cell and molecular biologists.

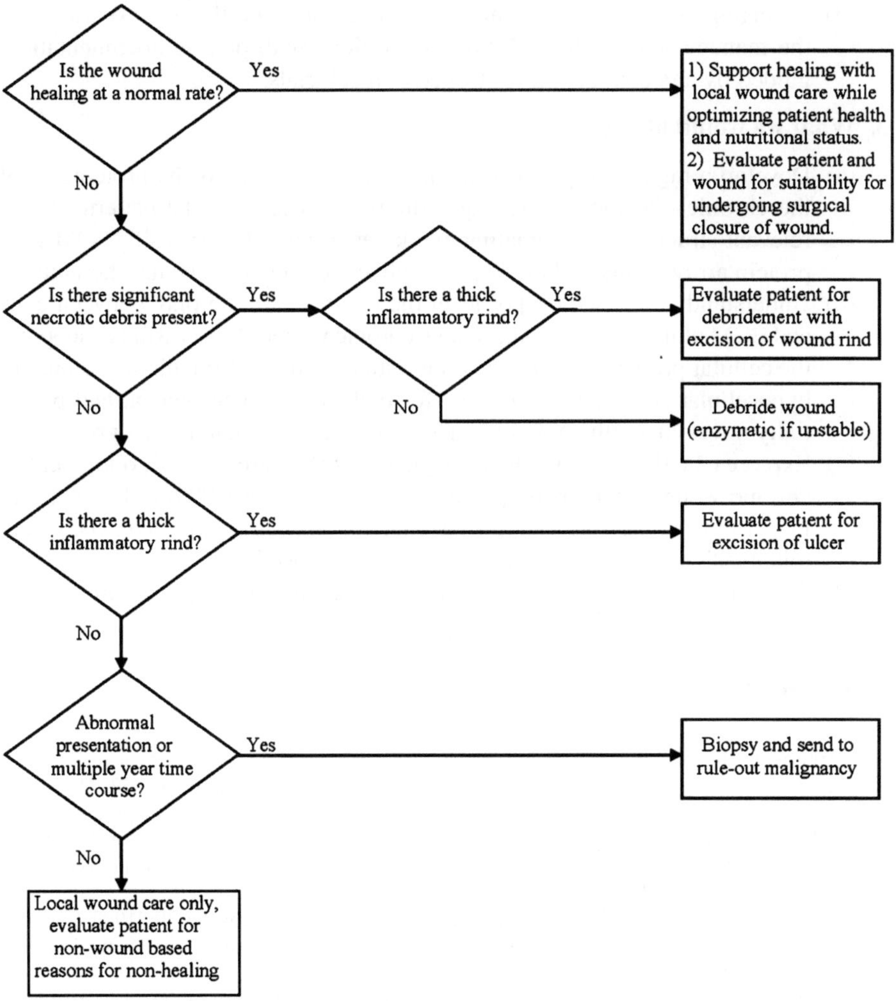

Fig. 9 The flow chart for surgical management of open wounds

2. **What we need to do:**
   (a) Reverse the current trend of hypercaloric diet (<1,800–2,000 Cal/day) and hypoactivity. Since 80% of the circulating glucose is used by muscle, exercise must be promoted from the regional to the global level.
   (b) Increase the amount of daily aerobic activity. Explore ways to increase the efficiency and effectiveness of exercises such as whole body vibrations and rest-inserted high-impact routines.
   (c) Decrease the obesity prevalence in children and adolescents by re-instituting physical education and fundamentally re-design the school lunch programs.

(d) Form regional specialized centers to concentrate expertise and experience in the management of these difficult conditions with precise documentation, guided by and conducting controlled clinical trials.

3. **What we would like to do:**

(a) Develop a high throughput inexpensive means to assess the proteomics of the chronic diabetic wounds, quantitatively measuring the pattern of 200 relevant structural and signaling cytokines, growth factors using solid state protein arrays. This will allow personalized, directed molecular therapies.
(b) Understand at a much deeper level the control and interplay between cells and ECM of the chronic wounds: how chronic wounds "successfully" suppress the cellular proliferation to convert granulation tissue from highly cellular to hypocellular state. This will allow for both wound management and potentially novel chemotherapy for cancer, where the problem is the opposite.
(c) Explore exactly what the role of oxygen and ROS are in the development of chronic wounds in diabetic patients, since there is a 1,000-fold increase in PO_2 at mitochondrial level in cells at edges of an open wound, and hyperglycemia itself can also induce sustained increase in ROS.
(d) Translate the clinical application of allogeneic and autologous stem cells.

References

1. Brolmann FE, Ubbink DT, Nelson EA, Munte K, van der Horst CM, Vermeulen H (2012) Evidence-based decisions for local and systemic wound care. Br J Surg 99:1172–1183
2. Driscoll P (2007) Wound management 2007–2016: established and emergine products, technologies and markets in the US, Europe, Japan and the Rest of the World. Report No.: S245
3. Cavanagh PR, Lipsky BA, Bradbury AW, Botek G (2005) Treatment for diabetic foot ulcers. Lancet 366:1725–1735
4. Guo S, Dipietro LA (2010) Factors affecting wound healing. J Dent Res 89:219–229
5. Cukjati D, Rebersek S, Karba R, Miklavcic D (2000) Modelling of chronic wound healing dynamics. Med Biol Eng Comput 38:339–347
6. Evidence-based Clinical Practice Guidelines (2006) Chronic wounds of the lower extremity. Evidence-based Clinical Practic Guidelines. http://www.plasticsurgery.org/Documents/medical-professionals/health-policy/evidence-practice/Evidence-based-Clinical-Practice-Guideline-Chronic-Wounds-of-the-Lower-Extremity.pdf. Accessed 26 Aug 2012
7. Cukjati D, Rebersek S, Miklavcic D (2001) A reliable method of determining wound healing rate. Med Biol Eng Comput 39:263–271
8. Sungen A (2006) Healing rate of chronic stage 3 and 4 pressure ulcers, evidence-based practice in wound care. Cleveland. United States Veteran's Affairs Cooperative Study No. 535 @. http://www.rehab.research.va.gov/meet/cleve06/spungen.pdf
9. Mustoe TA, O'Shaughnessy K, Kloeters O (2006) Chronic wound pathogenesis and current treatment strategies: a unifying hypothesis. Plast Reconstr Surg 117:35S–41S
10. Lyder CH (2003) Pressure ulcer prevention and management. JAMA 289:223–226
11. Heit JA, Rooke TW, Silverstein MD, Mohr DN, Lohse CM, Petterson TM, O'Fallon WM, Melton LJ 3rd (2001) Trends in the incidence of venous stasis syndrome and venous ulcer: a 25-year population-based study. J Vasc Surg 33:1022–1027
12. Linder-Ganz E, Gefen A (2004) Mechanical compression-induced pressure sores in rat hindlimb: muscle stiffness, histology, and computational models. J Appl Physiol 96:2034–2049

13. Defloor T, Schoonhoven L, Fletcher J, Furtado K, Heyman H, Lubbers M, Witherow A, Bale S, Bellingeri A, Cherry G, Clark M, Colin D, Dassen T, Dealey C, Gulacsi L, Haalboom J, Halfens R, Hietanen H, Lindholm C, Moore Z, Romanelli M, Soriano JV, Trustees of the European Pressure Ulcer Advisory Panel (2005) Statement of the European pressure ulcer advisory panel–pressure ulcer classification: differentiation between pressure ulcers and moisture lesions. J Wound Ostomy Continence Nurs 32:302–306
14. Brem H, Balledux J, Bloom T, Kerstein MD, Hollier L (2000) Healing of diabetic foot ulcers and pressure ulcers with human skin equivalent: a new paradigm in wound healing. Arch Surg 135:627–634
15. Greenhalgh DG (2003) Wound healing and diabetes mellitus. Clin Plast Surg 30:37–45
16. Jude EB, Oyibo SO, Chalmers N, Boulton AJ (2001) Peripheral arterial disease in diabetic and nondiabetic patients: a comparison of severity and outcome. Diabetes Care 24:1433–1437
17. Ubbink DT, Westerbos SJ, Evans D, Land L, Vermeulen H (2008) Topical negative pressure for treating chronic wounds. Cochrane Database Syst Rev 16:CD001898
18. Sheehan P, Jones P, Caselli A, Giurini JM, Veves A (2003) Percent change in wound area of diabetic foot ulcers over a 4-week period is a robust predictor of complete healing in a 12-week prospective trial. Diabetes Care 26:1879–1882
19. Liu ZJ, Velazquez OC (2008) Hyperoxia, endothelial progenitor cell mobilization, and diabetic wound healing. Antioxid Redox Signal 10:1869–1882
20. George B, Cebioglu M, Yeghiazaryan K (2010) Inadequate diabetic care: global figures cry for preventive measures and personalized treatment. EPMA J 1:13–18
21. Golubnitschaja O (2010) Time for new guidelines in advanced diabetes care: paradigm change from delayed interventional approach to predictive, preventative & personalized medicine. EPMA J 1:3–12
22. Pellettieri J, Sanchez Alvarado A (2007) Cell turnover and adult tissue homeostasis: from humans to planarians. Annu Rev Genet 41:83–105
23. Ahlfors JE, Billiar KL (2007) Biomechanical and biochemical characteristics of a human fibroblast-produced and remodeled matrix. Biomaterials 28:2183–2191
24. Bischoff JE, Arruda EM, Grosh K (2000) Finite element modeling of human skin using an isotropic, nonlinear elastic constitutive model. J Biomech 33:645–652
25. Reddy NP, Cochran GV (1981) Interstitial fluid flow as a factor in decubitus ulcer formation. J Biomech 14:879–881
26. Nash J (1950) The bargaining problem. Econometrica 18:155–162
27. Jensen K, Andersen HO, Olesen J, Lindblom U (1986) Pressure-pain threshold in human temporal region. Evaluation of a new pressure algometer. Pain 25:313–323
28. Reddien PW, Bermange AL, Murfitt KJ, Jennings JR, Sánchez Alvarado A (2005) Identification of genes needed for regeneration, stem cell function, and tissue homeostasis by systematic gene perturbation in planaria. Dev Cell 8:635–649
29. Brem H, Tomic-Canic M (2007) Cellular and molecular basis of wound healing in diabetes. J Clin Invest 117:1219–1222
30. Kassab A, Piwowar A (2012) Cell oxidant stress delivery and cell dysfunction onset in type 2 diabetes. Biochimie 94:1837–1848
31. Sivitz WI, Yorek MA (2010) Mitochondrial dysfunction in diabetes: from molecular mechanisms to functional significance and therapeutic opportunities. Antioxid Redox Signal 12:537–577
32. Mozaffari MS, Baban B, Abdelsayed R, Liu JY, Wimborne H, Rodriguez N, Abebe W (2012) Renal and glycemic effects of high-dose chromium picolinate in db/db mice: assessment of DNA damage. J Nutr Biochem 23:977–985
33. Mah LJ, El-Osta A, Karagiannis TC (2010) GammaH2AX: a sensitive molecular marker of DNA damage and repair. Leukemia 24:679–686
34. Evans JL, Goldfine ID, Maddux BA, Grodsky GM (2003) Are oxidative stress-activated signaling pathways mediators of insulin resistance and beta-cell dysfunction? Diabetes 52:1–8
35. Chung SS, Ho EC, Lam KS, Chung SK (2003) Contribution of polyol pathway to diabetes-induced oxidative stress. J Am Soc Nephrol 14:S233–S236

36. Ho EC, Lam KS, Chen YS, Yip JC, Arvindakshan M, Yamagishi S, Yagihashi S, Oates PJ, Ellery CA, Chung SS, Chung SK (2006) Aldose reductase-deficient mice are protected from delayed motor nerve conduction velocity, increased c-Jun NH2-terminal kinase activation, depletion of reduced glutathione, increased superoxide accumulation, and DNA damage. Diabetes 55:1946–1953
37. Nishikawa T, Kukidome D, Sonoda K, Fujisawa K, Matsuhisa T, Motoshima H, Matsumura T, Araki E (2007) Impact of mitochondrial ROS production on diabetic vascular complications. Diabetes Res Clin Pract 77(Suppl 1):S41–S45
38. Malhotra S, Bello E, Kominsky S (2012) Diabetic foot ulcerations: biomechanics, charcot foot, and total contact cast. Semin Vasc Surg 25:66–69
39. Badillo AT, Redden RA, Zhang L, Doolin EJ, Liechty KW (2007) Treatment of diabetic wounds with fetal murine mesenchymal stromal cells enhances wound closure. Cell Tissue Res 329:301–311
40. Blakytny R, Jude E (2006) The molecular biology of chronic wounds and delayed healing in diabetes. Diabet Med 23:594–608
41. Stadelmann WK, Digenis AG, Tobin GR (1998) Physiology and healing dynamics of chronic cutaneous wounds. Am J Surg 176:26S–38S
42. Osuka A, Hanschen M, Stoecklein V, Lederer JA (2012) A protective role for inflammasome activation following injury. Shock 37:47–55
43. Martin P, Leibovich SJ (2005) Inflammatory cells during wound repair: the good, the bad and the ugly. Trends Cell Biol 15:599–607
44. Ochoa O, Torres FM, Shireman PK (2007) Chemokines and diabetic wound healing. Vascular 15:350–355
45. Bishop AJ, Mudge E (2012) Diabetic foot ulcers treated with hyperbaric oxygen therapy: a review of the literature. Int Wound J. doi:10.1111/j.1742-481X.2012.01034.x
46. Iozzo RV (1998) Matrix proteoglycans: from molecular design to cellular function. Annu Rev Biochem 67:609–652
47. Lyons TJ, Basu A (2012) Biomarkers in diabetes: hemoglobin A1c, vascular and tissue markers. Transl Res 159:303–312
48. Morelli T, Neiva R, Nevins ML, McGuire MK, Scheyer ET, Oh TJ, Braun TM, Nör JE, Bates D, Giannobile WV (2011) Angiogenic biomarkers and healing of living cellular constructs. J Dent Res 90:456–462
49. Mah LJ, El-Osta A, Karagiannis TC (2010) GammaH2AX as a molecular marker of aging and disease. Epigenetics 5:129–136
50. Gazzaruso C, Coppola A, Montalcini T, Baffero E, Garzaniti A, Pelissero G, Collaviti S, Grugnetti A, Gallotti P, Pujia A, Solerte SB, Giustina A (2012) Lipoprotein(a) and homocysteine as genetic risk factors for vascular and neuropathic diabetic foot in type 2 diabetes mellitus. Endocrine 41:89–95
51. Thrailkill KM, Clay Bunn R, Fowlkes JL (2009) Matrix metalloproteinases: their potential role in the pathogenesis of diabetic nephropathy. Endocrine 35:1–10
52. Shen Y, Guo Y, Mikus P, Sulniute R, Wilczynska M, Ny T, Li J (2012) Plasminogen is a key proinflammatory regulator that accelerates the healing of acute and diabetic wounds. Blood 119:5879–5887
53. Brem H, Stojadinovic O, Diegelmann RF, Entero H, Lee B, Pastar I, Golinko M, Rosenberg H, Tomic-Canic M (2007) Molecular markers in patients with chronic wounds to guide surgical debridement. Mol Med 13:30–39
54. Kuo YR, Wang CT, Cheng JT, Wang FS, Chiang YC, Wang CJ (2011) Bone marrow-derived mesenchymal stem cells enhanced diabetic wound healing through recruitment of tissue regeneration in a rat model of streptozotocin-induced diabetes. Plast Reconstr Surg 128:872–880
55. Gill SE, Parks WC (2008) Metalloproteinases and their inhibitors: regulators of wound healing. Int J Biochem Cell Biol 40:1334–1347
56. Ladwig GP, Robson MC, Liu R, Kuhn MA, Muir DF, Schultz GS (2002) Ratios of activated matrix metalloproteinase-9 to tissue inhibitor of matrix metalloproteinase-1 in wound fluids are inversely correlated with healing of pressure ulcers. Wound Repair Regen 10:26–37

57. Zhang J, Dong J, Gu H, Yu S, Zhang X, Gou Y, Xu W, Burd A, Huang L, Miyado K, Huang Y, Chan HC (2012) CD9 is critical for cutaneous wound healing through JNK signaling. J Invest Dermatol 132:226–236
58. Loughlin DT, Artlett CM (2011) Modification of collagen by 3-deoxyglucosone alters wound healing through differential regulation of p38 MAP kinase. PLoS One 6:e18676
59. Park CW, Kim HW, Lim JH, Yoo KD, Chung S, Shin SJ, Chung HW, Lee SJ, Chae CB, Kim YS, Chang YS (2009) Vascular endothelial growth factor inhibition by dRK6 causes endothelial apoptosis, fibrosis, and inflammation in the heart via the Akt/eNOS axis in db/db mice. Diabetes 58:2666–2676
60. Broughton G 2nd, Janis JE, Attinger CE (2006) The basic science of wound healing. Plast Reconstr Surg 117:12S–34S
61. Lamkanfi M (2011) Emerging inflammasome effector mechanisms. Nat Rev Immunol 11:213–220
62. Gallagher KA, Liu ZJ, Xiao M, Chen H, Goldstein LJ, Buerk DG, Nedeau A, Thom SR, Velazquez OC (2007) Diabetic impairments in NO-mediated endothelial progenitor cell mobilization and homing are reversed by hyperoxia and SDF-1 alpha. J Clin Invest 117:1249–1259
63. Liu W, Shen SM, Zhao XY, Chen GQ (2012) Targeted genes and interacting proteins of hypoxia inducible factor-1. Int J Biochem Mol Biol 3:165–178
64. Nishimura Y, Ii M, Qin G, Hamada H, Asai J, Takenaka H, Sekiguchi H, Renault MA, Jujo K, Katoh N, Kishimoto S, Ito A, Kamide C, Kenny J, Millay M, Misener S, Thorne T, Losordo DW (2012) CXCR4 antagonist AMD3100 accelerates impaired wound healing in diabetic mice. J Invest Dermatol 132:711–720
65. Mahdipour E, Charnock JC, Mace KA (2011) Hoxa3 promotes the differentiation of hematopoietic progenitor cells into proangiogenic Gr-1+CD11b+ myeloid cells. Blood 117:815–826
66. American Diabetes Association (2012) Standards of medical care in diabetes- 2012. Diabetes Care 35(Suppl 1):S11–S63
67. Richard JL, Lavigne JP, Sotto A (2012) Diabetes and foot infection: more than double trouble. Diabetes Metab Res Rev 28(Suppl 1):46–53
68. American Diabetes Association (2003) Peripheral arterial disease in people with diabetes. Diabetes Care 26:3333–3341
69. Posthauer ME, Dorner B, Collins N (2010) Nutrition: a critical component of wound healing. Adv Skin Wound Care 23:560–572; quiz 73–74
70. Bader MS (2008) Diabetic foot infection. Am Fam Physician 78:71–79
71. Flynn MD, Tooke JE (1995) Diabetic neuropathy and the microcirculation. Diabet Med 12:298–301
72. Diabetic foot and ankle care physician performance measurement set (2007) Bethesda: American Podiatric Medical Association (APMA), American College of Foot and Ankle Surgeons, American Collage of Foot and Ankle Orthopedics and Medicine, Center for Medicare and Medicaid Services, United States Department of Health and Human Services
73. Edmonds M (2006) Diabetic foot ulcers: practical treatment recommendations. Drugs 66:913–929
74. Buchwald H, Estok R, Fahrbach K, Banel D, Jensen MD, Pories WJ, Bantle JP, Sledge I (2009) Weight and type 2 diabetes after bariatric surgery: systematic review and meta-analysis. Am J Med 122:248–256.e5
75. Lipsky BA, Berendt AR, Cornia PB, Pile JC, Peters EJ, Armstrong DG, Deery HG, Embil JM, Joseph WS, Karchmer AW, Pinzur MS, Senneville E, Infectious Diseases Society of America (2012) 2012 Infectious Diseases Society of America clinical practice guideline for the diagnosis and treatment of diabetic foot infections. Clin Infect Dis 54:e132–e173
76. Lipsky BA, Peters EJ, Berendt AR, Senneville E, Bakker K, Embil JM, Lavery LA, Urbančič-Rovan V, Jeffcoate WJ, International Working Group on Diabetic Foot (2012) Specific guidelines for the treatment of diabetic foot infections 2011. Diabetes Metab Res Rev 28(Suppl 1):234–235
77. Ambrosch A, Haefner S, Jude E, Lobmann R (2011) Diabetic foot infections: microbiological aspects, current and future antibiotic therapy focusing on methicillin-resistant Staphylococcus aureus. Int Wound J 8:567–577

78. Schade VL, Roukis TS (2008) Use of a surgical preparation and sterile dressing change during office visit treatment of chronic foot and ankle wounds decreases the incidence of infection and treatment costs. Foot Ankle Spec 1:147–154
79. Chiu CC, Huang CL, Weng SF, Sun LM, Chang YL, Tsai FC (2011) A multidisciplinary diabetic foot ulcer treatment programme significantly improved the outcome in patients with infected diabetic foot ulcers. J Plast Reconstr Aesthet Surg 64:867–872
80. Bekker A, Holland HD, Wang PL, Rumble D 3rd, Stein HJ, Hannah JL, Coetzee LL, Beukes NJ (2004) Dating the rise of atmospheric oxygen. Nature 427:117–120
81. Fike DA, Grotzinger JP, Pratt LM, Summons RE (2006) Oxidation of the Ediacaran ocean. Nature 444:744–747
82. Gnaiger E, Lassnig B, Kuznetsov A, Rieger G, Margreiter R (1998) Mitochondrial oxygen affinity, respiratory flux control and excess capacity of cytochrome c oxidase. J Exp Biol 201:1129–1139
83. Andersson SGE, Karlberg O, Canbäck B, Kurland CG (2003) On the origin of mitochondria: a genomic perspective. Philos Trans R Soc Lond B Biol Sci 358:165–179
84. Adkinson C (2011) Hyperbaric oxygen for treatment of problem wounds. Minn Med 94:41–46
85. Falanga V (2005) Wound healing and its impairment in the diabetic foot. Lancet 366: 1736–1743
86. Bao P, Kodra A, Tomic-Canic M, Golinko MS, Ehrlich HP, Brem H (2009) The role of vascular endothelial growth factor in wound healing. J Surg Res 153:347–358
87. Kranke P, Bennett MH, Martyn-St James M, Schnabel A, Debus SE (2012) Hyperbaric oxygen therapy for chronic wounds. Cochrane Database Syst Rev 4:CD004123
88. Niinikoski J (2003) Hyperbaric oxygen therapy of diabetic foot ulcers, transcutaneous oxymetry in clinical decision making. Wound Repair Regen 11:458–461
89. O'Reilly D, Linden R, Fedorko L, Tarride JE, Jones WG, Bowen JM, Goeree R (2011) A prospective, double-blind, randomized, controlled clinical trial comparing standard wound care with adjunctive hyperbaric oxygen therapy (HBOT) to standard wound care only for the treatment of chronic, non-healing ulcers of the lower limb in patients with diabetes mellitus: a study protocol. Trials 12:69
90. Mulder GD, Lee DK, Jeppesen NS (2012) Comprehensive review of the clinical application of autologous mesenchymal stem cells in the treatment of chronic wounds and diabetic bone healing. Int Wound J 9:595–600
91. Blumberg SN, Berger A, Hwang L, Pastar I, Warren SM, Chen W (2012) The role of stem cells in the treatment of diabetic foot ulcers. Diabetes Res Clin Pract 96:1–9
92. Hocking AM, Gibran NS (2010) Mesenchymal stem cells: paracrine signaling and differentiation during cutaneous wound repair. Exp Cell Res 316:2213–2219
93. Jones I, Currie L, Martin R (2002) A guide to biological skin substitutes. Br J Plast Surg 55: 185–193
94. Collins N (2003) Diabetes, nutrition, and wound healing. Adv Skin Wound Care 16: 291–294
95. Markuson M, Hanson D, Anderson J, Langemo D, Hunter S, Thompson P, Paulson R, Rustvang D (2009) The relationship between hemoglobin A(1c) values and healing time for lower extremity ulcers in individuals with diabetes. Adv Skin Wound Care 22:365–372
96. Smiley DD, Umpierrez GE (2006) Perioperative glucose control in the diabetic or nondiabetic patient. South Med J 99:580–589; quiz 90–91
97. Hunt TK, Pai MP (1972) The effect of varying ambient oxygen tensions on wound metabolism and collagen synthesis. Surg Gynecol Obstet 135:561–567
98. American Society of Plastic Surgeons (2007) Chronic wounds of the lower extremity. Evidence-based practic guidelines. American Society of Plastic Surgeons, Arlington Heights
99. Harlin SL, Willard LA, Rush KJ, Ghisletta LC, Meyers WC (2008) Chronic wounds of the lower extremity: a preliminary performance measurement set. Plast Reconstr Surg 121: 142–174

100. Strauss MB (2005) Surgical treatment of problem foot wounds in patients with diabetes. Clin Orthop Relat Res 439:91–96
101. Piaggesi A, Schipani E, Campi F, Romanelli M, Baccetti F, Arvia C, Navalesi R (1998) Conservative surgical approach versus non-surgical management for diabetic neuropathic foot ulcers: a randomized trial. Diabet Med 15:412–417
102. Webster J, Scuffham P, Sherriff KL, Stankiewicz M, Chaboyer WP (2012) Negative pressure wound therapy for skin grafts and surgical wounds healing by primary intention. Cochrane Database Syst Rev 4:CD009261
103. Ferguson MW, Herrick SE, Spencer MJ, Shaw JE, Boulton AJ, Sloan P (1996) The histology of diabetic foot ulcers. Diabet Med 13(Suppl 1):S30–S33
104. Robson MC, Stenberg BD, Heggers JP (1990) Wound healing alterations caused by infection. Clin Plast Surg 17:485–492
105. Edwards R, Harding KG (2004) Bacteria and wound healing. Curr Opin Infect Dis 17:91–96

Understanding Inflammation: The Key to Targeted Preventive Measures for Diabetes and Periodontitis

Ranjitha Krishna, Philip J. Hanes, and Christopher W. Cutler

Abstract Type 2 diabetes and periodontitis are both chronic inflammatory diseases. Common pathways of inflammation are responsible for these diseases and each appears to contribute to the severity and degree of control of the other. Periodontal treatment that reduces gingival inflammation aids in the control of hyperglycemia. Periodontitis is accompanied by gingival bleeding and the production of an inflammatory exudate termed gingival crevicular fluid (GCF) that arises from the inflamed gingival tissues surrounding the teeth. GCF contains byproducts of connective tissue degradation, enzymes from host and bacterial cells, cytokines and other inflammatory mediators, and has been studied for screening blood glucose and for biomarkers of both diabetes and periodontitis. This chapter focuses on the bidirectional nature of diabetes and periodontitis and the biomarkers common to both these diseases that may enable earlier detection, targeted preventive measures and individualized therapeutic intervention of these chronic conditions.

Keywords Periodontitis • Diabetes • Inflammation • Biomarkers • Gingival crevicular fluid • Personalized treatment

R. Krishna (✉) • C.W. Cutler
Department of Periodontics, Georgia Regents University, College of Dental Medicine, 1120 15th Street, GC 1339, Augusta, GA 30912, USA
e-mail: rkrishna@georgiahealth.edu

P.J. Hanes
Dental Academic Administration, Georgia Regents University, College of Dental Medicine, 1430 John Wesley Gilbert Drive, GC 1339, Augusta, GA 30912, USA

M.S. Mozaffari (ed.), *New Strategies to Advance Pre/Diabetes Care: Integrative Approach by PPPM*, Advances in Predictive, Preventive and Personalised Medicine 3, DOI 10.1007/978-94-007-5971-8_12,

1 Introduction

The identification and utilization of diagnostic biomarkers of various disease states has been advocated as a means of simplifying or improving the accuracy and efficiency of diagnostic procedures. In 2001, an NIH working group standardized the definition of a biomarker as "a characteristic that is objectively measured and evaluated as an indicator of normal biological processes, pathogenic processes, or pharmacologic responses to a therapeutic intervention" [1]. Improved diagnostic efforts resulting from the utilization of novel biomarkers should enable optimal treatment planning and a more focused prevention of common human pathologies. Such biomarkers could also be used to screen healthy patients and identify those who may be predisposed to potential pathologies later in life and thereby enable the clinician to implement well-timed preventive measures. For the last two decades efforts have been made to identify biomarkers for diabetes mellitus and periodontitis. Diabetes and periodontitis are chronic inflammatory diseases which share a number of pathogenic mechanisms with common inflammatory mediators which have been investigated as possible biomarkers of disease status. This review describes the relationship between diabetes mellitus and periodontitis and summarizes the research aimed at the identification of biomarkers for these two diseases.

Diabetes mellitus (DM) is a group of disorders associated with a quantitative reduction in insulin production or a qualitative reduction in the action of insulin leading to changes in carbohydrate, protein, and lipid metabolism [2] and accumulation of glucose in the bloodstream. Increased glucose in the bloodstream, or hyperglycemia, affects a variety of tissues and organ systems including eyes, nerves, kidneys, and blood vessels. The periodontal tissues are also affected by hyperglycemia. Periodontitis has been described as the sixth complication of diabetes mellitus [3].

The public health burden of diabetes in the world is enormous and many of the people suffering from diabetes are unaware of the problem. The most common types of diabetes include type 1, type 2, and gestational diabetes, although some other forms also exist. Type 1 DM is caused by the destruction of β-cells in the pancreatic islets of Langerhans [4] and most often occurs in children and young adults. Type 2 DM usually occurs in adults, but may also been seen in children and young adults and is the more common form of DM. This is also the form of DM that is most commonly encountered in the dental office. Type 2 DM is a chronic condition and unlike type 1 DM, can remain undiagnosed for a long time and may in fact be first recognized by a dental care provider. Type 2 DM is characterized by increased glucose production in the liver and increased peripheral resistance to insulin in the muscles and a subsequent reduction in insulin secretion. Exercise and weight loss improve insulin sensitivity in these patients and decrease insulin sensitivity in the peripheral tissues. Medications are often used to improve carbohydrate metabolism, increase pancreatic insulin production, or decrease insulin resistance. Sometimes, insulin is administered to improve glycemic state in type 2 DM as well. Gestational diabetes occurs mostly in women in their third trimester of pregnancy and results in significant perinatal morbidity and mortality. Most patients with gestational diabetes return to their normal glycemic state after delivery although these women have been shown to be more likely to develop diabetes at a later age.

The major systemic complications and the resulting morbidity and mortality associated with diabetes has been attributed to a sustained state of hyperglycemia in the diabetic subject which leads to micro- and macrovascular complications and eventual damage to the structure and function within several major organ systems. In the large blood vessels, diabetes is associated with an increase in lipid deposition, thickening of arterial walls, and atheroma formation. Microvascular complications are due to damage of endothelial cells and alterations in endothelial basement membranes [2].

Diabetes has several effects on the oral cavity including xerostomia (reduced salivary flow), burning mouth, altered wound healing, and increased incidence of periodontal diseases such as gingivitis and periodontitis. Clinical studies have established that periodontal diseases are more prevalent and of greater severity in patients with poorly controlled type 1 or type 2 diabetes than in non-diabetic patients [5–9]. In addition, the prevalence and severity of gingival inflammation and bleeding within diabetic patients has been shown to be directly related to their level of metabolic control [10–15]. In fact, a strong positive association has been demonstrated between elevated HbA1c levels over a 2-year, pre-examination period and the presence and severity of periodontitis [16], and several studies have suggested a dose-response relationship between poor metabolic control and the severity of periodontitis [9, 17, 18]. In addition periodontal pockets with probing attachment loss occur more frequently in diabetics [19]; diabetics with more advanced systemic complications have more severe periodontitis [20]; and poorly controlled diabetics respond less favorably to periodontal treatment [21].

2 The Normal Human Periodontium

The tissues that anchor the teeth and support them during health and function constitute the periodontium (Greek *peri*, "around"; *odont*, "tooth") and include the gingiva (Fig. 1), which covers the maxillary and mandibular alveolar bone, and the periodontal attachment apparatus (Fig. 2) which attaches the tooth to the jaw and

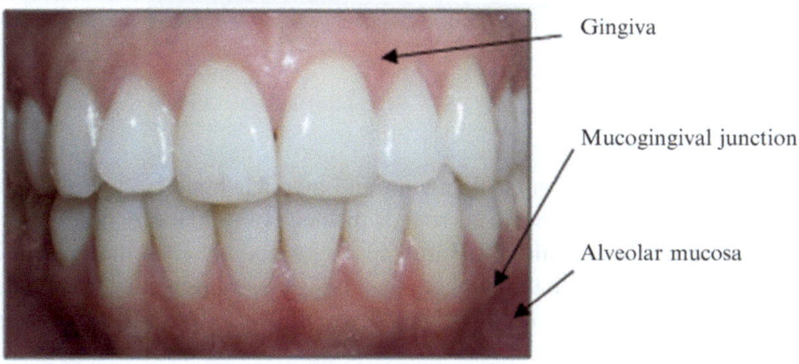

Fig. 1 Teeth surrounded by healthy periodontium

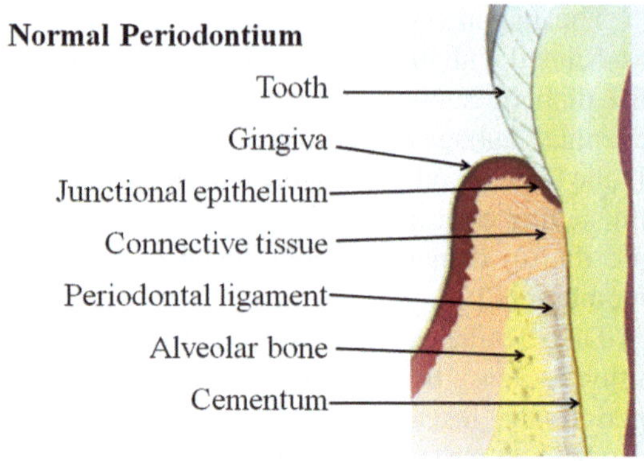

Fig. 2 Diagrammatic representation of the different components of the normal periodontium

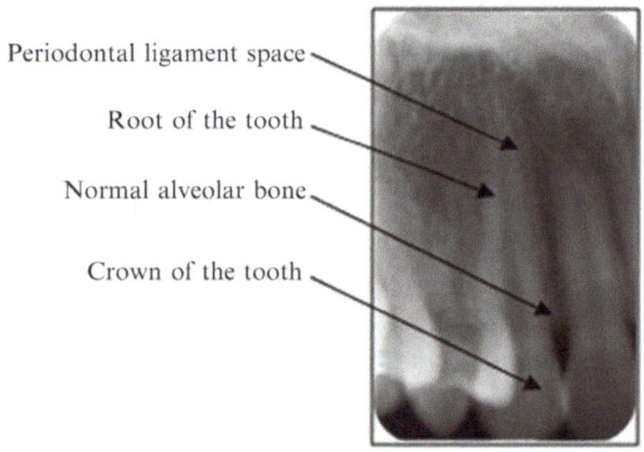

Fig. 3 Radiograph of healthy teeth and periodontium

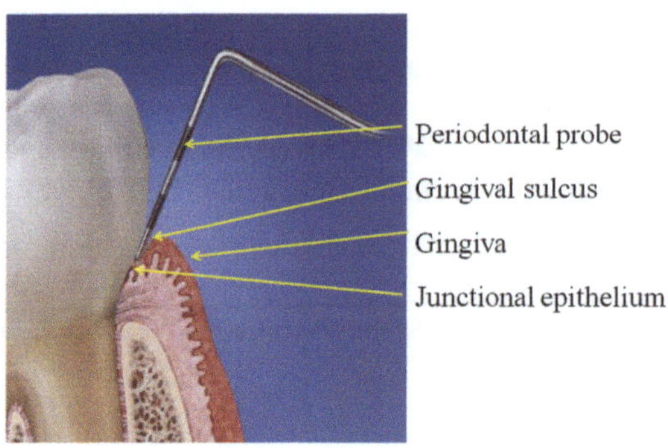

Fig. 4 Gingival sulcus

includes cementum, periodontal ligament, and alveolar bone [22] (Fig. 3). In the normal periodontium, the gingiva surrounds the neck of each tooth, much like a collar. There is normally a shallow 2–3 mm space formed between the margin of the gingiva and the crown of the tooth where the gingiva overlaps a portion of the crown of the tooth [22]. This space between the crown of the tooth and the gingival tissue is the gingival sulcus (Fig. 4).

The periodontal attachment apparatus is the means by which the tooth is attached to the jaw and consists of the cementum on the root of the tooth, the alveolar bone lining the tooth socket and the periodontal ligament (Figs. 2 and 3). Cementum is a mineralized connective tissue that covers the roots of the teeth and its main function is to attach the periodontal ligament fibers to the teeth. It is similar in structure to bone, but differs from it in its microscopic organization, lack of vascularity, and the absence of continuous remodeling. The periodontal ligament is a dense, fibrous connective tissue that attaches the cementum of the tooth to the alveolar bone. It occupies a narrow 0.1–0.25 mm [23] space between two calcified structures, cementum and alveolar bone and is mainly made up of collagen fibers (type 1 and 3), vascular elements, cells, and neural elements.

3 Periodontal Disease

Diseases of the periodontium can be broadly classified as either gingivitis or periodontitis. Gingivitis is an inflammation of the gingiva which manifests clinically with the usual signs and symptoms of inflammation including an increase in redness, swelling, bleeding, suppuration, and possible pain. Deposits of dental plaque and calculus are often found adherent to the tooth surface and in close contact with the gingival margin. The primary etiology of virtually all periodontal diseases is the bacterial biofilm termed dental plaque that forms on tooth surfaces and root surfaces below the gum line. Certain species in this biofilm infect the tissues lining the deepened periodontal pocket. These are predominantly anaerobic, gram-negative bacteria including *Aggregatibacter actinomycetemcomitans*, *Tannerella forsythensis*, *Porphyromonas gingivalis*, *Prevotella intermedia*, *Fusobacterium nucleatum*, *Campylobacter rectus, and Treponema denticola*. The disease is a result of a complex interplay between the bacterial products and the host response that are modified by behavioral and/or systemic factors [24]. About 50% of the tissue destruction associated with periodontal disease is attributed to the host response. The immune response or inflammation elicited by the chronic infectious burden in the pockets results in destruction of structural components of the periodontium, leading to the clinical signs of periodontitis. Page and Schroeder [25] described the histopathogenesis of the lesion of gingivitis in three stages: Initial (2–4 days after bacterial insult), early (4–7 days after) and established (14–21 days after). The initial lesion is an acute vasculitis characterized by vascular changes consisting essentially of dilation of capillaries and increased vascular permeability with an infiltration of polymorphonuclear leukocytes (PMNs) in the space between the gingival margin and the tooth, in the epithelium lining this space and in the subjacent connective tissue. As the inflammatory lesion progresses to the next stage, clinical signs of erythema may appear due to the proliferation of capillaries and increased formation of capillary loops. The epithelium lining the space between the tooth and the gingival margin begins to proliferate into the subjacent infiltrated connective tissue, and the inflammatory infiltrate consists largely of lymphocytes and PMNs. Further alterations

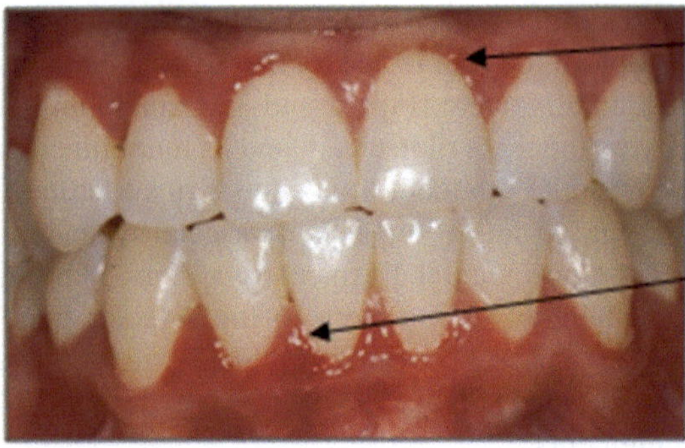

Fig. 5 Clinical presentation of gingivitis. Inflammation is localized to the gingiva and has not spread to the bone

in the histopathology of the lesion associated with established gingivitis are consistent with the development of a chronic inflammatory lesion, with the appearance of numerous plasma cells, increased size of the inflammatory infiltration of the connective tissue and increased vascular changes wherein gingival blood vessels become dilated and congested. The established lesion can be observed clinically as moderate to severely inflamed gingiva (Fig. 5). If left untreated, the established lesion progresses to a fourth stage, which is the advanced lesion in which destruction of connective tissue attachment to the tooth and loss of supporting bone occurs.

Gingivitis is a common disease of the oral cavity. In the third National Health and Nutrition Examination Survey (NHANES III, 1988-94), 50% of the US adult population was shown to have gingivitis on at least four teeth [26]. Although gingivitis is a fairly common disease and gingivitis is a pre-requisite for periodontitis, not all sites with gingivitis progress to periodontitis [27–30]. There are several factors other than the amount of plaque accumulation that have been shown to play a role in the progression of gingivitis to periodontal disease. These include but are not limited to age, pronounced inflammatory response, more pathogenic bacteria, stress, genetic predisposition (IL-1 positive genotype individuals), and smoking status [31–33].

Periodontitis is an inflammatory disease of the gingiva and the deeper periodontal tissues. Although periodontitis is preceded and accompanied by gingivitis, gingivitis may persist for indefinite periods without progressing to periodontitis. Periodontitis involves destruction of gingival and periodontal connective tissue fibers, resorption of tooth-supporting bone, and apical proliferation of the junctional epithelium resulting in the formation of deep pockets between the gingiva and the tooth surface. These pockets harbor pathogenic periodontal bacteria which are the primary etiology of the periodontal lesion. If left untreated this inflammatory lesion progresses, eventually leading to loss of supporting structures of the teeth to the point that the tooth is lost.

Clinically, patients with periodontitis demonstrate all or some of the classical signs of inflammation seen in gingivitis along with pocket formation and radiographic bone loss (Fig. 6a, b). It is well documented that moderate periodontal

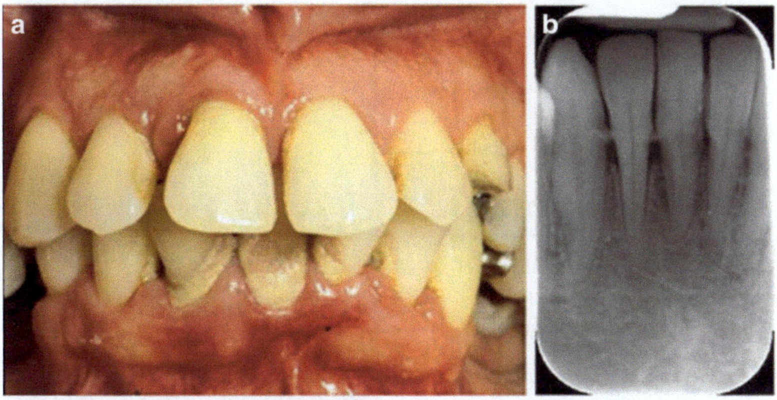

Fig. 6 (**a**) Clinical appearance of periodontitis. Note gingival inflammation, heavy deposits of dental plaque and calculus and gingival recession. (**b**) The radiograph from the same patient shows advanced bone loss and radiographic appearance of calculus deposits

disease affects a majority of the adult population, and is more severe in populations which do not receive regular dental care [34, 35].

The histopathologic lesion of periodontitis is characterized by apical proliferation of junctional epithelium onto cementum, lysis of collagen fibers of the periodontal ligament, and bone destruction at the alveolar crest (Fig. 7a).

There is an increase in the volume of the connective tissue inflammatory infiltrate characterized by the presence of PMNs, lymphocytes, macrophages, and plasma cells [25]. Updated immunofluorescence analyses of the established lesion indicate the presence of organized oral lymphoid foci or OLF. OLF contains a highly cellular infiltrate, consisting of various dendritic cell (DCs) subsets (Langerhans cells, dermal DCs), engaged with CD4+ T cells [36]. These DCs are apparently mobilized towards the microvasculature [37] and was reviewed in by Cutler and Jotwani [38] (Fig. 7b). The influence of diabetes on the content of this inflammatory infiltrate, and in particular the DC subsets are not clear.

4 The Inter-Relationship Between Periodontal Disease and Diabetes

Epidemiologic studies have shown a three to fourfold increased risk for progressive periodontal destruction in diabetic patients compared with individuals without diabetes [7]. Although the biological basis for the relationship between periodontitis prevalence and severity and the level of metabolic control of diabetes has not been clearly elucidated [39], there is ample clinical evidence to support the view that poor glycemic control increases the risk in the diabetic patient for more severe periodontal disease with greater periodontal attachment loss, bone loss and deeper periodontal probing depths [17, 40–45] (Fig. 8).

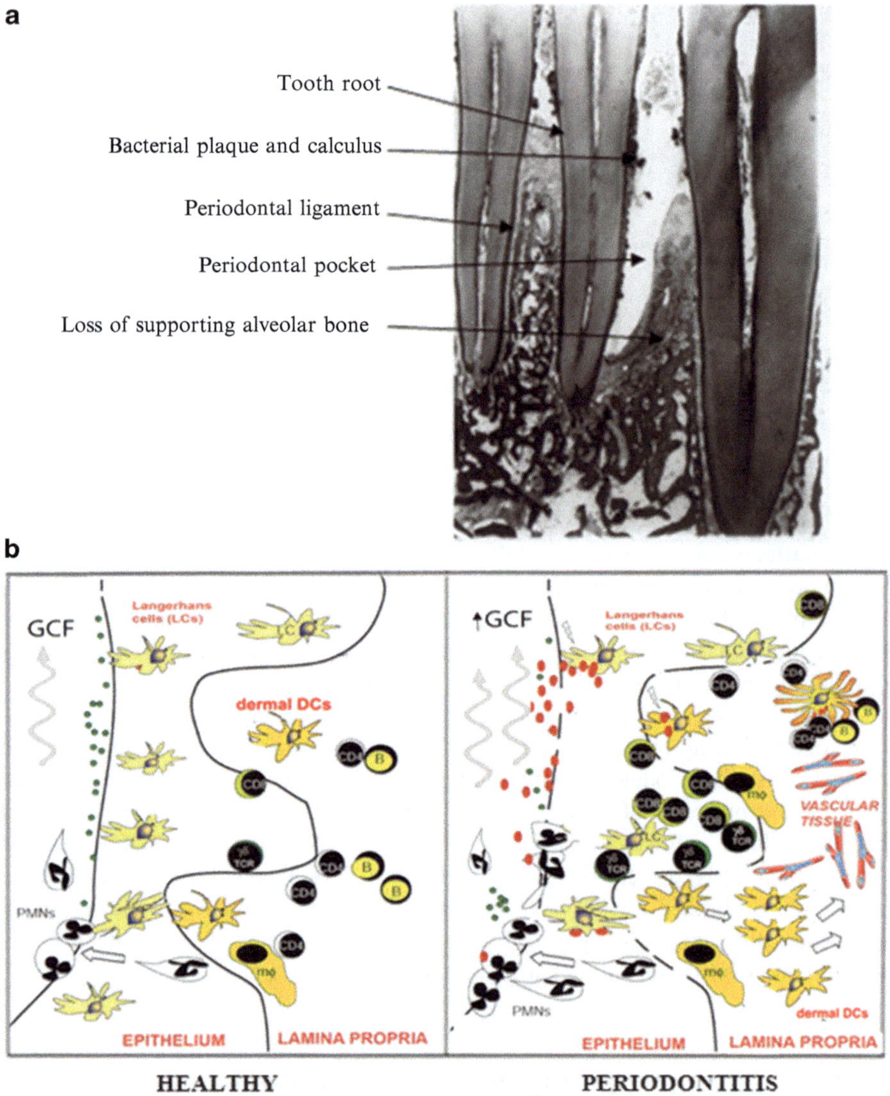

Fig. 7 (**a**) Histology of periodontitis lesion. Bacteria-induced inflammation has resulted in loss of connective tissue attachment to the tooth, epithelial migration, pocket formation and loss of supporting bone. (**b**) Oral lymphoid foci and GCF. The *top left panel* is a representation of a healthy biofilm comprised primarily of gram positive bacterial (*green dots*). Healthy gingival tissue has many Langerhans cells in the gingival epithelium as well as a few dermal dendritic cells. As disease progresses the oral biofilm is colonized by more gram negative bacteria (*red dots*). In response to this we see a dramatic loss (efflux) of Langerhans cells from the epithelium and an influx of dermal dendritic cells into the lamina propria as well as neutrophils, macrophages, B cells and T cells. The volume of gingival crevicular fluid (GCF) increases and the inflammatory cytokine content increases

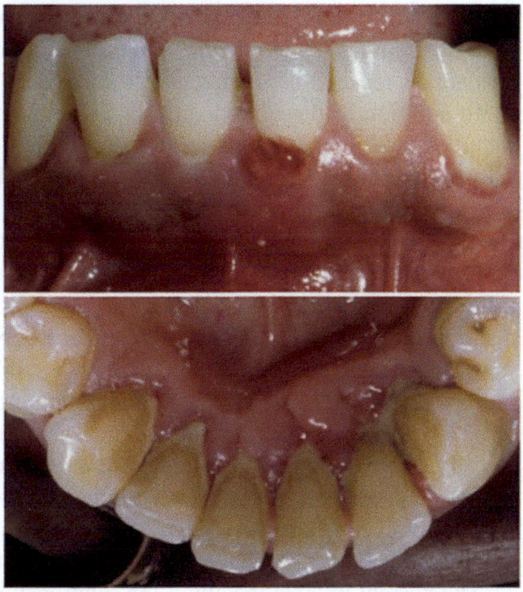

Fig. 8 Clinical presentation of periodontitis in patient with undiagnosed diabetes mellitus (type 1). History of multiple recurring periodontal abscesses

Early research into the relationship between diabetes and increased risk for periodontitis focused on potential differences in the bacterial microflora in diabetics versus non-diabetics. The majority of this research shows, however that the subgingival microbiota is very similar in diabetics and non-diabetics with periodontal disease [39, 46]. Subsequent research has been focused on differences in the immunoinflammatory response to periodontal pathogens in diabetic versus non-diabetic subjects.

Multiple factors have been suggested to explain the relationship between diabetes mellitus and periodontal diseases including (1) altered immunoinflammatory response to bacterial pathogens, (2) diminution of the formative aspects of connective tissue metabolism, (3) impaired wound healing, (4) microvascular changes and (5) formation of advanced glycation end products [4]. Although the components of the bacterial biofilm are similar in comparisons of dental plaque samples from diabetic and non-diabetic patients [39, 46, 47], the immunoinflammatory responses to these bacterial pathogens are altered in the diabetic patient. Immune cell functions such as adherence, chemotaxis and phagocytosis have been shown to be impaired in neutrophils from diabetic patients [48, 49]. In addition the monocyte/macrophage cell line has been shown to be hyper-responsive to bacterial antigens, particularly during episodes of hyperglycemia, resulting in increased production of the proinflammatory cytokines IL-1β and TNF-α [50, 51].

In addition to these altered immunoinflammatory responses, the formative aspects of connective tissue metabolism appear to be diminished in the diabetic patient and to be related to the level of metabolic control. Bone turnover and osseous healing have been shown to be impaired under the influence of hyperglycemia [52–55] due to inhibition of osteoblastic cell proliferation and collagen production

[56–58]. In addition, studies have shown an increased apoptotic response of fibroblasts and osteoblasts to *P. gingivalis* infection in the presence of hyperglycemia, resulting in a net decrease in the numbers of these matrix-producing cells [59–61].

In addition to these inadequacies in the formative aspects of connective tissue metabolism, hyperglycemia-induced inhibition of the attachment and spreading of fibroblasts [62] and microvascular changes [12, 13] all contribute to impaired wound healing which is a common problem in the diabetic patient. Characteristic microvascular changes associated with diabetic complications, such as abnormal growth and regeneration of blood vessels, have also been described in the periodontal tissues [12–14]. Advanced glycation end products (AGEs) have been shown to accumulate in periodontal tissues in diabetic subjects [63] and to play a role in these microvascular alterations. AGEs bind irreversibly to collagen forming macromolecules which accumulate in the basement membranes of endothelial cells resulting in increased basement membrane thickness and alteration in the normal homeostatic transport across the membrane [43]. AGE production has also been associated with increased production of vascular endothelial growth factor (VEGF), a cytokine that is also involved in the microvascular complications of diabetes [64, 65] and which has been reported to be elevated in the gingival tissues of diabetic patients [66].

The current knowledge about the relationship between diabetes and periodontal diseases suggests a bidirectional relationship between the two. Both diabetes mellitus and periodontitis can be classified as chronic inflammatory diseases although of different etiologic origin. Poorly controlled diabetic subjects tend to present with more severe periodontitis [18, 67–69], and diabetic subjects with untreated periodontitis generally have poorer metabolic control of their diabetes [69–71]. Such studies suggest that severe, long-standing inflammatory lesions in the untreated patient with periodontitis could contribute to the exacerbation of diabetes. In contrast, several studies have reported a positive association between periodontal treatment and improved metabolic control of diabetes mellitus [72–75].

Studies have suggested that an abnormal inflammatory response known as a "hyper-inflammatory trait" occurs in subjects with diabetes [76]. This hyper-inflammatory trait is characterized by an exaggerated secretion of inflammatory mediators such as TNFα and IL-6 as well as systemic markers of inflammation. Over time hyperglycemia in the diabetic patient leads to glycation of structural proteins and lipids in the extracellular matrix and connective tissues as well as the vascular tissues [77–79]. These vascular changes lead to disturbances in capillary function, impaired perfusion of tissues and the release of reactive oxygen species resulting in a systemic inflammatory challenge [80, 81]. In addition monocytes and endothelial cells can interact with advanced glycation end products (AGEs) via monocytic receptors (RAGE) to "prime" neutrophils and macrophages resulting in increased secretion of cytokines and inflammatory mediators by these cells. AGEs and elevated unsaturated fatty acids which are often associated with diabetes, are both capable of inducing a hypersecretory monocytic state leading to an increased production of inflammatory cytokines in response to antigenic or bacterial stimulus [82–85]. Both periodontitis and diabetes mellitus result in a systemic inflammatory challenge associated with a chronic elevation of inflammatory mediators including IL-1, TNFα, IL-6, PGE_2, C-reactive protein, and fibrinogen [81, 86, 87]. This hyper-inflammatory trait and the AGE-RAGE upregulation of the inflammatory response to bacterial antigens in the diabetic patient may result in a more severe systemic inflammatory challenge and overexpression of inflammatory mediators.

Given the fact that periodontal infections are associated with a predominately gram negative bacterial flora, cytokines may also be secreted by cells in response to stimulation by bacterial lipopolysaccharide. In this case cytokine production occurs via a Toll-like receptor mediated response resulting in secretion of cytokines IL-1β, IL-6, TNF-α, and PGE_2, all of which are implicated in the tissue destruction associated with periodontitis [88, 89]. Infections such as periodontitis result in elevated levels of circulating cytokines including IL-1, TNFα, and IL-6. These cytokines block lipoprotein lipase activity and induce hyperlipemia [81, 90, 91]. TNFα promotes glycogenolysis and impairs glucose uptake by cells leading to hyperglycemia [81, 91]. TNFα and IL-6 also target the hepatocyte and lead to production of C-reactive protein [81, 90, 91]. The systemic inflammatory challenge associated with both diabetes mellitus and periodontitis link diabetic metabolic dysregulation and the severity of periodontal disease and shed light upon the interactions between chronic oral and systemic inflammatory diseases.

5 Potential Diagnostic Components of Gingival Crevicular Fluid

As chronic inflammatory diseases, both diabetes and periodontitis share certain inflammatory markers indicative of level of disease control [92]. The initiation and progression of an inflammatory lesion in the periodontal tissues is positively associated with the flow of gingival crevicular fluid (GCF) from the gingival sulcus or periodontal pocket [93, 94] (Fig. 9). GCF is a serum exudate that originates from the

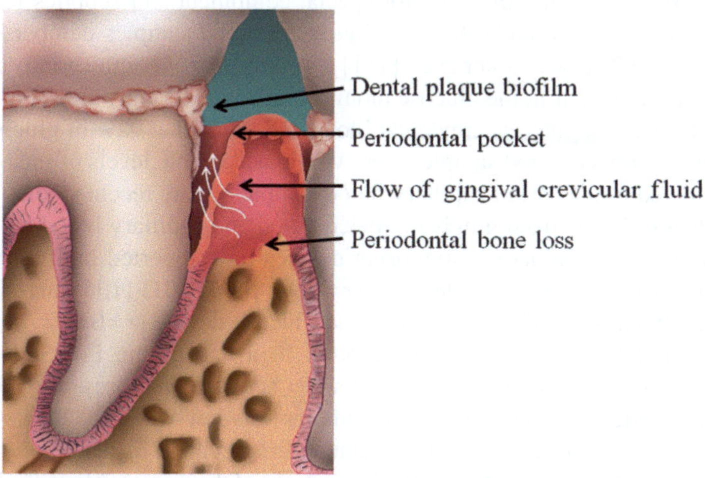

Fig. 9 Gingival crevicular fluid flow in a periodontal pocket

microvasculature in the lamina propria [95]. The content of GCF is reflective of the cellular infiltrate and its response to the subgingival microflora (Fig. 7b). Since the 1980s numerous studies have attempted to identify components in the GCF which could aid in the identification of specific periodontally involved sites which are in active stages of tissue destruction [96], or to identify GCF components which could be used to screen patients for periodontitis or otherwise simplify periodontal examination and diagnosis [97–100]. The composition of GCF somewhat mirrors that of plasma [97] and has, therefore been of interest to clinicians as a less invasive medium to evaluate for various systemic conditions including diabetes.

5.1 Glucose in GCF

Efforts to evaluate glucose levels in samples of blood from the gingival sulcus or periodontal pocket during dental examinations have been advocated as a method of expanding the range of venues for diabetic screening due to the fact that this method does not require a "finger stick" and takes advantage of the presence of a blood sample that would otherwise be discarded in the dental office [101]. Gingival bleeding is a common, virtually ubiquitous symptom of gingivitis and periodontitis -a finding supported by the high frequency of dental patients who report that their gingival tissues routinely bleed when they brush and/or floss their teeth. Likewise the gingival tissues of these patients bleed during the gentle provocation of the tissues associated with a routine dental and periodontal examination. Several studies have shown that in diabetic subjects glucose levels in capillary blood obtained from "finger sticks" correlate with glucose levels from blood obtained from gingival sulci or periodontal pockets during routine periodontal examinations [102–110]. Although early studies reported moderate to good correlations between the glucose levels in gingival crevice blood and capillary blood, the length of time required for collection of samples and the need for special equipment and supplies for sample collection and analysis made the use of gingival blood samples cumbersome and inconvenient [102–104]. Parker et al. [111] tested the use of commercially available reagent strips and an in-home glucose monitoring device for the purpose of measuring glucose levels in gingival crevicular blood and found that 92% of glucose levels from gingival crevice blood samples were within acceptable levels of measurement error compared to 90% from capillary blood samples. More recent studies comparing the glucose levels from gingival crevice blood and capillary blood from finger sticks using in-home glucose monitoring devices have reported high intra-patient correlations using the Bayer Elite 2000 ($r=0.98$; $p<0.0001$) [106], and the Xitux Diagnostics Smart-X self-monitoring blood glucose device ($r=0.997$; $p<0.0001$) [105]. Contradictory findings were reported in a test of the FreeStyle® Blood Glucose Monitoring System wherein agreement between measures of glucose level in gingival crevice and capillary blood samples was low when the data was compared using statistical measures of agreement/disagreement instead of correlation/ regression [107, 108]. Although it is possible that part of the discrepancy in the literature over the validity of glucose measurements in gingival crevice blood may be due to the smaller volume of blood sampled from the gingival crevice, studies have shown that glucose levels can be reliably determined in sub-microliter volumes (0.2–0.8 μl) of GCF and that GCF glucose levels are significantly higher in periodontitis sites [112]. However, no effort was made in this study to correlate or to determine the level of agreement between measures of glucose levels from GCF and serum glucose levels [112]. In the study Strauss et al. measured glucose levels from gingival crevice and capillary finger-stick blood in an OneTouch UltraMini Blood Glucose Monitoring System and compared these measurements using both measures of correlation and limits of agreement [101]. Samples were taken from patients with established gingivitis and bleeding on probing and compared with samples from patients with minimal inflammation and little or no bleeding. In the presence of frank gingivitis glucose levels in gingival crevice and capillary blood were highly correlated ($r = 0.89$) and the limits of agreement were acceptable (−27.1 to 29.7). On the other hand in the absence of gingivitis, correlations between the measures of glucose level were lower (0.78) and the limits of agreement were much wider (−25.1 to 80.5). The authors concluded that in the absence of frank gingivitis it was difficult to obtain enough blood from the gingival crevice for measurements without touching the tooth as the tissues were manipulated to stimulate bleeding, possibly resulting in contamination of the sample. Taken together these studies suggest that samples of gingival crevice blood taken during an initial, pre-treatment dental examination may be used to screen previously undiagnosed patients for diabetes using commercially available glucose monitoring devices. However following dental and periodontal treatment when gingivitis and bleeding on probing have been successfully controlled, it will be difficult to obtain a sufficient volume of gingival crevice blood for reliable assessment of glucose levels.

5.2 Structural Proteins and Proteolytic Enzymes

The identification of biomarkers of periodontal disease presence or active disease status has been a somewhat elusive goal in periodontal research. Considerable effort has been expended on the identification of some component of the GCF which could identify a site or a patient at risk or predict an impending episode of disease progression. Research efforts have focused on the presence of tissue specific structural molecules including collagen fragments such as type 1 collagen carboxyterminal telopeptide, proteoglycans such as chondroitin sulfate, hyaluronan, heparin sulfate, osteonectin and bone sialoproteins [100], and on regulators of tissue destruction such as collagenase and matrix metalloproteinases including matrix metalloproteinase-8 [99]. High levels of collagenase activity in GCF are associated with degradation of periodontal tissues in destructive periodontitis. In this regard, Safkan-Seppälä [113] evaluated GCF collagenase activity in poorly controlled and well-controlled diabetics and healthy controls and found that diabetic subjects had higher GCF collagenase activity than healthy controls and that the collagenase activity was related to the level of metabolic control. GCF levels of lysosomal enzymes such as beta-glucuronidase and elastase, and cytoplasmic enzymes such as lactate dehydrogenase have also been studied for their potential as a biomarker of periodontal destruction. Lamster et al. [97] reported a clinical study wherein GCF levels of beta-glucuronidase, but not lactate dehydrogenase were predictive of periodontal breakdown. Palcanis et al. [98] showed that GCF levels of elastase were significantly higher at sites which demonstrated periodontal bone loss in subsequent examinations over a 6 month period. More recently, Gursoy et al. [114] investigated the periodontal status and aspartate aminotransferase and lactate dehydrogenase enzyme activities in GCF of type 2 diabetic and/or obese chronic periodontitis patients, comparing these to healthy controls. Although the periodontal status was worse in subjects with diabetes, no significant differences in aspartate aminotransferase and lactate dehydrogenase enzyme activities were observed. Although the search continues for an appropriate biomarker in the GCF associated with periodontal disease presence and progression, most of the current research is focused on studies of various immunoinflammatory proteins including cytokines, inflammatory mediators and growth factors.

5.3 *Inflammatory Mediators in GCF*

GCF levels of IL-1β are known to be high in subjects with periodontitis [115–118], and even more so in subjects with diabetes [119–123]. Salvi et al. [119] compared the GCF levels of IL-1β, PGE_2 and TNF-α in diabetics and systemically healthy subjects with varying degrees of periodontal disease severity and found that diabetics had significantly higher GCF levels of both PGE_2 and IL-1β when compared to non-diabetic controls with similar periodontal status. In addition GCF levels of IL-1β and PGE_2 increased in diabetics as the severity of periodontal disease increased. Engebretson et al. [121] investigated the effects of glycemic control on GCF levels of IL-1β in patients with chronic periodontitis and type 2 diabetes and found that clinical periodontal measures and measures of glycemic control (HbA1c, random glucose) were significantly correlated with GCF IL-1β. Patients with greater than 8% HbA1c had significantly higher mean GCF IL-1β levels than patients with less than 8% HbA1c. Andriankaja et al. [123] compared the levels of serum IL-6 and GCF IL-1β and PGE_2 in subjects with gingivitis and type 2 diabetes. GCF IL-1β was significantly elevated in the diabetic compared to the non-diabetic group but serum IL-6 and GCF PGE_2 were not. GCF IL-1β and PGE_2 levels were significantly elevated in subjects with gingivitis compared to subjects with gingival health regardless of diabetic status. However, serum IL-6 was elevated in subjects with gingivitis compared to healthy subjects only among those subjects with diabetes. Kardeşler et al. [124] investigated the relationship of periodontal disease in type 2 diabetics with GCF levels of PGE_2 and IL-1β and in contrast to most studies, found lower IL-1β levels in the type 2 diabetic group, and similar levels of PGE_2 in diabetic and non-diabetic periodontitis patients. In general most studies report elevated levels of

IL-1β in GCF from diabetic patients and increased expression of IL-1β as the severity of periodontal inflammation increases which is consistent with the hypothesis that hyperglycemia contributes to an heightened inflammatory response, and suggests a mechanism to account for the association between poor glycemic control and periodontal destruction (Table 1).

Hyperglycemia-induced activation of the protein kinase C pathway leads to increased expression of vascular endothelial growth factor (VEGF) which can alter vascular permeability and angiogenesis. The increased concentration of VEGF in periodontitis [140, 141] may be one of the reasons for the increased vascularization and permeability associated with periodontitis, and it may be regarded as a marker for disease severity. Prapulla et al. [141] compared GCF VEGF levels in gingivitis and periodontitis patients and healthy controls and found that VEGF levels in GCF increased from health to periodontitis, and that periodontal treatment resulted in a reduction of GCF VEGF levels. The authors suggested that VEGF could be considered a biomarker of periodontal disease progression. Sakallioğlu et al. [142] studied gingival tissue and GCF VEGF levels in diabetic periodontitis patients with good metabolic control compared to that of systemically healthy patients with periodontitis. This study found significantly higher VEGF levels in the gingival tissue supernatants of the diabetic group, but GCF VEGF levels were similar between the controlled diabetic and the systemically healthy subjects.

Prostaglandin E_2 (PGE_2) is associated with periodontal disease progression and alveolar bone resorption [143]. Levels of GCF PGE_2 are elevated in the presence of periodontitis and perhaps even more so in diabetic patients [119]. Offenbacher et al. [144] reported significant increases in GCF PGE_2 levels a few months prior to clinically evident progression of periodontal attachment loss and suggested that monitoring of GCF PGE_2 levels could be of value in predicting episodes of periodontal disease progression enabling the administration of preventive therapy before the onset of tissue destruction. More recent studies of GCF PGE_2 levels in diabetic patients have, however cast some doubt on the usefulness of this agent as biomarker of disease progression. Kardesler et al. [124] observed that GCF PGE_2 levels in diabetics were higher than healthy controls when levels were expressed as total amount of PGE_2. When GCF PGE_2 levels were expressed as concentration of PGE_2 the concentration of PGE_2 was lower in GCF from diabetic subjects. These findings suggest that the differences in GCF PGE_2 levels are more a function of increased volume of GCF in the diabetic rather than an increased expression of PGE_2. Andriankaja et al. [123] compared GCF PGE_2 levels in diabetics and healthy controls during the development of experimental gingivitis, and found that GCF PGE_2 levels increased with the presence of gingivitis, but that the amount of GCF PGE_2 was similar in diabetics and healthy controls (Table 1).

Hyperglycemia-induced activation of the protein kinase C pathway also results in increased expression of transforming growth factor (TGF)-β which leads to changes in collagen and fibronectin thereby promoting capillary occlusion. Gurkan et al. [139] measured GCF TGF-β1 in several forms of periodontal disease and found that GCF levels of TGF-β1 were significantly elevated in subjects with periodontitis compared to healthy controls and that GCF TGF-β1 levels were correlated

Table 1 Inflammatory mediators in GCF as a function of periodontal disease and diabetes status

Cytokine	Study	Health	Gingivitis	Periodontitis	Effect of treatment	Effect of diabetes
IL-1α	Masada et al. [125]	—	—	Present	Reduced	—
	Mathur et al. [126]	Present	—	Amount and concentration higher	—	—
	Tsalikis et al. [127]	Present	Increased	—	Reduced	—
IL-1β	Masada et al. [125]	—	—	Present	Reduced	—
	Hou et al. [116]	Present	—	Amount three times higher than controls. (not concentration) Correlated with severity	Reduced	—
	Tsai et al. [117]	Present	—	Higher amount; correlated with severity	Reduced	—
	Salvi et al. [119]	Present	Present	Present	—	Concentration higher compared to Non-diabetic with similar level of periodontitis. Concentration increases with severity of periodontitis
	Figueredo et al. [128]	—	Present	Concentration higher than gingivitis. Not related to pocket depths.	—	—
	Bulut et al. [120]	Present	—	Concentration higher than heathy controls	—	Concentration higher than periodontitis
IL-1β	Gonzales et al. [129]	Present	Concentration increased	—	—	—
	Tsalikis et al. [127]	Present	Amount increased	—	Reduced	—

	Engebretson et al. [121]	Present	Amount higher than healthy sites	Amount higher than healthy and gingivitis sites	Reduced	—
	Faizuddin et al. [130]	Present	Concentration higher than healthy sites	Concentration higher than healthy and gingivitis sites	—	—
	Giannopoulou et al. [131]	Present	Increased amount	—	—	—
	Engebretson et al. [117]	—	—	Present	—	Amount correlated with periodontitis severity; and with HbA1c
	Orozco et al. [118]	—	Present	Concentration higher than gingivitis	—	—
	Zhong et al. [122]	—	—	Positively related to pocket depths and bleeding	—	Concentration higher than periodontitis
	Kardesler et al. [124]	Present	—	Present	—	Amount lower than periodontitis
	Schierano et al. [132]	Present	Amount higher than healthy sites	—	Reduced	—
	Andriankaja et al. [123]	Present	Amount higher than healthy sites	—	—	Amount higher than non-diabetic
IL-6	Kurtis et al. [133]	Present	—	Concentration higher than heal thy controls. No correlations with clinical parameters	—	Concentration higher than periodontitis

(continued)

Table 1 (continued)

Cytokine	Study	Health	Gingivitis	Periodontitis	Effect of treatment	Effect of diabetes
IL-8	Tsai et al. [117]	Present	—	Amount higher than healthy controls. Positive relationship between amount and clinical parameters.	Reduced	—
	Mathur et al. [126]	Present	—	Amount was significantly higher in diseased sites. Concentration was lower compared to healthy sites.	—	—
	Giannopoulou et al. [131]	Present	Increased amount. Higher amount in smokers.	—	—	—
IL-18	Orozco et al. [118]	—	Present	Higher concentration than gingivitis	—	—
Interferon-α	Mathur et al. [126]	Present	—	Amount higher but concentration lower in diseased sites	—	—
PGE_2	Salvi et al. [119]	Present	Present	Present	—	Concentration higher compared to non-diabetic with similar level of periodontitis. Concentration increases with severi ty of periodontitis
	Kardesler et al. [124]	Present	—	Present	—	Amount higher than healthy, but similar to periodontitis. Concentration lower than periodontitis

PGE_2	Andriankaja et al. [123]	Present	Amount higher than healthy sites	—	—	Amount similar to non-diabetic
TGFβ-1	Skaleric et al. [134]	—	Present	Higher concentration in sites with deeper pockets	—	—
	Buduneli et al. [135]	Similar levels as Gingivitis	Similar levels as healthy controls. Higher levels with cyclosporine therapy	—	—	—
	Wright et al. [136]	Present	Amount increased but similar concentration	—	Amount decreased	—
	Kuru et al. [137]	—	—	Present	Transient increase following periodontal surgery	
	Gurkan et al. [138]	—	—	Present	Increased over 3 month post-treatment period. More so with systemic anti-inflammatory treatment.	—
	Gurkan et al. [139]	Present	—	Higher amount in chronic and aggressive periodontitis. Amount correlated with clinical parameters.	—	—
	Schierano et al. [132]	Present	Similar amount as healthy sites	—	Similar amount as healthy sites	—

(continued)

Table 1 (continued)

Cytokine	Study	Health	Gingivitis	Periodontitis	Effect of treatment	Effect of diabetes
TNFα	Schierano et al. [132]	Present	Similar amount as healthy sites	—	Similar amount as healthy sites	—
VEGF	Güneri et al. [140]	Present	Similar amount as healthy sites	Similar amount as healthy sites	—	Amount higher than non-diabetics, but similar between healthy and periodontitis sites
	Prapulla et al. [141]	Present	Similar concentration as healthy sites	Higher concentration than gingivitis	Reduced	—
	Sakallioğlu et al. [142]	—	—	Present in concentration similar to controlled diabetic	—	Present in concentration similar to non-diabetic

with all clinical periodontal parameters. Schierano et al. [132] measured TNF-α, TGF-β2 and IL-1β levels in GCF before and after the onset of experimental gingivitis in non-diabetic subjects. Although the volume of the crevicular fluid and the GCF levels of IL-1β increased significantly with the onset of gingivitis, TNF-α and TGF-β2 levels in GCF did not change significantly. GCF expression of transforming growth factor in diabetic periodontitis patients has not been reported (Table 1).

Other inflammatory mediators which have been studied as potential biomarkers of periodontal disease include IL-1α, IL-6, IL-8, IL-18, Interferon–α, and TNF-α (Table 1). GCF levels of IL-1α have been shown to be increased in the presence of gingivitis and periodontitis [125–127] and to decrease following periodontal treatment [125, 127]. No studies of GCF levels of IL-1α in diabetic subjects have been reported. Kurtis et al. [133] compared GCF levels of IL-6 in diabetics and healthy controls with and without periodontitis. The concentration of IL-6 in GCF was higher in periodontitis than in healthy gingival sulci, and was higher in diabetic patients than that of both healthy controls and non-diabetic patients with periodontitis. GCF levels of IL-8 [117, 126, 131], IL-18 [118] and Interferon-α [126] have also been shown to increase in the presence of gingivitis and periodontitis. In addition the GCF levels of IL-8 have been shown to decrease following periodontal treatment [117]. However, no studies of GCF levels of IL-8, IL-18 or Interferon-α in diabetics have been reported (Table 1).

Studies have also investigated the potential of various components of saliva for diagnosing and evaluating the prognosis and evolution of diabetes mellitus and its oral manifestations, although it should be mentioned that whole saliva also contains GCF. Borg Andersson et al. [145] studied the glucose content of parotid saliva and found that salivary levels of glucose peaked at 2 h following a glucose challenge and was significantly elevated in subjects with impaired glucose tolerance or diabetes mellitus, but not in healthy controls. Arana et al. [146] compared markers of oxidative stress in saliva from patients with type 2 diabetes mellitus with that of healthy controls and found significantly higher glutathione peroxidase and glutathione reductase activity, and significantly lower reduced glutathione in diabetic subjects. Gumus et al. [147] also reported significantly lower reduced glutathione in diabetic subjects, but only in type 1 diabetics, and no other differences in the levels of other salivary antioxidants were observed. Possible explanations for the differences in the findings of these studies may include the fact that stimulated saliva samples were employed in one study [146], while unstimulated samples were evaluated in the other [147]. In addition, the prevalence and severity of periodontal disease differed somewhat between the two studies, as well as the methods used to diagnose and describe the degree of periodontal involvement.

5.4 *Oral Pathogens Secrete Enzymes That Degrade and Inactivate Cytokines/Chemokines*

It should be mentioned in this context that the cluster of pathogens that inhabit the deep periodontal pockets in periodontitis secrete proteolytic enzymes, in order to provide for their sustenance. These proteolytic enzymes have a dramatic effect on

Table 2 Degradation of cytokines by *P. gingivalis* in serum

Cytokine degraded	Enzyme responsible	Reference
IFN-γ, TNF-α, IL-6, IL-8, IL-12	RgpA or RgpB	[148, 152, 155–157]
IFN-γ, TNF-α, RANTES, IL-12	Kgp	[148, 152, 155, 157]
VCAM-1, VEGF, IL-17, IL-1β, IL-1ra, CD14	LIVE *P. gingivalis*: not clear which if any gingipains	[153, 155, 158–160]

the immune detection and functionality of the cytokines that permeate tissues and GCF. This suggests that infection with species such as *P. gingivalis* and other proteolytic species impairs our ability to accurately quantitate these cytokines. More importantly, these proteases impair the immune response and wound healing required for disease resolution. *P. gingivalis* has three promiscuous proteinases, termed gingipains, that cleave natural or synthetic substrates after arginine (RgpA, RgpB) or lysine (Kgp) residues [148–153]. These gingipains are associated with the bacterium and get secreted into the extracellular matrix or biofilm in a more soluble form. The gingipains are responsible for 85% of all general proteolytic activity of *P. gingivalis* and 100% of its "trypsin-like activity" [154]. The gingipains' hemagglutinin activity is not the only adhesive role of these gingipains; they are also implicated in binding to connective tissue components like fibronectin and fibrinogen followed by the subsequent degradation of these proteins [154]. The gingipains have also been associated with degradation of host extracellular matrix components such as laminin, fibronectin, and collagen type III, IV and V [154].

The gingipains are also active in degrading cytokines that are either associated with the bacterium or present in the biofilm. So how does *P. gingivalis* make use of these gingipains to mediate progression of infection? Table 2 is a review of the various cytokine substrates of the gingipains already described in the literature. As seen in Table 2, most of the cytokines that are digested by the gingipains are proinflammatory cytokines. One must also consider that *P. gingivalis* is able to degrade the complement peptides C3 and C5 with its RgpA and RgpB gingipains [158, 161]. This degradation produces a functional C5a and short lived C3a chemotactic factors. The gingipains also target the C5aR receptors from the PMN surface for degradation [161]. Overall this suggests that the organism may benefit from recruitment of activated PMN's to the site of infection.

6 Conclusion

In view of the fact that diabetes mellitus and periodontitis share many features consistent with chronic inflammatory disorders, it may be prudent to evaluate the biomarkers common to these diseases, to determine of they can be used to evaluate a patient's disease status and better design intervention strategies. Studies in diabetic patients have indicated that the ex vivo quantification of gelatinase activity in serum and the expression of recoverin in mononuclear blood cells were useful for the prediction of active proliferative retinopathy, suggesting that diabetic patients might be monitored for the expressions of these molecules for more effective timing of appropriate therapy [162]. Studies conducted to date suggest that gingival crevicular fluid may be an acceptable substitute for "finger stick" blood for determination of serum glucose levels and may therefore be a useful screening tool in the dental office. Although considerable effort has been applied in the attempt to identify an immunoinflammatory biomarker for these diseases, the ideal biomarker has yet to be found. Some of the promising inflammatory cytokines and growth factors that are expressed in the GCF as a part of the host response and that have been shown to be elevated in both periodontitis and diabetes include IL-1β, IL-6, PGE_2, and VEGF. However, none of these agents have been shown to be a valid biomarker due to inconclusive, conflicting or insufficient data. Although we have known for some time that the presence of these cytokines are increased in the GCF during infection and inflammation, it has been hard to quantify this association in terms of determining the extent to which the quantity of the expressed cytokine goes up or down with changes in inflammation. Further research is necessary to clarify the usefulness of these potential biomarkers and to likewise investigate additional cytokines which have yet to be studied in this regard.

References

1. Biomarkers Definitions Working Group (2001) Biomarkers and surrogate endpoints: preferred definitions and conceptual framework. Clin Pharmacol Ther 69:89–95
2. Mealey B, Rees T, Rose L (2004) Systemic factors impacting the periodontium. In: Rose L, Mealey B (eds) Periodontics medicine, surgery, and implants. Elsevier Mosby, St. Louis
3. Löe H (1993) Periodontal disease: the sixth complication of diabetes mellitus. Diabetes Care 16(Suppl 1):329–334
4. Mealey BL, Oates TW, American Academy of Periodontology (2006) Diabetes mellitus and periodontal diseases. J Periodontol 77:1289–1303
5. Papapanou PN (1996) Periodontal diseases: epidemiology. Ann Periodontol 1:1–36
6. Bacić M, Plancak D, Granić M (1988) CPITN assessment of periodontal status in diabetic patients. J Periodontol 59:816–822
7. Emrich LJ, Shlossman M, Genco RJ (1991) Periodontal disease in non-insulin dependent diabetes mellitus. J Periodontol 62:123–131
8. Shlossman M, Knowler WC, Pettitt DJ, Genco RJ (1990) Type 2 diabetes mellitus and periodontal disease. J Am Dent Assoc 121:532–536
9. Tervonen T, Oliver RC (1993) Long-term control of diabetes mellitus and periodontitis. J Clin Periodontol 20:431–435
10. Gusberti FA, Syed SA, Bacon G, Grossman N, Loesche WJ (1983) Puberty gingivitis in insulin-dependent diabetic children. I. Cross-sectional observations. J Periodontol 54:714–720
11. Sastrowijoto SH, van der Velden U, van Steenbergen TJ, Hillemans P, Hart AA, de Graaff J, Abraham-Inpijn L (1990) Improved metabolic control, clinical periodontal status, and subgingival microbiology in insulin-dependent diabetes mellitus. A prospective study. J Clin Periodontol 17:233–242
12. Karjalainen KM, Knuuttila ML (1996) The onset of diabetes and poor metabolic control increases gingival bleeding in children and adolescents with insulin-dependent diabetes mellitus. J Clin Periodontol 23:1060–1067

13. Ervasti T, Knuuttila M, Pohjamo L, Haukipuro K (1985) Relation between control of diabetes and gingival bleeding. J Periodontol 56:154–157
14. Cutler CW, Machen RL, Jotwani R, Iacopino AM (1999) Heightened gingival inflammation and attachment loss in type 2 diabetics with hyperlipidemia. J Periodontol 70:1313–1321
15. Lalla E, Park DB, Papapanou PN, Lamster IB (2004) Oral disease burden in Northern Manhattan patients with diabetes mellitus. Am J Public Health 94:755–758
16. Lalla E, Cheng B, Lal S, Kaplan S, Softness B, Greenberg E, Goland RS, Lamster IB (2007) Diabetes-related parameters and periodontal conditions in children. J Periodontal Res 42:345–349
17. Guzman S, Karima M, Wang HY, Van Dyke TE (2003) Association between interleukin-1 genotype and periodontal disease in a diabetic population. J Periodontol 74:1183–1190
18. Taylor GW, Burt BA, Becker MP, Genco RJ, Shlossman M, Knowler WC, Pettitt DJ (1998) Non-insulin dependent diabetes mellitus and alveolar bone loss progression over 2 years. J Periodontol 69:76–83
19. Westfelt E, Rylander H, Blohmé G, Jonasson P, Lindhe J (1996) The effect of periodontal therapy in diabetics. Results after 5 years. J Clin Periodontol 23:92–100
20. Karjalainen KM, Knuuttila ML, von Dickhoff KJ (1994) Association of the severity of periodontal disease with organ complications in type 1 diabetic patients. J Periodontol 65:1067–1072
21. Tervonen T, Karjalainen K (1997) Periodontal disease related to diabetic status. A pilot study of the response to periodontal therapy in type 1 diabetes. J Clin Periodontol 24:505–510
22. Ryder M (2004) Anatomy, development, and physiology of the periodontium. In: Rose L, Mealey B (eds) Periodontics medicine, surgery, and implants. Elsevier Mosby, St. Louis
23. Grant DA, Stern IB, Everett FG (1972) Orban's periodontics: a concept – theory and practice. Mosby, St. Louis
24. Page RC, Offenbacher S, Schroeder HE, Seymour GJ, Kornman KS (1997) Advances in the pathogenesis of periodontitis: summary of developments, clinical implications and future directions. Periodontol 2000 14:216–248
25. Page RC, Schroeder HE (1976) Pathogenesis of inflammatory periodontal disease. A summary of current work. Lab Invest 34:235–249
26. Oliver RC, Brown LJ, Löe H (1998) Periodontal diseases in the United States population. J Periodontol 69:269–278
27. Lindhe J, Okamoto H, Yoneyama T, Haffajee A, Socransky SS (1989) Longitudinal changes in periodontal disease in untreated subjects. J Clin Periodontol 16:662–670
28. Kornman KS (2001) Patients are not equally susceptible to periodontitis. Does this change dental practice and dental curriculum? J Dent Educ 65:777–784
29. Machtei EE, Hausmann E, Dunford R, Grossi S, Ho A, Davis G, Chandler J, Zambon J, Genco RJ (1999) Longitudinal study of predictive factors for periodontal disease and tooth loss. J Clin Periodontol 26:374–380
30. Burt B, Research, Science and Therapy Committee of the American Academy of Periodontology (2005) American academy of periodontology position paper: epidemiology of periodontal diseases. J Periodontol 76:1406–1419
31. Lang NP, Tonetti MS, Suter J, Sorrell J, Duff GW, Kornman KS (2000) Effect of interleukin-1 gene polymorphisms on gingival inflammation assessed by bleeding on probing in a periodontal maintenance population. J Periodontal Res 35:102–107
32. Müller HP, Stadermann S, Heinecke A (2002) Longitudinal association between plaque and gingival bleeding in smokers and non-smokers. J Clin Periodontol 29:287–294
33. Deinzer R, Forster P, Fuck L, Herforth A, Stiller-Winkler R, Idel H (1999) Increase of crevicular interleukin 1- beta under academic stress at experimental gingivitis sites and at sites of perfect oral hygiene. J Clin Periodontol 26:1–8
34. Baelum V (1987) Pattern of periodontal breakdown in adult Tanzanians. Scand J Dent Res 95:221–228
35. Baelum V, Fejerskov O, Manji F (1988) Periodontal diseases in adult Kenyans. J Clin Periodontol 15:445–452

36. Jotwani R, Palucka AK, Al-Quotub M, Nouri-Shirazi M, Kim J, Bell D, Banchereau J, Cutler CW (2001) Mature dendritic cells infiltrate the T cell-rich region of oral mucosa in chronic periodontitis: in situ, in vivo, and in vitro studies. J Immunol 167(8):4693–4700
37. Jotwani R, Muthukuru M, Cutler CW (2004) Increase in HIV receptors/co-receptors/alpha-defensins in inflamed human gingiva. J Dent Res 83:371–377
38. Cutler CW, Jotwani R (2006) Dendritic cells at the oral mucosal interface. J Dent Res 85:678–689
39. Zambon JJ, Reynolds H, Fisher JG, Shlossman M, Dunford R, Genco RJ (1988) Microbiological and immunological studies of adult periodontitis in patients with non-insulin dependent diabetes mellitus. J Periodontol 59:23–31
40. Tervonen T, Knuuttila M (1986) Relation of diabetes control to periodontal pocketing and alveolar bone level. Oral Surg Oral Med Oral Pathol 61:346–349
41. Safkan-Seppälä B, Ainamo J (1992) Periodontal conditions in insulin-dependent diabetes mellitus. J Clin Periodontol 19:24–29
42. Seppälä B, Seppälä M, Ainamo J (1993) A longitudinal study on insulin-dependent diabetes mellitus and periodontal disease. J Clin Periodontol 20:161–165
43. Seppälä B, Ainamo J (1994) A site-by-site follow-up study on the effect of controlled versus poorly controlled insulin-dependent diabetes mellitus. J Clin Periodontol 21:161–165
44. Tsai C, Hayes C, Taylor GW (2002) Glycemic control of type 2 diabetes and severe periodontal disease in the U.S. adult population. Community Dent Oral Epidemiol 30:182–192
45. Taylor GW, Burt BA, Becker MP, Genco RJ, Shlossman M (1998) Glycemic control and alveolar bone loss progression in type 2 diabetes. Ann Periodontol 3:30–39
46. Sastrowijoto SH, Hillemans P, van Steenbergen TJ, Abraham-Inpijn L, de Graaff J (1989) Periodontal condition and microbiology of healthy and diseased periodontal pockets in type 1 diabetes mellitus patients. J Clin Periodontol 16:316–322
47. Sbordone L, Ramaglia L, Barone A, Ciaglia RN, Iacono VJ (1998) Periodontal status and subgingival microbiota of insulin-dependent juvenile diabetics: a 3-year longitudinal study. J Periodontol 69:120–128
48. Manouchehr-Pour M, Spagnuolo PJ, Rodman HM, Bissada NF (1981) Comparison of neutrophil chemotactic response in diabetic patients with mild and severe periodontal disease. J Periodontol 52:410–415
49. McMullen JA, Van Dyke TE, Horoszewicz HU, Genco RJ (1981) Neutrophil chemotaxis in individuals with advanced periodontal disease and a genetic predisposition to diabetes mellitus. J Periodontol 52:167–173
50. Schmidt AM, Yan SD, Wautier JL, Stern D (1999) Activation of receptor for advanced glycation end products. A mechanism for chronic vascular dysfunction in diabetic vasculopathy and atherosclerosis. Circ Res 84:489–497
51. Schmidt AM, Hori O, Cao R, Yan SD, Brett J, Wautier JL, Ogawa S, Kuwabara K, Matsumoto M, Stern D (1996) RAGE: a novel cellular receptor for advanced glycation end products. Diabetes 45(Suppl 3):S77–S80
52. Loder RT (1988) The influence of diabetes mellitus on the healing of closed fractures. Clin Orthop 232:210–216
53. Kumeda Y, Emoto M, Kawagishi T, Ishimura E, Nakatsuka K, Shioi A, Morii H (1999) Poor glycemic control impairs the response of biochemical parameters of bone formation and resorption to exogenous 1,25-dihydroxyvitamin D3 in patients with type 2 diabetes. Osteoporos Int 9:525–531
54. Tisdel CL, Marcus RE, Heiple KG (1995) Triple arthrodesis for diabetic peritalar neuroarthropathy. Foot Ankle Int 16:332–338
55. White CB, Turner NS, Lee GC, Haidukewych GJ (2003) Open ankle fractures in patients with diabetes mellitus. Clin Orthop 414:37–44
56. Beam HA, Parsons JR, Lin SS (2002) The effects of blood glucose control upon fracture healing in the BB Wistar rat with diabetes mellitus. J Orthop Res 20:1210–1216
57. Gooch HL, Hale JE, Fujioka H, Balian G, Hurwitz SR (2000) Alterations of cartilage and collagen expression during fracture healing in experimental diabetes. Connect Tissue Res 41:81–85

58. Lu H, Kraut D, Gerstenfeld LC, Graves DT (2003) Diabetes interferes with bone formation by affecting the expression of transcription factors that regulate osteoblast differentiation. Endocrinology 144:346–352
59. He H, Liu R, Desta T, Leone C, Gerstenfeld LC, Graves DT (2004) Diabetes causes decreased osteoclastogenesis, reduced bone formation, and enhanced apoptosis of osteoblastic cells in bacteria stimulated bone loss. Endocrinology 145:447–452
60. Liu R, Desta T, He H, Graves DT (2004) Diabetes alters the response to bacteria by enhancing fibroblast apoptosis. Endocrinology 145:2997–3003
61. Liu R, Bal HS, Desta T, Krothapalli N, Alyassi M, Luan Q, Graves DT (2006) Diabetes enhances periodontal bone loss through enhanced resorption and diminished bone formation. J Dent Res 85:510–514
62. Nishimura F, Takahashi K, Kurihara M, Takashiba S, Murayama Y (1998) Periodontal disease as a complication of diabetes mellitus. Ann Periodontol 3:20–29
63. Schmidt AM, Weidman E, Lalla E, Yan SD, Hori O, Cao R, Brett JG, Lamster IB (1996) Advanced glycation end products (AGEs) induce oxidant stress in the gingiva: a potential mechanism underlying accelerated periodontal disease associated with diabetes. J Periodontal Res 31:508–515
64. Paques M, Massin P, Gaudric A (1997) Growth factors and diabetic retinopathy. Diabetes Metab 23:125–130
65. Chiarelli F, Santilli F, Mohn A (2000) Role of growth factors in the development of diabetic complications. Horm Res 53:53–67
66. Unlü F, Güneri PG, Hekimgil M, Yeşilbek B, Boyacioğlu H (2003) Expression of vascular endothelial growth factor in human periodontal tissues: comparison of healthy and diabetic patients. J Periodontol 74:181–187
67. Taylor GW, Burt BA, Becker MP, Genco RJ, Shlossman M, Knowler WC, Pettitt DJ (1996) Severe periodontitis and risk for poor glycemic control in patients with non-insulin-dependent diabetes mellitus. J Periodontol 67(10 Suppl):1085–1093
68. Grossi SG, Genco RJ (1998) Periodontal disease and diabetes mellitus: a two-way relationship. Ann Periodontol 3:51–61
69. Lalla E, Lamster IB, Drury S, Fu C, Schmidt AM (2000) Hyperglycemia, glycoxidation and receptor for advanced glycation end products: potential mechanisms underlying diabetic complications, including diabetes-associated periodontitis. Periodontol 2000 23:50–62
70. Soskolne WA, Klinger A (2001) The relationship between periodontal diseases and diabetes: an overview. Ann Periodontol 6:91–98
71. Taylor GW (2001) Bidirectional interrelationships between diabetes and periodontal diseases: an epidemiologic perspective. Ann Periodontol 6(1):99–112
72. Grossi SG, Skrepcinski FB, DeCaro T, Robertson DC, Ho AW, Dunford RG, Genco RJ (1997) Treatment of periodontal disease in diabetics reduces glycated hemoglobin. J Periodontol 68:713–719
73. Kiran M, Arpak N, Unsal E, Erdoğan MF (2005) The effect of improved periodontal health on metabolic control in type 2 diabetes mellitus. J Clin Periodontol 32:266–272
74. Faria-Almeida R, Navarro A, Bascones A (2006) Clinical and metabolic changes after conventional treatment of type 2 diabetic patients with chronic periodontitis. J Periodontol 77:591–598
75. Stewart JE, Wager KA, Friedlander AH, Zadeh HH (2001) The effect of periodontal treatment on glycemic control in patients with type 2 diabetes mellitus. Clin Periodontol 28:306–310
76. Salvi GE, Brown CE, Fujihashi K, Kiyono H, Smith FW, Beck JD, Offenbacher S (1998) Inflammatory mediators of the terminal dentition in adult and early onset periodontitis. J Periodontol 33:212–225
77. Iehara N, Takeoka H, Yamada Y, Kita T, Doi T (1996) Advanced glycation end products modulate transcriptional regulation in mesangial cells. Kidney Int 50:1166–1172
78. Brownlee M (1995) The pathological implications of protein glycation. Clin Invest Med 18:275–281

79. Tanimoto K, Ohno S, Fujimoto K, Honda K, Ijuin C, Tanaka N, Doi T, Nakahara M, Tanne K (2001) Proinflammatory cytokines regulate the gene expression of hyaluronic acid synthetase in cultural rabbit synovial membrane cells. Connect Tissue Res 42:187–196
80. Lalla E, Lamster IB, Stern DM, Schmidt AM (2001) Receptor for advanced glycation end products, inflammation, and accelerated periodontal disease in diabetes: mechanisms and insights into therapeutic modalities. Ann Periodontol 6:113–118
81. Tan KC, Chow WS, Tam S, Bucala R, Betteridge J (2004) Association between acute-phase reactants and advanced glycation end products in type 2 diabetes. Diabetes Care 27: 223–228
82. Jambou D, Dejour N, Bayer P, Poirée JC, Frédenrich A, Issa-Sayegh M, Adjovi-Desouza M, Lapalus P, Harter M (1993) Effect of human native low-density and high-density lipoproteins on prostaglandin production by mouse macrophage cell line P338D1: possible implication in pathogenesis of atherosclerosis. Biochim Biophys Acta 1168:115–121
83. Salbach PB, Specht E, von Hodenberg E, Kossmann J, Janssen-Timmen U, Schneider WJ, Hugger P, King WC, Glomset JA, Habenicht AJ (1992) Differential low density lipoprotein receptor-dependent formation of eicosanoids in human blood-derived monocytes. Proc Natl Acad Sci USA 89:2439–2443
84. Offenbacher S (1996) Periodontal diseases: pathogenesis (review). Ann Periodontol 1:821–878
85. Schmidt AM, Hori O, Brett J, Yan SD, Wautier JL, Stern D (1994) Cellular receptors for advanced glycation end products. Implications for induction of oxidant stress and cellular dysfunction in the pathogenesis of vascular lesions. Arterioscler Thromb 14:1521–1528
86. Raynes JG (1994) The acute phase response. Biochem Soc Trans 22:69–74
87. Paz K, Hemi R, LeRoith D, Karasik A, Elhanany E, Kanety H, Zick Y (1997) A molecular basis for insulin resistance–elevated serine/threonin phosphorylation of IRS-1 and IRS-2 inhibits their binding to the juxtamembrane region of the insulin receptor and impairs their ability to undergo insulin-induced tyrosine phosphorylation. J Biol Chem 272:29911–29918
88. Yoshimura A, Kaneko T, Kato Y, Golenbock DT, Hara Y (2002) Lipopolysaccharides from periodontopathic bacteria *Porphyromonas gingivalis* and *Capnocytophaga ochracea* are antagonists for human Toll-like receptor 4. Infect Immun 70:218–225
89. Ogawa T, Asai Y, Hashimoto M, Takeuchi O, Kurita T, Yoshikai Y, Miyake K, Akira S (2002) Cell activation by *Porphyromonas gingivalis* lipid A molecule through Toll-like receptor 4- and myeloid differentiation factor 88-dependent signaling pathway. Int Immunol 14:1325–1332
90. Streja D, Cressey P, Rabkin SW (2003) Associations between inflammatory markers, traditional risk factors, and complications in patients with type 2 diabetes mellitus. J Diabetes Complications 17:120–127
91. Black PH (2003) The inflammatory response in an integral part of the stress response: implications for atherosclerosis, insulin resistance, type 2 diabetes and metabolic syndrome X. Brain Behav Immun 17:350–364
92. Soory M (2002) Hormone mediation of immune responses in the progression of diabetes, rheumatoid arthritis and periodontal diseases. Curr Drug Targets Immune Endocr Metabol Disord 2:13–25
93. Suppipat N, Suppipat N (1977) Evaluation of an electronic device for gingival fluid quantitation. J Periodontol 48:388–394
94. Hinrichs JE, Bandt CL, Smith JA, Golub LM (1984) A comparison of 3 systems for quantifying gingival crevicular fluid with respect to linearity and the effects of qualitative differences in fluids. J Clin Periodontol 11:652–661
95. Williams R, Pacquette D (1997) Advances in periodontal diagnosis. In: Lindhe J, Karring T, Lang NP (eds) Clinical periodontology and implant dentistry, 3rd edn. Munksgaard, Copenhagen
96. Offenbacher S, Farr DH, Goodson JM (1981) Measurement of prostaglandin E in crevicular fluid. J Clin Periodontol 8:359–367

97. Lamster IB, Oshrain RL, Harper DS, Celenti RS, Hovliaras CA, Gordon JM (1988) Enzyme activity in crevicular fluid for detection and prediction of clinical attachment loss in patients with chronic adult periodontitis. Six month results. J Periodontol 59:516–523
98. Palcanis KG, Larjava IK, Wells BR, Suggs KA, Landis JR, Chadwick DE, Jeffcoat MK (1992) Elastase as an indicator of periodontal disease progression. J Periodontol 63: 237–242
99. Kinane DF (2000) Regulators of tissue destruction and homeostasis as diagnostic aids in periodontology. Periodontol 2000 2000(24):215–225
100. Embery G, Waddington RJ, Hall RC, Last KS (2000) Connective tissue elements as diagnostic aids in periodontology. Periodontology 2000 24:193–214
101. Strauss SM, Wheeler AJ, Russell SL, Brodsky A, Davidson RM, Gluzman R, Li L, Malo RG, Salis B, Schoor R, Tzvetkova K (2009) The potential use of gingival crevicular blood for measuring glucose to screen for diabetes: an examination based on characteristics of the blood collection site. J Periodontol 80:907–914
102. Yamaguchi M, Kambe S, Yamazaki K, Kobayashi M (2005) Error grid analysis of noninvasive glucose monitoring via gingival crevicular fluid. IEEE Trans Biomed Eng 52:1796–1798
103. Yamaguchi M, Takada R, Kambe S, Hatakeyama T, Naitoh K, Yamazaki K, Kobayashi M (2005) Evaluation of time-course changes of gingival crevicular fluid glucose levels in diabetics. Biomed Microdevices 7:53–58
104. Yamaguchi M, Kawabata Y, Kambe S, Wårdell K, Nystrom FH, Naitoh K, Yoshida H (2004) Non-invasive monitoring of gingival crevicular fluid for estimation of blood glucose level. Med Biol Eng Comput 42:322–327
105. Khader YS, Al-Zu'bi BN, Judeh A, Rayyan M (2006) Screening for type 2 diabetes mellitus using gingival crevicular blood. Int J Dent Hyg 4:179–182
106. Beikler T, Kuczek A, Petersilka G, Flemmig TF (2002) In-dental-office screening for diabetes mellitus using gingival crevicular blood. J Clin Periodontol 29:216–218
107. Müller H, Behbehani E (2004) Screening of elevated glucose levels in gingival crevice blood using a novel, sensitive self-monitoring device. Med Princ Pract 13:361–365
108. Müller H, Behbehani E (2005) Methods for measuring agreement: glucose levels in gingival crevice blood. Clin Oral Investig 9:65–69
109. Ardakani MR, Moeintaghavi A, Haerian A, Ardakani MA, Hashemzadeh M (2009) Correlation between levels of sulcular and capillary blood glucose. J Contemp Dent Pract 10:10–17
110. Ficara AJ, Levin MP, Grower MF, Kramer GD (1975) A comparison of the glucose and protein content of gingival crevicular fluid from diabetics and nondiabetics. J Periodontal Res 10:171–175
111. Parker RC, Rapley JW, Isley W, Spencer P, Killoy WJ (1993) Gingival crevicular blood for assessment of blood glucose in diabetic patients. J Periodontol 64(7):666–672
112. Ciantar M, Spratt DA, Newman HN, Wilson M (2002) Development of an in vitro microassay for glucose quantification in submicrolitre volumes of biological fluid. J Periodontal Res 37:79–85
113. Safkan-Seppälä B, Sorsa T, Tervahartiala T, Beklen A, Konttinen YT (2006) Collagenases in gingival crevicular fluid in type 1 diabetes mellitus. J Periodontol 77:189–194
114. Gursoy UK, Marakoglu I, Ersan S (2006) Periodontal status and cytoplasmic enzyme activities in gingival crevicular fluid of type 2 diabetic and/or obese patients with chronic periodontitis. J Int Acad Periodontol 8:2–5
115. Engebretson SP, Hey-Hadavi J, Ehrhardt FJ, Hsu D, Celenti RS, Grbic JT, Lamster IB (2004) Gingival crevicular fluid levels of interleukin-1β and glycemic control in patients with chronic periodontitis and type 2 diabetes. J Periodontol 75:1203–1208
116. Hou LT, Liu CM, Rossomando EF (1995) Crevicular interleukin-1 beta in moderate and severe periodontitis patients and the effect of phase I periodontal treatment. J Clin Periodontol 22:162–167
117. Tsai CC, Ho YP, Chen CC (1995) Levels of interleukin-1 beta and interleukin-8 in gingival crevicular fluids in adult periodontitis. J Periodontol 66:852–859

118. Orozco A, Gemmell E, Bickel M, Seymour GJ (2006) Interleukin-1beta, interleukin-12 and interleukin-18 levels in gingival fluid and serum of patients with gingivitis and periodontitis. Oral Microbiol Immunol 21:256–260
119. Salvi GE, Beck JD, Offenbacher S (1998) PGE_2, IL-1 β, and TNF-α responses in diabetics as modifiers of periodontal disease expression. Ann Periodontol 3:40–50
120. Bulut U, Develioglu H, Taner IL, Berker E (2001) Interleukin-1 beta levels in gingival crevicular fluid in type 2 diabetes mellitus and adult periodontitis. J Oral Sci 43:171–177
121. Engebretson SP, Grbic JT, Singer R, Lamster IB (2002) GCF IL-1b profiles in periodontal disease. J Clin Periodontol 29:48–53
122. Zhong Y, Slade GD, Beck JD, Offenbacher S (2007) Gingival crevicular fluid interleukin-1beta, prostaglandin E_2 and periodontal status in a community population. J Clin Periodontol 34:285–293
123. Andriankaja OM, Barros SP, Moss K, Panagakos FS, DeVizio W, Beck J, Offenbacher S (2009) Levels of serum interleukin (IL)-6 and gingival crevicular fluid of IL-1β and prostaglandin E_2 among non-smoking subjects with gingivitis and type 2 diabetes. J Periodontol 80:307–316
124. Kardeşler L, Buduneli N, Biyikoğlu B, Cetinkalp S, Kütükçüler N (2008) Gingival crevicular fluid PGE_2, IL-1β, t- PA, PAI-2 levels in type 2 diabetes and relationship with periodontal disease. Clin Biochem 41:863–868
125. Masada MP, Persson R, Kenney JS, Lee SW, Page RC, Allison AC (1990) Measurement of interleukin-1 alpha and -1 beta in gingival crevicular fluid: implications for the pathogenesis of periodontal disease. J Periodontal Res 25:156–163
126. Mathur A, Michalowicz B, Castillo M, Aeppli D (1996) Interleukin-1 alpha, interleukin-8 and interferon-alpha levels in gingival crevicular fluid. J Periodontal Res 31:489–495
127. Tsalikis L, Parapanisiou E, Bata-Kyrkou A, Polymenides Z, Konstantinidis A (2002) Crevicular fluid levels of interleukin-1alpha and interleukin-1beta during experimental gingivitis in young and old adults. J Int Acad Periodontol 4:5–11
128. Figueredo CM, Ribeiro MS, Fischer RG, Gustafsson A (1999) Increased interleukin-1β concentration in gingival crevicular fluid as a characteristic of periodontitis. J Periodontol 70:1457–1463
129. Gonzáles JR, Herrmann JM, Boedeker RH, Francz PI, Biesalski H, Meyle J (2001) Concentration of interleukin-1β and neutrophil elastase activity in gingival crevicular fluid during experimental gingivitis. J Clin Periodontol 28:544–549
130. Faizuddin M, Bharathi SH, Rohini NV (2003) Estimation of interleukin-1β levels in the gingival crevicular fluid in health and in inflammatory periodontal disease. J Periodontal Res 38:111–114
131. Giannopoulou C, Bernard JP, Buser D, Carrel A, Belser UC (2003) Effect of intracrevicular restoration margins on peri-implant health: clinical, biochemical, and microbiologic findings around esthetic implants up to 9 years. Int J Oral Maxillofac Implants 18:173–181
132. Schierano G, Pejrone G, Brusco P, Trombetta A, Martinasso G, Preti G, Canuto RA (2008) TNF-α, TGF-β2 and IL-1β levels in gingival and peri-implant crevicular fluid before and after de novo plaque accumulation. J Clin Periodontol 35:532–538
133. Kurtiş B, Develioğlu H, Taner IL, Baloş K, Tekin IO (1999) IL-6 levels in gingival crevicular fluid (GCF) from patients with non-insulin dependent diabetes mellitus (NIDDM), adult periodontitis and healthy subjects. J Oral Sci 41:163–167
134. Skaleric U, Kramar B, Petelin M, Pavlica Z, Wahl SM (1997) Changes inTGFβ1levels in gingiva, crevicular fluid and serum associated with periodontal inflammation in humans and dogs. Eur J Oral Sci 105:136–142
135. Buduneli N, Kütükçüler N, Aksu G, Atilla G (2001) Evaluation of transforming growth factor –β1 level in crevicular fluid of cyslosporin A-treated patients. J Periodontol 72:526–531
136. Wright HJ, Chapple IL, Matthews JB (2003) Levels of TGFβ1 in gingival crevicular fluid during a 21-day experimental model of gingivitis. Oral Dis 9:88–94
137. Kuru L, Griffiths GS, Petrie A, Olsen I (2004) Changes in transforming growth factor-β1 in gingival crevicular fluid following periodontal surgery. J Clin Periodontol 31:527–533

138. Gürkan A, Cinarcik S, Hüseyinov A (2005) Adjunctive subantimicrobial dose doxycycline: effect on clinical parameters and gingival crevicular fluid transforming growth factor b1levels in severe, generalized chronic periodontitis. J Clin Periodontol 32:244–253
139. Gürkan A, Emingil G, Cinarcik S, Berdeli A (2006) Gingival crevicular fluid transforming growth factor-β1 in several forms of periodontal disease. Arch Oral Biol 51:906–912
140. Güneri P, Unlü F, Yeşilbek B, Bayraktar F, Kokuludağ A, Hekimgil M, Boyacioğlu H (2004) Vascular endothelial growth factor in gingival tissues and crevicular fluids of diabetic and healthy periodontal patients. J Periodontol 75:91–97
141. Prapulla DV, Sujatha PB, Pradeep AR (2007) Gingival crevicular fluid VEGF levels in periodontal health and disease. J Periodontol 78:1783–1787
142. Sakallioğlu EE, Aliyev E, Lütfioğlu M, Yavuz U, Açikgöz G (2007) Vascular endothelial growth factor (VEGF) levels of gingiva and gingival crevicular fluid in diabetic and systemically healthy periodontitis patients. Clin Oral Investig 11:115–120
143. Heasman PA, Collins JG, Offenbacher S (1993) Changes in crevicular fluid levels of interleukin-1, beta leukotriene B4, prostaglandin E2, thromboxane B2 and tumour necrosis factor alpha in experimental gingivitis in humans. J Periodontal Res 28:241–247
144. Offenbacher S, Odle BM, Van Dyke TE (1986) The use of crevicular fluid prostaglandin E2 levels as a predictor of periodontal attachment loss. J Periodontal Res 21:101–112
145. Borg Andersson A, Birkhed D, Berntorp K, Lindgärde F, Matsson L (1998) Glucose concentration in parotid saliva after glucose/food intake in individuals with glucose intolerance and diabetes mellitus. Eur J Oral Sci 106:931–937
146. Arana C, Cutando A, Ferrera MJ, Gómez-Moreno G, Worf CV, Bolaños MJ, Escames G, Acuña-Castroviejo D (2006) Parameters of oxidative stress in saliva from diabetic and parenteral drug addict patients. J Oral Pathol Med 35:554–559
147. Gümüş P, Buduneli N, Cetinkalp S, Hawkins SI, Renaud D, Kinane DF, Scott DA (2009) Salivary antioxidants in patients with type 1 or type 2 diabetes and inflammatory periodontal disease: a case-control study. J Periodontol 80:1440–1446
148. Calkins CC, Platt K, Potempa J, Travis J (1998) Inactivation of tumor necrosis factor –alpha by proteinases (gingipains) from the periodontal pathogen. Porphyromonas gingivalis. Implications of immune evasion. J Biol Chem 273:6611–6614
149. Yonezawa H, Kato T, Kuramitsu HK, Okuda K, Ishihara K (2005) Immunization by Arg-gingipain A DNA vaccine protects mice against an invasive Porphyromonas gingivalis infection through regulation of interferon-gamma production. Oral Microbiol Immunol 20:259–266
150. Yun LW, Decarlo AA, Collyer C, Hunter N (2003) Enhancement of Th2 pathways and direct activation of B cells by the gingipains of Porphyromonas gingivalis. Clin Exp Immunol 134:295–302
151. Yun PL, DeCarlo AA, Collyer C, Hunter N (2002) Modulation of an interleukin-12 and gamma interferon synergistic feedback regulatory cycle of T-cell and monocyte co-cultures by Porphyromonas gingivalis lipopolysacchardie in the absence or presence of cysteine proteinases. Infect Immun 70:5695–5705
152. Yun PL, DeCarlo AA, Hunter N (1999) Modulation of major histocompatibility complex protein expression by human gamma interferon mediated by cysteine proteinase-adhesin polyproteins of Porphyromonas gingivalis. Infect Immun 67:2986–2995
153. Fletcher J, Reddi K, Poole S, Nair S, Henderson B, Tabona P, Wilson M (1997) Interactions between periodontopathogenic bacteria and cytokines. J Periodontal Res 32:200–205
154. Potempa J, Banbula A, Travis J (2000) Role of bacterial proteinases in matrix destruction and modulation of host responses. Periodontol 2000 24:153–192
155. Bodet C, Chandad F, Grenier D (2005) Modulation of cytokine production by porphyromonas gingivalis in a macrophage and epithelial cell co-culture model. Microbes Infect 7:448–456
156. Darveau RP, Belton CM, Reife RA, Lamont RJ (1998) Local chemokine paralysis, a novel pathogenic mechanism for Porphyromonas gingivalis. Infect Immun 66:1660–1665

157. Yun PL, Decarlo AA, Collyer C, Hunter N (2001) Hydrolysis of interleukin-12 by Porphyromonas gingivalis major cysteine proteinases may affect local gamma interferon accumulation and the Th1 or Th2 T-cell phenotype in periodontotis. Infect Immun 69:5650–5660
158. Teng YT (2006) Protective and destructive immunity in the periodontium: part 1- innate and humoral immunity and the periodontium. J Dent Res 85:198–208
159. Zhou Q, Desta T, Fenton M, Graves DT, Amar S (2005) Cytokine profiling of macrophages exposed to Porphyromonas gingivalis, its lipopolysaccharide, or its FimA protein. Infect Immun 73:935–943
160. Wilson M, Seymour R, Henderson B (1998) Bacterial perturbation of cytokine networks. Infect Immun 66:2401–2409
161. Jagels MA, Travis J, Potempa J, Pike R, Hugli TE (1996) Proteolytic inactivation of the leukocyte C5a receptor by proteinases and Porphyromonas gingivalis. Infect Immun 64:1984–1991
162. Golubnitschaja O, Jaksche A, Moenkemann H, Yeghiazaryan K, Karl SE, Trog D, Schild HH, Löffler KU (2005) Molecular imaging system for possible prediction of active retinopathy in patients with diabetes mellitus. Amino Acids 28:229–237

Cancer Predisposition in Diabetic Patients: Risk Assessment and Targeted Preventive Measures

Melanie Cebioglu, Hans H. Schild, and Olga Golubnitschaja

Abstract *Diabetes mellitus* (DM) is a lifelong progressive disease with high morbidity and mortality worldwide, in particular due to severe complications secondary to metabolic disorders. Increasing evidence indicates that diabetic patients are highly predisposed to a spectrum of oncologic diseases. Understanding of molecular pathomechanisms of the cancer-provocation in DM is of great importance for the properly organised diabetes care. This chapter focuses on the epidemiology and cancer aetiology in DM and provides expert recommendations for the targeted prevention in the diabetic patient cohort.

Keywords Cancer • Diabetes • Complications • Predictive, preventive and personalised medicine • Aetiology • Epidemiology • Advanced technologies • Molecular mechanism • Patient profile • Medical record • Biomarker pattern • Integrative medicine • Economy • Education • Healthcare

1 Diabetes Mellitus as the General Risk Factor for Almost All Kinds of Cancer with Ethnic- and Gender-Dependent Preferences

Diabetes mellitus (DM) is a group of metabolic disorders, mainly characterised by impaired glucose metabolism and consequent hyperglycaemia as the common feature. Whereas cardiovascular complications are well described for DM [1, 2], it is a

---

M. Cebioglu • H.H. Schild • O. Golubnitschaja (✉)
Department of Radiology, Rheinische Friedrich-Wilhelms-University of Bonn, Sigmund-Freud-Str. 25, Bonn D-53105, Germany

M.S. Mozaffari (ed.), *New Strategies to Advance Pre/Diabetes Care: Integrative Approach by PPPM*, Advances in Predictive, Preventive and Personalised Medicine 3,
DOI 10.1007/978-94-007-5971-8_13,

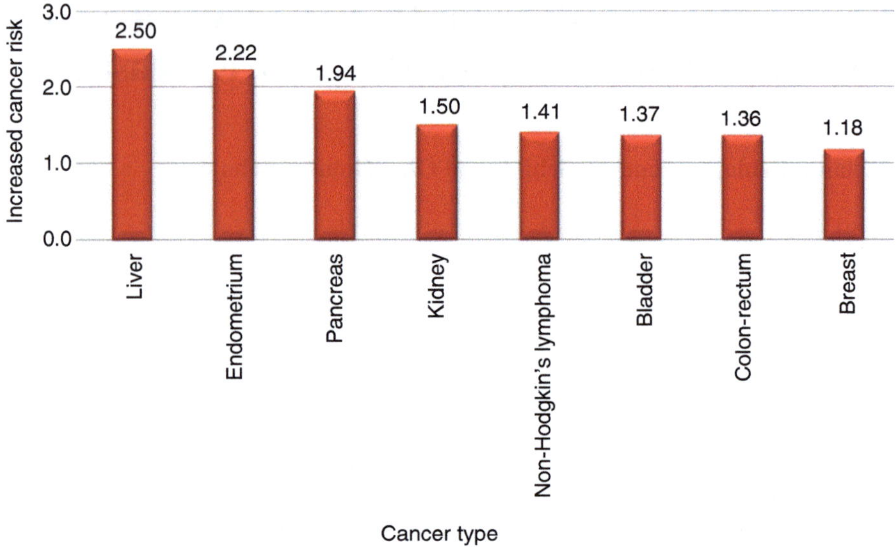

Fig. 1 Increased site-specific cancer risks of DM-patients compared to non-diabetic counterparts as registered worldwide in years 1999 until 2008 [3–12]

relatively new consideration that diabetic patients are highly predisposed to oncologic diseases. In spite of some ethnic-, gender-, and age-specific differences, significantly increased risks of liver, pancreas, bladder and digestive tract cancer types are worldwide recognised for DM-patients now (Fig. 1).

Series of population studies indicate increased risks for almost all kinds of cancer in the cohort of DM patients with ethnic- as well as gender-dependent preferences summarised here for the Japanese population (see Fig. 2).

2 The Ethnicity, Type of DM, Age and Gender as Factors Modifying the Cancer Risks in Diabetic Patients

With respect to single DM-types, population studies have been carried out in different countries showing geographic particularities in cancer incidence. Thus, the long-term monitoring of a cancer-specific incidence in the subpopulation of type 1 diabetic patients was performed in Sweden in the years 1965–1999 [14]. Noteworthy, as the particular risks in childhood, endometrial, buccal-cavity, bladder and cervix cancer types have been recognised as demonstrated in Fig. 3. For the type 1 diabetic patients aged above 14 years, the highest risks have been registered for cancer of stomach, endometrium, cervix, and bladder.

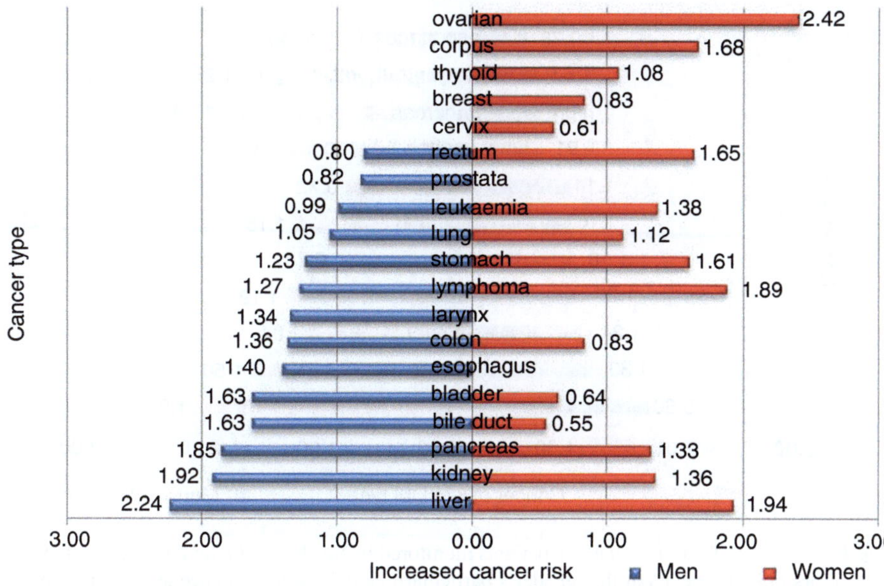

Fig. 2 Population studies in Japan to the general risk as an example of cancer predisposition of diabetic patients in years 1990–2003 [13]

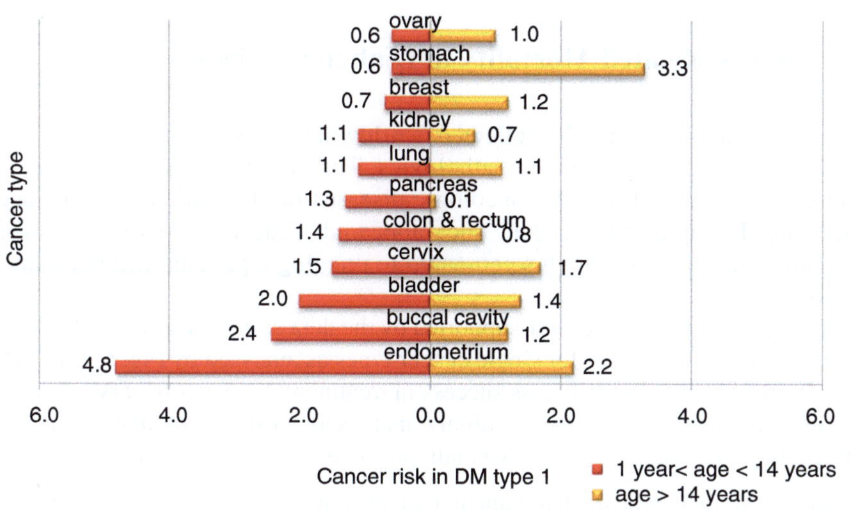

Fig. 3 The long-term population study performed for diabetes type 1 in Sweden with respect to the cancer predisposition in two age categories; the time-frame of the monitoring was 1965–1999 years [14]

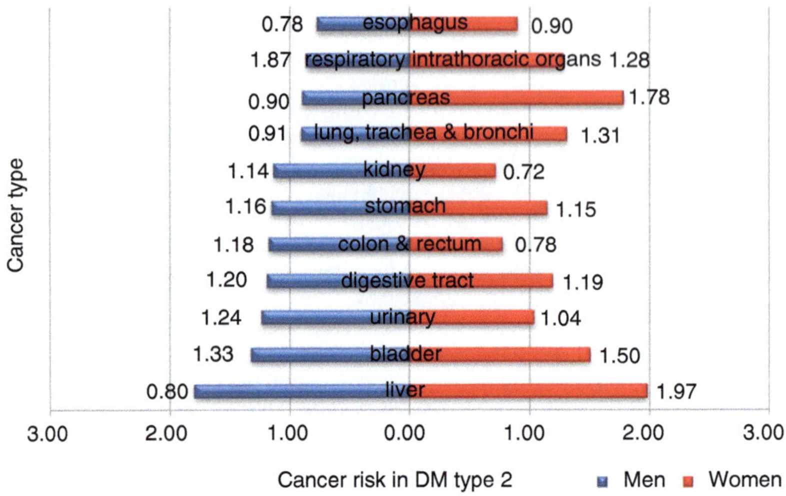

Fig. 4 Cancer risk for type 2 diabetic patients monitored in North Italy in the time-frame of 1987–1996 years [15]. It is evident that gender's particularities in cancer incidence are spread over all organs investigated with the exception of digestive tract with the incidence rated equal for both sexes

Similar studies carried out for type 2 diabetes demonstrate liver, pancreas and bladder cancer to be the leading types in terms of the incidence risk (Fig. 4).

3 Cancer-Related Mortality in Diabetic Patients

According to the currently collected worldwide statistics, every 10 sec one patient dies due to DM-related pathologies. Thereby, DM is the fourth leading cause of deaths [16]. For the oncologic diseases in DM patient-cohorts, accumulating data demonstrate that, once appeared, cancer outcomes have worse prognosis for diabetic patients compared to oncologic patients without diabetic history (Fig. 5).

This general figures are well-supported by the data collected specifically in USA to the treatment outcomes for single cancer types as analysed by Fig. 6. These data clearly indicate significantly less success in treatment of cancer for DM-patients.

Summarising the reasons for highly increased mortality of diabetic patients affected by oncologic diseases, two main problems should be pointed, namely

- Recent studies indicate that diabetic patients are generally predisposed to cancer development [7–9, 11, 13–15, 17–37];
- The hindered treatment of oncologic diseases in diabetes care.

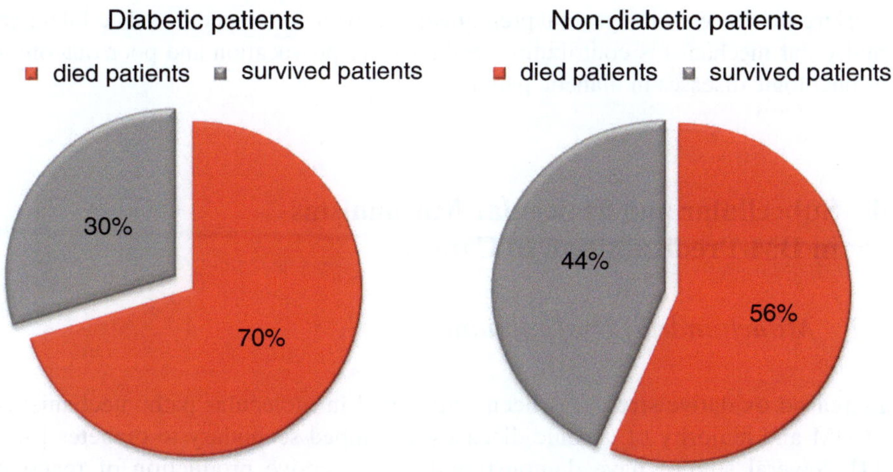

Fig. 5 Cancer related mortality within the cohort of diabetic patients compared to oncologic patients without DM-history as registered in the Netherlands during 8 years of monitoring in the years 1995–2002 [17]. Corresponding mortality risk is estimated to be at least 1.5-times high for treated diabetic patients compared to treated non-diabetic patients

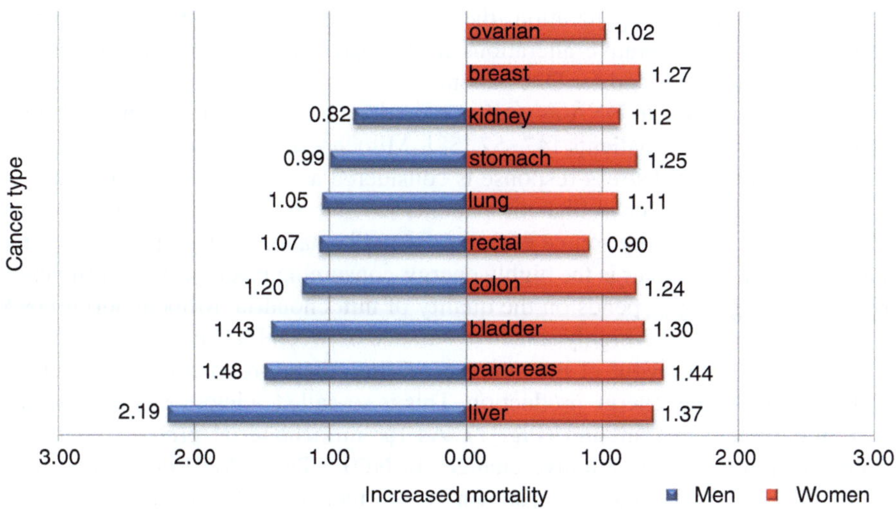

Fig. 6 Increased mortality of diabetic patients *versus* non-diabetic patients for single cancer types as documented for patients treated in USA in years 1982–1998 [18]

Here we overview the general prerequisites as well as the specific subcellular and molecular mechanisms contributing to the cancer-provocation and poor outcomes of oncologic diseases in diabetic patients.

4 Subcellular and Molecular Mechanisms in DM-Predisposition to Cancer

4.1 Mitochondrial Dysfunction

Increased oxidative stress has been implicated in molecular pathomechanisms of DM and majority of chronic diseases developed secondary to diabetes [38–44]. Several studies have demonstrated the excessive production of reactive oxygen species (ROS) to be the common feature of both diabetes type 1 and 2 [16, 37, 45–49]. Disturbed glucose/insulin homeostasis initiates overall cellular stress leading to severe and irreversible damage to subcellular structures [47, 49, 50].

Oxidative damage to DNA is well documented for cells isolated from diabetic patients [51–53]. These findings indicate an imbalance between the increased production of ROS and decreased DNA-repair capacity. Depending on a quality of cell-cycle controlling machinery, this imbalance can lead either to extensive apoptotic cell loss or proliferation of damaged cells [54]. Whereas the first process causes mainly tissue-degeneration, the latter predisposes to cancer development [55, 56]. Both degenerative alterations in damaged organs and predisposition to cancer have been reported for DM-patients.

Extensive damage to DNA as well as mutations has been demonstrated also for mitochondria in DM-patients [47, 57, 58]. Mitochondrial genetic background as well as mitochondrial stress-response is considered as an important contributor in predisposition to cancer in diabetic patients [57, 59, 60]. Electron transfer chain (ETC) is the main energy source essential for performance of all cellular functions. Particularly, DNA-repair is the highly energy-consuming machinery, the efficiency of which obligatory depends on the quality of mitochondria. Mitochondrial DNA variations may affect a highly sensitive balance between ETC-efficiency and production of ROS in favour of the latter increasing, therefore, mutagenic effects of ROS and decreasing energy production. This is so-called "vicious circle" resulting in mitochondrial dysfunction (Fig. 7). As for mitochondrial protein repertoire, both quantitative and qualitative changes in ND1-ND6 (NADH-dehydrogenases of the ETC-complex I) have been shown to contribute to the "vicious circle" resulting in premature ageing and multiple pathologies including neurodegeneration and cancer [61]. In patients with a history of diabetes, the mitochondrial "vicious circle" has been implicated specifically in individual predisposition to breast cancer [56, 62].

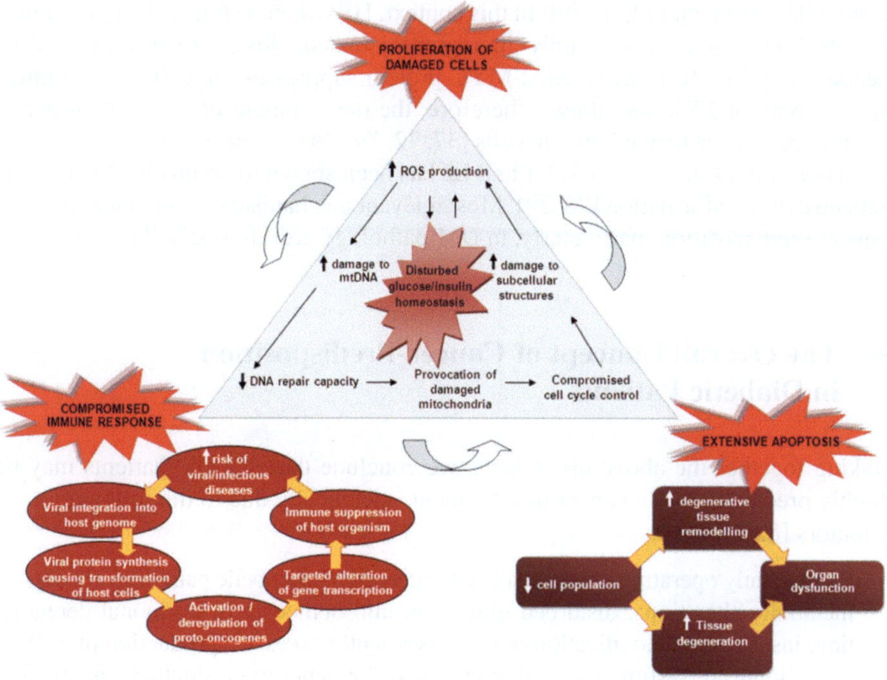

Fig. 7 Mitochondrial "vicious circle" causes a dangerous imbalance between highly increased production of ROS on one side and low energy production on the other side. Highly increased damage to chrDNA and mtDNA, remarkably decrease repair capacity as well as compromise cell-cycle control are direct consequences of stress-provoked mitochondrial dysfunction [37–41, 45, 47, 58, 61, 63–76]

5 Concept of Viral Aetiology in DM-Provoked Cancer

Patients with diabetic history are at significantly increased risk of infection [63–65]. There is a growing body of evidence for compromised immune response in this patient cohort as a consequence of metabolic syndrome [66–68]. This imbalance can lead to highly increased incidence of viral infection, further potentiating cancer risk in diabetic patients [68–70, 77].

It is already well-acknowledged now that more than 15% of viral infections are able to provoke cancer development in humans [78]. Hence, human papilloma virus (HPV) infection is attributed to 80% of all human cancer types and has been proposed to play a central role in molecular pathomechanisms of breast cancer [79–81]. Integration of viral particles in human genome frequently results in activation of several proto-oncogenes, which in turn trigger tumourigenic mechanisms in affected cells [37, 71–73, 79]. Thus, a viral activation of proto-oncogene c-MYC is well described in literature [79]. Thereby, a targeted integration of viral particles in human genome plays the crucial role for consequent cancer development [74,78, 79]. In this context, HPV demonstrates clear preference for its integration sites, mainly in regions situated closely to c-MYC coding sequences [79]. Once activated, c-MYC protein suppresses the cell-cycle controlling activity of P53, and allows, therefore, the development of new tumourigenic phenotype of transformed human cells [37, 72, 79]. In consensus, the activated synthesis of viral proteins E6, E7, E1 and E2 has been shown to be involved in cancer-related cell transformation [75, 79]. Most relevant mechanisms for viral aetiology of cancer predisposition, particularly, in DM-pathology are summarised in Fig. 7.

6 The Overall Concept of Cancer-Predisposition in Diabetic Patients

Taking together the above given facts, we conclude that diabetic patients may be highly predisposed to a cancer development, specifically due to the following contributors [82]:

- Permanently operating stress factors characteristic for diabetic pathology (excessive metabolic alterations, disturbed glucose/insulin homeostasis, hormonal deregulation, insufficient detoxification) with consequently excessive production of ROS;
- mitochondrial dysfunction with consequent low energy production, insufficient repair capacity and accumulating damage to both chromosomal and mitochondrial DNA;
- highly increased risk of infectious disorders with consequently induced (viral) proto-oncogenic activity as well as activity of particular pathogenic bacterial forms such as *Helicobacter pylori.*

Adequacy of stress response, repair capacity as well as immune defence are highly individual for each patient and strongly depend on risk factors such as genetic background, age, environmental factors, nutrition, body culture, life style, etc. [83, 84]. Figure 8 summarises the overall risk factors contributing to the individual outcomes in diabetes and secondary pathologies. Thus, varying breast cancer risk in different ethnic and social groups is well documented [32, 85–87]. Breast fat deposits and distribution increase risk for breast cancer in female and even in male patients [32, 88–92]. In contrast, breast cancer development is significantly reduced in people with regular physical activity [93]. Alcohol abuse and tobacco consumption are further contributors which remarkably increase risk of cancer development [33, 35, 94–98].

7 Outlook and Expert Recommendations

Current biotechnology possesses sufficient power to estimate the severity of damage to subcellular structures, individual stress reactions and repair capacity. For example, by stress proteome profiling in peripheral leukocytes and blood plasma, individual

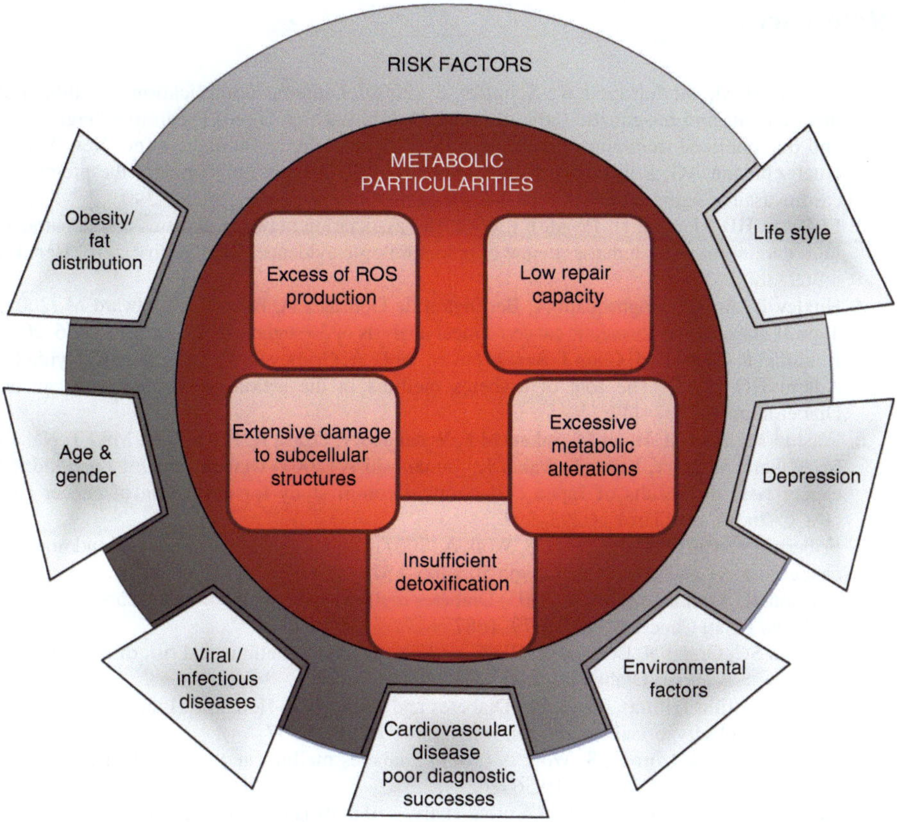

Fig. 8 Particularities of metabolic syndrome and factors contributing to cancer development. Corresponding references are listed below: Risk factors – Obesity/fat distribution [30, 32, 33, 88–92, 95, 97, 99–102], Age and gender [18, 20, 31, 86, 102–105], Viral/infectious diseases [38, 73, 78, 80, 81, 94, 95, 97, 106–108], Cardiovascular disease/poor diagnostic successes [1], Environmental factors [94, 95, 104, 109–111], Depression [112, 113], Life style [30, 33, 73, 85, 86, 92, 93, 95–99, 109, 114–121]; Metabolic particularities – Excess of ROS production [52, 53, 57, 71, 122–129], Extensive damage to subcellular structures [57, 58, 60, 109, 127–131], Insufficient detoxification [94, 109, 128, 129, 132], Excessive metabolic alterations [29, 30, 32–34, 37, 73, 89, 129, 133, 134], Low repair capacity [73, 114, 129, 135, 136]

stress reactions can be well estimated. Advanced predictive diagnostic approaches are currently close to clinical application and allow to select groups of risk and to estimate a predisposition to severe complications in diabetic patients [59, 137–139].

Much attention should be focused on preventive measures in diabetic healthcare, in order to restrict or even avoid severe secondary complications, such as cancer [140]. Potential groups of risk should be informed about good lifestyle choices. Nowadays, it is increasingly clear that a well-balanced individually created diet is considered as an effective preventive measure and treatment of majority of chronic complications in diabetic patients [141–143].

References

1. Mozaffari M, Abdelsayed R, Schaffer S (2009) Diabetic complications: pathogenic mechanisms and prognostic indicators. In: Golubnitschaja O (ed) Predictive diagnostics and personalized treatment: dream or reality? Nova Science Publishers Inc, New York
2. Huebschmann AG, Kohrt WM, Regensteiner JG (2011) Exercise attenuates the premature cardiovascular aging effects of type 2 diabetes mellitus. Vasc Med 16:378–390
3. El-Serag HB, Hampel H, Javadi F (2006) The association between diabetes and hepatocellular carcinoma: a systematic review of epidemiologic evidence. Clin Gastroenterol Hepatol 4:369–380
4. Huxley R, Ansary-Moghaddam A, Berrington de González A, Barzi F, Woodward M (2005) Type-II diabetes and pancreatic cancer: a meta-analysis of 36 studies. Br J Cancer 92:2076–2083
5. Lindblad P, Chow WH, Chan J, Bergström A, Wolk A, Gridley G, McLaughlin JK, Nyrén O, Adami HO (1999) The role of diabetes mellitus in the aetiology of renal cell cancer. Diabetologia 42:107–112
6. Washio M, Mori M, Khan M, Sakauchi F, Watanabe Y, Ozasa K, Hayashi K, Miki T, Nakao M, Mikami K, Ito Y, Kubo T, Wakai K, Tamakoshi A (2007) Diabetes mellitus and kidney cancer risk: the results of Japan collaborative cohort study for evaluation of cancer risk (JACC study). Int J Urol 14:393–397
7. Friberg E, Orsini N, Mantzoros CS, Wolk A (2007) Diabetes mellitus and risk of endometrial cancer: a meta-analysis. Diabetologia 50:1365–1374
8. Larsson SC, Orsini N, Wolk A (2005) Diabetes mellitus and risk of colorectal cancer: a meta-analysis. J Natl Cancer Inst 97:1679–1687
9. Larsson SC, Orsini N, Brismar K, Wolk A (2006) Diabetes mellitus and risk of bladder cancer: a meta-analysis. Diabetologia 49:2819–2823
10. Mitri J, Castillo J, Pittas AG (2008) Diabetes and risk of Non-Hodgkin's lymphoma: a meta-analysis of observational studies. Diabetes Care 31:2391–2397
11. Larsson SC, Mantzoros CS, Wolk A (2007) Diabetes mellitus and risk of breast cancer: a meta-analysis. Int J Cancer 121:856–862
12. Kasper JS, Giovannucci E (2006) A meta-analysis of diabetes mellitus and the risk of prostate cancer. Cancer Epidemiol Biomarkers Prev 15:2056–2062
13. Inoue M, Iwasaki M, Otani T, Sasazuki S, Noda M, Tsugane S (2006) Diabetes mellitus and the risk of cancer: results from a large-scale population-based cohort study in Japan. Arch Intern Med 166:1871–1877
14. Zendehdel K, Nyrén O, Ostenson CG, Adami HO, Ekbom A, Ye W (2003) Cancer incidence in patients with type 1 diabetes mellitus: a population-based cohort study in Sweden. J Natl Cancer Inst 95:1797–1800
15. Verlato G, Zoppini G, Bonora E, Muggeo M (2003) Mortality from site-specific malignancies in type 2 diabetic patients from Verona. Diabetes Care 26:1047–1051
16. Kowluru RA, Chan P-S (2007) Oxidative stress and diabetic retinopathy. Exp Diabetes Res 2007:43603
17. van de Poll-Franse LV, Houterman S, Janssen-Heijnen MLG, Dercksen MW, Coebergh JWW, Haak HR (2007) Less aggressive treatment and worse overall survival in cancer patients with diabetes: a large population based analysis. Int J Cancer 120:1986–1992
18. Coughlin SS, Calle EE, Teras LR, Petrelli J, Thun MJ (2004) Diabetes mellitus as a predictor of cancer mortality in a large cohort of US adults. Am J Epidemiol 159:1160–1167
19. Pandey A, Forte V, Abdallah M, Alickaj A, Mahmud S, Asad S, McFarlane SI (2011) Diabetes mellitus and the risk of cancer. Minerva Endocrinol 36:187–209
20. Luo W, Cao Y, Liao C, Gao F (2011) Diabetes mellitus and the incidence and mortality of colorectal cancer: a meta-analysis of twenty four cohort studies. Colorectal Dis. doi:10.1111/j.1463-1318.2011.02875.x
21. Jiang Y, Ben Q, Shen H, Lu W, Zhang Y, Zhu J (2011) Diabetes mellitus and incidence and mortality of colorectal cancer: a systematic review and meta-analysis of cohort studies. Eur J Epidemiol 26:863–876

22. Wang C, Wang X, Gong G, Ben Q, Qiu W, Chen Y, Li G, Wang L (2012) Increased risk of hepatocellular carcinoma in patients with diabetes mellitus: a systematic review and meta-analysis of cohort studies. Int J Cancer 130:1639–1648
23. Ren H-B, Yu T, Liu C, Li YQ (2011) Diabetes mellitus and increased risk of biliary tract cancer: systematic review and meta-analysis. Cancer Causes Control 22:837–847
24. Jing W, Jin G, Zhou X, Zhou Y, Zhang Y, Shao C, Liu R, Hu X (2012) Diabetes mellitus and increased risk of cholangiocarcinoma: a meta-analysis. Eur J Cancer Prev 21:24–31
25. Ge Z, Ben Q, Qian J, Wang Y, Li Y (2011) Diabetes mellitus and risk of gastric cancer: a systematic review and meta-analysis of observational studies. Eur J Gastroenterol Hepatol 23:1127–1135
26. Ben Q, Xu M, Ning X, Liu J, Hong S, Huang W, Zhang H, Li Z (2011) Diabetes mellitus and risk of pancreatic cancer: a meta-analysis of cohort studies. Eur J Cancer 47:1928–1937
27. Wang F, Gupta S, Holly EA (2006) Diabetes mellitus and pancreatic cancer in a population-based case–control study in the San Francisco Bay Area, California. Cancer Epidemiol Biomarkers Prev 15:1458–1463
28. Adami HO, McLaughlin J, Ekbom A, Berne C, Silverman D, Hacker D, Persson I (1991) Cancer risk in patients with diabetes mellitus. Cancer Causes Control 2:307–314
29. Jee SH, Ohrr H, Sull JW, Yun JE, Ji M, Samet JM (2005) Fasting serum glucose level and cancer risk in Korean men and women. JAMA 293:194–202
30. Xue F, Michels KB (2007) Diabetes, metabolic syndrome, and breast cancer: a review of the current evidence. Am J Clin Nutr 86:s823–s835
31. Swerdlow AJ, Laing SP, Qiao Z, Slater SD, Burden AC, Botha JL, Waugh NR, Morris AD, Gatling W, Gale EA, Patterson CC, Keen H (2005) Cancer incidence and mortality in patients with insulin-treated diabetes: a UK cohort study. Br J Cancer 92:2070–2075
32. Rose DP, Haffner SM, Baillargeon J (2007) Adiposity, the metabolic syndrome, and breast cancer in African-American and white American women. Endocr Rev 28:763–777
33. Mori M, Saitoh S, Takagi S, Obara F, Ohnishi H, Akasaka H, Izumi H, Sakauchi F, Sonoda T, Nagata Y, Shimamoto K (2000) A review of cohort studies on the association between history of diabetes mellitus and occurrence of cancer. Asian Pac J Cancer Prev 1:269–276
34. Chung YW, Han DS, Park KH, Eun CS, Yoo K-S, Park CK (2008) Insulin therapy and colorectal adenoma risk among patients with type 2 diabetes mellitus: a case–control study in Korea. Dis Colon Rectum 51:593–597
35. Suba Z, Ujpál M (2007) Disorders of glucose metabolism and risk of oral cancer. Fogorv Sz 100(250–257):243–249
36. Wideroff L, Gridley G, Mellemkjaer L, Chow WH, Linet M, Keehn S, Borch-Johnsen K, Olsen JH (1997) Cancer incidence in a population-based cohort of patients hospitalized with diabetes mellitus in Denmark. J Natl Cancer Inst 89:1360–1365
37. Vairaktaris E, Kalokerinos G, Goutzanis L, Spyridonidou S, Vassiliou S, Derka S, Nkenke E, Yapijakis C, Vylliotis A, Lazaris A, Patsouris E (2007) Diabetes alters expression of p53 and c-myc in different stages of oral oncogenesis. Anticancer Res 27:1465–1473
38. Khan S, Raghuram GV, Bhargava A, Pathak N, Chandra DH, Jain SK, Mishra PK (2011) Role and clinical significance of lymphocyte mitochondrial dysfunction in type 2 diabetes mellitus. Transl Res 158:344–359
39. Watanabe K, Thandavarayan RA, Harima M, Sari FR, Gurusamy N, Veeraveedu PT, Mito S, Arozal W, Sukumaran V, Laksmanan AP, Soetikno V, Kodama M, Aizawa Y (2010) Role of differential signaling pathways and oxidative stress in diabetic cardiomyopathy. Curr Cardiol Rev 6:280–290
40. Thandavarayan RA, Giridharan VV, Watanabe K, Konishi T (2011) Diabetic cardiomyopathy and oxidative stress: role of antioxidants. Cardiovasc Hematol Agents Med Chem 9:225–230
41. Schiffelers RM, Fens MH, van Blijswijk JM, Bink DI, Storm G (2007) Targeting the retinal microcirculation to treat diabetic sight problems. Expert Opin Ther Targets 11:1493–1502
42. Drechsler K, Fikenzer S, Sechtem U, Blank E, Breithardt G, Zeymer U, Niebauer J (2008) The Euro heart survey – Germany: diabetes mellitus remains unrecognized in patients with coronary artery disease. Clin Res Cardiol 97:364–370

43. Hofmann MA, Schiekofer S, Isermann B, Kanitz M, Henkels M, Joswig M, Treusch A, Morcos M, Weiss T, Borcea V, Abdel Khalek AK, Amiral J, Tritschler H, Ritz E, Wahl P, Ziegler R, Bierhaus A, Nawroth PP (1999) Peripheral blood mononuclear cells isolated from patients with diabetic nephropathy show increased activation of the oxidative-stress sensitive transcription factor NF-kappaB. Diabetologia 42:222–232
44. Bierhaus A, Schiekofer S, Schwaninger M, Andrassy M, Humpert PM, Chen J, Hong M, Luther T, Henle T, Klöting I, Morcos M, Hofmann M, Tritschler H, Weigle B, Kasper M, Smith M, Perry G, Schmidt AM, Stern DM, Häring HU, Schleicher E, Nawroth PP (2001) Diabetes-associated sustained activation of the transcription factor nuclear factor-kappaB. Diabetes 50:2792–2808
45. Palmeira CM, Rolo AP, Berthiaume J, Bjork JA, Wallace KB (2007) Hyperglycemia decreases mitochondrial function: the regulatory role of mitochondrial biogenesis. Toxicol Appl Pharmacol 225:214–220
46. Baynes JW (1991) Role of oxidative stress in development of complications in diabetes. Diabetes 40:405–412
47. Ricci C, Pastukh V, Mozaffari M, Schaffer SW (2007) Insulin withdrawal induces apoptosis via a free radical-mediated mechanism. Can J Physiol Pharmacol 85:455–464
48. Esper RJ, Vilariño JO, Machado RA, Paragano A (2008) Endothelial dysfunction in normal and abnormal glucose metabolism. Adv Cardiol 45:17–43
49. Karasu C (2010) Glycoxidative stress and cardiovascular complications in experimentally-induced diabetes: effects of antioxidant treatment. Open Cardiovasc Med J 4:240–256
50. Hou N, Torii S, Saito N, Hosaka M, Takeuchi T (2008) Reactive oxygen species-mediated pancreatic beta-cell death is regulated by interactions between stress-activated protein kinases, p38 and c-Jun N-terminal kinase, and mitogen-activated protein kinase phosphatases. Endocrinology 149:1654–1665
51. Frustaci A, Kajstura J, Chimenti C, Jakoniuk I, Leri A, Maseri A, Nadal-Ginard B, Anversa P (2000) Myocardial cell death in human diabetes. Circ Res 87:1123–1132
52. Sliwinska A, Blasiak J, Kasznicki J, Drzewoski J (2008) In vitro effect of gliclazide on DNA damage and repair in patients with type 2 diabetes mellitus (T2DM). Chem Biol Interact 173:159–165
53. Sliwinska A, Sliwinski T, Kasznicki J, Drzewoski J (2010) Effect of gliclazide on nucleotide excision repair (NER) and non-homologous DNA end joining (NHEJ) in normal and cancer cells. J Physiol Pharmacol 61:347–353
54. Golubnitschaja O (2009) Diabetes mellitus. In: Golubnitschaja O (ed) Predictive diagnostics and personalized treatment: dream or reality? Nova Science Publishers Inc, New York
55. Cebioglu M, Schild H, Golubnitschaja O (2010) Cancer predisposition in diabetics: risk factors considered for predictive diagnostics and targeted preventive measures. EPMA J 1: 130–137
56. Yeghiazaryan K, Cebioglu M, Braun M, Kuhn W, Schild HH, Golubnitschaja O (2011) Noninvasive subcellular imaging in breast cancer risk assessment: construction of diagnostic windows. Personalized Med 8:321–330
57. Ohta S (2003) A multi-functional organelle mitochondrion is involved in cell death, proliferation and disease. Curr Med Chem 10:2485–2494
58. Rolo AP, Palmeira CM (2006) Diabetes and mitochondrial function: role of hyperglycemia and oxidative stress. Toxicol Appl Pharmacol 212:167–178
59. Golubnitschaja O (2008) Clinical proteomics in application to predictive diagnostics and personalized treatment of diabetic patients. Curr Proteomics 5:35–44
60. Xu L, Hu Y, Chen B, Tang W, Han X, Yu H, Xiao C (2006) Mitochondrial polymorphisms as risk factors for endometrial cancer in southwest china. Int J Gynecol Cancer 16:1661–1667
61. Alexeyev MF, Ledoux SP, Wilson GL (2004) Mitochondrial DNA and aging. Clin Sci 107:355–364
62. Bai R-K, Leal SM, Covarrubias D, Liu A, Wong LJC (2007) Mitochondrial genetic background modifies breast cancer risk. Cancer Res 67:4687–4694

63. Bener A, Micallef R, Afifi M, Derbala M, Al-Mulla HM, Usmani MA (2007) Association between type 2 diabetes mellitus and Helicobacter pylori infection. Turk J Gastroenterol 18:225–229
64. Movahed MR, Hashemzadeh M, Jamal MM (2007) Increased prevalence of infectious endocarditis in patients with type II diabetes mellitus. J Diabetes Complicat 21:403–406
65. Ojetti V, Migneco A, Silveri NG, Ghirlanda G, Gasbarrini G, Gasbarrini A (2005) The role of H. pylori infection in diabetes. Curr Diabetes Rev 1:343–347
66. Foss NT, Foss-Freitas MC, Ferreira MAN, Cardili RN, Barbosa CMC, Foss MC (2007) Impaired cytokine production by peripheral blood mononuclear cells in type 1 diabetic patients. Diabetes Metab 33:439–443
67. Hopps E, Camera A, Caimi G (2008) Polimorphonuclear leukocytes and diabetes mellitus. Minerva Med 99:197–202
68. Barnea M, Madar Z, Froy O (2008) Glucose and insulin are needed for optimal defensin expression in human cell lines. Biochem Biophys Res Commun 367:452–456
69. Ali SS, Ali IS, Aamir AH, Jadoon Z, Inayatullah S (2007) Frequency of hepatitis C infection in diabetic patients. J Ayub Med Coll Abbottabad 19:46–49
70. Sangiorgio L, Attardo T, Gangemi R, Rubino C, Barone M, Lunetta M (2000) Increased frequency of HCV and HBV infection in type 2 diabetic patients. Diabetes Res Clin Pract 48:147–151
71. Hemann MT, Narita M (2007) Oncogenes and senescence: breaking down in the fast lane. Genes Dev 21:1–5
72. Ben-Yosef T, Yanuka O, Halle D, Benvenisty N (1998) Involvement of Myc targets in c-myc and N-myc induced human tumors. Oncogene 17:165–171
73. Pontén J, Guo Z (1998) Precancer of the human cervix. Cancer Surv 32:201–229
74. Johnson CN, Levy LS (2005) Matrix attachment regions as targets for retroviral integration. Virol J 2:68
75. Hebner CM, Laimins LA (2006) Human papillomaviruses: basic mechanisms of pathogenesis and oncogenicity. Rev Med Virol 16:83–97
76. Golubnitschaja O (2007) Cell cycle checkpoints: the role and evaluation for early diagnosis of senescence, cardiovascular, cancer, and neurodegenerative diseases. Amino Acids 32:359–371
77. Banerjee S, Banerjee M (2006) Hepatitis C and diabetes mellitus. J Indian Med Assoc 104:86–89
78. Shera KA, Shera CA, McDougall JK (2001) Small tumor virus genomes are integrated near nuclear matrix attachment regions in transformed cells. J Virol 75:12339–12346
79. Popescu NC, Zimonjic DB (2002) Chromosome-mediated alterations of the MYC gene in human cancer. J Cell Mol Med 6:151–159
80. de Villiers E-M, Sandstrom RE, zur Hausen H, Buck CE (2005) Presence of papillomavirus sequences in condylomatous lesions of the mamillae and in invasive carcinoma of the breast. Breast Cancer Res 7:R1–R11
81. Lawson JS, Günzburg WH, Whitaker NJ (2006) Viruses and human breast cancer. Future Microbiol 1:33–51
82. Cebioglu M, Schild HH, Golubnitschaja O (2008) Diabetes mellitus as a risk factor for cancer: stress or viral etiology? Infect Disord Drug Targets 8:76–87
83. Golubnitschaja O, Costigliola V (2010) Common origin but individual outcomes: time for new guidelines in personalized healthcare. Personalized Med 7:561–568
84. Golubnitschaja O (2010) Personalisierte Therapie – erforderliche Maßnahmen für eine wirksame Diabetes-Vorsorge: Zeit für neue Richtlinien? In: Niederlag W, Lemke HU, Golubnitschaja O, Rienhoff O (eds) Personalisierte Medizin. Health Academy, Dresden
85. Berger FG (2004) The interleukin-6 gene: a susceptibility factor that may contribute to racial and ethnic disparities in breast cancer mortality. Breast Cancer Res Treat 88:281–285
86. Di Pietro PF, Medeiros NI, Vieira FGK, Fausto MA, Belló-Klein A (2007) Breast cancer in southern Brazil: association with past dietary intake. Nutr Hosp 22:565–572

87. Fejerman L, Ziv E (2008) Population differences in breast cancer severity. Pharmacogenomics 9:323–333
88. Celis JE, Moreira JMA, Cabezón T, Gromov P, Friis E, Rank F, Gromova I (2005) Identification of extracellular and intracellular signaling components of the mammary adipose tissue and its interstitial fluid in high risk breast cancer patients: toward dissecting the molecular circuitry of epithelial-adipocyte stromal cell interactions. Mol Cell Proteomics 4:492–522
89. Xue Y, Guo X-T, Liu W-C (2007) Clinical research advancement on male breast cancer. Ai Zheng 26:1148–1152
90. Benchellal Z, Wagner A, Harchaoui Y, Huten N, Body G (2002) Male breast cancer: 19 case reports. Ann Chir 127:619–623
91. Riedl M, Kotzmann H, Luger A (2001) Growth hormone in the elderly man. Wien Med Wochenschr 151:426–429
92. Abu-Abid S, Szold A, Klausner J (2002) Obesity and cancer. J Med 33:73–86
93. Kruk J (2007) Physical activity in the prevention of the most frequent chronic diseases: an analysis of the recent evidence. Asian Pac J Cancer Prev 8:325–338
94. Irigaray P, Newby JA, Clapp R, Hardell L, Howard V, Montagnier L, Epstein S, Belpomme D (2007) Lifestyle-related factors and environmental agents causing cancer: an overview. Biomed Pharmacother 61:640–658
95. Danaei G, Vander Hoorn S, Lopez AD, Murray CJL, Ezzati M (2005) Causes of cancer in the world: comparative risk assessment of nine behavioural and environmental risk factors. Lancet 366:1784–1793
96. Lowenfels AB, Maisonneuve P (2005) Risk factors for pancreatic cancer. J Cell Biochem 95:649–656
97. Hart AR, Kennedy H, Harvey I (2008) Pancreatic cancer: a review of the evidence on causation. Clin Gastroenterol Hepatol 6:275–282
98. Hermansen K, Jørgensen K, Schmidt EB, Tjønneland A, Tolstrup JS, Grønbaek MN (2007) Alcohol and lifestyle diseases. Ugeskr Laeger 169:3404–3406
99. Hopps E, Caimi G (2011) Exercise in obesity management. J Sports Med Phys Fitness 51:275–282
100. Conroy KP, Davidson IM, Warnock M (2011) Pathogenic obesity and nutraceuticals. Proc Nutr Soc 70:426–438
101. Kim M-C, Kim C-S, Chung T-H, Park H-O, Yoo C-I (2011) Metabolic syndrome, lifestyle risk factors, and distal colon adenoma: a retrospective cohort study. World J Gastroenterol 17:4031–4037
102. Etemadi A, Golozar A, Kamangar F, Freedman ND, Shakeri R, Matthews C, Islami F, Boffetta P, Brennan P, Abnet CC, Malekzadeh R, Dawsey SM (2012) Large body size and sedentary lifestyle during childhood and early adulthood and esophageal squamous cell carcinoma in a high-risk population. Ann Oncol 23:1593–1600
103. Mönkemann H, De Vriese AS, Blom HJ, Kluijtmans LAJ, Heil SG, Schild HH, Golubnitschaja O (2002) Early molecular events in the development of the diabetic cardiomyopathy. Amino Acids 23:331–336
104. Mousavi SM, Fallah M, Sundquist K, Hemminki K (2012) Age- and time-dependent changes in cancer incidence among immigrants to Sweden: colorectal, lung, breast and prostate cancers. Int J Cancer 131:E122–E128
105. Thompson CL, Larkin EK, Patel S, Berger NA, Redline S, Li L (2011) Short duration of sleep increases risk of colorectal adenoma. Cancer 117:841–847
106. Mohamed AA, Loutfy SA, Craik JD, Hashem AGM, Siam I (2011) Chronic hepatitis c genotype-4 infection: role of insulin resistance in hepatocellular carcinoma. Virol J 8:496
107. Daley E, DeBate R, Dodd V, Dyer K, Fuhrmann H, Helmy H, Smith SA (2011) Exploring awareness, attitudes, and perceived role among oral health providers regarding HPV-related oral cancers. J Public Health Dent 71:136–142

108. Omland LH, Jepsen P, Krarup H, Christensen PB, Weis N, Nielsen L, Obel N, Sørensen HT, Stuver SO (2012) Liver cancer and non-Hodgkin lymphoma in hepatitis C virus-infected patients: results from the DANVIR cohort study. Int J Cancer 130:2310–2317
109. Saikali J, Picard C, Freitas M, Holt P (2004) Fermented milks, probiotic cultures, and colon cancer. Nutr Cancer 49:14–24
110. Chakraborty J (2012) Cancer risk from exposure to hazardous air pollutants: spatial and social inequities in Tampa Bay, Florida. Int J Environ Health Res 22:165–183
111. Fenton SE, Reed C, Newbold RR (2012) Perinatal environmental exposures affect mammary development, function, and cancer risk in adulthood. Annu Rev Pharmacol Toxicol 52:455–479
112. Gross AL, Gallo JJ, Eaton WW (2010) Depression and cancer risk: 24 years of follow-up of the Baltimore epidemiologic catchment area sample. Cancer Causes Control 21:191–199
113. Peters JL, Kubzansky LD, Ikeda A, Spiro A 3rd, Wright RO, Weisskopf MG, Kim D, Sparrow D, Nie LH, Hu H, Schwartz J (2011) Childhood and adult socioeconomic position, cumulative lead levels, and pessimism in later life: the VA normative aging study. Am J Epidemiol 174:1345–1353
114. Mahabir S, Wei Q, Barrera SL, Dong YQ, Etzel CJ, Spitz MR, Forman MR (2008) Dietary magnesium and DNA repair capacity as risk factors for lung cancer. Carcinogenesis 29:949–956
115. Howard RA, Freedman DM, Park Y, Hollenbeck A, Schatzkin A, Leitzmann MF (2008) Physical activity, sedentary behavior, and the risk of colon and rectal cancer in the NIH-AARP diet and health study. Cancer Causes Control 19:939–953
116. Toledo FGS, Menshikova EV, Ritov VB, Azuma K, Radikova Z, DeLany J, Kelley DE (2007) Effects of physical activity and weight loss on skeletal muscle mitochondria and relationship with glucose control in type 2 diabetes. Diabetes 56:2142–2147
117. Gerbi GB, Habtemariam T, Tameru B, Nganwa D, Robnet V (2011) Psychosocial responses associated with perceived risk of cancer in an African-American community. Psychol Rep 109:93–98
118. Cheetham PJ, Katz AE (2011) Diet and prostate cancer – a holistic approach to management. Arch Esp Urol 64:720–734
119. Pedersen BK (2011) Exercise-induced myokines and their role in chronic diseases. Brain Behav Immun 25:811–816
120. Mäkitie A, Keski-Säntti H, Saarilahti K, Bäck L, Kontio R, Grénman R, Koivunen P, Laranne J (2011) Therapeutic results in oral cancer–what has been learned? Duodecim 127:1986–1991
121. Kampman E (2008) Nutrition and oropharyngeal cancer. Ned Tijdschr Tandheelkd 115:558–562
122. Dröge W (2002) Free radicals in the physiological control of cell function. Physiol Rev 82:47–95
123. Dranka BP, Benavides GA, Diers AR, Giordano S, Zelickson BR, Reily C, Zou L, Chatham JC, Hill BG, Zhang J, Landar A, Darley-Usmar VM (2011) Assessing bioenergetic function in response to oxidative stress by metabolic profiling. Free Radic Biol Med 51:1621–1635
124. Zhang Y, Du Y, Le W, Wang K, Kieffer N, Zhang J (2011) Redox control of the survival of healthy and diseased cells. Antioxid Redox Signal 15:2867–2908
125. Jiang Z, Hu Z, Zeng L, Lu W, Zhang H, Li T, Xiao H (2011) The role of the Golgi apparatus in oxidative stress: is this organelle less significant than mitochondria? Free Radic Biol Med 50:907–917
126. Wojcik M, Burzynska-Pedziwiatr I, Wozniak LA (2010) A review of natural and synthetic antioxidants important for health and longevity. Curr Med Chem 17:3262–3288
127. Stadler K (2011) Peroxynitrite-driven mechanisms in diabetes and insulin resistance – the latest advances. Curr Med Chem 18:280–290
128. Valko M, Leibfritz D, Moncol J, Cronin MTD, Mazur M, Telser J (2007) Free radicals and antioxidants in normal physiological functions and human disease. Int J Biochem Cell Biol 39:44–84
129. Sedelnikova OA, Redon CE, Dickey JS, Nakamura AJ, Georgakilas AG, Bonner WM (2010) Role of oxidatively induced DNA lesions in human pathogenesis. Mutat Res 704:152–159

130. Gago-Dominguez M, Jiang X, Castelao JE (2007) Lipid peroxidation, oxidative stress genes and dietary factors in breast cancer protection: a hypothesis. Breast Cancer Res 9:201
131. Bonassi S, Znaor A, Norppa H, Hagmar L (2004) Chromosomal aberrations and risk of cancer in humans: an epidemiologic perspective. Cytogenet Genome Res 104:376–382
132. Munday R, Zhang Y, Fahey JW, Jobson HE, Munday CM, Li J, Stephenson KK (2006) Evaluation of isothiocyanates as potent inducers of carcinogen-detoxifying enzymes in the urinary bladder: critical nature of in vivo bioassay. Nutr Cancer 54:223–231
133. Szendroedi J, Phielix E, Roden M (2012) The role of mitochondria in insulin resistance and type 2 diabetes mellitus. Nat Rev Endocrinol 8:92–103
134. Phielix E, Szendroedi J, Roden M (2011) Mitochondrial function and insulin resistance during aging: a mini-review. Gerontology 57:387–396
135. Loizidou MA, Michael T, Neuhausen SL, Newbold RF, Marcou Y, Kakouri E, Daniel M, Papadopoulos P, Malas S, Kyriacou K, Hadjisavvas A (2008) Genetic polymorphisms in the DNA repair genes XRCC1, XRCC2 and XRCC3 and risk of breast cancer in Cyprus. Breast Cancer Res Treat 112:575–579
136. Yu D, Zhang X, Liu J, Yuan P, Tan W, Guo Y, Sun T, Zhao D, Yang M, Liu J, Xu B, Lin D (2008) Characterization of functional excision repair cross-complementation group 1 variants and their association with lung cancer risk and prognosis. Clin Cancer Res 14:2878–2886
137. Gahan PB (2008) Circulating nucleic acids in plasma and serum: roles in diagnosis and prognosis in diabetes and cancer. Infect Disord Drug Targets 8:100–108
138. Golubnitschaja O (2009) Advanced technologies for prediction of secondary complications in diabetes mellitus. In: Golubnitschaja O (ed) Predictive diagnostics and personalized treatment: dream or reality. Nova Science Publishers Inc, New York
139. Cebioglu M, Schild HH, Golubnitschaja O (2009) Diabetes mellitus as a risk factor for cancer: is predictive diagnosis possible? In: Predictive diagnostics and personalized treatment: dream or reality. Nova Science Publishers Inc, New York
140. Golubnitschaja O (2010) Advanced diabetes care: three levels of prediction, prevention & personalized treatment. Curr Diabetes Rev 6:42–51
141. Page NM, Kemp CF, Butlin DJ, Lowry PJ (2002) Placental peptides as markers of gestational disease. Reproduction 123:487–495
142. Schweigert FJ (2007) Nutritional proteomics: methods and concepts for research in nutritional science. Ann Nutr Metab 51:99–107
143. Mathers JC, Hesketh JE (2007) The biological revolution: understanding the impact of SNPs on diet-cancer interrelationships. J Nutr 137:253S–258S

Recognition of Individual Risks by Analysis of Subcellular Imaging Insights into Chromium Picolinate Therapy in Pre/Diabetes Care, Search into Its Safety and Opinion Controversy

Kristina Yeghiazaryan, Hans H. Schild, and Olga Golubnitschaja

Abstract Due to the important physiologic function of trivalent chromium in glucose/insulin homeostasis, some commercial organisations promote (Cr^{3+}) supplements for

- maintaining proper carbohydrate and lipid metabolism;
- regulation to reduce carbohydrate carvings and appetite;
- prevention of insulin resistance and glucose intolerance;
- to regulate body composition, including reducing fat mass and increasing lean body mass;
- optimal body building for athletes;
- a means of losing weight;
- an antidepressant in treatment of atypical depression;
- prevention of obesity and type 2 diabetes mellitus.

The ubiquitous agent is the chromium picolinate (CrPic) has become very popular for treating type 2 diabetics as well as obesity- and diabetes-predisposed individuals: it is reported that CrPic generates sales for more than 100 million dollars annually. The supplements by CrPic are freely available in numerous forms including chewing gums, pills, sport drinks and nutrition bars. A series of studies cautions against potential risks of unguided nutritional application of CrPic. Furthermore, a growing body of case reports warn against adverse health outcomes assigned to CrPic dietary application, whereas by others they are commented as non-evidence based anecdotes.

K. Yeghiazaryan • H.H. Schild • O. Golubnitschaja (✉)
Department of Radiology, Rheinische Friedrich-Wilhelms-University of Bonn, Sigmund-Freud-Str. 25, Bonn D-53105, Germany
e-mail: olga.golubnitschaja@ukb.uni-bonn.de

M.S. Mozaffari (ed.), *New Strategies to Advance Pre/Diabetes Care: Integrative Approach by PPPM*, Advances in Predictive, Preventive and Personalised Medicine 3,
DOI 10.1007/978-94-007-5971-8_14,

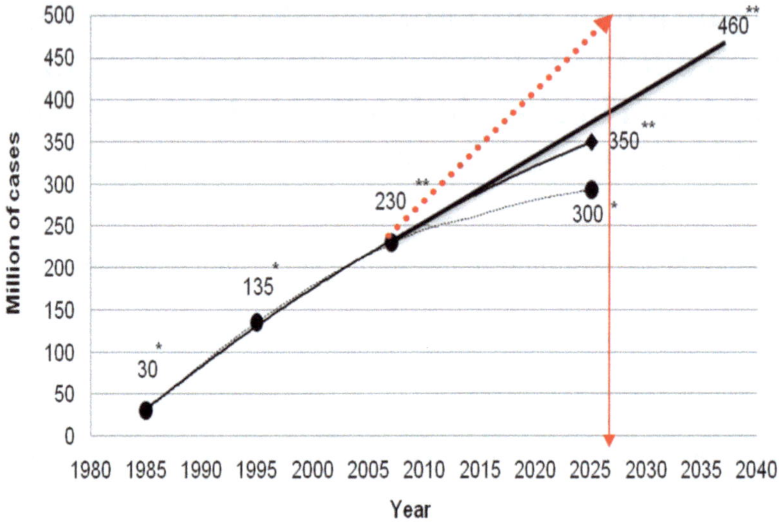

Fig. 1 Worldwide prognosis of the epidemic scale in diabetes; *estimations as published around the year 2000; **stepwise worsening prognosis as published in 2003–2008; current prognoses are marked in *red color* [1, 2]

This chapter analyses opinion controversies in the research field of trivalent chromium therapy, demonstrates highly individual reactions towards CrPic dietary supplements and highlights individual risks, when dietary supplements used freely as therapeutic agents, without application of advanced diagnostic tools to predict individual outcomes.

Keywords Diabetes • Predisposition • Healthcare • Predictive diagnostics • Therapy monitoring • Guidelines • Prevention • Personalised medicine • Nutrition • Dietary supplements • Drug • Safety • Individual reaction • Patient profile • Medical record • Biomarker pattern • Pandemy • Education

1 Changing Long-Held Beliefs Is Never Easy

As a consequence of the accumulating clinical data and knowledge about the epidemiology and pathological mechanisms of the most frequent causes of morbidity and mortality, we are currently reconsidering our view of the origins and progression of cardiovascular, oncologic and neurodegenerative diseases. The majority of these pathologies are of chronic nature: they progress from precursor lesions over one or even several decades of life until a diagnosis is made that is often too late for effective therapeutic intervention. An excellent example is the epidemic scale of diabetes mellitus type 2 witnessed in the European Union. In most industrialised countries and

Fig. 2 Severe complications developed secondary to diabetes: from "upstream" (up) to "downstream" (bottom) in the cascade of pathologic processes (Adapted from [4])

countries with large populations, the permanently growing cohort of diabetics creates a serious healthcare problem and dramatic health economic burden. Estimates for diabetes prevalence in the years 20-25-2030 is half a billion patients worldwide – see Fig. 1.

Moreover, the contemporary onset of the dominant diabetes type 2 is already observed in subpopulations of teenagers [3]. Severe complications secondary to early onset of diabetes mellitus, such as retinopathy, nephropathy, silent ischaemia, dementia, and cancer (Fig. 2), soon may lead to collapsing healthcare systems [4].

Optimistic *versus* pessimistic prognosis in future developments of the healthcare sector depends much on diagnostic, preventive and treatment approaches which diabetes care will preferably adopt in the near future. Without innovation in healthcare, diabetes related complications may cause such a dramatic level of burden that any performance of personalised medicine will not be feasible from economical point of view. In contrast, effective utilisation of advanced early/predictive diagnostics and targeted prevention could enable rapidly ageing populations in Europe and elsewhere to be economically effective for the society. Therefore, it is timely to consider the integrative medical approaches to advance pre/diabetes care utilising the innovation by predictive diagnostics as the basis for concomitant targeted prevention, patient profiling as the basis for individualised treatment algorithms and medical approaches tailored to the patient [5–8].

Our message is – new guidelines should create the robust juristic and economic platform for advanced medical services utilising the cost-effective models of risk assessment followed by tailored treatments focussed on the precursor stages of chronic disease [9].

2 Potential Role of Biologically Active Chromium in Diabetes Prevention

Chromium is an essential trace element in human physiology. The physiologic value of the chromium compounds depends on the oxidative states of the chromium ions that vary from bivalent to hexavalent ones. The best investigated forms are trivalent and hexavalent chromium compounds. The latter is frequently used in a variety of industrial technologies, chronic exposure to which has been shown to be responsible for a spectrum of severe health problems affecting skin and inner organs [10]. Investigations of its carcinogenic effects are in progress at molecular and cellular levels [11, 12]. In contrast, trivalent chromium is considered physiologically safe essential trace metal functionally involved in several metabolic pathways in human beings as published elsewhere. The recommended daily allowance of trivalent chromium corresponds to the range of 50–200 μg [13]. Dietary deficiency of chromium leads to disturbances in carbohydrate metabolisms, increases risk of glucose intolerance, insulin resistance, hyperglycaemia and predisposes to obesity and type 2 diabetes mellitus [14, 15].

3 Controversial Benefits by Cr^{3+} Nutrient Supplement in Foods

Biologically valuable forms of chromium are compounds of trivalent chromium bound to low molecular weight organic complexes such as oligopeptides. This recognition led to the development of a commercial branch with Cr^{3+} nutrient supplements which suppose to promote or even improve health status.

What are the currently available "for" *versus* "against" arguments to approve/disprove potential benefits of Cr^{3+} nutrient supplement?

Chromium is the essential metal but only traces of trivalent chromium are utilised in human body for the glucose conversion by insulin-driven reactions in carbohydrate metabolic pathways. Furthermore, naturally available food sources used for well-balanced diets are capable to cover the requirement of trivalent chromium completely: fruits, vegetables, meat, fish, grains, Brewer's yeast, etc. [16]. Cr deficiency cases are extremely rare, and evidence-based biochemical justification for artificial source needed additionally to standard foods has not been provided in the literature. Nevertheless, due to the important physiologic function of trivalent chromium in maintaining proper carbohydrate and lipid metabolism and glucose/insulin homeostasis, some commercial organisations promote Cr^{3+} supplements as

- regulation to reduce carbohydrate carvings and appetite;
- prevention of insulin resistance and glucose intolerance;
- to regulate body composition, including reducing fat mass and increasing lean body mass;

- optimal body building for athletes;
- a means of losing weight;
- an antidepressant in treatment of atypical depression;
- prevention of obesity and type 2 diabetes mellitus.

Widely spread chemical compound used as the trivalent chromium nutrient supplement is chromium-picolinate (CrPic). CrPic has become a very popular nutritional supplement for treating type 2 diabetics, obese and diabetes-predisposed individuals [17]. However, over a decade of human studies with CrPic indicate that the supplement has not demonstrated effects on the body composition of healthy individuals, even when taken in combination with an exercise training programme [18]. Also a potential weight loss by CrPic therapy has not been confirmed [19].

After the evidence-based search into the issue, U.S. Food and Drug Administration concluded that the "relationship between chromium picolinate intake and insulin resistance is highly uncertain" [20]. Also some later studies have concluded that "CrPic does not improve key features of metabolic syndrome in obese nondiabetic adults" [21]. Nevertheless, it is reported that CrPic generates sales for more than 100 million dollars annually [22]. The supplements by CrPic are freely available in numerous forms including chewing gums, pills, sport drinks and nutrition bars [23]. CrPic is commonly supplied in the range of 200–500 μg as a daily dose [13]. Big advantage of chromium-picolinate is its high bioavailability due to increased absorbing capacity in human tissues compared with dietary chromium [24]. However, a mechanism of chromium-picolinate action at molecular level is not completely understood. In 1999, Speetjens et al. reported that CrPic cleaves DNA molecules by the radical mechanism: in the presence of oxygen from air and biologically available reductants, CrPic generates hydroxyl-radical damaging DNA molecules [25]. Further, organisms exposed to CrPic demonstrate the tendency to accumulate intracellular Cr^{3+} ions that may lead to long-term genotoxic effects by formation of covalent bonds to DNA [26]. Taken together, cytotoxic, genotoxic and mutagenic effects as well as activity damaging to mitochondria and induction of apoptosis have been reported for CrPic using mammalian cell cultures, *Drosophila* and animal models [27–30]. Potential toxicity of CrPic has raised the concerns about safety of CrPic-related nutrient supplements.

4 Animal Experiments to Model Treatment Effects by CrPic on Human Beings

To simulate treatment affects by CrPic a spectrum of models has been used. To simulate DM type 2 a well-acknowledged animal model of db/db-mice was utilised in our recent project [31]. The experimental design is summarised in Fig. 3.

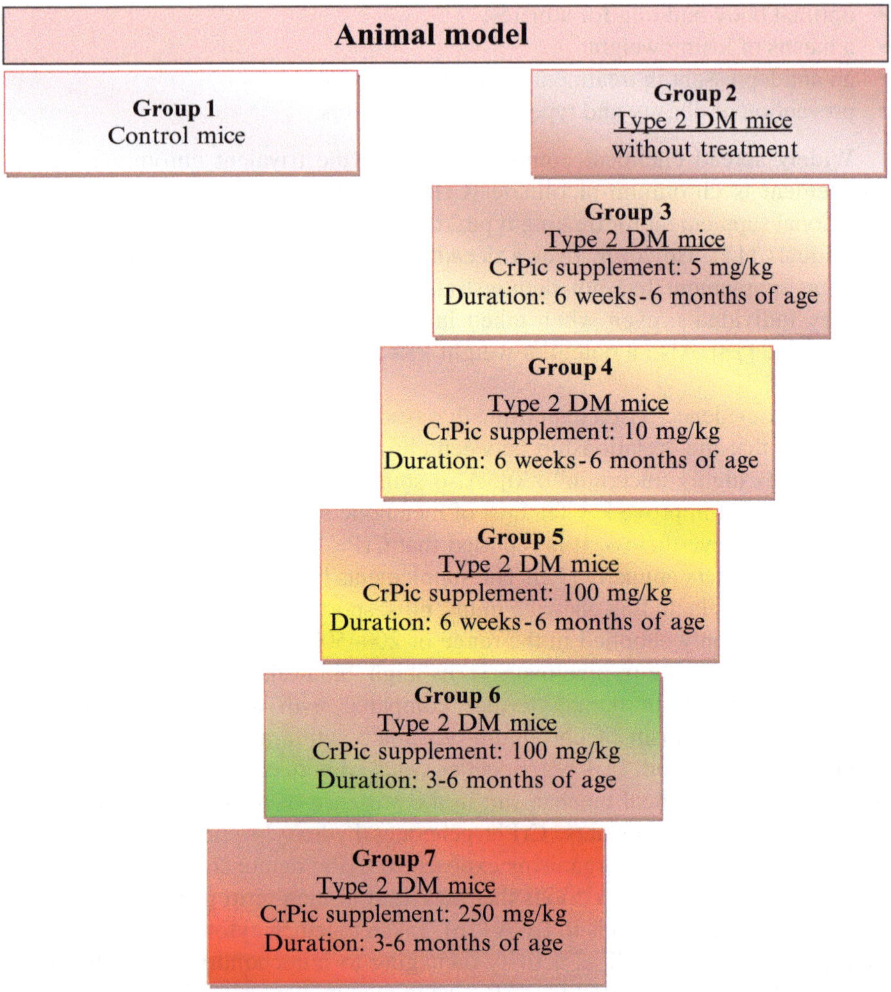

Fig. 3 Experimental design to model CrPic treatment effects. Animals were grouped into eight groups. Group 1 represented the healthy control. Group 2 was the model mice for type 2 DM. Groups 3–7 were diabetes-model mice treated under individual algorithms: 5 mg/kg CrPic from 6 weeks to 6 months of age (Group 3), 10 mg/kg CrPic from 6 weeks to 6 months of age (Group 4), 100 mg/kg CrPic from 6 weeks to 6 months of age (Group 5), 100 mg/kg CrPic from 3 to 6 months of age (Group 6), and 250 mg/kg CrPic from 3 to 6 months of age (Group 7). Kidney tissue samples were collected and stored at – 80°C till the below described analysis were performed

5 Subcellular Imaging Insights into Chromium Picolinate Therapy

5.1 Quantitative Subcellular Imaging by "Comet Assay"-Analysis

The single cell gel electrophoresis assay is a simple and effective method for evaluating DNA damage in cells. It is based upon the ability of denatured, cleaved DNA fragments to migrate out of the cell under the influence of an electric field, whereas undamaged DNA remains within the confines of the nucleoid and migrates slower. The assessment of DNA damage is done via the evaluation of the DNA "comet" tail shape and migration pattern. The cells are immobilised in a bed of low melting point agarose, on a Trevigen CometSlide™. After cell lysis, samples are treated with alkali to unwind and denature the DNA and hydrolyse sites of damage. After performing electrophoresis, staining with a fluorescent DNA intercalating dye (SYBR® Green I) is done and the sample is visualised by epifluorescence microscopy. The alkaline electrophoresis is very sensitive and detects small amounts of damage.

In our experiments, DNA damage was designated to four classes based on the visual aspect considering the extent of DNA migration as published earlier [32]. Comets with a bright head and almost no tail were classified as class I indicating minimal DNA damage. Whereas comets with no visible head and a long diffuse tail were classified as class IV revealing complete DNA fragmentation. Comets with intermediate characteristics were assigned to classes II and III based on the ratio $R=T/r$, in which T represents the comet's tail length and r is the radius of the comet's head. The characteristic value of R for class I is 1 and for class IV is ∞ ($r=0$). Comet with R values ranging between $1<R>3$ were designated as class II. The comet classification is demonstrated in Fig. 4a. Figure 4b demonstrates comet patterns typical for experimental group 1 (left) with the lowest level of comets class IV (apoptosis) and experimental groups 5–7 (right) with the highest level of apoptosis.

5.2 Statistical Analysis Reveals Increased Heterogeneity of Individual Reactions Towards Highest Doses of CrPic

Statistical analysis of comets in classes I (intact DNA) and IV (apoptosis) is demonstrated in Fig. 5a, b, respectively. With the red frame, the experimental groups are marked with highest rates of statistical deviations towards mean value. The highest heterogeneity is obvious for the groups 5, 6 and 7, i.e. the experimental groups with highest doses of CrPic applied for the treatment.

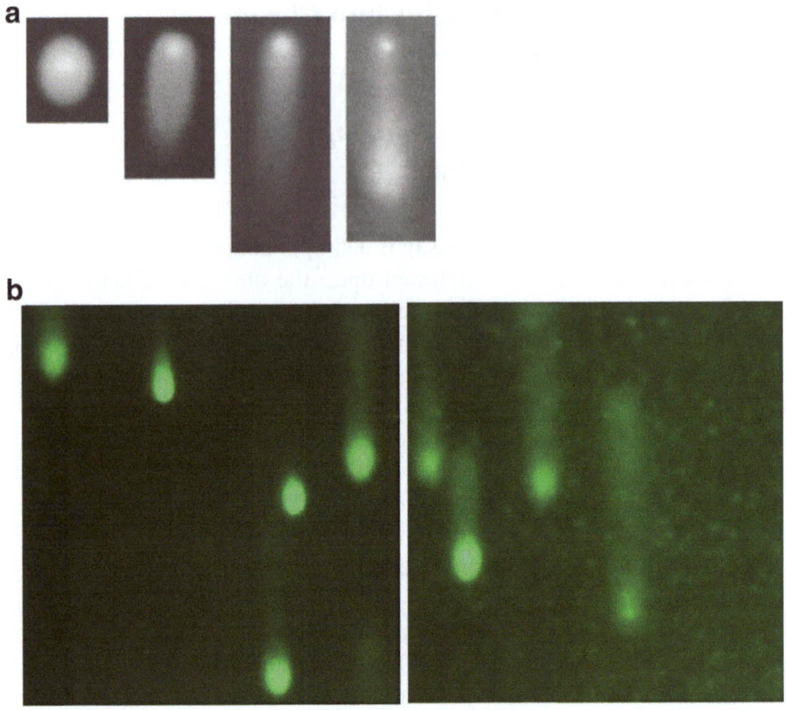

Fig. 4 (**a**) Comet classes I, II, III and IV (from *left to right*); (**b**) Comet pattern typical for the experimental group 1 (*left*, low apoptosis rates) and experimental groups 5–7 (*right*, high apoptosis rates)

5.2.1 DNA-Damage in Untreated Diabetic Is Significantly Higher Compared to Healthy Control Group

Quantitative subcellular imaging by "Comet assay"-analysis revealed significantly higher DNA-damage in untreated diabetic group compared to healthy controls (see Fig. 6) that is well in agreement with data published and reviewed for diabetic patients [4, 8].

5.2.2 Individuality Within Untreated Diabetic Group

Although subcellular imaging generally revealed group-specific patterns when diabetic animals were compared with controls, however, high heterogeneity was demonstrated within untreated diabetic group: in some animals intact DNA (class I comets) was monitored, whereas only comet classes with damaged DNA was monitored for others (see Fig. 7). This heterogeneity clearly demonstrates individualised reaction of organism towards diabetic condition. This is a very important observation that well corresponds with the clinic picture of diabetes [8]. Hence, we conclude that the animal model used is suitable for issue-related studies to simulate medical condition of diabetes and investigate individualised therapeutic effects.

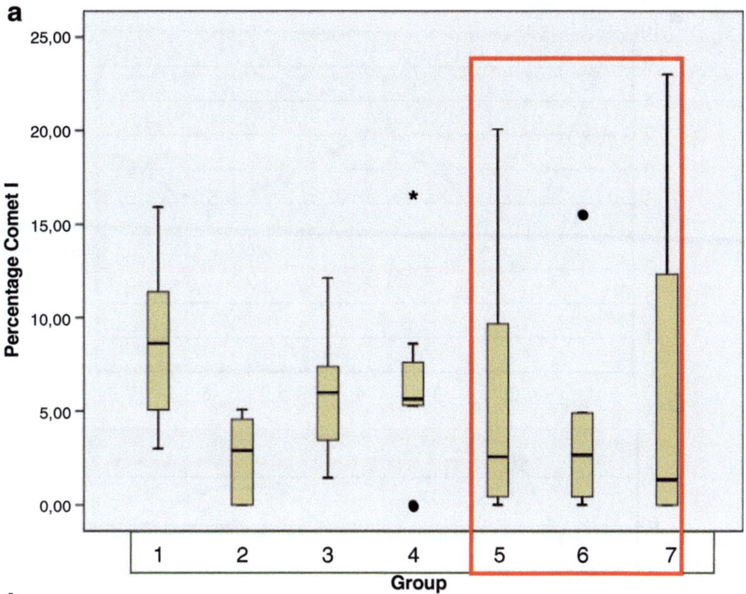

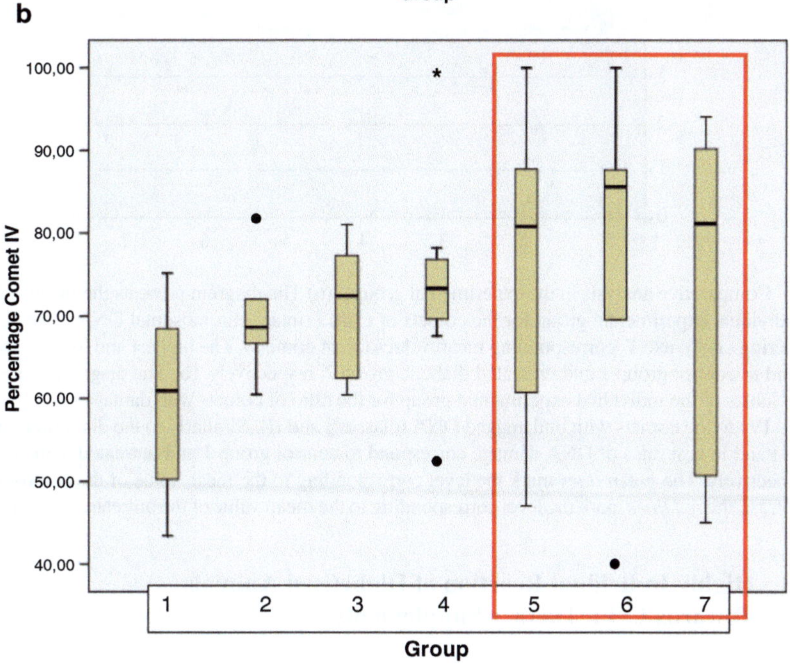

Fig. 5 Statistical analyses for the comet types (**a**) I and (**b**) IV (see "comet classification" in Fig. 4a) in groups of comparison as described in Fig. 3 ("experimental design"). Analyses were carried out using SPSS 17.0 software (SPSS, Chicago, IL, USA) by the application of univariable variance analysis with Bonferroni. The *red box* marks the groups with the highest deviation towards the mean value, i.e. the greatest heterogeneity within corresponding groups. The individual values which go overboard by statistical calculations are marked with *black circles* (for the standard deviation in groups 4 and 6) and *asterisk* (for the mean value in group 4)

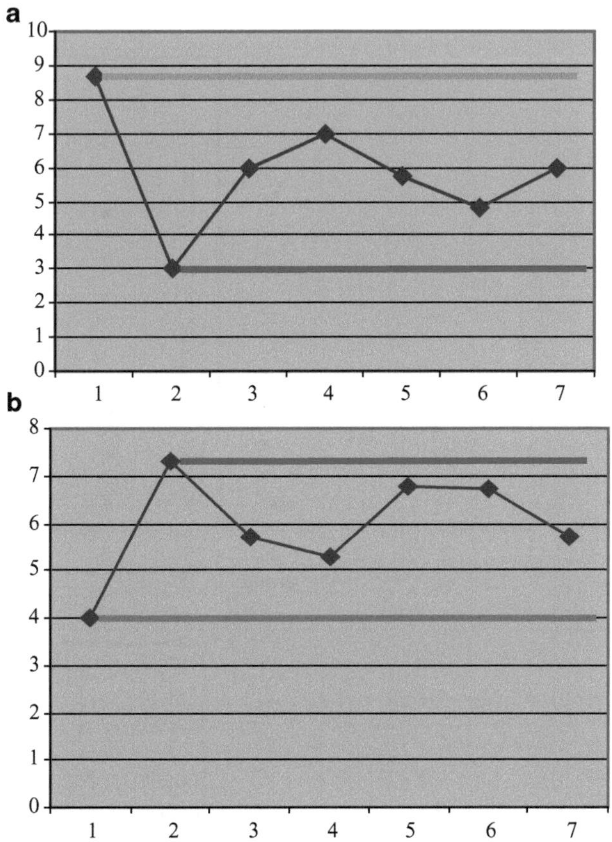

Fig. 6 Comparative analysis in the experimental groups. (**a**) The diagram presents the mean values of the individual experimental group for the comets of class I (intact chromosomal DNA): axe *X*, group numbering (*1–7*); axe *Y*, corresponding mean values (% of comets). The highest and lowest rates correspond to control group 1 and untreated diabetic group 2, respectively. (**b**) The diagram presents the mean values of the individual experimental group for the ratio of comets with damaged DNA (classes III and IV) to the comets with undamaged DNA (classes I and II). Similarly to the diagram in (**a**), the highest and lowest rates of DNA damage correspond to control group 1 and untreated diabetic group 2, respectively. The *green lines* mark the level corresponding to the mean value of the control group; in contrast, the *red lines* mark the level corresponding to the mean value of the untreated diabetic group

5.2.3 Highly Individual Reaction of Diabetised Animals Towards CrPic Dietary Supplements

Despite group-specific patterns, the subcellular imaging of comet assay indicated that each animal within a group responded individually to the same dosage of CrPic administered and treatment duration. Within diabetic groups, experimental animals demonstrated highly individual comet patterns with respect to single dosages and treatment algorithms. This observation is important in regard of highly individual response to the treatment algorithms applied. We conclude individual reaction of diabetised animals towards doses and duration of CrPic treatment. This observation

Fig. 7 Diagram demonstrates individual reactions towards CrPic treatments in the groups of comparison (1–7, group numbering is marked in the middle). Individual levels of intact DNA (axe **X**, individual numbering; axe Y, comet class I in %) are highly heterogeneous in diabetic groups compared to control group 1. The highest level of heterogeneity is evident in groups 5, 6 and 7 with the highest doses of CrPic supplements (see "experimental design" in Fig. 3) The *turquoise lines* mark the level corresponding to the mean value of the control group; in contrast, the *red arrows* mark the level corresponding to the mean of the untreated diabetic group

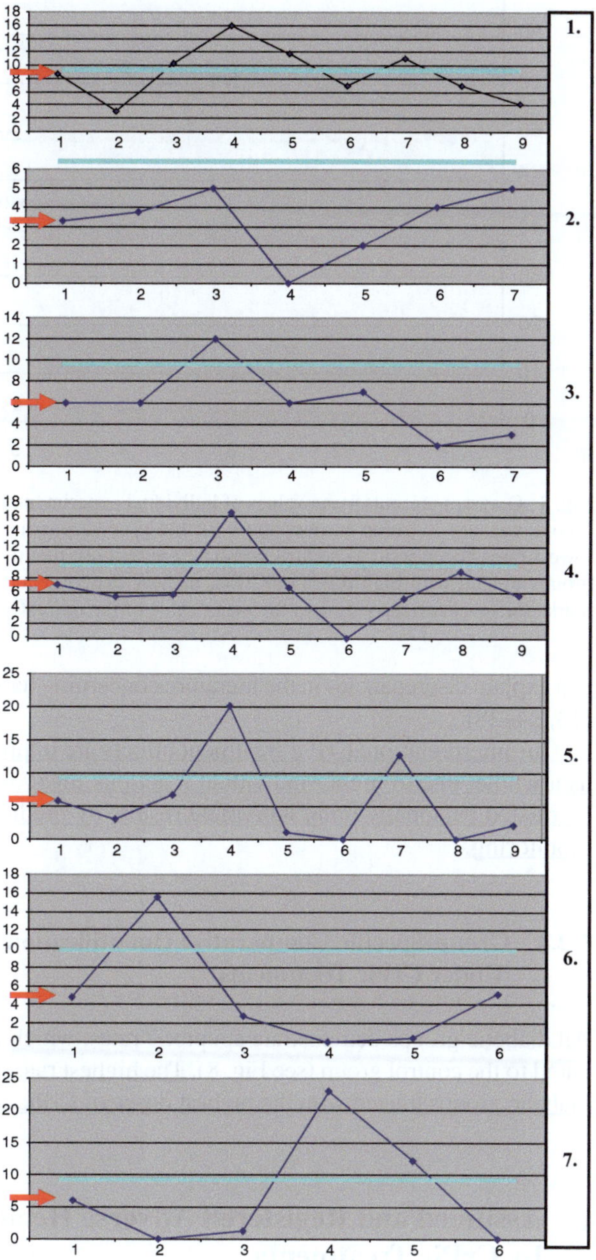

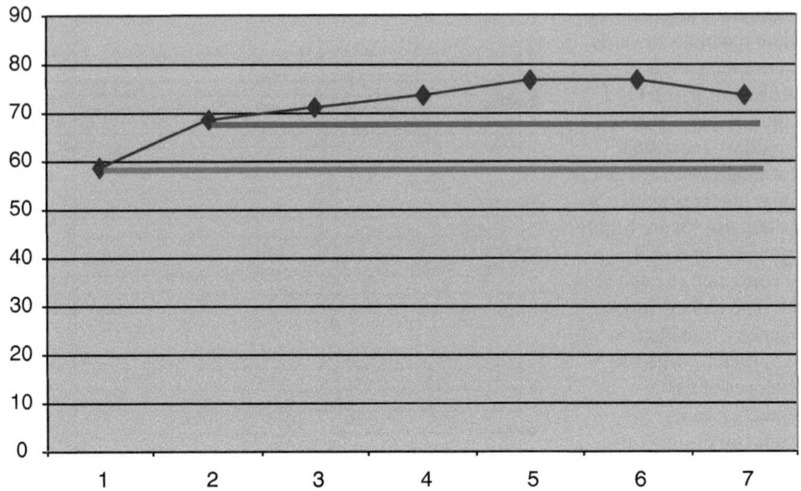

Fig. 8 Diagram presents mean values of individual experimental groups for the comets of class IV (apoptosis): axe X, group numbering (*1–7*); axe Y, corresponding mean values (% of comets). The lowest rates correspond to control group 1 (*green line*); the highest rates correspond to diabetic groups treated with high doses of CrPic, namely experimental groups 5, 6 and 7. The *red line* marks the level corresponding to the mean value of the untreated diabetic group

can explain discrepancies in the literature concerning harmful effects of CrPic-therapy [13, 24–28].

Our interpretation: CrPic-treatment effects are unpredictable for a patient-cohort as a whole, due to highly individual reactions towards therapy. Individuals should be treated personally upon individual results by predictive diagnostics and therapy monitoring.

5.2.4 Group-Specific Comparative Quantification of Apoptosis Under CrPic Treatments

All diabetic groups demonstrate apoptotic rates, which are significantly higher compared to the control group (see Fig. 8). The highest rate of apoptosis was monitored in diabetic groups treated with the highest doses of CrPic namely in groups 5, 6 and 7.

6 Assumed and Registered Adverse Health Outcomes by CrPic Treatments

Taken the above observations together we conclude possible risks for individual long-term effects, when CrPic is being used freely as a therapeutic nutritional modality agent without application of advanced diagnostic tools to predict individual outcomes.

6.1 *In Experimental Models*

In terms of potential treatments of gestational diabetes, there is very little information about the safety of chromium picolinate applied as dietary supplement in pregnant women. However, experiments performed with CrPic administered to pregnant mice resulted in skeletal birth defects in the developing fetus [33].

Comparative cytotoxic and genotoxic studies of trivalent chromium demonstrated that the compound safety depends on the ligand bound to chromium-ion [34]. Thereby, CrPic produces significantly more oxidative stress and DNA damage compared, for example, to niacin-bound chromium(III). The implicated toxicity of CrPic may result in renal impairment, severe biochemical, histological and morphological changes in eye, skin blisters and pustules, anaemia, haemolysis, tissue oedema, liver dysfunction; neuronal cell injury, impaired cognitive, perceptual and motor activity; enhanced production of hydroxyl radicals, chromosomal aberration, depletion of antioxidant enzymes, oxidative stress and DNA damage [23, 29, 34–38]. Increased apoptotic effects by CrPic are demonstrated by the authors in this chapter for kidney and for circulating leukocytes elsewhere [17, 39]. Potential carcinogenic effects are assumed [40, 41].

6.2 *In Man*

A growing body of case reports warns against adverse health outcomes assigned to CrPic dietary application, whereas by others they are interpreted as "anecdotal reports" [42].

Hence, the case reports have described with acute kidney failure, liver damage and anaemia by taking high dosage of CrPic as the dietary supplement [43, 44]. Adverse cutaneous reactions to CrPic supplements have been described [45].

There are some concerns that CrPic may affect levels of neurotransmitters with potential risks for patients treated due to depression, bipolar disorder and schizophrenia [18].

In the broader sense, CrPic supplements are hormone-related and may influence hormone secretion through its function in the endocrine/metabolic system [46]. Chromium supplements taken with medications that block the formation of prostaglandins, such as ibuprofen, indomethacin, naproxen, and aspirin, may increase the absorption of chromium in the body followed by unpredictable consequences such as long-term genotoxic effects by formation of covalent bonds to DNA [26]. Some additive medication effects may be expected, if CrPic therapy is combined with diabetes treatments causing blood glucose levels to dip too low.

Here we are wishing to stress the point that according to our analysis, the above listed adverse heath effects are not expected to happen to everybody. They may occur individually with a severity grade depending on individual predispositions, but can be effectively avoided by application of advanced diagnostic tools for therapy monitoring and predicting individual outcomes. Further, we appeal to react effectively towards the case reports demonstrating adverse health outcomes by artificial dietary supplements often regarded as harmless by the public and lay media.

7 Artificial Supplements for Diabetes Prevention: Hype or Hope?

7.1 Conclusions and Recommendations

Definitely CrPic-therapy is solely one example of several therapy forms which currently are applied "across-the-board" in diabetes care. An administration of artificial supplements is an attempt to prevent or at least to postpone the onset of type 2 diabetes in the group of risk with individuals who are unwilling to make prudent changes in their diets and sedentary habits. Should this approach be considered as the hype or the hope? Considering diabetes type 2 as the multi factorial disease, our answer is - the above question is rather of rhetoric nature.

High efficacy of balanced diet, individually optimised life-style and personalised treatment regiments hardly can be substituted by a limited number of single supplements to cover all the multi factorial risks such as the upward trends of the population ageing, environmental risk factors, urbanisation, additive effects of diverse stress factors, incorrectly chosen life-style including unfavourable nutritional habits, increasing prevalence of obesity, low physical activity, etc.

Although ageing is the well-acknowledged factor contributing to the disease development, there are completely new epidemiologic factors characteristic for the twenty-first century that speed up the disease progression particularly in youth and young adults. Hence, it has been demonstrated that the prognosed DM-rates progression will be inversely increasing with the age (younger age – higher progression), and the youngest group of 20–39 years old people will be delivering the highest rates of diabetic progression that will double the DM-cohort of this age group by the year 2030 compared to 2010 [47]. This is a completely new situation and very big challenge for the societies around the globe requiring special competencies of several groups of professionals as well as innovative approaches in healthcare and health economy.

Population at-risk for diabetes is huge and increasing in pandemic scale. One of the reasons might be the failed attempt to prevent the disease by application of artificial supplements and drugs with hardly recognised individual risks. Consequently, a multimodal approach of integrative medicine by predictive diagnostics, targeted prevention and individually created treatment algorithms is highly desirable.

As discussed and reviewed earlier more individualised treatments are desirable in effective protection against diabetic retinopathy and poly neuropathy, diabetes-related cardiovascular complications and cancer [8]. Further, field-related research is needed to establish simplified non-invasive diagnostic approaches for routine medical practice which would allow for an accurate prediction of individualised therapy risks and outcomes. A promising technological platform has been recently created using approaches with the detection of circulating nucleic acids in blood plasma [48] and clinical proteomics of body fluids [49].

Targeted measures require a creation of new guidelines essential to regulate (renoprotective) therapy approaches and application of more individualised therapeutic modalities for advanced Diabetes care. These measures should provide a legitimate regulation for well-timed predictive diagnostics, an effective prevention and creation of individualised treatment algorithms in pre/diabetes [50].

Acknowledgements The project has been granted by NIH ("Is chromium-picolinate renoprotective in diabetes?", National Institutes of Health, study no. 1R21 AT003012-01A1). The authors thank to Prof. Dr. M.S. Mozaffari (Georgia Regents University, USA) for providing tissue samples in the project. For the statistical analysis and assistance in preparation of the original paper [31], the authors thank to V. Peeva, M.Sc. and A. Shenoy, M.Sc., University of Bonn, Germany.

References

1. George B, Cebioglu M, Yeghiazaryan K (2010) Inadequate diabetic care: global figures cry for preventive measures and personalized treatment. EPMA J 1:13–18
2. Golubnitschaja O (2012) Time for new guidelines in advanced healthcare: the mission of the EPMA journal to promote an integrative view in predictive, preventive and personalized medicine. EPMA J 3:5
3. Golubnitschaja O (2010) Time for new guidelines in advanced diabetes care: paradigm change from delayed interventional approach to predictive, preventive & personalized medicine. EPMA J 1:3–12
4. Golubnitschaja O (2009) Diabetes mellitus. In: Golubnitschaja O (ed) Predictive diagnostics and personalized treatment: dream or reality? Nova Science Publishers Inc, New York
5. Golubnitschaja O, Jaksche A, Moenkemann H, Yeghiazaryan K, Karl SE, Trog D, Schild HH, Löffler KU (2005) Molecular imaging system for possible prediction of active retinopathy in patients with diabetes mellitus. Amino Acids 28:229–237
6. Golubnitschaja O (2009) Advanced technologies for prediction of secondary complications in diabetes mellitus. In: Golubnitschaja O (ed) Predictive diagnostics and personalized treatment: dream or reality? Nova Science Publishers Inc, New York
7. Mozaffari M, Abdelsayed R, Schaffer S (2009) Diabetic complications: pathogenic mechanisms and prognostic indicators. In: Golubnitschaja O (ed) Predictive diagnostics and personalized treatment: dream or reality? Nova Science Publishers Inc, New York
8. Golubnitschaja O (2010) Advanced diabetes care: three levels of prediction, prevention & personalized treatment. Curr Diabetes Rev 6:42–51
9. Golubnitschaja O, Swanton C, Danesi R, Costigliola V (2011) Promoting predictive, preventive and personalised medicine: European event of global importance. EPMA J 2:131–136
10. van Heerden PV, Jenkins IR, Woods WP, Rossi E, Cameron PD (1994) Death by tanning–a case of fatal basic chromium sulphate poisoning. Intensive Care Med 20:145–147
11. Fan Y, Ovesen JL, Puga A (2012) Long-term exposure to hexavalent chromium inhibits expression of tumor suppressor genes in cultured cells and in mice. J Trace Elem Med Biol 26:188–191

12. Urbano AM, Ferreira LMR, Alpoim MC (2012) Molecular and cellular mechanisms of hexavalent chromium-induced lung cancer: an updated perspective. Curr Drug Metab 13:284–305
13. Rhodes MC, Hébert CD, Herbert RA, Morinello EJ, Roycroft JH, Travlos GS, Abdo KM (2005) Absence of toxic effects in F344/N rats and B6C3F1 mice following subchronic administration of chromium picolinate monohydrate. Food Chem Toxicol 43:21–29
14. Jeejeebhoy KN, Chu RC, Marliss EB, Greenberg GR, Bruce-Robertson A (1977) Chromium deficiency, glucose intolerance, and neuropathy reversed by chromium supplementation, in a patient receiving long-term total parenteral nutrition. Am J Clin Nutr 30:531–538
15. Brown RO, Forloines-Lynn S, Cross RE, Heizer WD (1986) Chromium deficiency after long-term total parenteral nutrition. Dig Dis Sci 31:661–664
16. Nielsen F (1994) Chromium. In: Shils M, Olson J, Shike M (eds) Modern nutrition in health and disease. Lea & Febiger, Philadelphia
17. Jana M, Rajaram A, Rajaram R (2009) Chromium picolinate induced apoptosis of lymphocytes and the signaling mechanisms thereof. Toxicol Appl Pharmacol 237:331–344
18. Vincent JB (2003) The potential value and toxicity of chromium picolinate as a nutritional supplement, weight loss agent aond muscle development agent. Sports Med 33:213–230
19. Yazaki Y, Faridi Z, Ma Y, Ali A, Northrup V, Njike VY, Liberti L, Katz DL (2010) A pilot study of chromium picolinate for weight loss. J Altern Complement Med 16:291–299
20. Trumbo PR, Ellwood KC (2006) Chromium picolinate intake and risk of type 2 diabetes: an evidence-based review by the United States Food and Drug Administration. Nutr Rev 64:357–363
21. Iqbal N, Cardillo S, Volger S, Bloedon LT, Anderson RA, Boston R, Szapary PO (2009) Chromium picolinate does not improve key features of metabolic syndrome in obese nondiabetic adults. Metab Syndr Relat Disord 7:143–150
22. Vincent JB (2000) Elucidating a biological role for chromium at a molecular level. Acc Chem Res 33:503–510
23. Andersson MA, Petersson Grawé KV, Karlsson OM, Abramsson-Zetterberg LAG, Hellman BE (2007) Evaluation of the potential genotoxicity of chromium picolinate in mammalian cells in vivo and in vitro. Food Chem Toxicol 45:1097–1106
24. Bailey MM, Sturdivant J, Jernigan PL, Townsend MB, Bushman J, Ankareddi I, Rasco JF, Hood RD, Vincent JB (2008) Comparison of the potential for developmental toxicity of prenatal exposure to two dietary chromium supplements, chromium picolinate and [Cr3O(O2CCH2CH3)(6(H2O)3]+, in mice. Birth Defects Res B Dev Reprod Toxicol 83:27–31
25. Speetjens JK, Collins RA, Vincent JB, Woski SA (1999) The nutritional supplement chromium(III) tris(picolinate) cleaves DNA. Chem Res Toxicol 12:483–487
26. Kareus SA, Kelley C, Walton HS, Sinclair PR (2001) Release of Cr(III) from Cr(III) picolinate upon metabolic activation. J Hazard Mater 84:163–174
27. Manygoats KR, Yazzie M, Stearns DM (2002) Ultrastructural damage in chromium picolinate-treated cells: a TEM study. Transmission electron microscopy. J Biol Inorg Chem 7:791–798
28. Stearns DM, Silveira SM, Wolf KK, Luke AM (2002) Chromium(III) tris(picolinate) is mutagenic at the hypoxanthine (guanine) phosphoribosyltransferase locus in Chinese hamster ovary cells. Mutat Res 513:135–142
29. Hepburn DDD, Xiao J, Bindom S, Vincent JB, O'Donnell J (2003) Nutritional supplement chromium picolinate causes sterility and lethal mutations in Drosophila melanogaster. Proc Natl Acad Sci USA 100:3766–3771
30. Stallings DM, Hepburn DDD, Hannah M, Vincent JB, O'Donnell J (2006) Nutritional supplement chromium picolinate generates chromosomal aberrations and impedes progeny development in Drosophila melanogaster. Mutat Res 610:101–113
31. Yeghiazaryan K, Peeva V, Shenoy A, Schild HH, Golubnitschaja O (2011) Chromium-picolinate therapy in diabetes care: molecular and subcellular profiling revealed a necessity for individual outcome prediction, personalised treatment algorithms and new guidelines. Infect Disord Drug Targets 11:188–195

32. Golubnitschaja O, Moenkemann H, Kim K, Mozaffari MS (2003) DNA damage and expression of checkpoint genes p21(WAF1/CIP1) and 14-3-3 sigma in taurine-deficient cardiomyocytes. Biochem Pharmacol 66:511–517
33. Bailey MM, Boohaker JG, Sawyer RD, Behling JE, Rasco JF, Jernigan JJ, Hood RD, Vincent JB (2006) Exposure of pregnant mice to chromium picolinate results in skeletal defects in their offspring. Birth Defects Res B Dev Reprod Toxicol 77:244–249
34. Bagchi D, Stohs SJ, Downs BW, Bagchi M, Preuss HG (2002) Cytotoxicity and oxidative mechanisms of different forms of chromium. Toxicology 180:5–22
35. Mahmoud AA, Karam SH, Abdel-Wahhab MA (2006) Chromium-picolinate induced ocular changes: protective role of ascorbic acid. Toxicology 226:143–151
36. Coryell VH, Stearns DM (2006) Molecular analysis of hprt mutations induced by chromium picolinate in CHO AA8 cells. Mutat Res 610:114–123
37. Tan G-Y, Zheng S-S, Zhang M-H, Feng J-H, Xie P, Bi J-M (2008) Study of oxidative damage in growing-finishing pigs with continuous excess dietary chromium picolinate intake. Biol Trace Elem Res 126:129–140
38. Refaie FM, Esmat AY, Mohamed AF, Aboul Nour WH (2009) Effect of chromium supplementation on the diabetes induced-oxidative stress in liver and brain of adult rats. Biometals 22:1075–1087
39. Jana M, Rajaram R, Rajaram A (2010) Autoschizis of T-cells is induced by the nutritional supplement, Cr(III)picolinate. Toxicol In Vitro 24:586–596
40. Stout MD, Nyska A, Collins BJ, Witt KL, Kissling GE, Malarkey DE, Hooth MJ (2009) Chronic toxicity and carcinogenicity studies of chromium picolinate monohydrate administered in feed to F344/N rats and B6C3F1 mice for 2 years. Food Chem Toxicol 47:729–733
41. Collins BJ, Stout MD, Levine KE, Kissling GE, Melnick RL, Fennell TR, Walden R, Abdo K, Pritchard JB, Fernando RA, Burka LT, Hooth MJ (2010) Exposure to hexavalent chromium resulted in significantly higher tissue chromium burden compared with trivalent chromium following similar oral doses to male F344/N rats and female B6C3F1 mice. Toxicol Sci 118:368–379
42. Lamson DW, Plaza SM (2002) The safety and efficacy of high-dose chromium. Altern Med Rev 7:218–235
43. Cerulli J, Grabe DW, Gauthier I, Malone M, McGoldrick MD (1998) Chromium picolinate toxicity. Ann Pharmacother 32:428–431
44. Wani S, Weskamp C, Marple J, Spry L (2006) Acute tubular necrosis associated with chromium picolinate-containing dietary supplement. Ann Pharmacother 40:563–566
45. Young PC, Turiansky GW, Bonner MW, Benson PM (1999) Acute generalized exanthematous pustulosis induced by chromium picolinate. J Am Acad Dermatol 41:820–823
46. Di Luigi L (2008) Supplements and the endocrine system in athletes. Clin Sports Med 27: 131–151, ix
47. Yeghiazaryan K, Cebioglu M, Golubnitschaja O (2012) Global figures argue in favour of preventive measures and personalised treatment to optimise inadequate diabetes care. In: Mozaffari M (ed) New strategies to advance pre/diabetes care: integrative approach by PPPM. In book series: Golubnitschaja O (ed) Advances in PPPM. Springer, Dordrecht/Heidelberg/New York/London
48. Gahan PB (2008) Circulating nucleic acids in plasma and serum: roles in diagnosis and prognosis in diabetes and cancer. Infect Disord Drug Targets 8:100–108
49. Golubnitschaja O (2008) Clinical proteomics in application to predictive diagnostics and personalized treatment of diabetic patients. Curr Proteomics 5:35–44
50. Golubnitschaja O, Costigliola V (2010) Common origin but individual outcomes: time for new guidelines in personalized healthcare. Personalized Med 7:561–568

Remote Control in Diabetes

Dimiter V. Dimitrov

Abstract The demonstration that near-normal glucose control substantially lowers microvascular and cardiovascular complications has heralded a new era of diabetes care. Major improvements in glucose monitoring and insulin delivery introduced in the past decade are now helping patients control their blood glucose more precisely and conveniently and reduce the risk of hypoglycemia. Tight glucose control requires almost continuous measurements and different sensors for continuous blood glucose measurement have been under development for the last two decades. Minimally invasive sensors able to measure glucose in interstitial fluid, and thus more suitable for self-monitoring, have also been developed. Real-time access to healthcare information on mobile devices such as iPad, Blackberry, iPhone, Windows PDA mobile and other popular smart phones enables healthcare providers and partners to have anytime-anywhere access to information while ensuring patient confidentiality and 24/7 access to critical care regardless of the location. Robust, clinically acceptable remote devices are widely expected to become available in the near years to come.

Keywords Diabetes management • Enhanced home care support • Remote control • Microchips • ICT intelligent platforms • Virtual clinical trials • Personal health systems

D.V. Dimitrov (✉)
Diavita Ltd, Bulgaria
e-mail: dimiter.v.dimitrov@gmail.com

M.S. Mozaffari (ed.), *New Strategies to Advance Pre/Diabetes Care: Integrative Approach by PPPM*, Advances in Predictive, Preventive and Personalised Medicine 3,
DOI 10.1007/978-94-007-5971-8_15, © Springer Science+Business Media Dordrecht 2013

1 Preventive Monitoring: A Revolution in E-Health

Information technology plays an important role in many aspects of our society and is an emerging force in the healthcare industry. One area of health IT that shows great promise is remote health monitoring, which during the past 10 years has risen from humble beginnings to become a significant player in healthcare.

Remote health monitoring is designed to help empower patients in becoming more actively involved with their personal health management [1]. Patients with a chronic illness, such as diabetes, asthma, or hypertension, need to be motivated to follow management plans and make necessary lifestyle modifications in order to minimize the chances that they will develop additional complications that could further jeopardize their health, thus requiring expensive treatment [2]. The feedback provided by constant monitoring can help reinforce adherence and good health practices. Researchers in remote health monitoring have even found that patients who use the technology make fewer unnecessary doctor visits and are less likely to require hospitalization. Remote health monitoring can also help reduce healthcare costs [3]. That is the first degree of personalization for the patient – it's personalized to their condition. Herewithin some examples of available platforms for remote monitoring in diabetes care are given.

2 ePatch: Wireless Body Measurements

ePatch is a breakthrough technology platform ready to carry a wide range of sensors that monitor physiological parameters direct from the wearer's skin (Fig. 1). These sensors provide real-time information about the user's health, movement and body

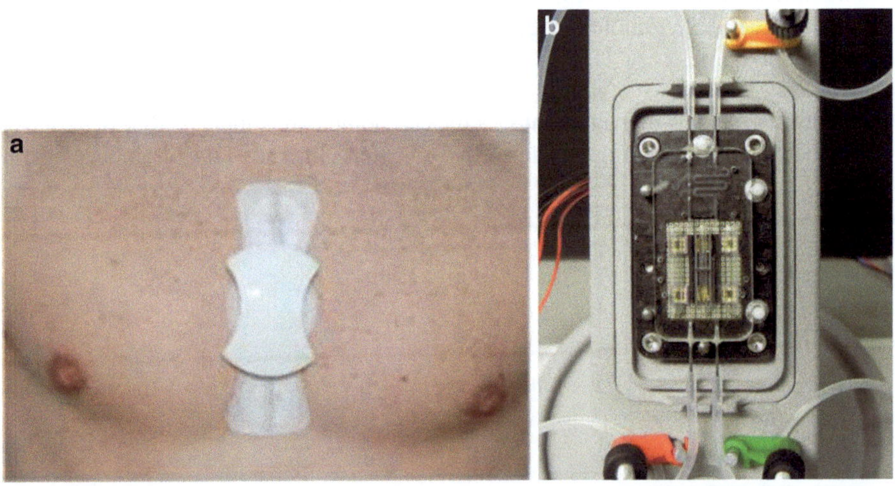

Fig. 1 ePatch skin sensors. One type of sensor is non-invasive with measuring technique based on impedance spectroscopy. Changes in glucose level cause changes in the skin's impedance which is measured by electrodes in direct contact with the patient's skin (**a**). The second type is minimally invasive with measuring technique based on infrared spectroscopy. Using a subcutaneous microneedle, absorption spectroscopy is performed on the interstitial liquid or blood (**b**)

status. This valuable data is wirelessly transmitted to any chosen type of receiver, and can be put to use in countless different ways [4].

ePatches are ideal for monitoring people who're at risk of developing a particular condition, or whose health or handicap makes certain activities a risk. Intelligent filtering of the monitoring data from the ePatch platform can be made to set off an alarm if the wearer exceeds any pre-determined physiological parameters, e.g. hypoglycemia. This paves the way to prompt remedial action that can limit harm or save lives. The sensors connect through a wireless network to various hospital systems and advanced mathematical models that can assess the patients' condition and provide directions for insulin therapy and possible crisis situations. The platform offers closed-loop feedback in both hospital environments and in outpatient clinics to patients, carers and healthcare professionals at the point of care with important information for effectively control of blood glucose levels in all situations so that health risks are minimised and the lives of patients improved.

The REACTION project is a 4-year research project funded by the European Commission, as part of the objective of the 7th Framework Programme for Personal Health Systems [5]. The project aims to develop an integrated ICT platform that supports improved long term management of diabetes based on wearable, continuous blood glucose monitoring sensors, such as ePatch and automated closed-loop delivery of insulin.

MicroCHIPS drug delivery and sensor technology both leverage the ability to microfabricate multiple wells into a microchip, and integrate that into an implantable device that uses these wells to deliver drugs or sense various analytes [6]. Much like existing pacemakers and other implantable devices, the MicroCHIPs device also communicates wirelessly, allowing both control of the device and data transmission (Fig. 2). MicroCHIPS' first biochemical sensor product in development is a continuous glucose sensing platform designed to transform how people with diabetes manage the condition. The development of a Continuous Glucose Monitor (CGM) has been attempted and abandoned by a number of companies and researchers [7]. The MicroCHIPS approach solves several of the more vexing problems associated with an implantable sensor including the biological fouling of the sensor and the need for an oxygen source to drive the chemical reaction that detects glucose levels. MicroCHIPS' continuous glucose monitor will be a discreet, long-term implanted monitor designed to wirelessly deliver continuous, convenient, reliable and accurate glucose measurements. Sophisticated alerts and alarms will be designed to provide additional peace of mind and security — if glucose levels fluctuate beyond pre-set boundaries, alerts can warn the user, or his/her caregivers, of the condition.

Dynamic Electrochemistry is an innovative technology which extracts a spectrum of information from the blood that is inaccessible to traditional electrochemical methods [8]. This new method compensates for many interfering factors that can often distort blood glucose results, helping to ensure accurate and reliable blood glucose readings (Fig. 3). As hematocrit variations occur in daily routine (e.g., because of smoking, exercise, hypermenorrhea, pregnancy, stay in mountains, etc.),

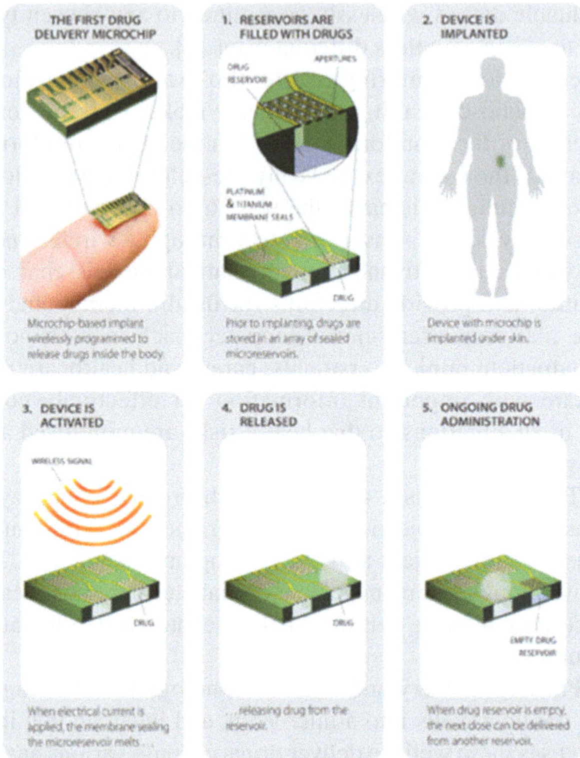

Fig. 2 An implantable microchip device that provides real-time dose schedule tracking, and as part of a network, physicians can remotely adjust treatment schedules as necessary

dynamic electrochemistry may be useful for stable measures under these conditions. It is now used in the BGStar device (Sanofi-Aventis), which appears to be an effective technology to correct for potential environmental influence on blood glucose results [9].

3 Mobile Applications for Diabetes

Various strategies using electronic technologies or educational programs have been proposed to improve the quality and efficiency of care for people with diabetes [10]. Recently, mobile phones have been developed that can measure blood glucose level and transmit data directly to doctors. The combination of a glucose meter with a

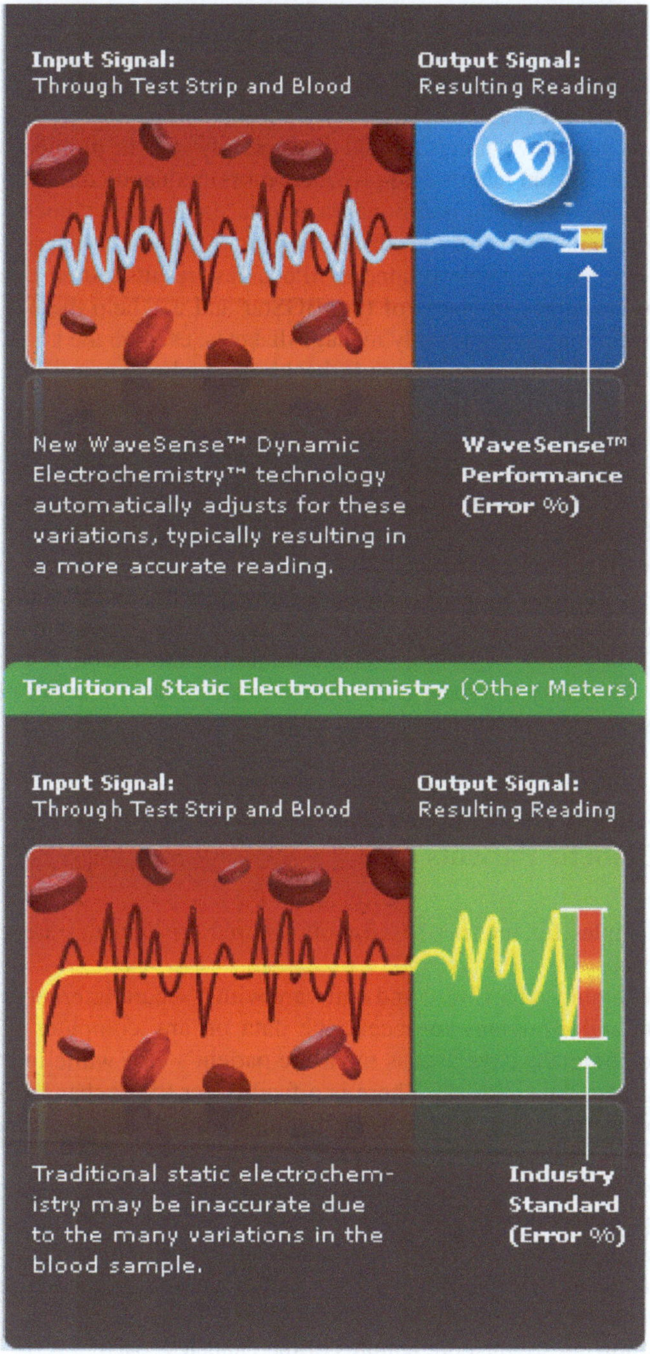

Fig. 3 Dynamic electrochemistry extracts a spectrum of information inaccessible to traditional static electrochemical methods. Error-correcting input signals from WaveSense™ enabled blood glucose meters provide more comprehensive data. This information can then be exploited by sophisticated digital signal processing algorithms to give more accurate glucose readings

mobile phone makes it easier for patients to report their self-monitored blood glucose (SMBG) levels and receive feedback from their doctors [11].

These systems allow patients to send data, identify specific problems and receive recommendations from health-care providers more frequently and with fewer time and place limitations than at traditional face-to-face outpatient clinics.

Sanofi – Aventis have also introduced iBgStar, a Diabetes Manager App, which can be accessed from Apple iPhone or iPod. Test results stored on the meter's memory are visualized in charts of the iBGStar and can help the patient analyze trends and variations to proactively manage diabetes, even on the go [12].

The need for mobile applications in healthcare and processes is at an all time high. Healthcare providers, doctors, clinicians and nurses need real-time access to patient, schedule, billing, drug and other practice related information to provide best patient care. On the other hand, insurance companies, drug suppliers and vendors of the healthcare ecosystem need 24/7 access to business critical information regardless of the location [13].

Happtique has found that there are 23,000 medical apps for iOS and Android devices alone and aims to build a standard for mobile health apps [14]. This is the first mobile application store developed by healthcare professionals and offers healthcare enterprises – like hospitals, continuing care facilities, and physician practices – the ability to create individually branded, secure, multi-platform app stores that support employee and patient mobile technology use.

4 Web-Based Communication Between Care Providers and Diabetic Patients

Specific examples for Internet based tools tailored for diabetics are Diasend and DiabetesCoach.

Diasend combines hardware and software with a secure internet server allowing patients and clinicians to access the data on any computer with internet access. The clinicians have access to all the patient's data while individuals can access their own data (Fig. 4). There is a facility for the exchange of messages. The suite of analysis tools allows graphical, tabular and statistical analysis of the data [15].

DiabetesCoach, is a supporting self-care among patients with type 2 diabetes and was developed to encourage patients to play a more active role in their own care. The web application is provided free of charge as a supplement to regular diabetes care. The application was developed in close collaboration with general practitioners, nurses, patients, behavioral scientists, and vendors (i.e. health insurance companies) [16]. The following are the core features of the DiabetesCoach:

- *My personal data*: documentation of personal details (e.g., treatment plan, medication use).
- *Online monitoring*: registration of metabolic values: weight, blood glucose level, blood pressure, and cholesterol.

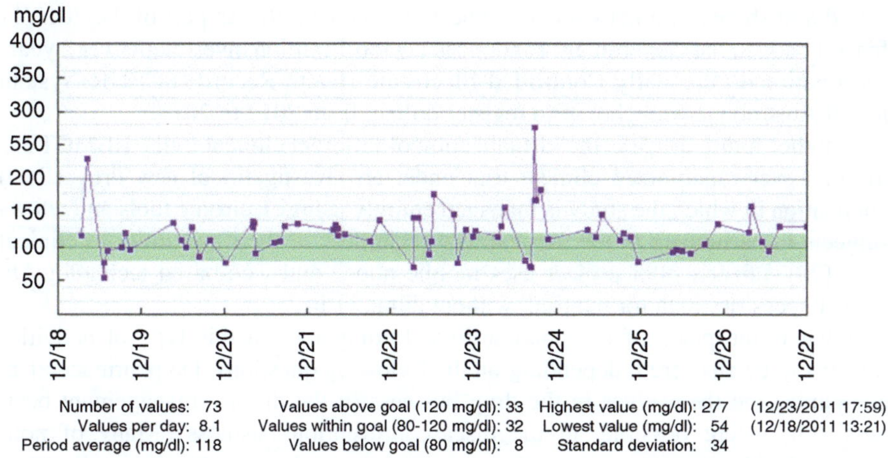

Fig. 4 Diasend patient weekly summary. Where 45% of blood glucose is above target, 11% is below and 44% within target

- *Email contact*: secured possibility for patient–nurse email communication (response within 5 working days).
- *Online education*: diabetes information and instructions.
- *Calendar*: a place to write down comments, appointments, and personal goals.
- *Lifestyle coach*: self-tests to support lifestyle changes.

5 Future Directions: Authorities, Pharma, Public Health

With the support of remote control devices the doctors with can manage appointments at their finger tips and make instant decisions. The doctors can also view the patient scheduling and alter appointments as necessary with access health record systems like Microsoft HealthVault and Google Health [17]. Real-time access to Electronic Health Records (EHR) and Electronic Medical Records (EMR) can stream home-based medical care by enabling two-way communications between the caregivers and healthcare administrative systems.

The U.S. Food and Drug Administration (FDA) advanced the federal effort to create electronic health records for Americans within the next decade by making it easier to share drug information electronically. FDA is moving the effort forward by adopting the Systematized Nomenclature of Medicine (SNOMED) as the standard computerized medical vocabulary system to be used to electronically code important terms in the Highlights section of prescription drug labeling. This move will allow healthcare professionals nationwide electronically to access and share critical health and treatment information more easily and efficiently [18].

Successful example of implementation is the Australian Personally Controlled Electronic Health Record (PCEHR) – a secure, electronic record of medical history, stored and shared in a network of connected systems [19]. As part of the 2010/11 federal budget, the Government announced a $466.7 million investment over 2 years for a national Personally Controlled Electronic Health Record (PCEHR) system for all Australians who choose to register online, from 2012 to 2013.

Another future usage is the virtualization of different clinical trails. REMOTE is the first-ever randomized clinical trial under an investigational new drug (IND) application in which the study is managed entirely using electronic tools and allows subjects to participate in the clinical trial regardless of their proximity to clinical sites [20]. Pfizer's pilot project uses mobile phone and web-based technology to collect necessary data for the trial, without clinic visits.

What is the place of the pharmaceutical companies in this type of activities remains to be answered, depending on the following questions. Do pharmaceutical companies see themselves in the drug business or the disease management business? Or, where possible, in the disease prevention business? Many of non-communicable diseases such as diabetes lend themselves to the use applications (mobile and web) or biometric devices. This is going to drive a greatly expanded focus on non-traditional partnerships. An example of how a pharmaceutical company could transition from being product centric to being customer centric is doing something like the company Ambucor [21]. Ambucor provides Ambulatory Electrocardiographic and Remote Device Monitoring services for cardiologists. Pharma companies already have deep relationships with cardiologists. Thus next widespread disease such as diabetes is just knocking on the door for open avenues towards remote monitoring.

In 2007, US Senate introduced the Remote Monitoring Access Act of 2007, which would "provide for coverage of remote patient management services for chronic healthcare conditions under the Medicare Program." The text included nine statements (or "findings"), such as: "Remote patient monitoring can make chronic disease management more effective and efficient for patients and the health care system" and "Utilization of these technologies not only improves the quality of care given to patients, it also reduces the need for frequent physician office appointments, costly emergency room visits, and unnecessary hospitalizations" [22].

For the over 35 million diabetics in the US, and over 300 million people worldwide, closely managing glucose levels is critical to avoiding the acute dangers of low blood sugar and the long term complications of prolonged hyperglycemia. Complications from poorly controlled diabetes include cardiovascular disease, renal failure, blindness and amputation. In 2007 US healthcare costs directly linked to diabetes were estimated at $174 billion [23].

The introduction of the Internet into clinical practice has brought about many opportunities for self-care, as it can be used as a powerful medium for promoting a healthy lifestyle and for increasing understanding about the condition. However, to be effective in empowering patients' self-awareness and engagement, remote control applications should be designed to allow individuals to tailor the program to their own specific needs, because patients are increasingly demanding convenient access to a high level of personalized health care.

References

1. Adibi S (2012) Link technologies and BlackBerry mobile health (mHealth) solutions: a review. IEEE Trans Inf Technol Biomed 16:586–597
2. Franc S, Daoudi A, Mounier S, Boucherie B, Dardari D, Laroye H, Neraud B, Requeda E, Canipel L, Charpentier G (2011) Telemedicine and diabetes: achievements and prospects. Diabetes Metab 37:463–476
3. Better Health Care Together (2008) Remote monitoring technologies could shave health care costs by $197 billion; broadband-based applications can improve care For chronic disease. Medical News Today. MediLexicon, Intl., Web: http://www.medicalnewstoday.com/releases/127982.php. Accessed 5 Dec 2012
4. Lobodzinski SS, Laks MM (2012) New devices for very long-term ECG monitoring. Cardiol J 19:210–214
5. Spat S, Höll B, Plank J, Schaupp L, Neubauer K, Chiarugi F, Beck P, Pieber TR (2012) A mobile in-hospital application supporting insulin dosing for patients with diabetes type 2. Oral presentation for ATTD 2012: advanced technologies and treatments for diabetes, Barcelona, Spain, 8–11 Feb 2012
6. Prescott JH, Lipka S, Baldwin S, Sheppard NF Jr, Maloney JM, Coppeta J, Yomtov B, Staples MA, Santini JT Jr (2006) Chronic, programmed polypeptide delivery from an implanted, multireservoir microchip device. Nat Biotechnol 24:437–438
7. DeVries JH (2012) Continuous glucose monitoring: coming of age? Eur J Endocrinol 166:1–4
8. Rice MJ (2011) Dynamic electrochemistry: a step in the right direction. J Diabetes Sci Technol 5:1176–1178
9. Pfützner A, Mitri M, Musholt PB, Sachsenheimer D, Borchert M, Yap A, Forst T (2012) Clinical assessment of the accuracy of blood glucose measurement devices. Curr Med Res Opin 28:525–531
10. Benhamou PY, Melki V, Boizel R, Benhamou PY, Melki V, Boizel R, Perreal F, Quesada JL, Bessieres-Lacombe S, Bosson JL, Halimi S, Hanaire H (2007) One-year efficacy and safety of web-based follow-up using cellular phone in type 1 diabetic patients under insulin pump therapy: the PumpNet study. Diabetes Metab 33:220–226
11. Carroll AE, Marrero DG, Downs SM (2007) The HealthPia GlucoPack Diabetes Phone: a usability study. Diabetes Technol Ther 9:158–164
12. iBGStar - Blood Glucose Meter for iPhone & iPod touch. http://www.ibgstar.us/. Accessed 5 Dec 2012
13. Holtz B, Lauckner C (2012) Diabetes management via mobile phones: a systematic review. Telemed J E Health 18:175–184
14. Peck AD (2011) App-solutely fabulous. Hundreds of new apps for iPAD and tablets make mHealth a reality and a lifestyle choice. Med Econ 88:S11–S14
15. Lamb W (2008) DIASEND®: one device to read them all and in the clinic guide them. Endocrine Abstracts 17:p48
16. Nijland N, van Gemert-Pijnen JE, Kelders SM, Brandenburg BJ, Seydel ER (2011) Factors influencing the use of a web-based application for supporting the self-care of patients with type 2 diabetes: a longitudinal study. J Med Internet Res 13:e71
17. Hsieh SH, Hsieh SL, Cheng PH, Lai F (2012) E-health and healthcare enterprise information system leveraging service-oriented architecture. Telemed J E Health 18:205–212
18. de Lusignan S (2011) SNOMED is coming, and more about using and interacting with technology in primary care. Inform Prim Care 19:1–2
19. Liaw ST, Chen HY, Maneze D, Taggart J, Dennis S, Vagholkar S, Bunker J (2012) Health reform: is routinely collected electronic information fit for purpose? Emerg Med Australas 24:57–63
20. Web-based methodology trial to evaluate the efficacy and safety of tolterodine ER in subjects with overactive bladder (REMOTE). Web: http://www.clinicaltrials.gov/ct2/show/NCT01302938?term=oab+pfizer&rank=24. Accessed 5 Dec 2012

21. Arrocha A, Klein GJ, Benditt DG, Sutton R, Krahn AD (2010) Remote electrocardiographic monitoring with a wireless implantable loop recorder: minimizing the data review burden. Pacing Clin Electrophysiol 33(11):1347–1352
22. Ubl SJ (2007) Public policy implications for using remote monitoring technology to treat diabetes. J Diabetes Sci Technol 1(3):436–439
23. Campbell AW (2011) The diabetes pandemic. Altern Ther Health Med 17:8–9

Drug Delivery Systems for Predictive Medicine: Polymers as Tools for Advanced Applications

Ana C. Fonseca, Paula Ferreira, Rosemeyre A. Cordeiro, Patrícia V. Mendonça, Joana R. Góis, Maria H. Gil, and Jorge F.J. Coelho

Abstract Predictive medicine represents a new philosophy in healthcare involving the detection of pathology-specific molecular patterns before the emergence of signs and symptoms, which in some diseases (*e.g.*, diabetes) surged accompanied by severe secondary complications. Advanced drug delivery systems (DDS) present important benefits for this new and fascinating area, since they enable to deploy active molecules in specific targeted regions of the body in a controlled manner. The exponential development achieved in DDS during the last decades can be used to develop effective solutions for the premature detection and effective treatment of different pathologies. Polymers provide the ideal opportunities for the development of new effective DDS due to the easy processing and to the control over the physical and chemical characteristics that can be accomplished during the polymerization.

This chapter provides an overview of the available DDS in view to understand the most important concepts of the area. An important focus is given to the sort of polymers commonly used and reported in the literature, while the advantages and disadvantages of each type of material is critically discussed. Finally, the miniaturization of DDS is described and the most important methods to prepare micro- and nanocarriers are presented.

A.C. Fonseca • R.A. Cordeiro • P.V. Mendonça • J.R. Góis • M.H. Gil
Department of Chemical Engineering, University of Coimbra, Coimbra 3030-290, Portugal

P. Ferreira
Department of Chemical Engineering, University of Coimbra, Coimbra 3030-290, Portugal

Department of Health Sciences, Portuguese Catholic University, Viseu 3504-505, Portugal

J.F.J. Coelho (✉)
Department of Chemical Engineering, University of Coimbra, Coimbra 3030-290, Portugal

Departamento de Engenharia Química, University of Coimbra, Polo II/Rua Silvio Lima, Coimbra 3030-790, Portugal
e-mail: jcoelho@eq.uc.pt

M.S. Mozaffari (ed.), *New Strategies to Advance Pre/Diabetes Care: Integrative Approach by PPPM*, Advances in Predictive, Preventive and Personalised Medicine 3,
DOI 10.1007/978-94-007-5971-8_16, © Springer Science+Business Media Dordrecht 2013

Keywords Drug delivery systems • Polymers • Miniaturization • Liposomes • Nanoparticles • Microparticles

Abbreviations

AAm	Acrylamide
bFGF	Basic fibroblast growth factor
BMP-2	Bone morphogenetic proteins
BSA	Bovine serum albumin
cfDNA	calf thymus DNA
CLRP	Controlled/living radical polymerization
DDS	Drug delivery systems
Dex-HEMA	Dextran-hydroxyethyl methacrylate
Dex-MA	Dextran-methacrylate
Dex-SA	Stearate-g-dextran
DNA	Deoxyribonucleic acid
DOX	Doxorubicin
EC	Ethylcellulose
FDA	Food and Drug Administration
FITC-Dextran	Fluorescein isothiocyanate-dextran
GUVs	Giant unilamellar vesicles
HeLa	Human cervical carcinoma
HEMA	2-hydroxyethyl methacrylate
HM	Heptamolybdate
HMQC	Hydrophobically modified quaternized cellulose
HPMC	Hydroxypropylmethylcellulose
HUVECs	Human umbilical vein endothelial cells
LCST	Lower critical solution temperature
LUVs	Large unilamellar vesicles
MLVs	Multilamellar vesicles
MMP	Matrix metalloproteinase
NDDAAm	N,N-Didodecylacrylamide
ODA	Octadecyl acrylate
PAA	Poly(acrylic acid)
PACA	Poly(alkylcyanoacrylates)
PC	Phosphatidylcholine
PCL	poly(ε-caprolactone)
PDLA	Poly(D-lactic acid)
PDLLA	Poly(D,L-lactic acid)
pDNA	Plasmid DNA
PE	Phosphatidylethanolamine
PEAA	Poly(2-ethyl acrylic acid)
PEG	Poly(ethylene glycol)
PEI	Poly(ethyleneimine)

PEO	Poly(ethylene oxide)
PGA	Poly(glycolic acid)
PHEMA	Poly(2-hydroxyethyl methacrylate)
PLA	Poly(lactic acid)
PLGA	Poly(lactic-co-glycolic acid)
PLL	Poly(L-lysine)
PLLA	Poly(L-lactic acid)
PMMA	Poly(methyl methacrylate)
PNIPAAm	Poly(N-isopropylacrylamide)
POE	Poly(ortho esters)
RESOLV	Rapid expansion of supercritical solution into liquid solvent
RESS	Rapid expansion of supercritical solution
rhBDNF	Recombinant human brain-derived neurotrophic factor
rhBMP-2	Recombinant human bone morphogenetic protein-2
rhVEGF	Recombinant vascular endothelial growth factor
SA	Stearic acid
SCF	Supercritical fluids
siRNA	Small interfering RNA
Sp H1	Sea urchin sperm histone H1
SucPG	Succinylated poly(glycidol)
SUVs	Small unilamellar vesicle
VEGF	Vascular endothelial growth factor

1 An Overview in Drug Delivery Systems

The introduction of drugs in the human body may be accomplished by several anatomic routes (Fig. 1) [1].

In order to achieve the therapeutic purpose, the choice of the most suitable administration route is of unquestionable importance. Therefore, several factors must be taken into consideration when administrating a drug, namely its own properties, the disease to be treated and the desired therapeutic time. The drugs can be administrated directly to the target tissue or organ or can be delivered by systemic routes [2]. Systemic drug delivery routes, as well as the time necessary to achieve therapeutic effect, are presented systematically in Table 1.

Pharmaceutical treatments started plenty of decades, or even centuries ago (*e.g.*, aspirin since 1828) both with the oral administration of solid pills (and liquids) [4] and injectable active chemical drugs [5]. When either of these methods is applied, drug dose maintenance in the body is achieved by repeated administrations. Despite the effectiveness of these treatments, dose peaks at administration times alternated with sub-therapeutic drug levels are unavoidable. Therefore, the impossibility of controlling the drug level over a long period of time constituted an important drawback. During the past two decades, new approaches and strategies have been developed to control several parameters considered essential for enhancing the

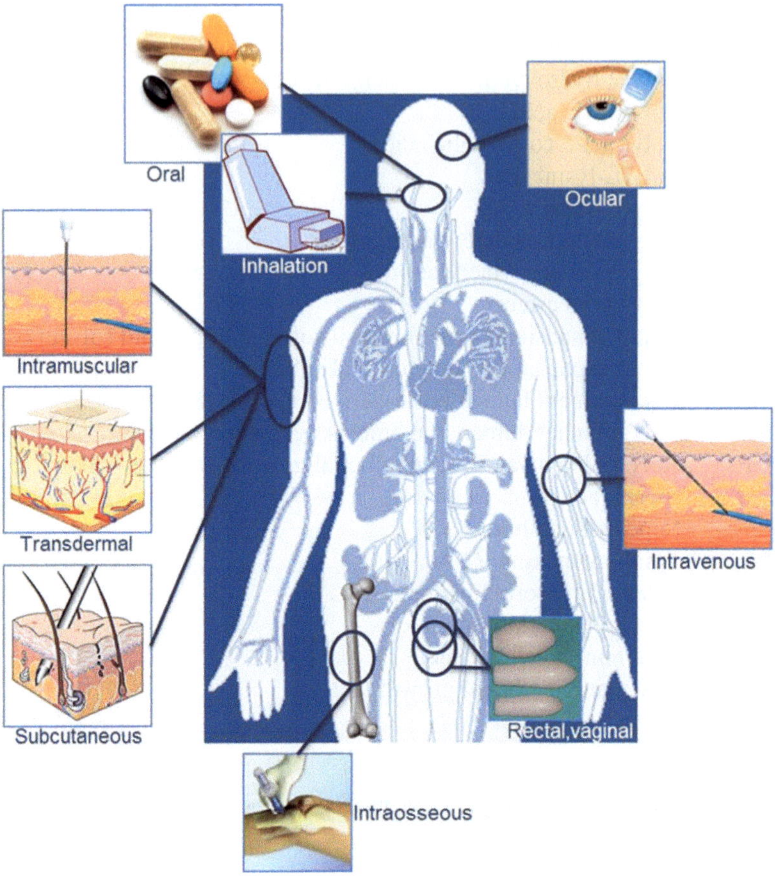

Fig. 1 Representation of the most common drug administration routes

Table 1 Classification of systemic drug delivery routes and correspondent time until therapeutic effect

Anatomic routes for systemic drug delivery	Time until therapeutic effect
Gastrointestinal systems	
Oral	30–90 min
Rectal	5–30 min
Parenteral	
Subcutaneous injection	15–30 min
Intramuscular injection	10–20 min
Intravenous injection	30–60 s
Intra-arterial injection	15–60 s
Transmucosal	15–30 min
Transnasal	15–30 min
Pulmonary (inhalation)	2–3 min
Transdermal	Variable (minutes to hours)
Intraosseous	30–60 s

Adapted from [3]

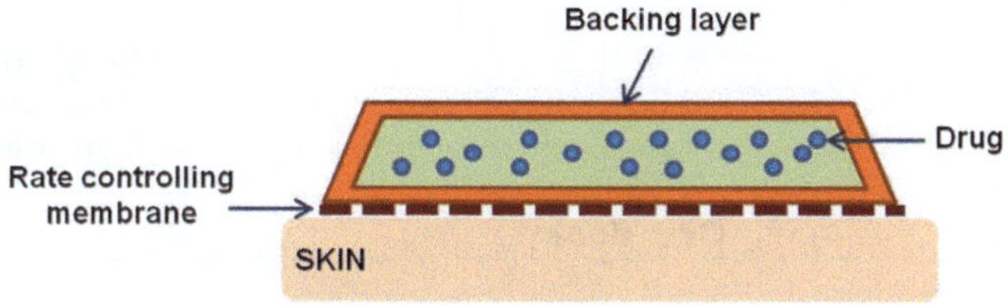

Fig. 2 Scheme of a transdermal drug delivery system

treatment performance such as the rate, period of time and targeting of delivery. This was the beginning of the so-called DDS [3].

Transdermal DDS (Fig. 2) are probably the most desired ones since skin provides a painless and compliant interface for systemic drug administration [6].

Human skin is an easily accessible surface for drug delivery and covers a surface of approximately 2 m^2 in a young adult. Also, it receives about one-third of the blood circulating through the body. For thousands of years, people have placed substances on the skin for therapeutic effects and, in the modern era, a variety of topical formulations have been developed for local treatments [7].

The first transdermal DDS was approved by the Food and Drug Administration (FDA) in 1979 and consisted in a patch with a 3 days release of scopolamine (an alkaloid used in the treatment of nausea and motion sickness). During following years, many systems were developed and some of them remain until now as real best-sellers [7]. Among them, nicotine patches are probably the most broadly used. Other systems available on the market include the ones containing: fentanyl (a synthetic narcotic analgesic), lidocaine (a local anesthetic) and hormones (either for contraception or hormone replacement). In fact, transdermal patches are so largely used nowadays, that it is estimated that more than one billion are currently manufactured each year.

Besides transdermal administration, DDS are also broadly used through other anatomic administration routes such as oral [8], ocular [9], nasal [10], pulmonary [11], rectal [12] and vaginal [13].

The main purpose of using a DDS is, as implied, not only to deliver a biologically active compound in a controlled manner (time period and releasing rate) but also to maintain drug level in the body within therapeutic window (Fig. 3). Besides, one can direct the drug towards a specific organ or tissue (targeted drug delivery) [14]. The first two features were addressed by using drug carriers, usually polymers (either biopolymers or synthetic polymers) whose properties could be manipulated in order to improve the DDS efficiency.

Although both natural (Table 2) and synthetic polymers (Table 3) are being used in the preparation of DDS, there are some advantages that can be pointed to synthetic macromolecules [15].

When the polymers are man-made, it becomes possible to control some aspects of their structure that allows the synthesis of tailor-made materials suitable to the desired biological application [16]. Also, three-dimensional structure as well as chemical composition can be controlled in order to adjust materials properties and orientation of specific functional groups that can interact with the drug.

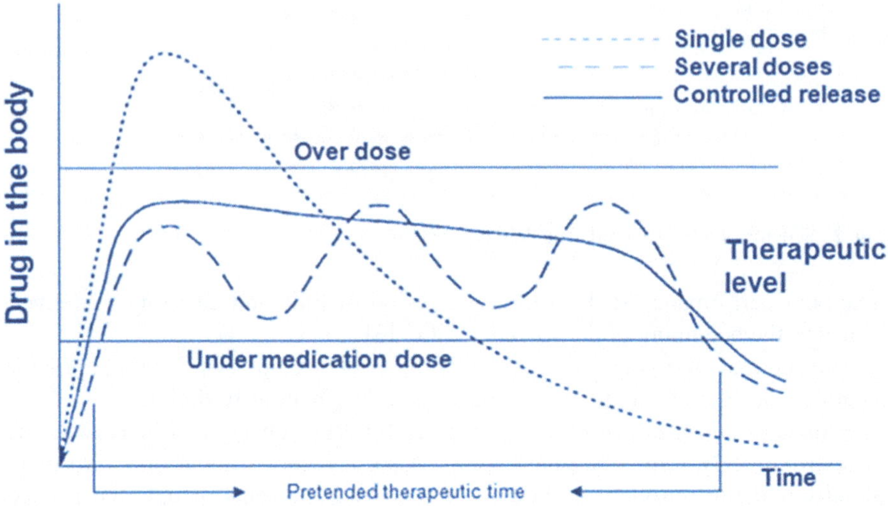

Fig. 3 Scheme of the effect in drug concentration in the body when using different administration methods (adapted from [14])

However, attention must be paid to molecular weight of synthetic polymers which are not biodegradable. Since biodegradation not always occur, synthetic polymers must be eliminated trough renal excretion [17]. Therefore, they should present a uniform molecular weight distribution that fits under the threshold of renal excretion. The new methods based on controlled/living radical polymerization (CLRP) are a very reliable and applied technology in order to obtain well defined macromolecular structures with narrow molecular weight distributions [18], resulting in systems with reproducible pharmacokinetic behavior.

DDS present several advantages, including important factors from decrease of drug side-effects to increased patient compliance. However, DDS disadvantages are also well-known, particularly the final price. DDS main advantages and disadvantages are summarized in Fig. 4.

Targeted drug delivery (Fig. 5) aroused the interest of the scientific community and consequently has witnessed tremendous developments over the last decade.

The active compounds targeting involves the conjunction of different areas related to active compounds design and carriers, biological systems, genetic approaches and precise design of new molecules.

In order to improve the effectiveness of the existing methods for drug delivery, several steps need to be accomplished. The main goal is generically related to deliver suitable active compounds at a desired target without any sign of degradation during the whole process. The development of a controlled delivery system that can dose orally, being less expensive and less painful for the patients and at the same time extremely effective considering a specific disease represents a final target for the research community [19]. DDS must possess some features, such as being

Table 2 Molecular structures of natural polymers used in drug delivery applications

Polymers name	Polymers structure
Collagen	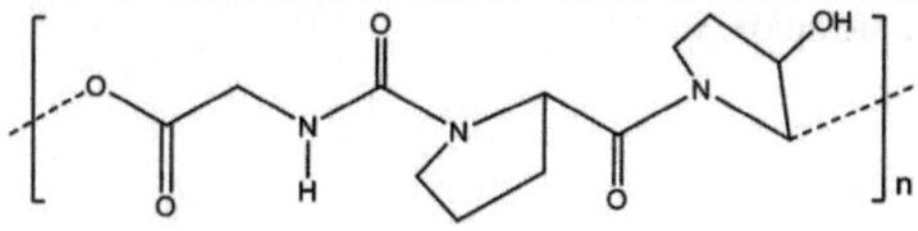
Alginic Acid	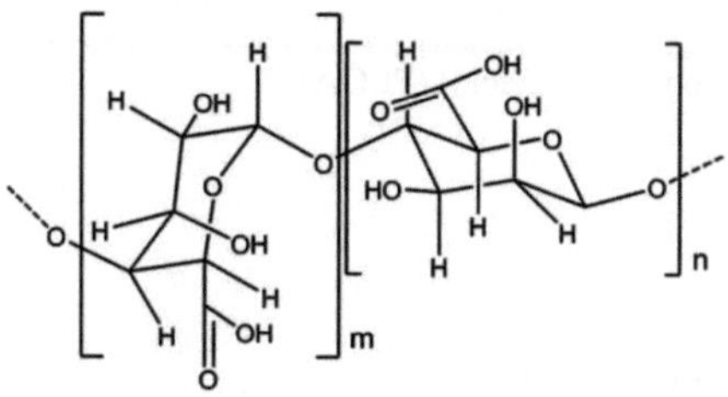
Chitosan	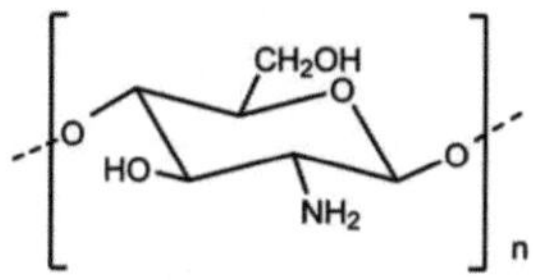
Dextran	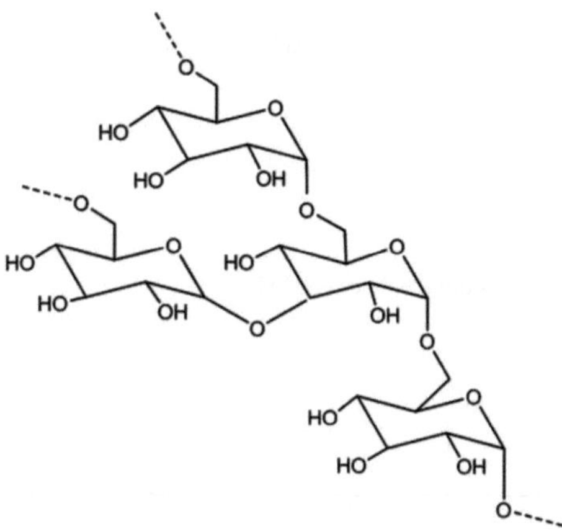
Cellulose	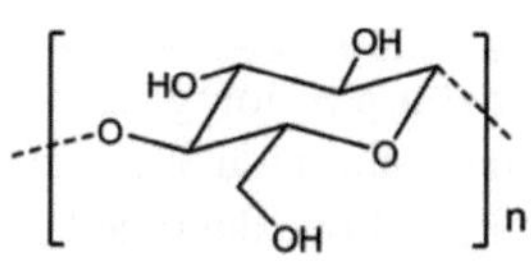

recognized by the specific target tissues [20]. In fact, the delivering of the drug in a specific area of the body is extremely important, in terms of lowering possible side effects of the active compounds, when enter non-targeted organs and tissues. The polymeric carrier itself, once in the targeting area should be able to control the drug

Table 3 Molecular structures of synthetic polymers used in drug delivery applications

Polymers name	Polymers structure
Poly(lactic acid) (PLA)	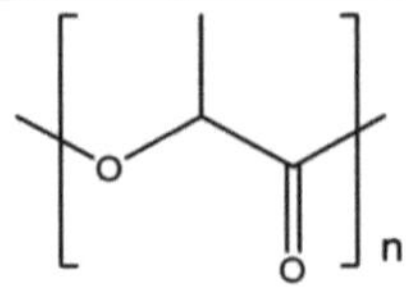
Poly(lactic-co-glycolic acid) (PLGA)	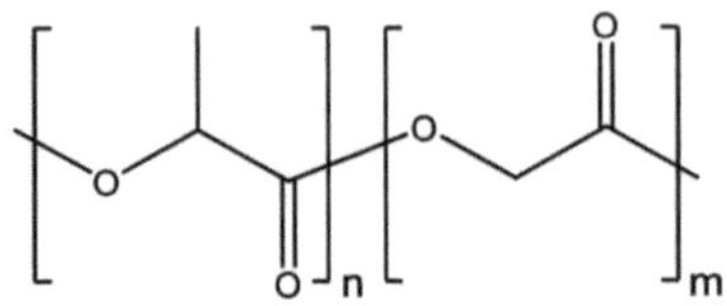
Poly(ε-caprolactone) (PCL)	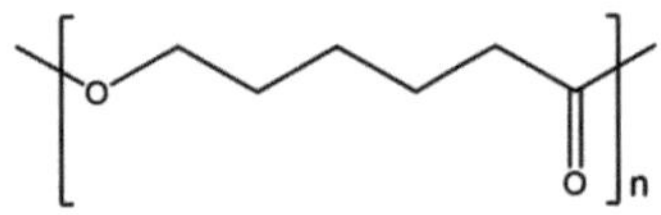
Poly(ethylene glycol) (PEG)	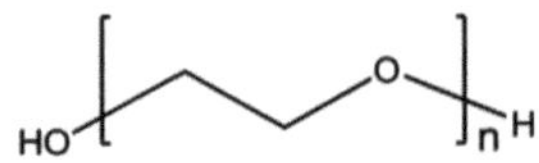
Poly(alkyl cyanoacrylates) (PACA)	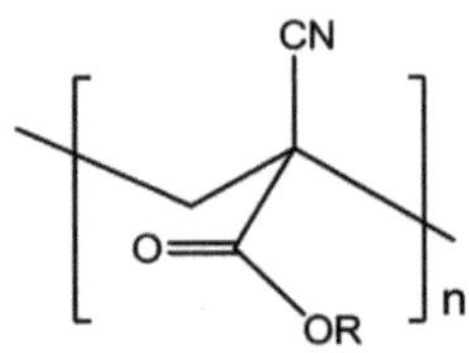
Poly(methyl methacrylate) (PMMA)	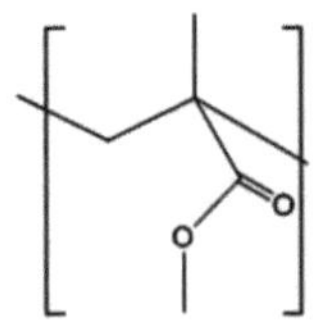

delivery by means of either a physiological or a chemical trigger. The design of a delivery system should be done in such a way that it will be suitable for specific areas of the body, where it could be degraded by environmental conditions, *e.g.*, pH of the stomach or the presence of some enzymes [3].

Polymeric based DDS currently available can be classified in four different categories: diffusion-controlled systems, chemically controlled systems, solvent-activated systems, and magnetically controlled systems [21]. The summary of this classification as well as drug release control mechanisms is presented in Table 4.

Diffusion-controlled systems enclose both matrix and reservoir systems (Fig. 6). The first type of system is based on a polymer matrix in which the drug is distributed homogeneously, while the second type is based in a polymeric membrane that

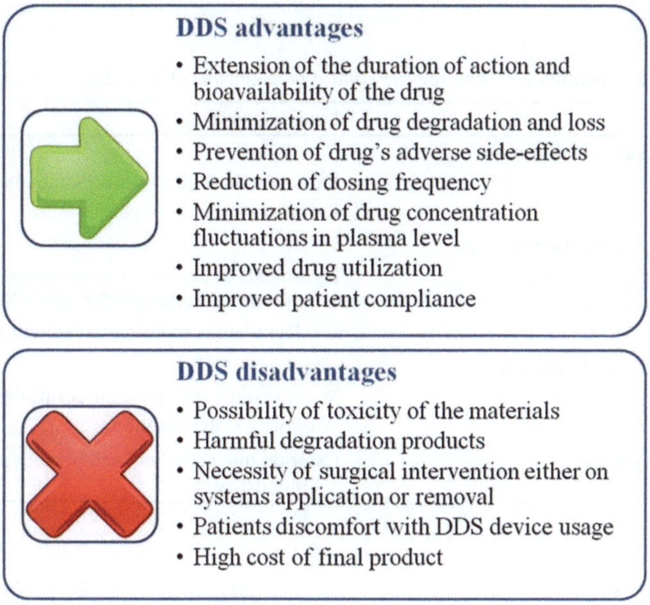

Fig. 4 DDS advantages and disadvantages

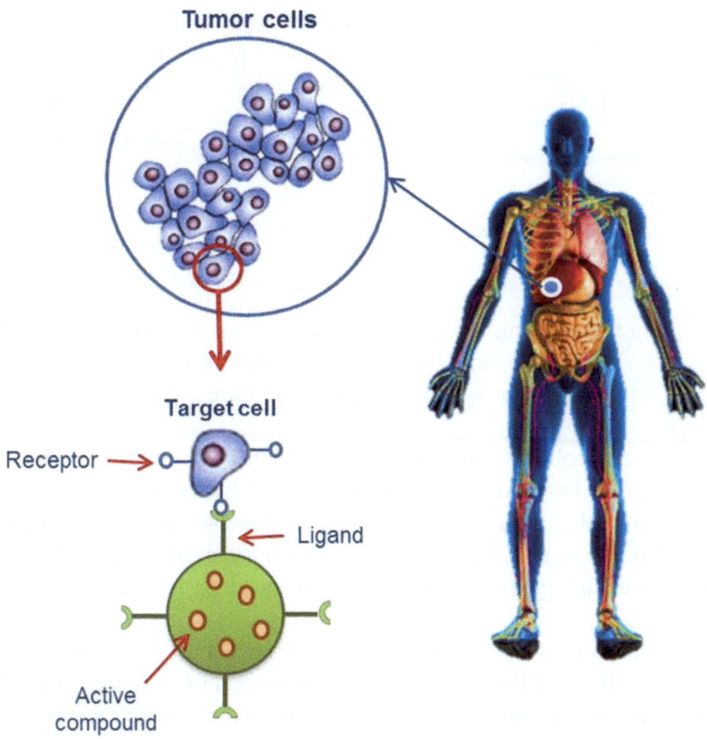

Fig. 5 Targeted drug delivery involves transporters of the active compound that are recognized by target cells. An example of such therapeutics is cancer treatment. Ligands, often antibodies, can be attached to drug carriers that are specific for receptors present on cancer cells membrane

Table 4 Classification and drug release control mechanisms of drug delivery systems

DDS type	Drug release control
Diffusion controlled systems	
Reservoir	Diffusion through systems membrane
Matrix	Diffusion through polymer's bulk
Chemically controlled systems	
Polymer-drug conjugates	Hydrolysis or enzymatic cleavage of pendant groups and diffusion though polymer's bulk
Matrix	Biodegradation and/or bioerosion of polymer
Solvent-activated systems	
Osmotically controlled	Water osmosis trough semi-permeable membrane
Swelling controlled	Polymer's water sorption
Magnetic systems	External magnetic field and hemodynamic force of bloodstream

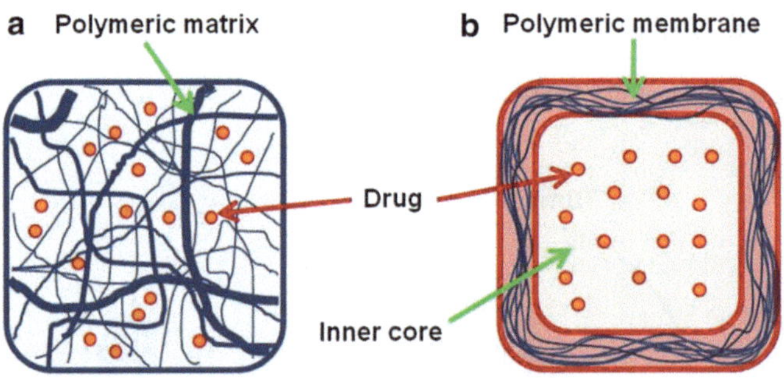

Fig. 6 Structure of a matrix (**a**) and a reservoir (**b**) DDS

surrounds a core containing the drug. Drug release is, in both cases, controlled by diffusion (Fig. 7). However, attention must be paid to the resistance of the polymeric membrane of the reservoir systems since its rupture will cause an abrupt drug release [22].

Chemically controlled systems include polymer-drug conjugates in which drug molecules are linked to a polymeric backbone often by means of a spacer molecule. Once inside the body, the linkage between polymer carrier and the drug is cleaved either by hydrolysis or enzyme cleavage. Different types of biodegradable or hydrolysable chemical linkages are used to attach the drug to the polymer backbone (Fig. 8) [23].

These polymer-drug conjugates usually possess a transport system which is responsible for directing the polymer to the target organs or tissues (Fig. 9).

Another type of chemically controlled system is the one that involves the use of biodegradable/bioerodible polymers. The distinction between these two concepts is based on how degradation occurs. The term biodegradation is usually applied when

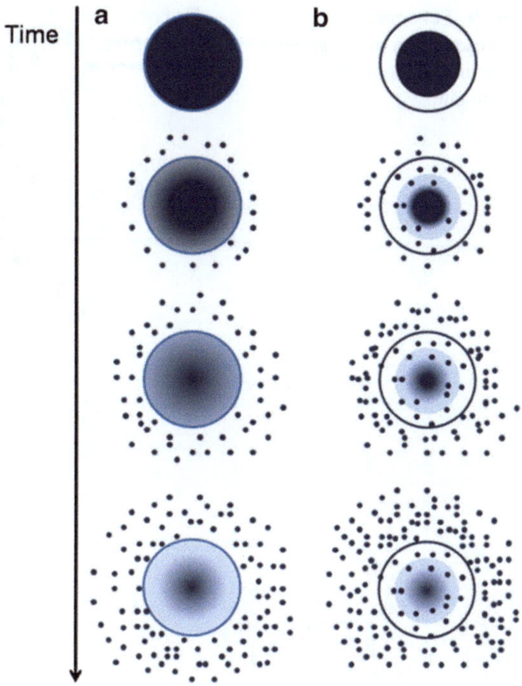

Fig. 7 Drug diffusion profile for both matrix (**a**) and reservoir systems (**b**)

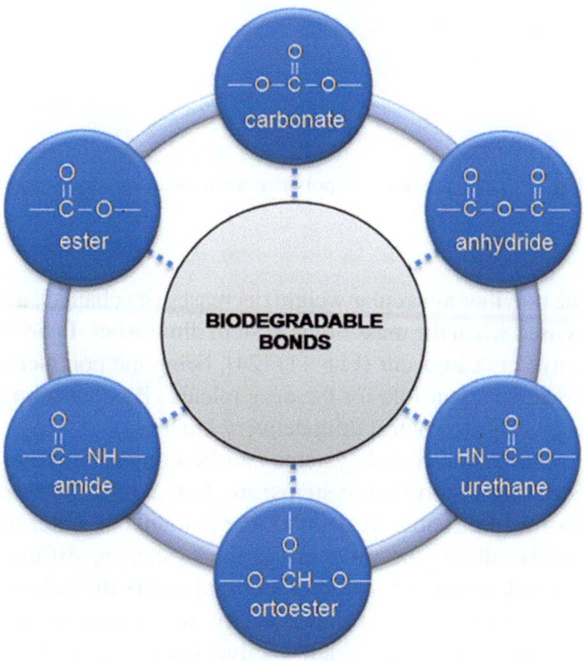

Fig. 8 Biodegradable chemical linkages

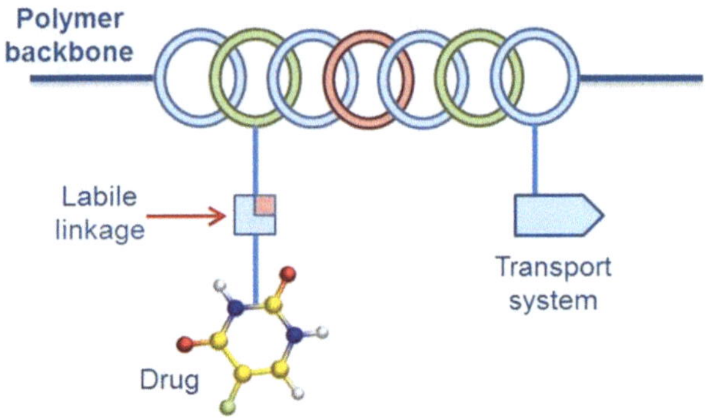

Fig. 9 Schematic representation of the general structure of a polymer-drug conjugate

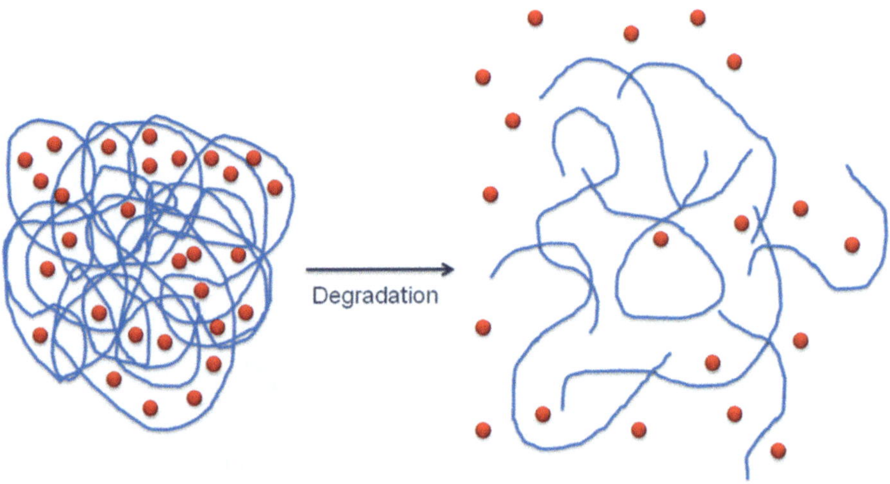

Fig. 10 Biodegradation of polymer chains with consequent drug release

the polymer molecular weight decreases (by chain cleavage, Fig. 10) while bioerosion is used when the mass of the system diminishes. In both cases surface or bulk eroding polymers can occur (Fig. 11) [24], being the polymeric chains matrix disruption the process responsible for the drug release. By controlling polymer degradation rates it is possible to control drug delivery kinetics.

Solvent-activated systems can be controlled either by swelling or by osmosis. Swelling controlled systems are based on a hydrophilic polymeric crosslinked chains that is able to absorb large amounts of water without dissolving. This water uptake allows the drug inside the system to diffuse outwards at a velocity that depends on the amount of water that enters the polymeric matrix (Fig. 12).

Osmotically controlled systems rely on a device containing a semipermeable membrane through which a solvent without or with small amount of drug flows

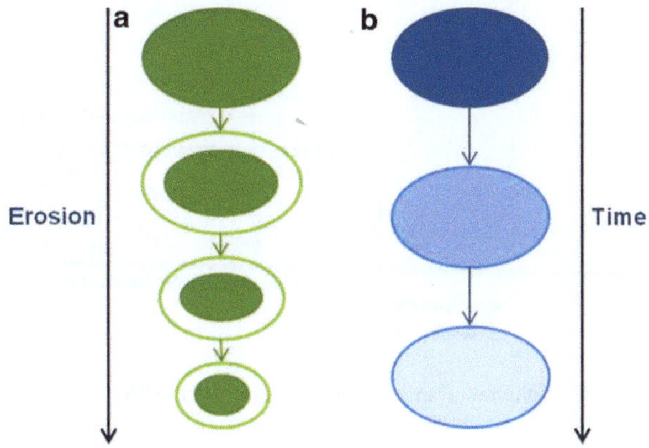

Fig. 11 Scheme representing surface (**a**) and bulk (**b**) bioerosion

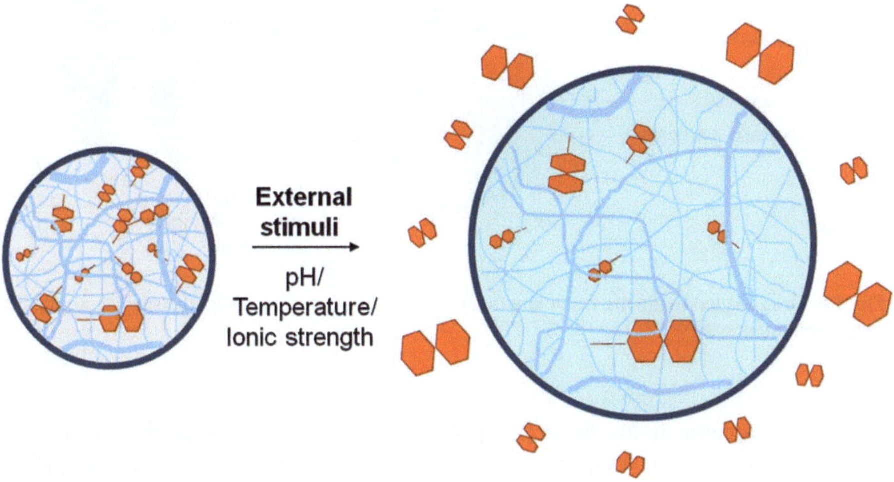

Fig. 12 Drug release resulting from swelling of a polymeric matrix

toward a chamber in which the drug is contained [25]. The solvent flow increases pressure inside the chamber containing the drug and forces the exit of the drug though an orifice present in the device (Fig. 13).

Finally, magnetically controlled systems have been developed mostly by combining a polymer with magnetic particles. Due to these magnetic properties, the particles movement inside the body can be influenced by an externally applied magnetic field. This specific force combines with the hemodynamic force of the bloodstream resulting in a final motion force. However, in order to obtain an effective control over the particles movement, the external magnetic force has to overcome the hemodynamic force. Therefore, when *in vivo* application is desired, low

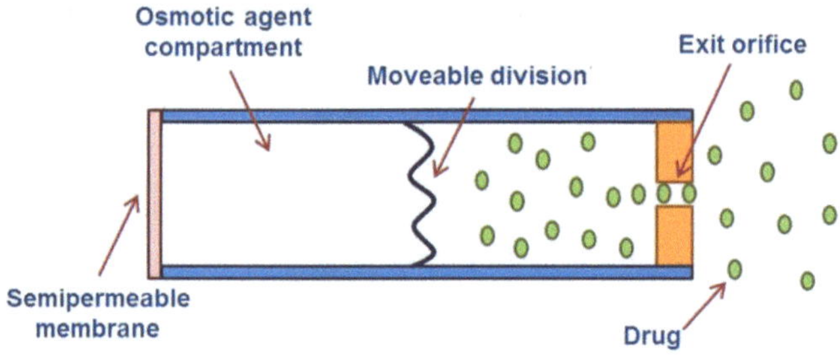

Fig. 13 Scheme of an osmotically controlled DDS

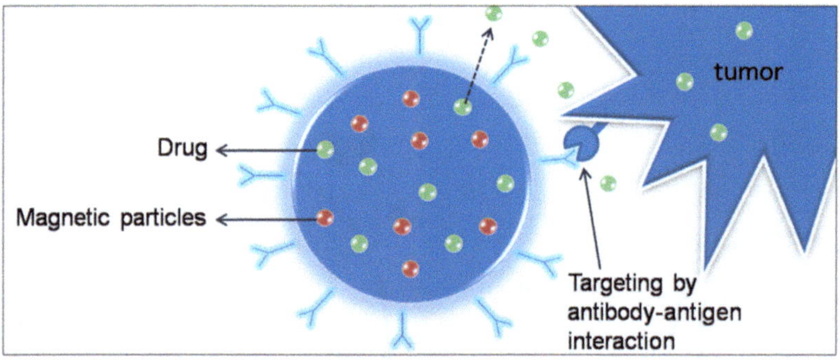

Fig. 14 Drug loaded magnetic particle with specific antibodies attached to the surface applied in cancer treatment

values of magnetic fields must be applied. For this reason, materials with high magnetization at room temperature must be used. Among these, the most applied ones are iron, cobalt and nickel [26]. A quite used strategy in developing DDS using magnetic particles is their association with a responsive polymer [27, 28]. The most studied of these responsive polymers for drug delivery applications are the thermo-responsive polymers since the temperature is an easily controlled parameter [29]. The most interesting point on using thermo-responsive hydrogels combined with magnetic particles results from the ability of these same particles to produce heat because of hysteresis energy loss when subjected to an external magnetic field [30]. Therefore, it is possible to produce a DDS that would be activated when a magnetic field is applied externally causing this energy release from the magnetic particles. This type of systems has been mainly applied in cancer treatment by attaching specific antibodies to their surface allowing a targeted delivery (Fig. 14) [31, 32].

In summary, advanced controlled DDS present indubitable advantages for pharmacologically active compounds administration. Owing to rapid advances in recent years, the application of polymers to drug delivery has grown noticeably. Different treatment methods aiming to control several diseases are currently available while some are still under development or even in researchers' imagination.

2 Polymers in Drug Delivery Systems

As already mentioned, DDS can be prepared by using natural or synthetic polymers, which can be biodegradable or non-biodegradable (Fig. 15).

These polymeric systems are suitable materials to prepare carriers for drugs, proteins, cells and genes. To be used as DDS, polymers should present a set of properties that allow them to be introduced in the human body. Biodegradable polymers are an attractive class of materials to be used in drug delivery field, since they can be broken down into products that are normal metabolites of the human body or into products that can be easily excreted by the body. Thus, once they are introduced in the human body, it is not necessary further manipulation or removal [33–35].

Some naturally occurring polymers have proven to be useful in the design of DDS because of their excellent biocompatibility, biodegradability and low cost. The use of this family of polymers has, however, some drawbacks, namely the strong immunogenic response, possibility of disease transmission, non-uniformity in properties from batch-to-batch and difficulties in their purification [33, 34].

Synthetic polymers, in turn, offer a wide variety of compositions with adjustable properties (chemical, mechanical and biological). By changing the building blocks or the preparation technique it is possible to develop a DDS with particular properties, for a very specific application. Since their preparation is very reproducible, it is possible to prepare DDS with the same specifications quite easily [33, 36].

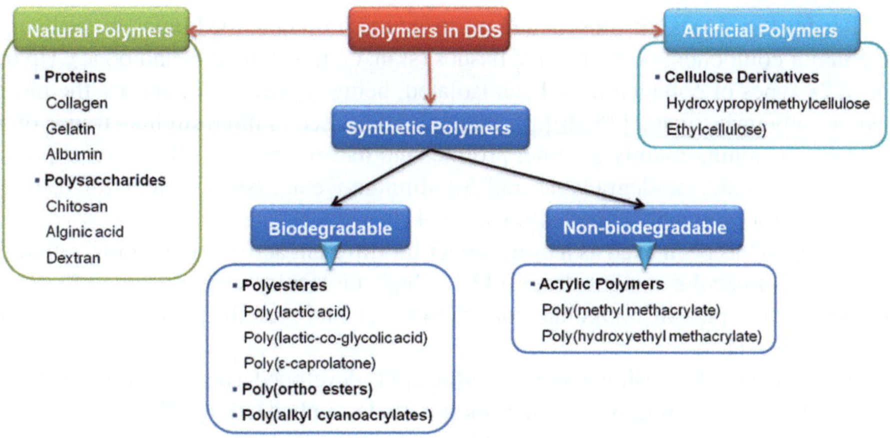

Fig. 15 Overview of polymers used in DDS

They can be biodegradable or non-biodegradable. The biodegradation can occur by hydrolysis, if the polymers present, for instance, ester, anhydride, ortho ester, urea or urethane linkages in the backbone, or by enzymatic pathways, if the polymers possess α-amino acids in the structure [35, 36]. Thus, it is possible to manipulate the biodegradability of non-biodegradable polymers by introducing such moieties in their backbone.

The design of DDS comprising natural and synthetic polymers is a very interesting alternative. The final aim is to combine in the same material the best properties of the natural polymer (*e.g.*, biodegradability and biocompatibility) and the synthetic polymer (*e.g.*, mechanical properties) [37, 38].

2.1 Natural Polymers

This sub section presents an overview of the most currently used natural biodegradable polymers in drug delivery devices. The source of materials and some of the main application as DDS are presented and described.

2.1.1 Proteins

Proteins are high molecular weight molecules composed by α-amino acid residues linked together by amide linkages and usually occur in tridimensional folded structures. They are the major structural components in human tissues. Various DDS using proteins have been designed and some contributions are hereafter presented.

Collagen

Collagen (Fig. 16) is the most abundant structural protein in the human body, being the major component of connective tissues (skin, cartilage, tendon and bone). Up to now, 28 types of collagen have been isolated, being type I, II, III, and IV the most extensively investigated [35, 40]. Collagen is composed of three similar strands of a protein containing mainly glycine, proline, and hydroxyproline. This natural polymer is non-toxic, biodegradable and low-immunogenic, being a very interesting candidate for biomedical applications [35, 41].

Collagen has been used as a drug carrier for different active compounds, ranging from low molecular weight drugs [42] to high molecular weight drugs [43, 44], growth factors [45–48], oligonucleotides [49–52], and proteins [35], in a variety of shapes [35, 41].

Very recently, Kawashima and co-authors [44] developed a device for transscleral drug delivery consisting of collagen microparticles embedded in a PEG membrane. Fluorescein isothiocyanate-dextran (FITC-dextran) (40 kg/mol) and recombinant human brain-derived neurotrophic factor (rhBDNF) were used as active compounds.

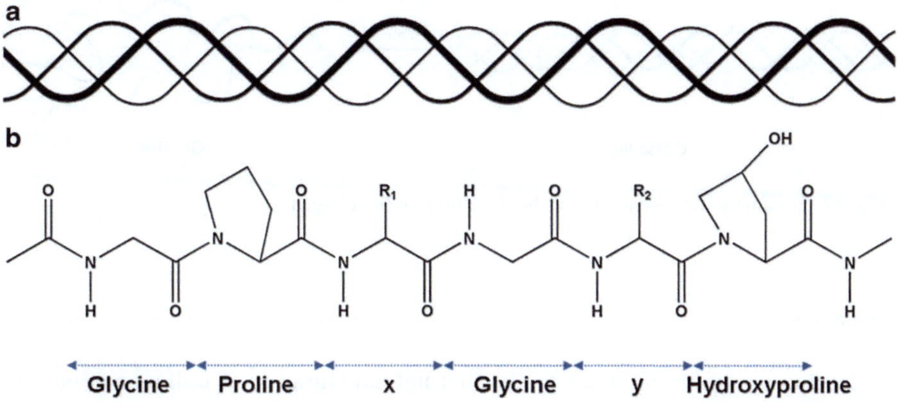

Fig. 16 Chemical structure of collagen type I: (**a**) secondary left handed helix and tertiary right handed triple-helix structure; (**b**) primary amino acid sequence (adapted from [39])

It was found that the release of drugs followed a zero-order kinetic and could be controlled by the concentration of collagen microparticles within the membrane.

Pastorino and co-authors [40] prepared a "smart" drug delivery device based on collagen microcapsules. The aim of the study was the evaluation of the behavior of such system in the presence of matrix metalloproteinase (MMP), the enzyme responsible for cartilage damage. The authors concluded that only in the presence of MMP the microcapsule's shape and morphology (formation of pores) changed significantly. Thus, the findings suggested that if some active compound is encapsulated in the capsule, it will be released only in the presence of MMP. A possible application of this system is the treatment of osteoarthritis.

Collagen based devices showed to be adequate carriers in the field of tissue engineering. Nagai and co-authors [46] prepared injectable collagen microspheres for the release of recombinant vascular endothelial growth factor (rhVEGF). rhVEGF was released in a sustained manner during 4 weeks, with the growth factor remaining bioactive after the release. The authors considered this system very promising for tissue engineering applications. An hydroxyapatite scaffold coated with a type I collagen film was prepared by Teixeira and co-authors [47] and tested as a drug delivery device for the release of bone morphogenetic proteins (BMP-2). Heparinized and non-heparinized scaffolds were prepared and the influence of heparin in the release of BMP-2 was studied. It was found that when heparin is present, the initial burst release phase is significantly reduced and BMP-2 is released in a sustained manner during 7 days. Wang and co-authors [48] prepared collagen and collagen-chondroitin sulfate carriers for the delivery of recombinant human bone morphogenetic protein-2 (rhBMP-2). The authors reported that the incorporation of chondroitin sulfate in the collagen scaffold led to an increase of the initial burst release of rhBMP-2, which is crucial for the acceleration of the bone regeneration process.

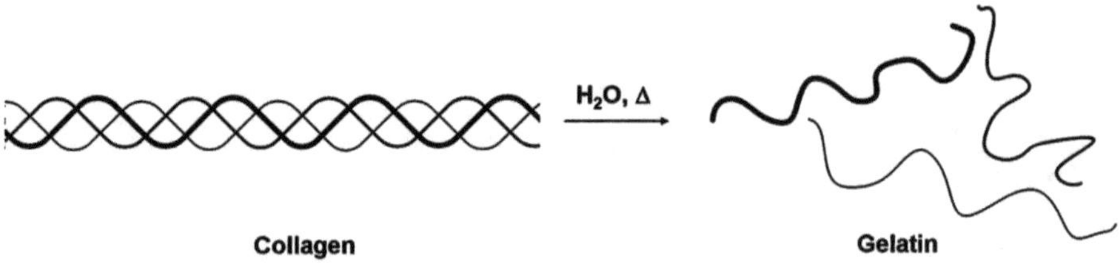

Fig. 17 Schematic representation of the denaturation of collagen

Gelatin

Gelatin is a protein obtained by the thermal denaturation of collagen (Fig. 17). In this process, the triple-helix of collagen is broken yielding single-strain molecules, which are water soluble [53, 54]. When collagen is subjected to an acidic treatment before denaturation, type A gelatin is obtained. On the other hand, if collagen is submitted to an alkaline treatment before denaturation, then type B gelatin is obtained. The isoelectric point of type A and type B gelatin is *ca.* 9 and *ca.* 4.8, respectively [53, 54].

In aqueous solution, gelatin undergoes a sol–gel transition when temperature is lowered below 35°C. This natural polymer being biodegradable, biocompatible and non-immunogenic is commonly used in the biomedical field (*e.g.*, drug delivery vehicles and wound dressings). Due to its high solubility in water and poor mechanical properties, crosslinking of gelatin with other materials is often necessary [53, 55].

Gelatin based drug delivery devices have been used in the delivery of different active compounds, namely anti-cancer drugs [54, 56–60], proteins [59–62] and growth factors [60, 63–65].

Some delivery systems based on gelatin and other polymers have been developed [55, 61, 62, 64–66]. Lin and co-authors [55] proposed a novel DDS composed of PCL particles embedded in a gelatin film, using ibuprofen as a model drug. This system showed a prolonged sustained drug release and an enhancement in the adhesion properties. Therefore it can be applied, for instance, in wound healing. Kim and co-authors [64] developed fibrous matrices of PCL/gelatin crosslinked with genipin for the controlled release of basic fibroblast growth factor (bFGF). Heparin was covalently conjugated to the matrices in order to reversibly immobilize bFGF. The sustained release of bFGF from the matrices was achieved during 4 weeks and the results obtained from both *in vitro* and *in vivo* tests demonstrated that the bFGF released from the matrices promoted the proliferation of human umbilical vein endothelial cells (HUVECs) and facilitated the formation of neo-blood vessels. Thus, this system showed to be promising in tissue engineering applications requiring angiogenesis.

PLGA/gelatin scaffolds were developed by Meng and co-authors [66] and their potential as drug carriers was evaluated. Fenbufen (a non-steroid anti-inflammatory drug) was used as model drug. It was found that the drug release depends on the amount of gelatin in the scaffold as well as on the nanofibers orientation within the scaffold.

Guo and co-authors [58] developed gelatin/heptamolybdate (HM) hybrid nanoparticles and tested their potential in cancer treatment. The nanoparticles were prepared in very mild conditions, using solely the electrostatic interaction between gelatin (zwiterionic molecule) and HM (anionic molecule). Up to 70% of HM was loaded in the nanoparticles. *In vitro* and *in vivo* tests showed that the nanoparticles showed better antitumor efficacy than plain HM solution.

Albumin

Albumin is the most abundant protein in the human blood plasma, representing 50% of the total plasma mass. It is hydrosoluble and presents a molecular weight of about 66 kg/mol. Albumin is a carrier of hydrophobic fatty acid molecules in the blood stream and has also the function of maintaining the blood pH [33, 34]. Characteristics like biodegradability, non-toxicity, non-immunogenicity and antioxidant properties makes albumin a very promising material for biomedical/pharmaceutical applications, including drug delivery purposes [67].

Albumin based drug carriers have proven to be efficient in applications related with cancer treatment [67–72]. This is mainly due to the fact that albumin is used by cancer cells as a source of nitrogen and energy, being taken up by tumor cells by a mechanism of fluid phase endocytosis, followed by lysosomal breakdown. Thus, the drugs entrapped in the albumin carriers are delivered in the specific site of action, minimizing systemic toxicity [67, 68]. An injectable formulation of paclitaxel-albumin nanoparticles for cancer therapy, known as Abraxane®, was approved by FDA in 2005. This system is being used, up to this time, in the treatment of breast cancer.

Albumin can also be used in gene therapy, which is a procedure that uses the delivery of genetic material to specific cells to prevent or treat a disease [69, 73–75]. Okoroukwu and co-authors [74] prepared a gene delivery system based on albumin microspheres containing plasmid deoxyribonucleic acid (pDNA) or sea urchin sperm histone H1 (Sp H1)-(pDNA) complexes. With the complex, the transfection efficiency of pDNA increased and a sustained release of pDNA from the microspheres was achieved during 20 days. This albumin based DDS showed to be suitable for controlled release of pDNA, while protecting it from physical stress and proteins action.

In another work, Eisele and co-authors [73] prepared different cationic polyelectrolytes based on albumin, that were subsequently complexed with pDNA by electrostatic interactions. It was found that the albumin polyelectrolyte containing a higher number of amino groups and consequently with higher cationic nature was the best non-viral DNA delivery agent, showing higher transfection efficiency.

Zöphel and co-authors [75] prepared a positively charged polyelectrolyte comprising two bovine serum albumin (BSA) molecules linked by a poly(ethylene

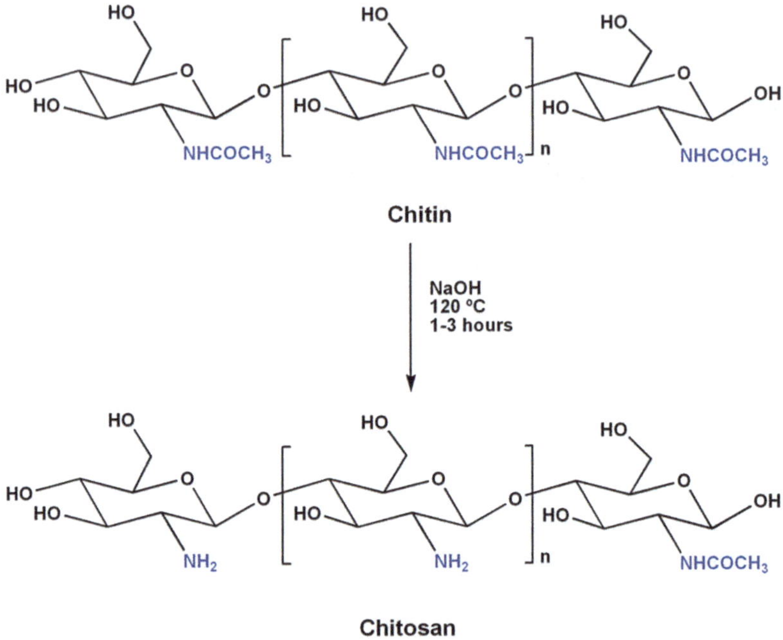

Fig. 18 Schematic representation of the alkaline deacetylation of chitin to obtain chitosan (adapted from [76])

oxide) (PEO) unit. This polyelectrolyte was then complexed with pDNA coding for luminescent enzyme luciferase and high transfection efficiency was obtained.

2.1.2 Polysaccharides

Polysaccharides are high molecular weight compounds, composed by monosaccharide repeating units. They present a wide range of properties and structures. The presence of reactive lateral groups allows the modification of their structure, further increasing their possible applications. Among these, chitosan, alginate and dextran have been widely used in DDS development.

Chitosan

Chitosan is a cationic polysaccharide obtained by the alkaline deacetylation of chitin (Fig. 18), which is the main constituent of the shells of marine crustaceans. It is composed by β-1,4-linked D-glucosamine groups with randomly located *N*-acetyl glucosamine residues [34, 35]. The ratio of N-acetyl glucosamine to D-glucosamine residues defines the deacetylation degree, which typically presents values between

70 and 95%. This is a very important characteristic since it determines chitosan's solubility, hydrophobicity and the ability of electrostatically interact with polyanions [76–78].

Chitosan is insoluble in water and in organic solvents, but can be dissolved in mildly acidic solutions [79]. The main reason for the difficulty in chitosan's solubilization is due to its rigid and highly crystalline structure, because of the intra and intermolecular hydrogen bonds established between acetamido and primary amino groups. The functionalization of chitosan primary amino groups with different chemical groups leads to the rupture of its crystalline structure augmenting its possibility of being dissolved in other solvents rather than only in acidic medium [76].

Chitosan is an attractive natural polymer for application in drug delivery because of its high biodegradability, low toxicity, good biocompatibility, mucoadhesiveness and bioresorbability [35, 76, 77, 79, 80]. Besides, it has the possibility of being functionalized with other moieties [76, 81, 82].

Micro- and nanoparticles of chitosan have been used in the encapsulation of different active compounds [59, 78, 83, 84]. Various routes of administration are known for chitosan carriers, namely ocular, oral, nasal, intravenous, and mucosal [83, 84]. Ionotropic gelation, polyelectrolyte complexation and complex coacervation are the simplest and mildest methods for the preparation of chitosan particles [59, 78]. They are particularly attractive methods since they do not require severe reaction conditions, which is essential to maintain the integrity of the encapsulated drug. Proteins, for instance, have been successfully encapsulated in chitosan particles by the ionotropic gelation method. The influence of some parameters (*e.g.*, chitosan molecular weight and deacetylation degree, chitosan concentration, chitosan/polyanion mass ratio, protein concentration, presence of stabilizer) in the encapsulation efficiency and release of these proteins was evaluated [85–89].

Chitosan particles have been investigated as drug carriers for ophthalmologic applications and the obtained results showed that such particles were well tolerated by the ocular surface tissues [90–92]. A review by de la Fuente and co-authors [93] gives an overview of the application of chitosan carriers in ocular therapeutics.

Carriers based on chitosan have been used as DDS in the field of cancer treatment [76, 84, 94, 95]. Injectable chitosan hydrogels prepared by different methods were used for the entrapment of anti-cancer drugs, as reviewed by Ta and co-authors [94].

Chitosan, due to its cationic nature, can easily bind nucleic acids, reason why it has been very used as a potential material to be used in gene therapy [35, 83, 96]. Research works dealing with the preparation of chitosan-nucleic acids complexes have been presented in literature and the main factors influencing the transfection efficiencies have been figured out. The chitosan molecular weight, chitosan degree of deacetylation, chitosan/nucleic acid ratio and type of cell transfected showed to be important factors [96–100]. The prepared complexes showed to be efficient in protecting the genetic material from nuclease attack, thus maintaining their bioactivity.

Malmo and co-authors [101] studied the influence of chitosan chain architecture in the transfection efficiency in human cervical carcinoma (HeLa) cells. Self-branched chitosan, self-branched trisaccharide-substituted chitosan and linear chitosan were

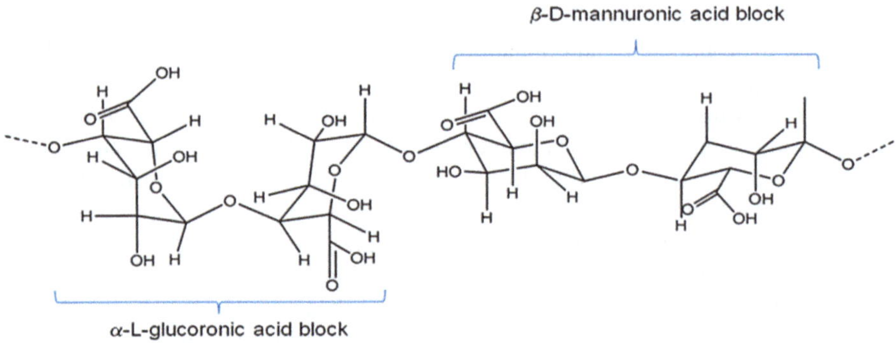

Fig. 19 Alginic acid molecular structure

used to complex pDNA, and the cellular uptake, transfection efficiency and citotoxicity of these complexes were evaluated. The results demonstrated that self-branched chitosans were the most efficient in transfecting the HeLa cells, exerting very low toxicity. In turn, the self-branched chitosan containing trisaccharides showed high intracellular uptake, but low transfection efficiency, whereas linear chitosan did not show the ability of transfecting the HeLa cellular line.

Recently, Opanosopit and co-authors [102] demonstrated that transfection efficiency of poly(L-arginine)-pDNA complex in HeLa cells can be enhanced by mixing chitosan with such complex.

Alginic Acid

Alginic acid is a cationic polysaccharide extracted from the cell walls of brown algae, by an alkaline treatment, followed by an acidic treatment. It is a linear polysaccharide composed of two uronic acid units, β-D-mannuronic acid and α-L-glucoronic acid, joined by (1→4) glycosidic bonds (Fig. 19). Their molecular weight can be higher than 500 kg/mol [33–35, 103]. Usually, alginic acid is used in its sodium salt form, usually named sodium alginate, or for some authors, simply alginate.

Due to the high content of acidic groups along the polymeric backbone, alginic acid readily forms gels when in contact with divalent cations (*e.g.*, Ca^{2+}), at ambient temperature. This is a very important characteristic for drug delivery, since it opens the possibility of encapsulating active substances, under mild conditions, while maintaining their full biological activity [34, 35, 104]. Furthermore, the great number of carboxylic groups along the alginate chain opens the possibility of further chemical modification, changing its properties and enlarging its applications [105]. Alginate is biocompatible and non-immunogenic. However, it is not enzymatically degraded by mammals, limiting its degradation *in vivo* and exhibits poor cell adhesion. These are the main drawbacks in the use of alginates as biomaterials [34, 35, 104, 106].

Alginate based gels crosslinked with divalent ions can be used in the sustained and localized release of low-molecular weight drugs and macromolecules. However, they

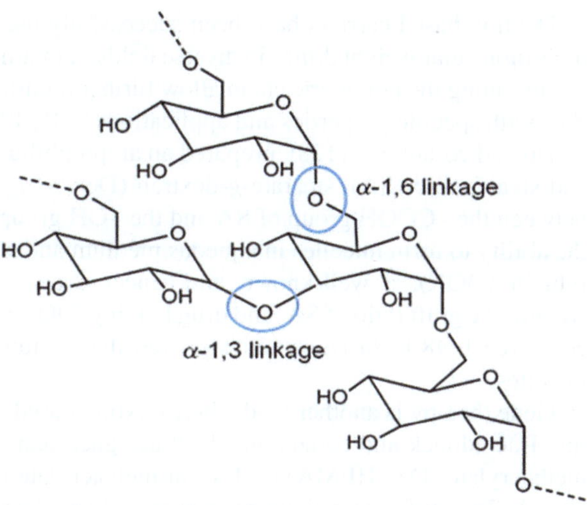

Fig. 20 Molecular structure of dextran (adapted from [125])

showed to be unstable in biological fluids, due to the exchange of the crosslinking divalent ions by the monovalent ions presented in the surrounding medium. As a consequence, the mechanical strength of the gel is weakened and the drugs release profiles are modified [105, 106]. To overcome this situation, a chemical modification of alginate gels with other molecules is needed [104–107].

Alginate based carriers have demonstrated their ability to be used in the field of cancer [108–110] and ocular diseases [111–113]. Angiogenic growth factors, like vascular endothelial growth factor (VEGF) and bFGF, have been entrapped in alginate gels [114–118]. These alginate based systems showed to be very promising for applications related with the formation/regeneration of tissues. Alginate based carriers also demonstrated their ability in encapsulating and delivering both pDNA and small interfering ribonucleic acid (siRNA) [52, 119, 120].

Systems based on alginate and other polyssacharides have also proved to be very efficient in the drug delivery field [121–124].

Dextran

Dextran is a water soluble homopolysaccharide of bacterial origin composed essentially by α-1,6-linked D-glucopyranose units, with branches extending mainly from the α-1,3- and occasionally from the α-1,2- and α-1,4- positions (Fig. 20). Usually, dextran degree of branching is about 5% [106, 125]. Dextran can present different molecular weights, molecular weight distributions and degrees of branching depending on the conditions and bacterial strain used for its production [106].

It is an adequate material for biomedical applications due to its biodegradability, biocompatibility, non-immunogenicity and non-antigenicity [126]. Besides, dextran resists to protein adsorption and has non-cell binding properties [106].

Dextran based carriers have been successfully used in the delivery of low- and high-molecular weight drugs, in diverse fields of treatment [107, 127]. The hydroxyl groups along the polymeric chain allow further modifications, leading to new materials with specific properties and applications [106, 127].

Du and co-authors [128], prepared an amphiphilic macromolecule from dextran and stearic acid (SA), stearate-*g*-dextran (Dex-SA), via an esterification reaction between the –COOH group of SA and the –OH group of dextran. Dex-SA showed the ability to form micelles in aqueous medium and was used to encapsulate doxorubicin (DOX), a well-known anti-cancer drug. Adjusting dextran molecular weight, the graft ratio of SA, and drug loading, DOX release from Dex-SA micelles could reach 48 h. *In vivo* studies proved the antitumor activity of Dex-SA/DOX micelles.

Gene therapy is another field where dextran based carriers showed to be promising. Raemdonck and co-authors [129] designed nanogels of dextran-hydroxyethyl methacrylate (Dex-HEMA) or dextran-methacrylate (Dex-MA) for the delivery of siRNA. Two different cationic monomers, [2-(methacryloyloxy)ethyl]trimethylammonium chloride and 2-aminoethyl methacrylate hydrochloride, were used in order to prepare different cationic dextran nanogels. High loading of siRNA through electrostatic interactions was achieved. The nanogels were internalized by the HuH-7 human cells without relevant citotoxicity. In order to augment the time of circulation of such nanogels in the blood stream, PEG was covalently attached to their surface. The PEGylation did not reveal to have any detrimental effect in the transfection efficiency [130]. A dextran-protamine polycation was prepared by Thomas and co-authors [131] to be used as a carrier for calf thymus DNA (cfDNA). Protamine, a cationic polypeptide containing L-arginine residues, was conjugated with dextran, turning it cationic due to the presence of amine groups. The results showed that dextran-protamine is hemocompatible and non-toxic. In addition, a high gene transfection and cellular uptake in HepG2 cells was observed.

Pro-drugs, as mentioned in Sect. 1, are another form of delivering active compounds. Horning and co-authors [126], prepared a pro-drug made from dextran and hydrophobic drugs. The pro-drug was amphiphilic and, when in contact with water or water miscible solvents, self-assembled into nanoparticles. These nanoparticles presented high loading efficiency and showed to be stable under a pH range from 4 to 11, for several months. Thus, this system showed to be reliable for the sustained release of hydrophobic drugs.

2.2 *Artificial Polymers*

2.2.1 Cellulose Derivatives

Cellulose is the most abundant occurring biopolymer in nature. It is a linear polymer composed of $\beta(1 \rightarrow 4)$ linked D-glucose units, each one presenting three hydroxyl groups. These hydroxyl groups are responsible for the strong

Fig. 21 HPMC structure

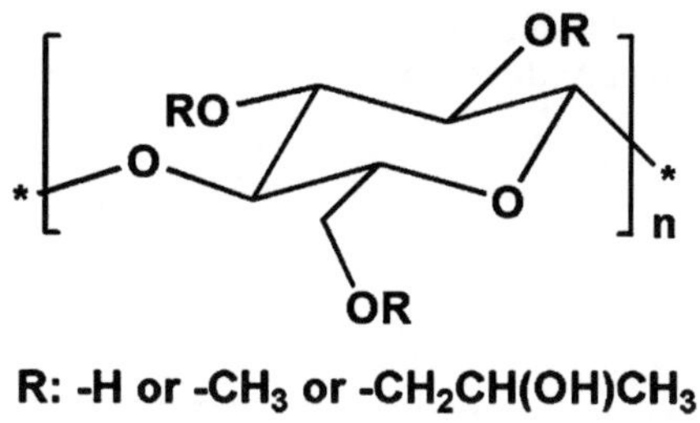

Fig. 22 EC structure

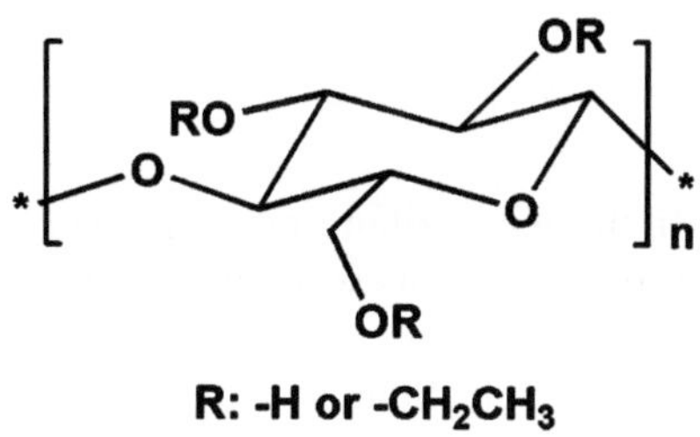

intermolecular and intramolecular H-bonds that are established between the cellulosic chains, making it insoluble in water and organic solvents. However, such hydroxyl groups are reactive and can be easily functionalized, leading to other cellulose based materials with distinct properties, spreading the fields of application of this biopolymer [33, 132]. Cellulose derivatives in the form of ethers, esters and acetals have been prepared and its applicability in biomedical applications has been investigated.

Hydroxypropylmethylcellulose (HPMC), also known as hypromellose (Fig. 21), is a cellulose ether widely used in the preparation of drug delivery devices.

When in contact with water or biological fluids, HPMC becomes hydrated, leading to a "disentanglement" of the polymeric matrix, forming a swelling gel layer. It is accepted that the drug release from a HPMC matrix comprises two steps: diffusion through the swelling gel layer and release due to the erosion of the swollen matrix [132, 133]. The drug release from this type of matrices can also be influenced by viscosity of the gel layer formed during the hydration of the polymer [132].

Along the years, HPMC has been used as carrier for several drugs. The different parameters that influence the drug release behavior have been studied, as documented by Kamel and co-authors [132]. Ethylcellulose (EC) (Fig. 22) is another cellulose derivative used in DDS.

EC is a non-ionic cellulose ether, insoluble in water, but soluble in some polar organic solvents. It has been used for the controlled release of various types of drugs: diclofenac sodium [134], ketoprofen [135], betamethazone [136], nimesulide [137], 5-fluorouracil [138] and cefpodoxime proxetil [139].

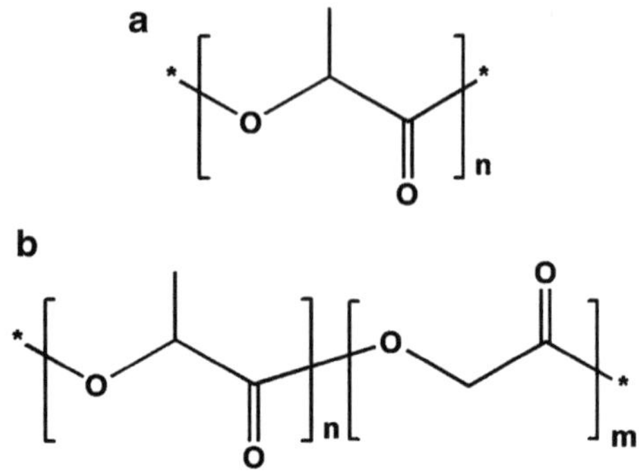

Fig. 23 Structures of: (**a**) PLA; (**b**) PLGA

In these contributions, some parameters like drug and polymer concentration or type of solvents were evaluated in order to get improved delivery systems based on this polymer.

Recently, Song and co-authors [140] prepared a DDS based on hydrophobically modified quaternized cellulose (HMQC). This derivative was amphiphilic: long alkyl chains constituted the hydrophobic part, whereas the hydrophilic moieties were quaternary ammonium groups. This polymer showed the ability of self-assembling in water and was used to encapsulate a poor water-soluble drug. In addition, HMQC showed to have a low toxicity.

2.3 Synthetic Polymers

This sub section presents the most common synthetic polymers used in DDS. As aforementioned, synthetic polymers offer the great advantage of being synthesized with specific properties for a given application.

2.3.1 Biodegradable Synthetic Polymers

Polyesters

PLA (Fig. 23) is a biocompatible and biodegradable material. It degrades by hydrolysis of the ester bonds, through a bulk eroding mechanism, releasing lactic acid, a normal metabolite in the human body. PLA can occur in four forms: poly(L-lactic acid) (PLLA), poly(D-lactic acid) (PDLA), poly(D,L-lactic acid) (PDLLA) and meso-poly(lactic acid). Only PLLA and PDLLA have been studied for biomedical applications, and both showed to take considerable time to be degraded *in vivo* [33–35]. Despite this fact, PLA nanoparticles have been successfully used as carriers for diverse active compounds [59].

PLGA is one of the most widely used polymer in DDS. It is biodegradable, biocompatible and low immunogenic. Upon hydrolytic degradation it releases lactic acid and glycolic acid, which are compounds effectively processed by the human body [35, 59, 141] . The properties of PLGA can be easily tailored by changing the ratio PLA to poly(glycolic acid) (PGA) [35, 103]. When compared to PLA, PLGA has a higher degradation rate.

Particles of PLGA have been used in the controlled release of several active compounds (proteins, vaccines, genes, antigens as well as growth factors). The different methods of preparation, factors influencing the encapsulation efficiency and the drug release from PLGA carriers were reviewed by Mundargi and co-authors [142]. These particles are also suitable for the application in the field of cancer treatment [59, 141]. Several studies have been done on the influence of certain parameters (presence or absence of stabilizers in the formulations [143], type of solvent [144] and molecular weight of the drug [145]) in the encapsulation efficiency or drug release profiles from these nano/microparticles.

New DDS with novel properties can be designed by using copolymers based on lactic acid, glycolic acid and other monomers. For instance, triblock copolymers based on PLGA and PEG have been prepared and showed to be thermosensitive. Such copolymer has demonstrated to be adequate for the drug delivery of different active compounds [146–149].

In the field of gene therapy, nanoparticles of PLA and PLGA showed to be very promising in delivering DNA and siRNA. In order to get an enhanced efficacy in gene delivery, the surface of the nanoparticles was modified with cationic moieties. Wang and co-authors [150] prepared four different cationic nanoparticulate systems based on methoxyPEG-PLA and PLGA, using poly(ethyleneimine) (PEI) and chitosan as the cationic moieties. The cationic nanoparticles were then tested as carriers for siRNA. The authors found that the mPEG-PLA-PEI nanoparticles were the most promising in binding siRNA, and were also characterized by higher transfection efficiencies. Recently, Mishra and co-authors [151] synthesized a triblock copolymer PLGA-PEI-PLGA that, when in contact with water, self-assembled into micelles, exhibiting positive charge at the surface. Such micelles were then used to complex pDNA. A subsequent addition of low molecular weight PEI to the previously formed complex, completely shielded the pDNA in the micelle, contributing to the maintenance of its bioactivity. The shielded system showed to have a transfection efficiency 50- to 100-fold higher than its non-shielded counterpart in a breast cancer cellular line. Fu and co-authors [152] used a triblock copolymer of PLA-PEG-Poly(L-lysine) (PLL) to be used as a non-viral vector for pDNA and its efficacy was compared with PEI. The triblock copolymer self-assembled into nanoparticles in aqueous medium and complexed pDNA through an electrostatic interaction between its negative groups and the positive amine groups of PLL. The PEG shell around the nanoparticle provided additional stability and reduced the possible interactions between the nanoparticle and proteins in biological fluids. The *in vitro* transfection efficiency was evaluated in HeLa and HepG2 cells and the results showed that the nanoparticles have a higher gene transfection than the PEI/pDNA complex. It is worth to mention that no cytotoxic effect was exerted by the nanoparticles.

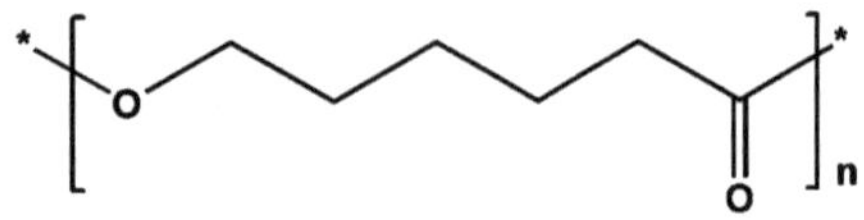

Fig. 24 Structure of PCL

PCL (Fig. 24) is a semicrystalline polymer with a low melting point (T_m = 55–60°C) and a low glass transition temperature (T_g = −60°C), which can be dissolved in a vast number of organic solvents. It possesses a very low degradation rate *in vivo*, reason why it has been mainly used for the preparation of long-term drug delivery devices [33, 35]. Upon degradation, PCL does not release acidic by-products, contrarily to what happens with PLA and PLGA. Additionally, PCL offers the possibility of being easily blended with other polymers [153]. PCL micro- and nanoparticles have been used as carriers for several drugs, as reviewed by Sinha and co-authors [153] and more recently by Kumari and co-authors [59].

Blending PCL with other polymers or copolymerizing ε-caprolactone with other monomers, led to PCL based materials with novel properties, enhancing its range of applications [35]. Blends of PCL with PEO based copolymers, processed by different techniques, were used as DDS for the prolonged release of anti-glaucoma drugs [154–156].

Block copolymers based on PCL and PEG, for instance, showed to be very promising materials for the encapsulation of various drugs, as reviewed by Wei and co-authors [157].

Micelles formed from PCL/PEI block copolymers were used to complex pDNA and, at same time, encapsulate an anti-cancer drug. It was shown that the drug loading did not have any effect on the interaction between PEI shell and pDNA. Moreover, a high transfection efficiency was achieved with this complex in HepG2 cells. The results obtained showed the potential of developing DDS that display the synergistic effect of combined drug and gene therapy in cancer treatment [158]. Making use of this synergy, two different research groups [159, 160] developed DDS based on PCL-PEO block copolymers modified with cationic moieties for carrying both anti-cancer drugs and siRNA. The results obtained were very promising and the use of combined therapies showed to be useful in preventing the tumor growth.

Poly(ortho esters)

Poly(ortho esters) (POE) are hydrophobic polymers that present three geminal ether bonds and degrade by a surface eroding mechanism [33–35].

Currently, four families of POE are known: POE I, POE II, POE III and POE IV (Fig. 25). POE I is obtained by a transesterification reaction between a diol and diethoxytetrahydrofuran. POE II is synthesized from diols and diketene acetal 3,9-bis(ethylidene 2,4,8,10-tetraoxaspiro[5] undecane). This polymer is highly hydrophobic, and to make it degradable under physiological conditions it is necessary the addition of an acid excipient. POE III can be obtained by a reaction

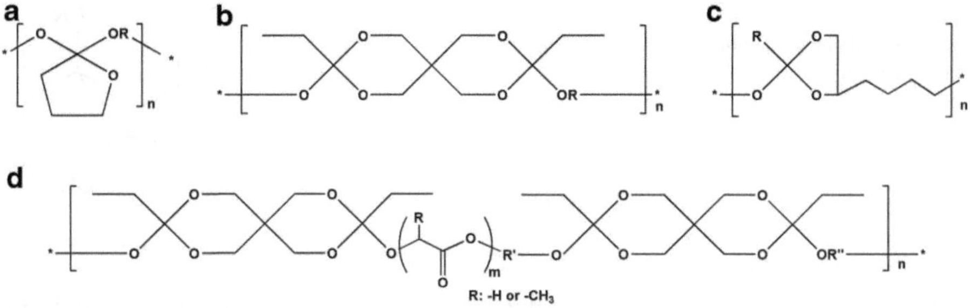

Fig. 25 Structures of the different families of POE: (**a**) POE I; (**b**) POE II; (**c**) POE III and (**d**) POE IV

between a triol and an ortho ester. The use of different triols enables the obtention of POEs with different chain flexibility. POE IV is a modification of POE II; in this specific case units of lactic acid or glycolic acid are incorporated in the polymer backbone, which enables the degradation of these polymers, without the addition of acidic excipients. Besides, the rate of degradation of these polymers can be tuned by the amount of lactic or glycolic acid present along the polymer chain [33, 34].

POE IV is the preferred POE for drug delivery purposes, since it presents several advantages over the other families of POE, namely the possibility of controlling the polymer properties and erosion rate, high stability at room temperature and drug release dependent on erosion mechanism [161].

The DDS based on POE, for different types of drugs, have been extensively described by Heller [161–164].

A possible application of POE carriers is in the delivery of DNA vaccines. An early work by Wang and co-authors [165] showed that POEs microspheres protected DNA from degradation, enabled uptake by antigen-presenting cells, and released DNA in response to phagosomal pH. Also they did not present any cytotoxic effect. Therefore, this system seemed to be suitable for the purpose of delivering DNA. More recently, Nguyen and co-authors [166] proposed a "modified" microparticulate system based on POE for the release of DNA vaccines. In this new system, PEI was blended in the POE matrix and even a small amount of PEI showed to have important influence in the microspheres properties. A higher control of the release rate and enhanced gene transfection efficiency were obtained.

POE based carriers have also proved to be useful in the sustained release of analgesic drugs [167] and antiproliferative drugs [168].

Poly(alkylcyanoacrylates)

PACA (Fig. 26) are biodegradable polymers, with a wide range of applications in the biomedical/pharmaceutical field (*e.g.*, surgical glues, skin adhesives, drug delivery devices). PACA are obtained by the anionic polymerization of alkyl cyanoacrylic

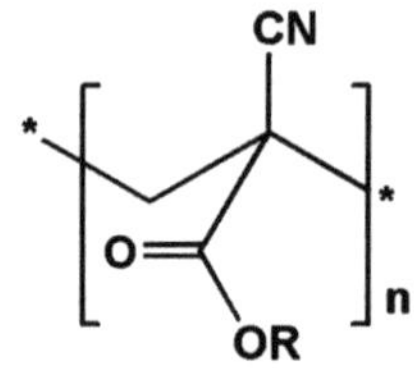

Fig. 26 Structure of PACA. R represents an alkyl chain of variable length

monomers initiated by traces of humidity [33, 34]. Their C-C bonds are hydrolytically instable, which can be ascribed to the high inductive activation of methylene hydrogen atoms by the electron-withdrawing neighboring groups.

The degradation rate of PACA can vary from hours to days, depending on the alkyl chain length [33, 34].

The development of PACA particles for drug delivery purposes has started two decades ago. Almost all type of substances (peptides, proteins, oligonucleotides, anti-cancer, anti-infeccious and anti-inflamatory drugs) have been successfully encapsulated in PACA particles (microparticles, nanoparticles or capsules) [59, 169]. Vauthier and co-authors [169] comprehensively reviewed the methods of preparation, potential applications of PACA carriers. Graf and co-authors [170] reviewed the different methods of preparation, the factors influencing the encapsulation efficiency and the drug release profiles from PACA particles. The results of some *in vivo* experiments were presented.

Nanoparticles from PACA have been successfully used in the field of cancer treatment. The main reason beyond this is related with the fact that PACA particles showed to overcome the multidrug resistance phenomenon, sometimes observed in cancer treatment [171]. For instance, PLGA and alginate nanoparticles failed in the delivery of DOX to drug resistant cancer cells, whereas PACA nanoparticles were able to do it. Thus, a considerable number of research works dealing with cancer therapies using PACA nanoparticles are presented in literature [172–175].

2.3.2 Non-biodegradable Synthetic Polymers

Acrylic Polymers

PMMA (Fig. 27), a biocompatible and biostable polymer, was the first acrylic polymer used in a biomedical application. Just after the World War II, PMMA was used in the fabrication of intraocular lenses, and is still used in the fabrication of rigid contact lenses [103].

PMMA is a versatile biocompatible polymer with applications as DDS in various areas of the biomedical field. PMMA has been used for over than 20 years in orthopedics (*e.g.*, artificial joints, bone cements). However, some of the applications have been impaired by its bio-inertness. To overcome this problem, it was

Fig. 27 Structure of PMMA

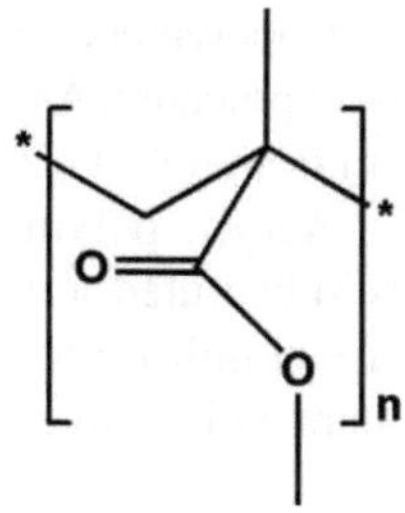

Fig. 28 Structure of PHEMA

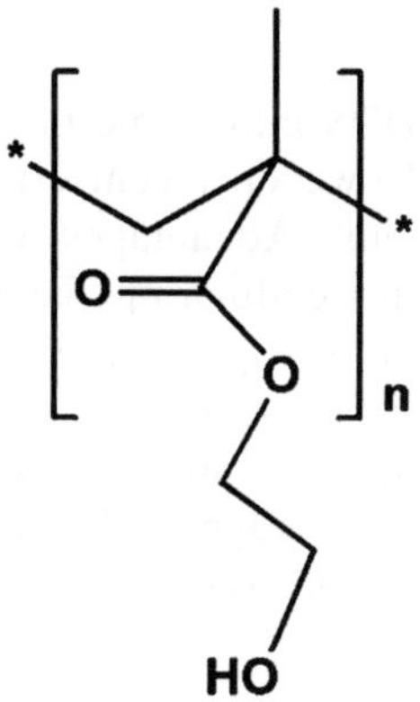

proposed the addition of bioactive glasses/ceramics fillers to the PMMA matrix [176]. PMMA microparticles have shown to be useful in the delivery of antimicrobial drugs to treat orthopedic infections [177] and in the delivery of drugs in the gastrointestinal tract [178, 179]. Microparticles based on PMMA copolymers exhibiting stimuli-responsive behavior have been prepared and demonstrated their potentialities for drug delivery purposes [180, 181].

Poly(2-hydroxyethyl methacrylate) (PHEMA) (Fig. 28) is structurally similar to PMMA; instead of a pendant methylester group it has a pendant hydroxyethyl ester group [103].

PHEMA is a biostable polymer that has the ability of forming hydrogels. The use of PHEMA based materials in controlled release applications is well known [182, 183]. In some cases, PHEMA delivery systems present an initial drug "burst release", immediately after being hydrated. Therefore, sometimes, it is necessary to perform structural modifications. Anderson and co-authors [184] developed a drug delivery device based on PHEMA, whose surface was hydrophobized by a reaction with octadecyl isocyanate. This system showed to be adequate in the prevention of post-operative infection (endophthalmitis), after cataract surgery. Other areas of the biomedical field have been accessed by PHEMA carriers, namely cancer treatment [182, 185] and neurologic diseases treatment [186].

Copolymers based on PHEMA presenting stimuli-responsive behavior have also been prepared. As example, Zhang and co-authors [187] prepared a pH-sensitive hydrogel from poly(L-glutamic acid-*g*-2-hydroxyethyl methacrylate) for oral drug delivery.

Acrylic polymers with pendant acid groups, *e.g.*, poly(acrylic acid) (PAA), and N-substituted acrylamide polymers, *e.g.*, poly(N-isopropylacrylamide) (PNIPAAm), are stimuli-responsive materials, commonly known as intelligent or "smart" polymers which have found wide use in biomedical applications [188].

3 Miniaturization of Drug Delivery Systems

DDS may come in every shape and size depending on the desired application. However, recent efforts have been made in trying to develop small-sized systems. Advantages of such small devices (including nano-scale) include: higher intracellular uptake, higher drug concentration transportation, easier permeation through physiological barriers and possible drug targeting. For all these reasons, the association of nanotechnology and medicine is, until now, most wanted. As a result, nanomedicine is an undergoing research subject. The combination of such knowledge fields aims to create miniature and multi-functionalized systems that can act as drug carriers improving drug delivery in both time and disease specific manner [189].

Among drug carriers, polymers have so far been the most tested materials. There are several forms of combining a polymer and a drug and still remain in a micro- or nano-scale. In Fig. 29 are presented some of the polymer-based nanocarriers that have been studied and that proved to be efficient until now.

During the next sections of this chapter an overview in some of these polymeric drug carriers will be presented. Recent research areas and main achieved results will be described and contextualized.

3.1 Polymeric Micro- and Nanoparticles

DDS often use polymeric carriers that act as "drug transporters". The use of a carrier allows overcoming some problems associated with the development and application of new drugs. Despite the fact that new biologically active molecules are being discovered constantly, some of them present serious limitations to their use such as: *in vivo* degradation and short circulating half-lives [190]; poor solubility and biodistribution which can compromise interaction of drug and site of action; low solubility at physiological pH [191] and insufficient cellular uptake [192]. The use of a DDS can be a very effective way of overcoming these barriers. However, when DDS are concerned, both design and engineering have to be considered in order to optimize drug efficiency as well as reduce their side effects [193].

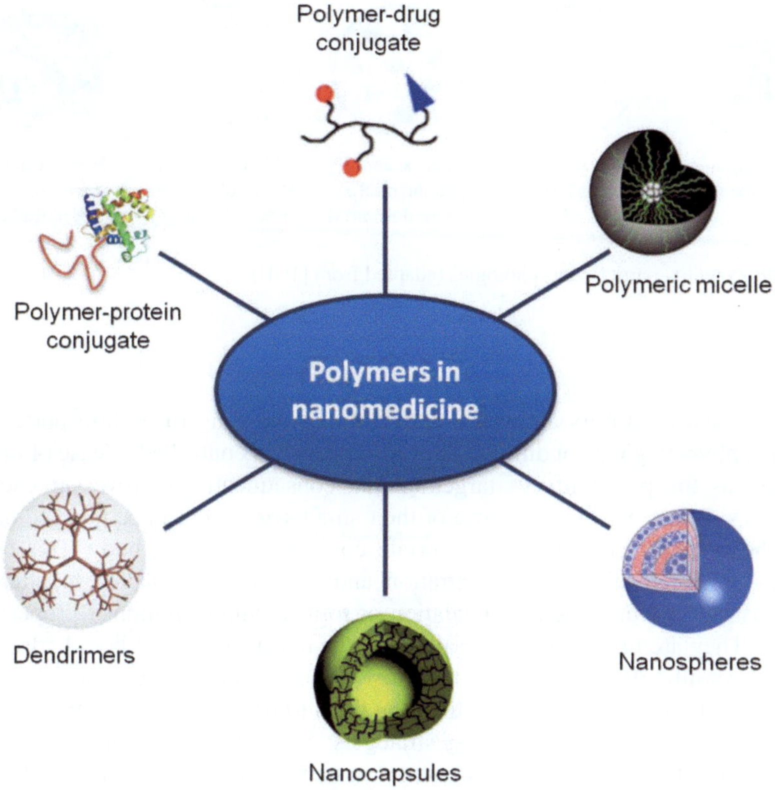

Fig. 29 Polymeric nanocarriers applied in DDS (adapted from [189])

Recently, researchers have developed a strong interest in particulate DDS since they have proved to be efficient drug carriers. Within particles, two different categories can be distinguished: spheres and capsules. The first type refers to spherical particles composed of homogeneous mixtures of active agent and polymer, while the second is applied when the particle presents a core (where the active agent is placed) that is delimited by a different material (usually the polymer). The core may present itself in distinct forms and states from solid to even gas. Several active agent domains may be found inside the capsule core. Some potential particles morphologies are presented in Fig. 30.

Another parameter used in particles classification refers to their size. Therefore, two categories are considered: micro- and nanoparticles. The distinction between both classes is often confusing and even polemic. Lately, the most accepted classification reports that particles up to 100 nm are considered nanoparticles while the ones from 1 μm up to 1,000 μm are classified as microparticles [195].

Micro- and nanoparticles are recently emerging as highly efficient vehicles in the delivery of several drugs, including insoluble and sensitive (such as DNA) ones.

Fig. 30 Potential particles morphologies (adapted from [194])

These types of carriers allow the protection of the drug to be transported from adverse physiological conditions. They also provide a controlled release of the drug as well as the possibility of targeting that consequently improves efficacy and reduces side effects. Also, because of their small size, both can be injected directly into the systemic circulation or a certain compartment in the body. This process allows a high local drug concentration and can also be adapted to alternative administration routes such as inhalation or topical administration if properly sustained. They are extremely stable allowing the encapsulation of both hydrophobic and hydrophilic drugs. However, the most explored ability of the particles during the last years is the possibility of directing them to specific places where the drug is to be delivered by using targeting strategies. Several factors can influence drug release from both micro- and nanoparticles, namely the nature of the carried drug (including charge and even size), particle size and composition and even surrounding conditions such as pH and temperature [196].

Particles to be applied as drug carriers can be prepared from different materials, namely polymers, lipids, peptides or even metals such as gold. The choice of the used material, as well as the method of preparation, allows obtaining a variety of morphologies, sizes and shapes that allows the tailoring of DDS for specific organs and diseases. Particles prepared from either natural or synthetic polymers have been extensively investigated to be applied as DDS. Among the polymers available for synthesis of micro- and nanoparticles, the ones generally used are chitosan [197], PCL [198], PLA [199], PLGA [141] and gelatin [200], because of their stated biodegradability and biocompatibility [59]. Polymeric particles may be synthesized by a variety of methods which can be divided in two different categories: the one that starts from preformed polymers and the one based in polymerization reactions between monomers [201] (Fig. 31). The choice of the method to be applied is conditioned by several requirements such as pretended polymeric system, adequate size and even final application of the systems (industrial or biomedical as examples).

As already stated by other authors [202], most of the methods applied in the synthesis of microparticles may also be used to prepare nanoparticles by adjusting experimental conditions. Therefore, during the next sections of this chapter, the most relevant methods of nanoparticles preparation based either in polymerization

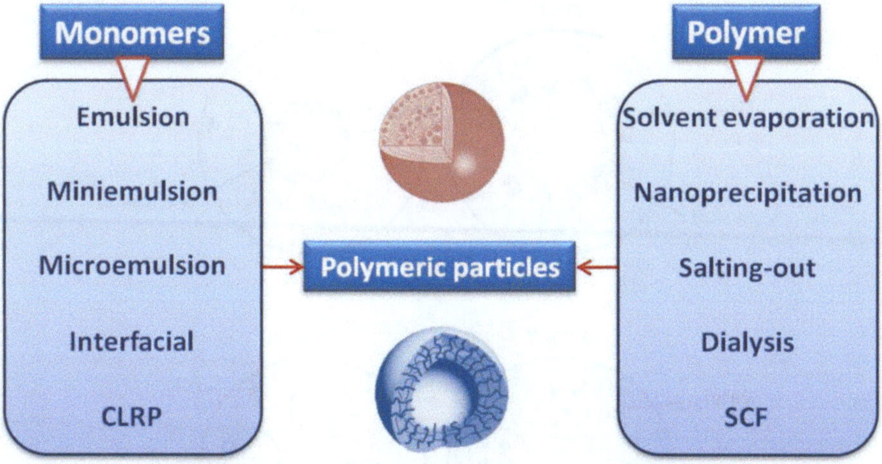

Fig. 31 Schematic summary of the available methods to produce polymeric particles (adapted from [201])

reaction or in preformed polymers will be addressed, keeping in mind that microparticles may also be prepared by the same techniques.

3.1.1 Micro- and Nanoparticles Prepared by Polymerization of Monomers

Emulsion Polymerization

The most commonly used technique to prepare nanoparticles is the emulsion polymerization followed by solvent evaporation [203]. An emulsion can be defined as a dispersion in which the dispersed phase is composed of small globules of a liquid distributed throughout a vehicle in which it is immiscible. There are several types of emulsion, from single emulsions (water in oil or W/O and oil in water or O/W) to multiple emulsions (Fig. 32).

Among these, the O/W emulsion polymerization is the most frequently applied to prepare nanoparticles. The necessary components to perform such reaction include: a monomer with low water solubility, water, a water-soluble initiator and a surfactant (which prevents particles aggregation during polymerization). During this process, the monomer is firstly added to an organic solvent. Emulsion is formed when this organic phase is added to the aqueous phase (water, initiator and the surfactant). This emulsion (oil in water or O/W) is then homogenized or placed under sonication and particles are formed. Finally, the organic solvent is evaporated and the particles remaining in the aqueous phase are isolated by centrifugation or lyophilization (Fig. 33).

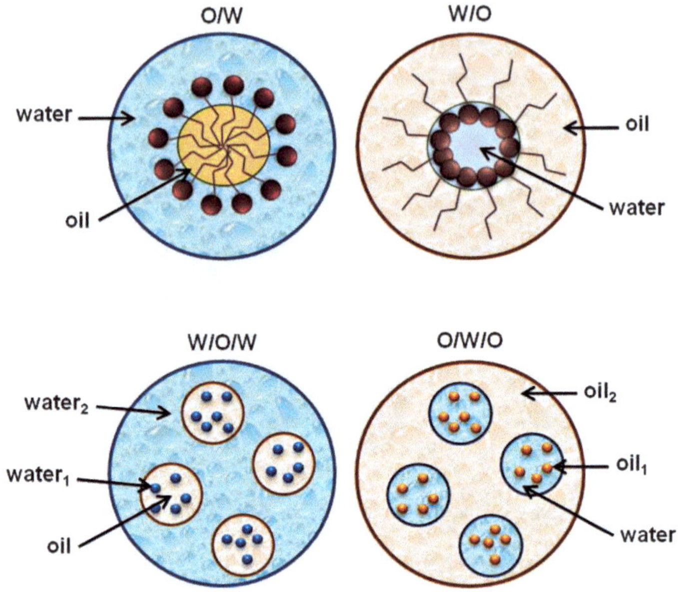

Fig. 32 Representation of different types of emulsions (single emulsions O/W and W/O and double emulsions W/O/W and O/W/O)

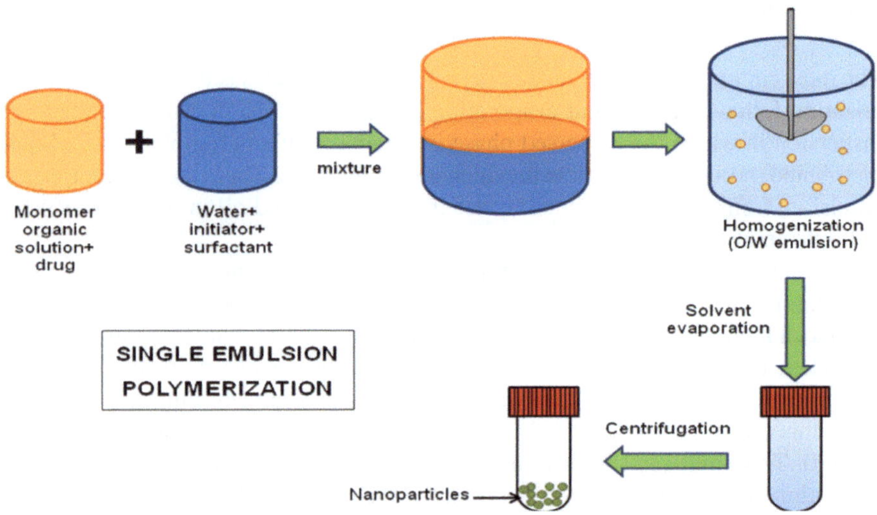

Fig. 33 Representation of nanoparticles preparation by single emulsion polymerization (O/W)

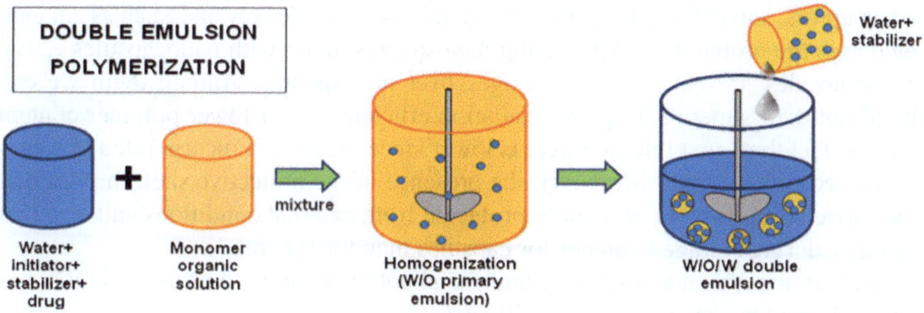

Fig. 34 Representation of nanoparticles preparation by double emulsion polymerization (W/O/W)

Although this method has proved to be highly efficient to encapsulate hydrophobic drugs, the same is not the case when encapsulating hydrophilic compounds [203]. When the objective is to encapsulate hydrophilic molecules, double (or multiple) emulsions have to be prepared. During this process, the hydrophilic drug is solubilized in water and a stabilizer is added to the solution (W_1). This aqueous solution is then added to an organic phase containing the organic solvent and the monomer (O), resulting in the formation of the primary emulsion. The double emulsion is prepared by adding the primary emulsion to another aqueous phase containing, again, the stabilizer ($W_1/O/W_2$). The obtained double emulsion then follows the same steps described for the single emulsion until collecting the nanoparticles (Fig. 34).

This type of emulsion polymerization involves the use of surfactants which need to be removed from the final systems. Therefore, the time and the costs associated to particles production are significantly affected. For this reason, new techniques that do not require the addition of a surfactant are now being applied and are generally referred to as "surfactant-free" emulsion polymerization [201]. The reagents involved in such reaction include water, water-soluble initiator and obviously, the monomer. Since no surfactant is to be added to the mixture, the stabilization of the particles is ensured by the presence of ionic co-monomers or ionizable initiators. Several authors have so far reported the preparation of nanoparticles using surfactant-free emulsions to be applied as DDS [204–206].

Interfacial Polymerization

Interfacial polymerization reaction is another common technique applied in nanoparticles preparation. It involves the condensation of two reactive monomers at the interface of the organic and aqueous phases [201]. This means that each of the monomers is dissolved in different phases of the system.

The main advantages of such method include: relative simplicity of performance and scale-up, high efficiency of drug encapsulation, and the possibility of producing nanocapsules. According to the literature, nanocapsules are more likely to be formed when aprotic solvents are used. Contrarily, the use of protic solvents, such as ethanol, increases the probability of preparing nanospheres along with nanocapsules [207]. Nanocapsules present some advantages over nanospheres. Among them we can highlight their superior drug encapsulation efficiency, their lower polymer content and the fact that burst release effect is lower since the drug is incorporated in a cavity protected by a shell. Moreover, the presence of the protective shell implies that the carried drug is more efficiently protected from external conditions and also that stability during storage is higher for capsules than for spheres.

Preparation of nanocapsules by interfacial polymerization to be used as DDS has been reported by several authors [170, 208].

3.1.2 Micro- and Nanoparticles Prepared by Dispersion of Preformed Polymers

Solvent Evaporation

Solvent evaporation technique is similar to emulsion polymerization described previously, with the difference that dispersed polymers are used instead of monomers. As an overall description, a polymer is dispersed in an organic solvent that can be for example chloroform or dichloromethane. The hydrophobic drug is then added to this solution and after dissolution the mixture is emulsified in an aqueous phase in the presence of a surfactant or emulsifier such as poly(vinyl alcohol). The resultant mixture is, therefore, an oil in water (O/W) emulsion and is prepared by high speed homogenization or sonication [209]. Once stabilized, the solvent is evaporated by stirring or by increasing temperature resulting in the precipitation of the polymer under the form of nanoparticles [202]. These particles are then washed with water in order to remove components such as the surfactant and finally lyophilized. This method is, however, only possible to be applied during encapsulation of hydrophobic drugs. As in emulsion polymerization, when hydrophilic drugs are to be used, multiple emulsions, such as double emulsion W/O/W have to be prepared. In this case the drug is primarily solubilized in water and this aqueous solution is afterwards added to the organic solvent containing the polymer. The following steps are similar to the O/W emulsion.

This technique has been extensively used to prepare nanoparticles based on several polymers, such as PLA, PLGA [142] and PCL. Some drugs, including anti HIV zidovudine or AZT [210] and lamivudine [199], antioxidants such as quercitrin [211] and antitumoral drugs [212] have already been entrapped in these particles for the preparation of DDS.

Salting-Out

Salting-out is a technique that results from the modification of emulsion and solvent evaporation methodologies. The main advantage of such modification is to avoid the

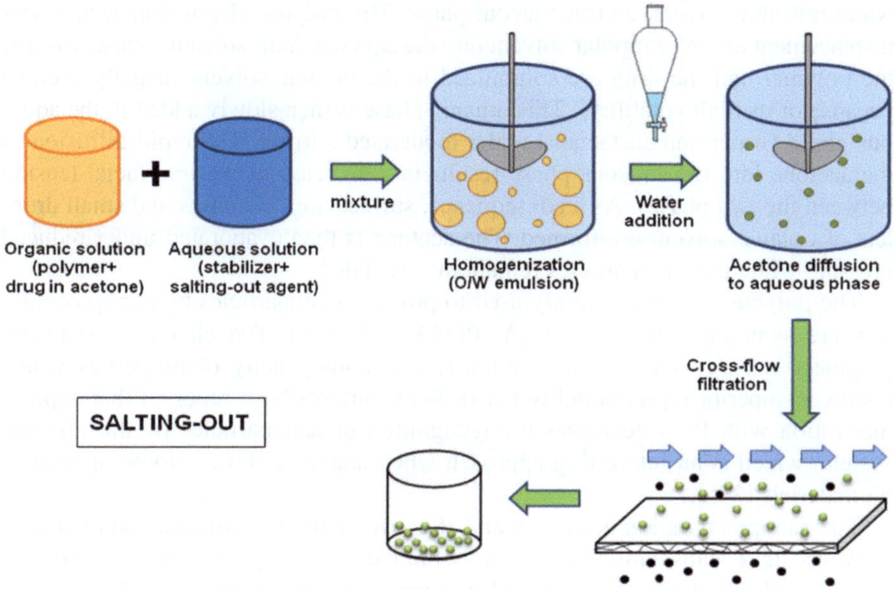

Fig. 35 Nanoparticles preparation by the salting-out technique

presence of surfactants and chloride compounds that need to be posteriorly removed from the system. This technique of nanoparticles preparation also begins with an emulsion. The organic phase of such emulsion is prepared by dissolving the polymer in a solvent which is totally miscible in water (like acetone). This organic solution is then added to the aqueous phase, which, in this case, is composed by water, the salting-out agent (an electrolyte such as calcium chloride or non-electrolytes such as sucrose) and a stabilizer (polyvinylpyrrolidone for example). The emulsion is obtained, unlike during previous techniques, with no application of high-shear forces [201]. After this process, an adequate volume of water is added to the mixture in order to allow the diffusion of acetone to the aqueous phase and therefore, the formation of the nanoparticles. Finally, the resultant nanoparticles are collected by cross-flow filtration, during which are eliminated both the solvent and the salting-out agent (Fig. 35).

Despite the absence of chlorinated compounds during the process, this technique presents a serious disadvantage that is related to the fact that several bioactive compounds are incompatible with the presence of salts [203].

Nanoprecipitation

The nanoprecipitation technique, also commonly named solvent displacement, was firstly developed by Fessi and co-authors [213]. This method is based on the deposition of a polymer at the interface of an organic phase (containing a semipolar water-miscible solvent) and an aqueous phase. This polymer deposition occurs after displacement of the semipolar solvent into the aqueous (non-solvent) phase. Briefly, the polymer and the drug are solubilized in the organic solvent (usually acetone because of its high volatility). This organic phase is then slowly added to the aqueous phase (water and surfactant) under moderated stirring. The rapid diffusion of the acetone into the aqueous phase results in a decrease of the interfacial tension between the two phases. As a consequence, surface area increases and small droplets of organic solvent are formed. The acetone is then evaporated under reduced pressure while the nanoparticles remain in suspended.

The polymers most frequently used to produce nanoparticles by nanoprecipitation are synthetic polyesters (PLA, PLGA and PCL). The choice of synthetic polymers is based on their higher purity and homogeneity of properties which results in superior reproducibility [214]. Some authors have reported that copolymerization with PEG decreases the recognition of nanoparticles by the immune system, which is an interesting approach when nanoparticles are to be applied as biomaterials [215].

The nanoprecipitation is an easy and effective method to produce nanoparticles. However, its encapsulation efficiency is limited to hydrophobic drugs. Although several studies have been conducted by researchers in encapsulating hydrophilic compounds, the general conclusions were that drug loading was significantly lower for this type of drugs [216, 217].

Supercritical Fluid Technology

The previously described methods used to prepare nanoparticles have in common the use of toxic solvents or surfactants. Therefore, concerns about environmental safety, as well as biological toxicity, are leading research in the direction of developing new techniques, that are simultaneously environmentally-friendly and that allow the synthesis of biocompatible drug-loaded micro- and nanoparticles. Supercritical fluid technology is recently being considered an interesting and quite effective method to achieve such goals [218].

There are two possible routes that make use of supercritical fluids (SCF) to achieve the preparation of micro- and nanoparticles. The first that will be addressed is called rapid expansion of supercritical solution (RESS). During this process, the drug is solubilized in a SCF (usually CO_2) and the obtained solution is then expanded across an orifice or a capillary nozzle into ambient air [201]. With this expansion, the solvent power of the SCF is highly reduced leading to the precipitation of the polymer (Fig. 36).

The main disadvantage of the RESS technique is related to the low solubility of most polymers in SCF. In fact, when preparing drug-loaded particles, a uniform drug distribution is only accomplished when polymers with low molecular mass (<10 kg/mol) are used. Another feature of such method, which may also be considered as a drawback, is the fact that particles prepared by RESS are usually microsized. When nanoparticles are desired, a modified version of RESS must be used.

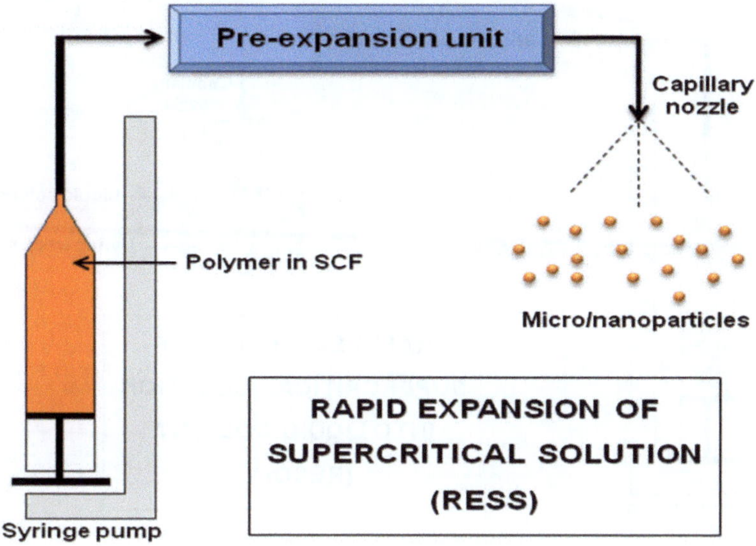

Fig. 36 Representation of the system for the preparation of micro- and nanoparticles by RESS (adapted from [201]). Polymer solution in SCF is placed in a syringe pump and before expansion through the nozzle is pumped into a pre-expansion unit where is heated isobarically to the necessary temperature

The technique resulting from this adaptation is generally called RESOLV, which stands for rapid expansion of supercritical solution into liquid solvent. In this case, solution expansion occurs into a liquid solution instead of into ambient air. The liquid solvent inhibits particles' growth, leading to the formation of nano-sized particles (Fig. 37).

As already mentioned, both variations of SCF technology are useful in obtained drug loaded particles in an environmental-friendly form. However, the problem associated with the poor polymer solubility in SCF is still a constraint that limits the use of such methodology in a more extended way.

3.2 *Liposomes*

Liposomes are self-assembled vesicles consisting of an aqueous core domain entrapped by a lipidic bilayer. They were reported for the first time during the 1960s, and their clinical use was approved in the mid-1990s [219].

The formation of liposomes is related to the hydrophilic/hydrophobic interaction between lipid/lipid and lipid/water molecules. In aqueous medium, lipids rearrange themselves, first, as bilayer sheets, with the headgroup of the lipid (hydrophilic) facing the water phase while the hydrocarbon chains (hydrophobic) are forced to face each other. Immediately afterwards, the formation of liposomes occurs (Fig. 38) [220, 221].

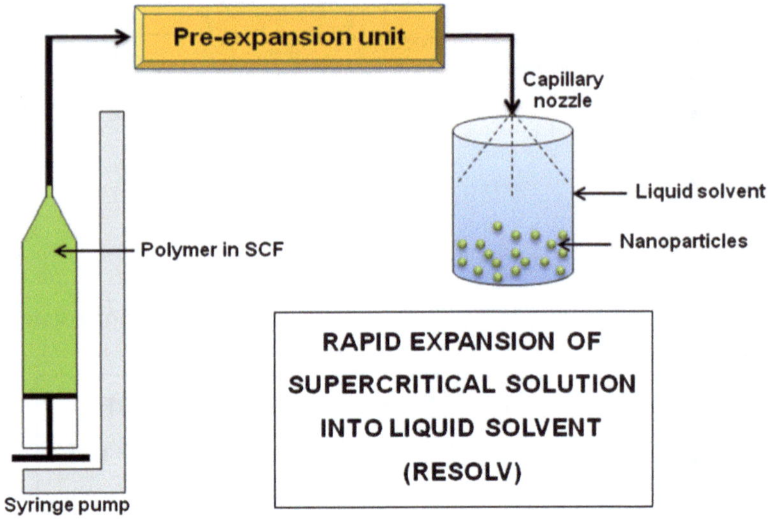

Fig. 37 Representation of the system for the preparation of nanoparticles by RESOLV (adapted from [201]). Polymer solution in SCF is placed in a syringe pump and before expansion through the nozzle is pumped into a pre-expansion unit where is heated isobarically to the necessary temperature. Unlike in RESS, expansion of polymeric solution is preformed into a liquid solvent that allows the control over particles growing, leading to nanoparticles formation

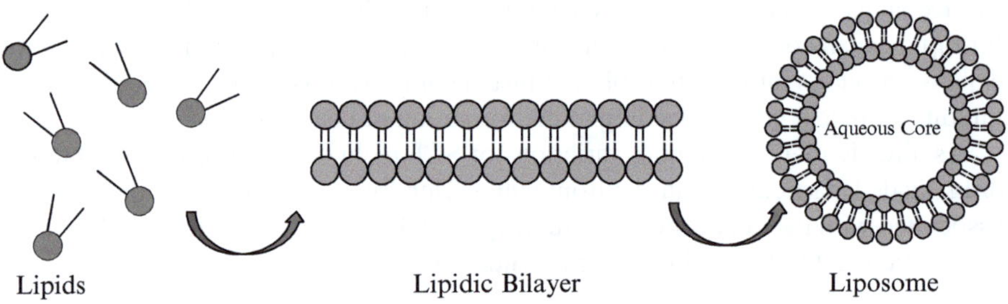

Fig. 38 Schematic representation of the liposomes' formation

Liposomes can be classified, according to their lamellarity (number of bilayer membranes) and size (Fig. 39) in [219, 220]: small unilamellar vesicle (SUVs) (~0.02 to ~0.2 μm), large unilamellar vesicles (LUVs) (~0.2 to ~1 μm), giant unilamellar vesicles (GUVs) (>1 μm) and multilamellar vesicles (MLVs) (>0.1 μm).

The simplest method for the preparation of liposomes is the thin-film hydration method, which yields mostly MLVs. Other methods, like reverse-phase evaporation technique and solvent injection are also used and produce mostly LUVs. Preparations containing MLVs or LUVs can, subsequently, be sonicated or extruded in order to give SUVs [219, 220].

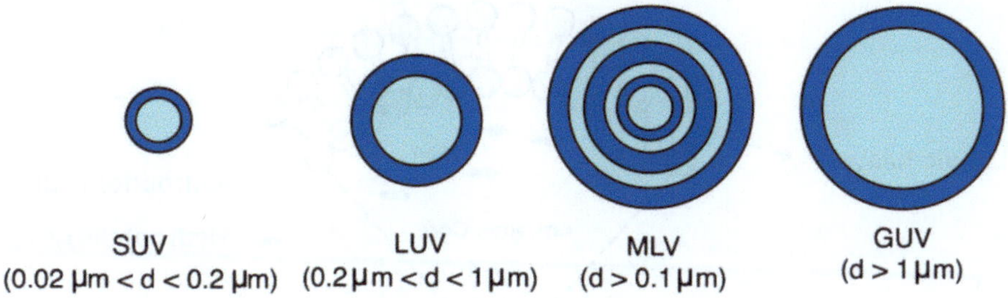

Fig. 39 Schematic representation of the different types of liposomes, depending on their size and number of lamellae

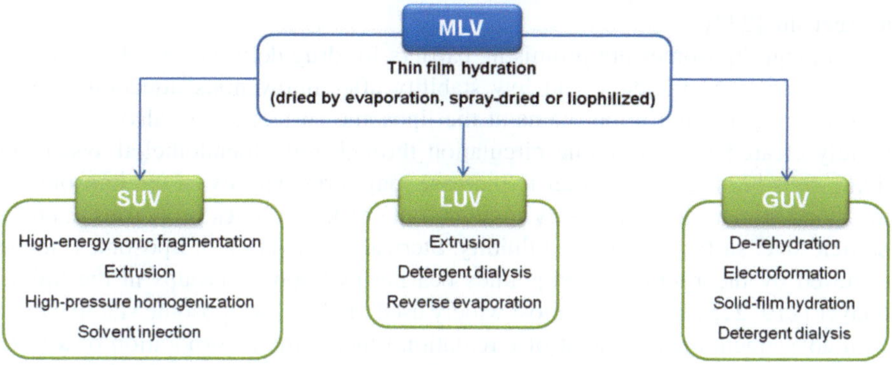

Fig. 40 Different methods used in liposomes preparation (adapted from [220])

In the thin-film hydration method, a dried lipidic film is hydrated in an aqueous buffer solution, at a temperature above the transition temperature (temperature at which a lipid passes from the gel state to a liquid state) of the lipids. As stated above, this method originates preferentially MLVs, with high heterogeneity in size, size distribution and lamellarity. In solvent injection method, an ethanolic or etheric solution of lipids is slowly injected into an aqueous phase, forming unilamellar liposomes. The vesicles obtained by this method present a well-defined size distribution and high encapsulation efficiency. Removal of residual solvent is done by dialysis against distilled water [220]. Figure 40 presents the different methods that can be used in the liposomes preparation.

Recently, due to their biocompatibility, non-immunogenicity, non-toxicity and ability of entrapping both hydrophilic and hydrophobic compounds, liposomes emerged as attractive materials for the development of DDS (Fig. 41) [219, 222]. Furthermore, a wide spectrum of liposomes properties can be obtained by just varying the lipid (natural or synthetic) or lipid composition used in their preparation. Chemical surface modification of liposomes is also possible. The incorporation of

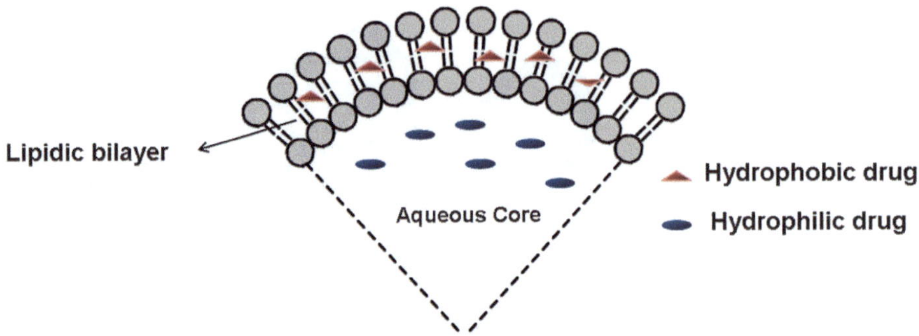

Fig. 41 Localization of hydrophilic and hydrophobic drugs within the liposome

specific molecules or macromolecules can enhance the interaction of liposome with a target site [223].

Although liposomes are promising carriers for drug delivery, they have shown low transfection efficiency and low stability after intravenous injection, due to absorption of plasma lipoproteins at the liposome surface. They also tend to be rapidly cleared from systemic circulation through reticuloendothelial system in living cells [224, 225]. In order to increase long-term stability of the liposome in the blood stream, it is necessary to alter some of the vesicles properties, namely particle size, surface charge and fluidity. Sterical stabilization of liposomes can be achieved by incorporating compounds bearing hydrophilic groups in the lipidic bilayer [219, 225]. One of the most widely used method for liposome stabilization, with consequent enhancement of circulation time, is the incorporation of a PEG phosphatidylethanolamine (PEG-PE) conjugate into its structure [226–228].

An additional stabilizing effect of the liposome bilayer can be achieved by including cholesterol in the formulation. Cholesterol is used as a membrane sealer, minimizing bilayers permeability and providing mechanical and cohesive strength [224, 229].

A possible strategy to increase the usefulness of liposomes in the drug delivery field is to prepare functional liposomes that can respond to changes, for example, in pH or temperature of the surrounding medium. These are useful for site-specific and/or cytoplasmatic drug delivery [230]. The most effective method for that purpose is the incorporation of functional polymers in the liposome. These polymers should be amphiphilic, with the polymers hydrophobic part acting as an anchor to the liposome surface. The stimulus-induced conformational changes of the polymers induce a partial or total arrangement of the liposome structure, which is believed to be responsible for the release of drugs from the liposome [231].

Temperature-sensitive liposomes are prepared from polymers that exhibit a low critical solution temperature (LCST) in aqueous medium. These kinds of polymers are highly hydrophilic below the LCST, but become hydrophobic above this temperature value. Therefore, incorporation of the polymer in the liposome is expected to destabilize the liposomal structure when temperature is raised above the LCST. Their copolymerization with other monomers, with varying hydrophilicity/

hydrophobicity, enables the adjustment of the LCST to a desired temperature [232, 233]. PNIPAAm, that exhibits a LCST around 32°C, is the most widely used polymer for the preparation of temperature-sensitive liposomes.

Takagishi and co-authors prepared thermo-responsive phosphatidylcholine (PC) liposomes, by coating them with a NIPAAm-octadecylacrylate (ODA) copolymer [234]. In other study, Takagishi and co-authors used a copolymer of NIPAAm and N,N-didodecylacrylamide (NDDAAm) to coat a Phosphatidylethanolamine (PE) based liposome. NDDAAm presents a long acyl chain, which allows the copolymer to be tightly bound to the liposome. In addition to hydrophobic interactions, the interaction between the membrane and the copolymer can be enhanced by the establishment of hydrogen bonds. During this work, it was shown that the incorporation of PE in liposome formulation is important for the development of more efficient DDS [232]. Han and co-authors studied the release of DOX from liposomes modified with poly(NIPAAm-co-acrylamide) P(NIPAAm-AAm) and PEG. They showed that incorporation of the copolymer and PEG enhanced drug release from liposomes and reduced protein adsorption in serum [235].

Recently, Paasonen and co-authors prepared a temperature-sensitive PC and PE based liposomes using poly(N-(2-hydroxypropyl) methacrylamide mono/dilactate). The copolymer was functionalized with a cholesterol moiety that acts as an anchor to the lipidic vesicle. When compared with the NIPAAm based copolymers, this kind of copolymer presents the advantage of being biodegradable, with a LCST ranging from 13°C to 65°C [222].

The sensitization of liposomes to pH by incorporation of a pH-sensitive polymer was reported by Tirrel and co-authors, in the early 80s. They showed that the attachment of poly(2-ethyl acrylic acid) (PEAA) onto liposome surface gave pH-sensitive content release properties. This polymer interacts with lipidic membrane in a pH dependent manner. At mildly acidic conditions, in high concentrations, PEAA induces micelization of the membranes, whereas at low concentrations, it permeabilizes them [236]. The pH at which permeabilization occurs can be modulated by adjusting the polymer molecular weight or substituting PEAA by another acrylic acid derivative [237].

Succinylated poly(glycidol) (SucPG), that presents –OH and –COOH groups on the side chain, can also be used to prepare pH-sensitive liposomes [237]. Kono and co-authors have incorporated SucPG bearing long alkyl chains in the structure of PC vesicles [238]. This complex is stable at neutral pH, being destabilized under mildly acidic conditions. It was shown that it could deliver calcein into the cytosol through fusion with endosome/lysosome membranes. It is believed that the mechanism responsible for the destabilization of liposome, when in contact with SucPG, is related to two phenomena: the hydrophobic interaction of SucPG backbone with lipid membrane and hydrogen bond formation between polymer carboxyl groups and phosphate groups of the phospholipid membrane [237]. More recently, Sakaguchi and co-authors demonstrated that pH sensitization of eggPC (PC obtained from egg) was enhanced with increasing side-chain hydrophobicity of poly(glycidol) derivatives [239]. The introduction of a titratable comonomer in the structure of hydrophobically-modified PNIPAAm can make the polymer pH-responsive. Leroux and co-authors have prepared a pH-sensitive liposome from eggPC/cholesterol, by incorporating a copolymer of NIPAAm, methacrylic acid and ODA into the liposomes lipidic layer [240].

Recently, Cho and co-authors prepared a pH-sensitive system composed of poly(methacrylic acid-co-stearyl methacrylate), PC and cholesterol. The complex was stable at pH above 6, becoming slightly disorganized when pH was 5 [231].

All the systems above referred are commonly known as polymer incorporated liposomes. Although they are able to respond to temperature or pH changes of the medium, it is known that they can dissociate from the liposome surface, returning them to the unstable state. This is related with the chain length and/or saturation degree of the anchor.

References

1. Zempsky WT (1998) Alternative routes of drug administration – advantages and disadvantages (subject review). Pediatrics 101(4):730–731
2. Golan DE, Armen H, Tashjian J, Armstrong EJ, Armstrong AW (2008) Principles of pharmacology: the pathophysiological basis of drug therapy, 2nd edn. Lippincott Williams & Wilkins, Philadelphia
3. Jain KK (2008) Drug delivery systems-an overview. In: Jain KK (ed) Drug delivery systems. Humana Press, New York
4. Jeffreys D (2005) Aspirin: the remarkable story of a wonder drug. Bloomsbury, New York
5. Magalhães PO, Lopes AM, Mazzola PG, Rangel-Yagui C, Penna TCV, Pessoa A Jr (2006) Methods of endotoxin removal from biological preparations: a review. J Pharm Sci 10:388–404
6. Mitragotri S (2008) Recent developments in needle-free drug delivery. In: Bugliarello G (ed) The bridge, vol 38. National Academy of Sciences, Washington DC
7. Prausnitz MR, Langer R (2008) Transdermal drug delivery. Nat Biotechnol 26:1261–1268
8. Sangalli ME, Maroni A, Zema L, Busetti C, Giordano F, Gazzaniga A (2001) In vitro and in vivo evaluation of an oral system for time and/or site-specific drug delivery. J Control Release 73:103–110
9. Rupenthal ID, Green CR, Alany RG (2011) Comparison of ion-activated in situ gelling systems for ocular drug delivery. Part 1: physicochemical characterisation and in vitro release. Int J Pharm 411:69–77
10. Nazar H, Fatouros DG, van der Merwe SM, Bouropoulos N, Avgouropoulos G, Tsibouklis J, Roldo M (2011) Thermosensitive hydrogels for nasal drug delivery: the formulation and characterisation of systems based on N-trimethyl chitosan chloride. Eur J Pharm Biopharm 77:225–232
11. El-Sherbiny IM, Smyth HDC (2010) Biodegradable nano-micro carrier systems for sustained pulmonary drug delivery: (I) self-assembled nanoparticles encapsulated in respirable/swellable semi-IPN microspheres. Int J Pharm 395:132–141
12. Zhao B-Q, Guo Y-R, Li X-L, Zang T, Qu H-Y, Zhou J-P, Li Q, Sun M-J (2011) Amelioration of dementia induced by A beta 22–35 through rectal delivery of undecapeptide-hEGF to mouse brain. Int J Pharm 405:1–8
13. Woolfson AD, Umrethia ML, Kett VL, Malcolm RK (2010) Freeze-dried, mucoadhesive system for vaginal delivery of the HIV microbicide, dapivirine: optimisation by an artificial neural network. Int J Pharm 388:136–143
14. Bajpai AK, Shukla SK, Bhanu S, Kankane S (2008) Responsive polymers in controlled drug delivery. Prog Polym Sci 33:1088–1118

15. Gilhotra RM, Mishra DN (2009) Polymeric systems for ocular inserts. http://www.pharmainfo.net/reviews/polymeric-systems-ocular-inserts. Accessed 21 June 2012
16. Kelner A, Schacht EH (2005) Tailor-made polymers for local drug delivery: Release of macromolecular model drugs from biodegradable hydrogels based on poly(ethylene oxide). J Control Release 101:13–20
17. Kabanov A, Okano T (2001) Challenges in polymer therapeutics, state of the art and prospects of polymer drugs. In: Maeda H, Kabanov A (eds) Polymer drugs in the clinical stage: advantages and prospects. Kluwer Academic/Plenum Publishers, New York
18. Braunecker WA, Matyjaszewski K (2007) Controlled/living radical polymerization: features, developments, and perspectives. Prog Polym Sci 32:93–146
19. Robinson JR, Lee VHL (1987) Controlled drug delivery: fundamentals and applications, 2nd edn. Marcel Dekker, New York
20. Koo OM, Rubinstein I, Onyuksel H (2005) Role of nanotechnology in targeted drug delivery and imaging: a concise review. Nanomed Nanotechnol Biol Med 1:193–212
21. Schmaljohann D (2006) Thermo- and pH-responsive polymers in drug delivery. Adv Drug Deliv Rev 58:1655–1670
22. Brannon-Peppas L (1997) Polymers in controlled drug delivery. Med Plast Biomater 4:34–44
23. Chasin M, Langer RS (1990) Biodegradable polymers as drug delivery systems. Marcel Dekker, New York
24. Burkersroda FV, Schedl L, Göpferich A (2002) Why degradable polymers undergo surface erosion or bulk erosion. Biomaterials 23:4221–4231
25. Sharma S, Singh SP, Bhardwaj S, Gaurave K, Gupta GD (2008) Osmotic controlled drug delivery system. http://www.pharmainfo.net/reviews/osmotic-controlled-drug-delivery-system. Accessed 21 June 2012
26. Urbina MC, Zinoveva S, Miller T, Sabliov CM, Monroe WT, Kumar C (2008) Investigation of magnetic nanoparticle-polymer composites for multiple-controlled drug delivery. J Phys Chem C 112:11102–11108
27. Liu T-Y, Hu S-H, Liu K-H, Liu D-M, Chen S-Y (2006) Preparation and characterization of smart magnetic hydrogels and its use for drug release. J Magn Magn Mater 304:397–399
28. Craciunescu I, Nan A, Turcu R, Kacso I, Bratu I, Leostean C, Vekas L (2009) Synthesis, characterization and drug delivery application of the temperature responsive pNIPA hydrogel. J Phys Conf Ser 182:1–4
29. Huang J, Huang Z, Bao Y, Weng Z (2006) Thermosensitive poly(N-isopropylacrylamide-co-acrylonitrile) hydrogels with rapid response. Chin J Chem Eng 14:87–92
30. Hergt R, Dutz S (2007) Magnetic particle hyperthermia–biophysical limitations of a visionary tumour therapy. J Magn Magn Mater 311:187–192
31. Hergt R, Dutz S, Muller R, Zeisberger M (2006) Magnetic particle hyperthermia: nanoparticle magnetism and materials development for cancer therapy. J Phys Condens Matter 18:S2919–S2934
32. Shinkai M, Ito A (2004) Functional magnetic particles for medical application. In: Scheper T, Kobayashi T (eds) Recent progress of biochemical and biomedical engineering in Japan II, vol 91. Springer, Berlin/Heidelberg
33. Nair LS, Laurencin CT (2006) Polymers as biomaterials for tissue engineering and controlled drug delivery. Adv Biochem Eng Biotechnol 102:47–90
34. Nair LS, Laurencin CT (2007) Biodegradable polymers as biomaterials. Prog Polym Sci 32:762–798
35. Ulery BD, Nair LS, Laurencin CT (2011) Biomedical applications of biodegradable polymers. J Polym Sci, Part B: Polym Phys 49:832–864
36. Pillai O, Panchagnula R (2001) Polymers in drug delivery. Curr Opin Chem Biol 5:447–451
37. Cascone MG, Sim B, Downes S (1995) Blends of synthetic and natural polymers as drug delivery systems for growth hormone. Biomaterials 16:569–574
38. Sahoo S, Sasmala A, Nanda R, Phani AR, Nayak PL (2010) Synthesis of chitosan–polycaprolactone blend for control delivery of ofloxacin drug. Carbohydr Polym 79:106–113

39. Friess W (1998) Collagen – biomaterial for drug delivery. Eur J Pharm Biopharm 45:113–136
40. Pastorino L, Erokhina S, Soumetz FC, Bianchini P, Konovalov O, Diaspro A, Ruggiero C, Erokhin V (2011) Collagen containing microcapsules: smart containers for disease controlled therapy. J Colloid Interface Sci 357:56–62
41. MaHam A, Tang ZW, Wu H, Wang J, Lin YH (2009) Protein-based nanomedicine platforms for drug delivery. Small 5:1706–1721
42. Berthold A, Cremer K, Kreuter J (1998) Collagen microparticles: carriers for glucocorticosteroids. Eur J Pharm Biopharm 45:23–29
43. Metzmacher I, Radu F, Bause M, Knabner P, Friess W (2007) A model describing the effect of enzymatic degradation on drug release from collagen minirods. Eur J Pharm Biopharm 67:349–360
44. Kawashima T, Nagai N, Kaji H, Kumasaka N, Onami H, Ishikawa Y, Osumi N, Nishizawa M, Abe T (2011) A scalable controlled-release device for transscleral drug delivery to the retina. Biomaterials 32:1950–1956
45. Kanematsu A, Yamamoto S, Ozeki M, Noguchi T, Kanatani I, Ogawa O, Tabata Y (2004) Collagenous matrices as release carriers of exogenous growth factors. Biomaterials 25:4513–4520
46. Nagai N, Kumasaka N, Kawashima T, Kaji H, Nishizawa M, Abe T (2010) Preparation and characterization of collagen microspheres for sustained release of VEGF. J Mater Sci Mater Med 21:1891–1898
47. Teixeira S, Yang L, Dijkstra PJ, Ferraz MP, Monteiro FJ (2010) Heparinized hydroxyapatite/collagen three-dimensional scaffolds for tissue engineering. J Mater Sci Mater Med 21:2385–2392
48. Wang Y, Zhang LH, Hu M, Wen WS, Xiao HX, Niu Y (2010) Effect of chondroitin sulfate modification on rhBMP-2 release kinetics from collagen delivery system. J Biomed Mater Res A 92A:693–701
49. Holladay C, Keeney M, Greiser U, Murphy M, O'Brien T, Pandit A (2009) A matrix reservoir for improved control of non-viral gene delivery. J Control Release 136:220–225
50. Vinas-Castells R, Holladay C, di Luca A, Diaz VM, Pandit A (2009) Snail1 down-regulation using small interfering RNA complexes delivered through collagen scaffolds. Bioconjugate Chem 20:2262–2269
51. Peng L, Cheng XR, Zhuo RX, Lan J, Wang YN, Shi B, Li SQ (2009) Novel gene-activated matrix with embedded chitosan/plasmid DNA nanoparticles encoding PDGF for periodontal tissue engineering. J Biomed Mater Res A 90A:564–576
52. Krebs MD, Jeon O, Alsberg E (2009) Localized and sustained delivery of silencing RNA from macroscopic biopolymer hydrogels. JACS 131:9204–9206
53. Kuijpers AJ, Engbers GHM, Feijen J, De Smedt SC, Meyvis TKL, Demeester J, Krijgsveld J, Zaat SAJ, Dankert J (1999) Characterization of the network structure of carbodiimide cross-linked gelatin gels. Macromolecules 32:3325–3333
54. Naidu BVK, Paulson AT (2011) A new method for the preparation of gelatin nanoparticles: encapsulation and drug release characteristics. J Appl Polym Sci 121:3495–3500
55. Lin M, Meng S, Zhong W, Cai R, Du QG, Tomasik P (2009) Novel drug-loaded gelatin films and their sustained-release performance. J Biomed Mater Res Part B 90B:939–944
56. GuhaSarkar S, Banerjee R (2010) Intravesical drug delivery: challenges, current status, opportunities and novel strategies. J Control Release 148:147–159
57. Muvaffak A, Gurhan I, Hasirci N (2004) Prolonged cytotoxic effect of colchicine released from biodegradable microspheres. J Biomed Mater Res Part B 71B:295–304
58. Guo R, Cheng Y, Ding D, Li X, Zhang L, Jiang X, Liu B (2011) Synthesis and antitumoral activity of gelatin/polyoxometalate hybrid nanoparticles. Macromol Biosci 11:839–847
59. Kumari A, Yadav SK, Yadav SC (2010) Biodegradable polymeric nanoparticles based drug delivery systems. Colloids Surf B Biointerfaces 75:1–18
60. Yadav SC, Kumari A, Yadav R (2011) Development of peptide and protein nanotherapeutics by nanoencapsulation and nanobioconjugation. Peptides 32:173–187

61. Kuijpers AJ, van Wachem PB, van Luyn MJA, Brouwer LA, Engbers GHM, Krijgsveld J, Zaat SAJ, Dankert J, Feijen J (2000) In vitro and in vivo evaluation of gelatin-chondroitin sulphate hydrogels for controlled release of antibacterial proteins. Biomaterials 21:1763–1772
62. Kuijpers AJ, van Wachem PB, van Luyn MJA, Engbers GHM, Krijgsveld J, Zaat SAJ, Dankert J, Feijen J (2000) In vivo and in vitro release of lysozyme from cross-linked gelatin hydrogels: a model system for the delivery of antibacterial proteins from prosthetic heart valves. J Control Release 67:323–336
63. Huang S, Fu XB (2010) Naturally derived materials-based cell and drug delivery systems in skin regeneration. J Control Release 142:149–159
64. Kim MS, Bhang SH, Yang HS, Rim NG, Jun I, Kim SI, Kim BS, Shin H (2010) Development of functional fibrous matrices for the controlled release of basic fibroblast growth factor to improve therapeutic angiogenesis. Tissue Eng Part A 16:2999–3010
65. Kim MS, Shin YM, Lee J-H, Kim SI, Nam YS, Shin CS, Shin H (2011) Release kinetics and in vitro bioactivity of basic fibroblast growth factor: effect of the thickness of fibrous matrices. Macromol Biosci 11:122–130
66. Meng ZX, Xu XX, Zheng W, Zhou HM, Li L, Zheng YF, Lou X (2011) Preparation and characterization of electrospun PLGA/gelatin nanofibers as a potential drug delivery system. Colloids Surf B Biointerfaces 84:97–102
67. Chuang VTG, Kragh-Hansen U, Otagiri M (2002) Pharmaceutical strategies utilizing recombinant human serum albumin. Pharm Res 19:569–577
68. Wunder A, Muller-Ladner U, Stelzer EHK, Funk J, Neumann E, Stehle G, Pap T, Sinn H, Gay S, Fiehn C (2003) Albumin-based drug delivery as novel therapeutic approach for rheumatoid arthritis. J Immunol 170:4793–4801
69. Kratz F (2008) Albumin as a drug carrier: design of prodrugs, drug conjugates and nanoparticles. J Control Release 132:171–183
70. Li QY, Liu C, Zhao XH, Zu YG, Wang Y, Zhang BY, Zhao DM, Zhao Q, Su L, Gao Y, Sun BH (2011) Preparation, characterization and targeting of micronized 10-hydroxycamptothecin-loaded folate-conjugated human serum albumin nanoparticles to cancer cells. Int J Nanomed 6:397–405
71. Sebak S, Mirzaei M, Malhotra M, Kulamarva A, Prakash S (2011) Human serum albumin nanoparticles as an efficient noscapine drug delivery system for potential use in breast cancer: preparation and in vitro analysis. Int J Nanomed 5:525–532
72. Zhao DM, Zhao XH, Zu YG, Li JL, Zhang Y, Jiang R, Zhang ZH (2011) Preparation, characterization, and in vitro targeted delivery of folate-decorated paclitaxel-loaded bovine serum albumin nanoparticles. Int J Nanomed 5:669–677
73. Eisele K, Gropeanu RA, Zehendner CM, Rouhanipour A, Ramanathan A, Mihov G, Koynov K, Kuhlmann CRW, Vasudevan SG, Luhmann HJ, Weil T (2010) Fine-tuning DNA/albumin polyelectrolyte interactions to produce the efficient transfection agent cBSA-147. Biomaterials 31:8789–8801
74. Okoroukwu ON, Green GR, D'Souza MJ (2010) Development of albumin microspheres containing Sp H1-DNA complexes: a novel gene delivery system. J Microencapsul 27:142–149
75. Zöphel L, Eisele K, Gropeanu R, Rouhanipour A, Koynov K, Lieberwirth I, Müllen K, Weil T (2010) Preparation of defined albumin–polymer hybrids for efficient cell transfection. Macromol Chem Phys 211:146–153
76. Rani M, Agarwal A, Negi YS (2010) Review: chitosan based hydrogel polymeric beads as drug delivery system. Bioresources 5:2765–2807
77. Hamman JH (2010) Chitosan based polyelectrolyte complexes as potential carrier materials in drug delivery systems. Mar Drugs 8:1305–1322
78. Nagpal K, Singh SK, Mishra DN (2010) Chitosan nanoparticles: a promising system in novel drug delivery. Chem Pharm Bull 58:1423–1430
79. Sinha VR, Singla AK, Wadhawan S, Kaushik R, Kumria R, Bansal K, Dhawan S (2004) Chitosan microspheres as a potential carrier for drugs. Int J Pharm 274:1–33
80. Denkbas EB (2006) Perspectives on: chitosan drug delivery systems based on their geometries. J Bioact Compat Polym 21:351–368

81. Carreira AS, Gonçalves FAMM, Mendonça PV, Gil MH, Coelho JFJ (2010) Temperature and pH responsive polymers based on chitosan: applications and new graft copolymerization strategies based on living radical polymerization. Carbohydr Polym 80:618–630
82. Ping Y, Liu CD, Tang GP, Li JS, Li J, Yang WT, Xu FJ (2010) Functionalization of chitosan via atom transfer radical polymerization for gene delivery. Adv Funct Mater 21:CP7–CP7
83. Peniche H, Peniche C (2011) Chitosan nanoparticles: a contribution to nanomedicine. Polym Int 60:883–889
84. Wang JJ, Zeng ZW, Xiao RZ, Xie TA, Zhou GL, Zhan XR, Wang SL (2011) Recent advances of chitosan nanoparticles as drug carriers. Int J Nanomed 6:765–774
85. Xu Y, Du Y (2003) Effect of molecular structure of chitosan on protein delivery properties of chitosan nanoparticles. Int J Pharm 250:215–226
86. Gan Q, Wang T (2007) Chitosan nanoparticle as protein delivery carrier–systematic examination of fabrication conditions for efficient loading and release. Colloids Surf B Biointerfaces 59:24–34
87. El-Sherbiny IM, Abdel-Bary EM, Harding DRK (2010) Preparation and in vitro evaluation of new pH-sensitive hydrogel beads for oral delivery of protein drugs. J Appl Polym Sci 115:2828–2837
88. Luangtana-anan M, Limmatvapirat S, Nunthanid J, Chalongsuk R, Yamamoto K (2010) Polyethylene glycol on stability of chitosan microparticulate carrier for protein. AAPS PharmSciTech 11:1376–1382
89. Sonia TA, Rekha MR, Sharma CP (2011) Bioadhesive hydrophobic chitosan microparticles for oral delivery of insulin: in vitro characterization and in vivo uptake studies. J Appl Polym Sci 119:2902–2910
90. Campos AM, Diebold Y, Carvalho ELS, Sánchez A, Alonso MJ (2004) Chitosan nanoparticles as new ocular drug delivery systems: in vitro stability, in vivo fate, and cellular toxicity. Pharm Res 21:803–810
91. Salamanca AE, Diebold Y, Calonge M, García-Vazquez C, Callejo S, Vila A, Alonso MJ (2006) Chitosan nanoparticles as a potential drug delivery system for the ocular surface: toxicity, uptake mechanism and in vivo tolerance. Invest Ophthalmol Vis Sci 47:1416–1425
92. Nagarwal RC, Singh PN, Kant S, Maiti P, Pandit JK (2011) Chitosan nanoparticles of 5-fluorouracil for ophthalmic delivery: characterization, in-vitro and in-vivo study. Chem Pharm Bull 59:272–278
93. de la Fuente M, Raviña M, Paolicelli P, Sanchez A, Seijo B, Alonso MJ (2010) Chitosan-based nanostructures: a delivery platform for ocular therapeutics. Adv Drug Del Rev 62:100–117
94. Ta HT, Dass CR, Dunstan DE (2008) Injectable chitosan hydrogels for localised cancer therapy. J Control Release 126:205–216
95. Bhattarai N, Gunn J, Zhang MQ (2010) Chitosan-based hydrogels for controlled, localized drug delivery. Adv Drug Del Rev 62:83–99
96. Rudzinski WE, Aminabhavi TM (2010) Chitosan as a carrier for targeted delivery of small interfering RNA. Int J Pharm 399:1–11
97. Mao H-Q, Roy K, Troung-Le VL, Janes KA, Lin KY, Wang Y, August JT, Leong KW (2001) Chitosan-DNA nanoparticles as gene carriers: synthesis, characterization and transfection efficiency. J Control Release 70:399–421
98. Yuan Y, Tan J, Wang Y, Qian C, Zhang M (2009) Chitosan nanoparticles as non-viral gene delivery vehicles based on atomic force microscopy study. Acta Biochim Biophys Sin 41:515–526
99. Gaspar VM, Sousa F, Queiroz JA, Correia IJ (2011) Formulation of chitosan-TPP-pDNA nanocapsules for gene therapy applications. Nanotechnology 22:015101
100. Sarkar K, Srivastava R, Chatterji U, Kundu PP (2011) Evaluation of chitosan and their self-assembled nanoparticles with pDNA for the application in gene therapy. J Appl Polym Sci 121:2239–2249
101. Malmo J, Varum KM, Strand SP (2011) Effect of chitosan chain architecture on gene delivery: comparison of self-branched and linear chitosans. Biomacromolecules 12:721–729

102. Opanasopit P, Tragulpakseerojn J, Apirakaramwong A, Ngawhirunpat T, Rojanarata T (2011) Chitosan enhances transfection efficiency of cationic polypeptides/DNA complexes. Int J Pharm 410:161–168
103. Hoffman AS (2006) Selecting the right polymer for biomaterial applications. In: Uchegbu IF, Schatzlein AG (eds) Polymers in drug delivery. Taylor & Francis, Boca Raton
104. Augst AD, Kong HJ, Mooney DJ (2006) Alginate hydrogels as biomaterials. Macromol Biosci 6:623–633
105. d'Ayala GG, Malinconico M, Laurienzo P (2008) Marine derived polysaccharides for biomedical applications: chemical modification approaches. Molecules 13:2069–2106
106. Baldwin AD, Kiick KL (2010) Polysaccharide-modified synthetic polymeric biomaterials. Pept Sci 94:128–140
107. Coviello T, Matricardi P, Marianecci C, Alhaique F (2007) Polysaccharide hydrogels for modified release formulations. J Control Release 119:5–24
108. Azhdarinia A, Yang DJ, Yu DF, Mendez R, Oh C, Kohanim S, Bryant J, Kim EE (2005) Regional radiochemotherapy using in situ hydrogel. Pharm Res 22:776–783
109. Hori Y, Stern PJ, Hynes RO, Irvine DJ (2009) Engulfing tumors with synthetic extracellular matrices for cancer immunotherapy. Biomaterials 30:6757–6767
110. Hori Y, Winans AM, Huang CC, Horrigan EM, Irvine DJ (2008) Injectable dendritic cell-carrying alginate gels for immunization and immunotherapy. Biomaterials 29:3671–3682
111. Cohen S, Lobel E, Trevgoda A, Peled Y (1997) A novel in situ-forming ophthalmic drug delivery system from alginates undergoing gelation in the eye. J Control Release 44:201–208
112. Sechoy O, Tissie G, Sebastian C, Maurin F, Driot JY, Trinquand C (2000) A new long acting ophthalmic formulation of carteolol containing alginic acid. Int J Pharm 207:109–116
113. Hassan MA (2007) A long acting ophthalmic gel formulations of atenolol. Drug Dev Ind Pharm 33:1192–1198
114. Tanihara M, Suzuki Y, Yamamoto E, Noguchi A, Mizushima Y (2001) Sustained release of basic fibroblast growth factor and angiogenesis in a novel covalently crosslinked gel of heparin and alginate. J Biomed Mater Res 56:216–221
115. Ohta M, Suzuki Y, Chou H, Ishikawa N, Suzuki S, Tanihara M, Mizushima Y, Dezawa M, Ide C (2004) Novel heparin/alginate gel combined with basic fibroblast growth factor promotes nerve regeneration in rat sciatic nerve. J Biomed Mater Res A 71A:661–668
116. Freeman I, Cohen S (2009) The influence of the sequential delivery of angiogenic factors from affinity-binding alginate scaffolds on vascularization. Biomaterials 30:2122–2131
117. Han YF, Han YQ, Pan YG, Chen YL, Chai JK (2010) Transplantation of microencapsulated cells expressing VEGF improves angiogenesis in implanted xenogeneic acellular dermis on wound. Transplant Proc 42:1935–1943
118. Silva EA, Mooney DJ (2010) Effects of VEGF temporal and spatial presentation on angiogenesis. Biomaterials 31:1235–1241
119. Tian J-Y, Sun X-Q, Chen X-G (2008) Formation and oral administration of alginate microspheres loaded with pDNA coding for lymphocystis disease virus (LCDV) to Japanese flounder. Fish Shellfish Immunol 24:592–599
120. Krebs MD, Salter E, Chen E, Sutter KA, Alsberg E (2010) Calcium phosphate-DNA nanoparticle gene delivery from alginate hydrogels induces in vivo osteogenesis. J Biomed Mater Res A 92A:1131–1138
121. Lucinda-Silva RM, Salgado HRN, Evangelista RC (2010) Alginate-chitosan systems: in vitro controlled release of triamcinolone and in vivo gastrointestinal transit. Carbohydr Polym 81:260–268
122. Kulkarni RV, Baraskar VV, Setty CM, Sa B (2011) Interpenetrating polymer network matrices of sodium alginate and carrageenan for controlled drug delivery application. Fiber Polym 12:352–358
123. Meng X, Li P, Wei Q, Zhang HX (2011) pH sensitive alginate-chitosan hydrogel beads for carvedilol delivery. Pharm Dev Technol 16:22–28
124. Zhu AM, Chen JH, Liu QL, Jiang YL (2011) Controlled release of berberine hydrochloride from alginate microspheres embedded within carboxymethyl chitosan hydrogels. J Appl Polym Sci 120:2374–2380

125. Chiellini E, Sunamoto J, Migliaresi C, Ottenbrite RM, Cohn D (2001) Biomedical polymers and polymer therapeutics. Kluwer Academic/Plenum Publishers, New York
126. Hornig S, Bunjes H, Heinze T (2009) Preparation and characterization of nanoparticles based on dextran-drug conjugates. J Colloid Interface Sci 338:56–62
127. Shrivastava PK, Shrivastava SK (2009) Dextran polysaccharides: successful macromolecular carrier for drug delivery. Int J Pharm Sci 1:353–368
128. Du Y-Z, Weng Q, Yuan H, Hu F-Q (2010) Synthesis and antitumor activity of stearate-g-dextran micelles for intracellular doxorubicin delivery. ACS Nano 4:6894–6902. doi:10.1021/nn100927t
129. Raemdonck K, Naeye B, Buyens K, Vandenbroucke RE, Hogset A, Demeester J, De Smedt SC (2009) Biodegradable dextran nanogels for RNA interference: focusing on endosomal escape and intracellular siRNA delivery. Adv Funct Mater 19:1406–1415
130. Naeye B, Raemdonck K, Remaut K, Sproat B, Demeester J, De Smedt SC (2010) PEGylation of biodegradable dextran nanogels for siRNA delivery. Eur J Pharm Sci 40:342–351
131. Thomas JJ, Rekha MR, Sharma CP (2010) Dextran-protamine polycation: an efficient nonviral and haemocompatible gene delivery system. Colloids Surf B Biointerfaces 81:195–205
132. Kamel S, Ali N, Jahangir K, Shah SM, El-Gendy AA (2008) Pharmaceutical significance of cellulose: a review. Express Polym Lett 2:758–778
133. Kavanagh N, Corrigan OI (2004) Swelling and erosion properties of hydroxypropylmethylcellulose (hypromellose) matrices–influence of agitation rate and dissolution medium composition. Int J Pharm 279:141–152
134. Murthy TEGK, Chowdary KPR (2005) Formulation and evaluation of ethyl cellulose-coated diclofenac sodium microcapsules: influence of solvents. Indian J Pharm Sci 67:216–219
135. Yamada T, Onishi H, Machida Y (2001) Sustained release ketoprofen microparticles with ethylcellulose and carboxymethylethylcellulose. J Control Release 75:271–282
136. Santhi K, Venkatesh D, Dhanaraj S, Sangeetha S, Suresh B (2005) Development and in-vitro evaluation of a tropical drug delivery system containing betamethazone loading ethyl cellulose nanopheres. Tropical J Pharm Res 4:495–500
137. Ravikumara NR, Madhusudhan B, Nagaraj TS, Hiremat SR, Raina G (2009) Preparation and evaluation of nimesulide-loaded ethylcellulose and methylcellulose nanoparticles and microparticles for oral delivery. J Biomater Appl 24:47–64
138. Vaghani S, Vasanti S, Chaturvedi K, Satish CS, Jivani NP (2011) Stomach-specific drug delivery of 5-fluorouracil using ethylcellulose floating microspheres. Pharm Dev Technol 15:154–161
139. Sharma AK, Keservani RK, Dadarwal SC, Choudhary Y, Ramteke S (2011) Formulation and in vitro characterization of cefpodoxime proxetil gastroretentive microballoons. Daru J Pharm Sci 19:33–40
140. Song YB, Zhang LZ, Gan WP, Zhou JP, Zhang LN (2011) Self-assembled micelles based on hydrophobically modified quaternized cellulose for drug delivery. Colloids Surf B Biointerfaces 83:313–320
141. Acharya S, Sahoo SK (2011) PLGA nanoparticles containing various anticancer agents and tumour delivery by EPR effect. Adv Drug Del Rev 63:170–183
142. Mundargi RC, Babu VR, Rangaswamy V, Patel P, Aminabhavi TM (2008) Nano/micro technologies for delivering macromolecular therapeutics using poly(D, L-lactide-co-glycolide) and its derivatives. J Control Release 125:193–209
143. Kumar PS, Saini TR, Chandrasekar D, Yellepeddi VK, Ramakrishna S, Diwan PV (2007) Novel approach for delivery of insulin loaded poly(lactide-co-glycolide) nanoparticles using a combination of stabilizers. Drug Deliv 14:517–523
144. Bilati U, Allémann E, Doelker E (2005) Poly(D, L-lactide-co-glycolide) protein-loaded nanoparticles prepared by the double emulsion method—processing and formulation issues for enhanced entrapment efficiency. J Microencapsul 22:205–214
145. Sandor M, Enscore D, Weston P, Mathiowitz E (2001) Effect of protein molecular weight on release from micron-sized PLGA microspheres. J Control Release 76:297–311
146. Liu P, Wang BC, Oiao WL, Li J (2009) Novel thermosensitive nano-micelles from triblock copolymer for drug delivery. J Biomed Nanotechnol 5:310–313

147. Tang Y, Singh J (2009) Biodegradable and biocompatible thermosensitive polymer based injectable implant for controlled release of protein. Int J Pharm 365:34–43
148. Gao YA, Sun Y, Ren FZ, Gao S (2010) PLGA-PEG-PLGA hydrogel for ocular drug delivery of dexamethasone acetate. Drug Dev Ind Pharm 36:1131–1138
149. Tang Y, Singh J (2010) Thermosensitive drug delivery system of salmon calcitonin: in vitro release, in vivo absorption, bioactivity and therapeutic efficacies. Pharm Res 27:272–284
150. Wang J, Feng S-S, Wang S, Chen Z-Y (2010) Evaluation of cationic nanoparticles of biodegradable copolymers as siRNA delivery system for hepatitis B treatment. Int J Pharm 400:194–200
151. Mishra D, Kang HC, Bae YH (2011) Reconstitutable charged polymeric (PLGA)2-b-PEI micelles for gene therapeutics delivery. Biomaterials 32:3845–3854
152. Fu CH, Sun XL, Liu DH, Chen ZJ, Lu ZJ, Zhang N (2011) Biodegradable tri-block copolymer poly(lactic acid)-poly(ethylene glycol)-poly(L-lysine)(PLA-PEG-PLL) as a non-viral vector to enhance gene transfection. Int J Mol Sci 12:1371–1388
153. Sinha VR, Bansal K, Kaushik R, Kumria R, Trehan A (2004) Poly-[epsilon]-caprolactone microspheres and nanospheres: an overview. Int J Pharm 278:1–23
154. Natu MV, de Sousa HC, Gil MH (2010) Effects of drug solubility, state and loading on controlled release in bicomponent electrospun fibers. Int J Pharm 397:50–58
155. Natu MV, Gaspar MN, Ribeiro CAF, Correia IJ, Silva D, de Sousa HC, Gil MH (2011) A poly(epsilon-caprolactone) device for sustained release of an anti-glaucoma drug. Biomed Mater 6:025003
156. Natu MV, Gil MH, de Sousa HC (2008) Supercritical solvent impregnation of poly(epsilon-caprolactone)/poly(oxyethylene-b-oxypropylene-b-oxyethylene) and poly(epsilon-caprolactone)/poly(ethylene-vinyl acetate) blends controlled release applications. J Supercrit Fluids 47:93–102
157. Wei X, Gong C, Gou M, Fu S, Guo Q, Shi S, Luo F, Guo G, Qiu L, Qian Z (2009) Biodegradable poly([var epsilon]-caprolactone)-poly(ethylene glycol) copolymers as drug delivery system. Int J Pharm 381:1–18
158. Qiu LY, Bae YH (2007) Self-assembled polyethylenimine-graft-poly(epsilon-caprolactone) micelles as potential dual carriers of genes and anticancer drugs. Biomaterials 28:4132–4142
159. Sun T-M, Du J-Z, Yao Y-D, Mao C-Q, Dou S, Huang S-Y, Zhang P-Z, Leong KW, Song E-W, Wang J (2011) Simultaneous delivery of siRNA and paclitaxel via a 'two-in-one' micelleplex promotes synergistic tumor suppression. ACS Nano 5:1483–1494
160. Xiong XB, Lavasanifar A (2011) Traceable multifunctional micellar nanocarriers for cancer-targeted co-delivery of MDR-1 siRNA and doxorubicin. ACS Nano 5:5202–5213
161. Heller J, Barr J, Ng SY, Shen HR, Schwach-Abdellaoui K, Emmahl S, Rothen-Weinhold A, Gurny R (2000) Poly(ortho esters) – their development and some recent applications. Eur J Pharm Biopharm 50:121–128
162. Heller J (2005) Ocular delivery using poly(ortho esters). Adv Drug Del Rev 57:2053–2062
163. Heller J, Barr J (2004) Poly(ortho esters) – from concept to reality. Biomacromolecules 5:1625–1632
164. Heller J, Barr J, Ng SY, Abdellauoi KS, Gurny R (2002) Poly(ortho esters): synthesis, characterization, properties and uses. Adv Drug Del Rev 54:1015–1039
165. Wang C, Ge Q, Ting D, Nguyen D, Shen HR, Chen JZ, Eisen HN, Heller J, Langer R, Putnam D (2004) Molecularly engineered poly(ortho ester) microspheres for enhanced delivery of DNA vaccines. Nat Mater 3:190–196
166. Nguyen DN, Raghavan SS, Tashima LM, Lin EC, Fredette SJ, Langer RS, Wang C (2008) Enhancement of poly(orthoester) microspheres for DNA vaccine delivery by blending with poly(ethylenimine). Biomaterials 29:2783–2793
167. Qi M, Li X, Yang Y, Zhou S (2008) Electrospun fibers of acid-labile biodegradable polymers containing ortho ester groups for controlled release of paracetamol. Eur J Pharm Biopharm 70:445–452
168. Polak MB, Valamanesh F, Felt O, Torriglia A, Jeanny JC, Bourges JL, Rat P, Thomas-Doyle A, BenEzra D, Gurny R, Behar-Cohen F (2008) Controlled delivery of 5-chlorouracil using poly(ortho esters) in filtering surgery for glaucoma. Invest Ophthalmol Vis Sci 49:2993–3003

169. Vauthier C, Labarre D, Ponchel G (2007) Design aspects of poly(alkylcyanoacrylate) nanoparticles for drug delivery. J Drug Target 15:641–663
170. Graf A, Jack KS, Whittaker AK, Hook SM, Rades T (2008) Protein delivery using nanoparticles based on microemulsions with different structure-types. Eur J Pharm Sci 33:434–444
171. Vauthier C, Dubernet C, Chauvierre C, Brigger I, Couvreur P (2003) Drug delivery to resistant tumors: the potential of poly(alkyl cyanoacrylate) nanoparticles. J Control Release 93:151–160
172. Arias JL, Linares-Molinero F, Gallardo V, Delgado AV (2008) Study of carbonyl iron/poly(butylcyanoacrylate) (core/shell) particles as anticancer drug delivery systems – loading and release properties. Eur J Pharm Sci 33:252–261
173. Arias JL, Reddy LH, Couvreur P (2009) Polymeric nanoparticulate system augmented the anticancer therapeutic efficacy of gemcitabine. J Drug Target 17:586–598
174. Arias JL, Ruiz MA, López-Viota M, Delgado ÁV (2008) Poly(alkylcyanoacrylate) colloidal particles as vehicles for antitumour drug delivery: a comparative study. Colloids Surf B Biointerfaces 62:64–70
175. Yordanov G, Kaneva N, Dushkin C (2009) Synthesis and characterization of novel core-shell colloidal particles ZnO/poly(ethylcyanoacrylate). Colloid Polym Sci 287:733–738
176. Lin M, Wang H, Meng S, Zhong W, Li Z, Cai R, Chen Z, Zhou X, Du Q (2007) Structure and release behavior of PMMA/silica composite drug delivery system. J Pharm Sci 96:1518–1526
177. Anguita-Alonso P, Giacometti A, Cirioni O, Ghiselli R, Orlando F, Saba V, Scalise G, Sevo M, Tuzova M, Patel R, Balaban N (2007) RNAIII-inhibiting-peptide-loaded in vivo polymethylmethacrylate prevents in vivo Staphylococcus aureus biofilm formation. Antimicrob Agents Chemother 51:2594–2596
178. Tao SL, Lubeley MW, Desai TA (2003) Bioadhesive poly(methyl methacrylate) microdevices for controlled drug delivery. J Control Release 88:215–228
179. Yuksel N, Baykara M, Shirinzade H, Suzen S (2011) Investigation of triacetin effect on indomethacin release from poly(methyl methacrylate) microspheres: evaluation of interactions using FT-IR and NMR spectroscopies. Int J Pharm 404:102–109
180. Fundueanu G, Constantin M, Stanciu C, Theodoridis G, Ascenzi P (2009) pH- and temperature-sensitive polymeric microspheres for drug delivery: the dissolution of copolymers modulates drug release. J Mater Sci Mater Med 20:2465–2475
181. Dalmoro A, Lamberti G, Titomanlio G, Barba AA, d'Amore M (2010) Enteric microparticles for targeted oral drug delivery. AAPS PharmSciTech 11:1500–1507
182. Chouhan R, Bajpai AK (2009) An in vitro release study of 5-fluoro-uracil (5-FU) from swellable poly-(2-hydroxyethyl methacrylate) (PHEMA) nanoparticles. J Mater Sci Mater Med 20:1103–1114
183. Singhal R, Datta M (2009) Studies on the development of biodegradable poly(HEMA)/cloisite nanocomposites. Polym Compos 30:887–890
184. Anderson EM, Noble ML, Garty S, Ma HY, Bryers JD, Shen TT, Ratner BD (2009) Sustained release of antibiotic from poly(2-hydroxyethyl methacrylate) to prevent blinding infections after cataract surgery. Biomaterials 30:5675–5681
185. Nyangoga H, Zecheru T, Filmon R, Basle MF, Cincu C, Chappard D (2009) Synthesis and use of pHEMA microbeads with human EA.hy 926 endothelial cells. J Biomed Mater Res B Appl Biomater 89:501–507
186. Jhaveri SJ, Hynd MR, Dowell-Mesfin N, Turner JN, Shain W, Ober CK (2009) Release of nerve growth factor from HEMA hydrogel-coated substrates and its effect on the differentiation of neural cells. Biomacromolecules 10:174–183
187. Zhang Z, Chen L, Deng MX, Bai YY, Chen XS, Jing XB (2011) Biodegradable thermo- and pH-responsive hydrogels for oral drug delivery. J Poly Sci Part A-Poly Chem 49:2941–2951

188. Minoo-Rabeeh–Hobabi, Hassanzadeh D, Azarmi S, Entezami AA (2007) Effect of synthesis method and buffer composition on the LCST of a smart copolymer of N-isopropylacrylamide and acrylic acid. Poly Adv Technol 18:986–992
189. Tong R, Cheng J (2007) Anticancer polymeric nanomedicines. Poly Rev 47:345–381
190. Parveen S, Misra R, Sahoo SK (2012) Nanoparticles: a boon to drug delivery, therapeutics, diagnostics and imaging. Nanomedicine 8:147–166
191. Persson EM, Gustafsson AS, Carlsson AS, Nilsson RG, Knutson L, Forsell P, Hanisch G, Lennernas H, Abrahamsson B (2005) The effects of food on the dissolution of poorly soluble drugs in human and in model small intestinal fluids. Pharm Res 22:2141–2151
192. Lysik MA, Wu-Pong S (2003) Innovations in oligonucleotide drug delivery. J Pharm Sci 92:1559–1573
193. Kingsley JD, Dou HY, Morehead J, Rabinow B, Gendelman HE, Destache CJ (2006) Nanotechnology: a focus on nanoparticles as a drug delivery system. J Neuroimmune Pharmacol 1:340–350
194. Birnbaum DT, Brannon-Peppas L (2004) Microparticle drug delivery systems. In: Brown DM (ed) Drug delivery systems in cancer therapy. Humana Press, Totowa
195. PAS 71:2005, Vocabulary — nanoparticles. British Standards Institution, Accessed 21 June 2011
196. Tirelli N (2006) (Bio)Responsive nanoparticles. Curr Opin Colloid Interface Sci 11:210–216
197. Li PW, Wang YC, Peng Z, She FH, Kong LX (2011) Development of chitosan nanoparticles as drug delivery systems for 5-fluorouracil and leucovorin blends. Carbohydr Polym 85:698–704
198. Wang T, Li M, Gao H, Wu Y (2011) Nanoparticle carriers based on copolymers of poly(epsilon-caprolactone) and hyperbranched polymers for drug delivery. J Colloid Interface Sci 353:107–115
199. Dev A, Binulal NS, Anitha A, Nair SV, Furuike T, Tamura H, Jayakumar R (2010) Preparation of poly(lactic acid)/chitosan nanoparticles for anti-HIV drug delivery applications. Carbohydr Polym 80:833–838
200. Ofokansi K, Winter G, Fricker G, Coester C (2010) Matrix-loaded biodegradable gelatin nanoparticles as new approach to improve drug loading and delivery. Eur J Pharm Biopharm 76:1–9
201. Rao JP, Geckeler KE (2011) Polymer nanoparticles: preparation techniques and size-control parameters. Prog Polym Sci 36:887–913
202. Pinto Reis C, Neufeld RJ, Ribeiro AJ, Veiga F (2006) Nanoencapsulation I. Methods for preparation of drug-loaded polymeric nanoparticles. Nanomed Nanotechnol Biol Med 2:8–21
203. Hans ML, Lowman AM (2002) Biodegradable nanoparticles for drug delivery and targeting. Curr Opin Solid State Mater Sci 6:319–327
204. Jeon H-J, Jeong Y-I, Jang M-K, Park Y-H, Nah J-W (2000) Effect of solvent on the preparation of surfactant-free poly(dl-lactide-co-glycolide) nanoparticles and norfloxacin release characteristics. Int J Pharm 207:99–108
205. Buyukserin F, Camli ST, Yavuz MS, Budak GG (2011) Novel antifouling oligo(ethylene glycol) methacrylate particles via surfactant-free emulsion polymerization. J Colloid Interface Sci 355:76–80
206. Yanasarn N, Sloat BR, Cui Z (2009) Nanoparticles engineered from lecithin-in-water emulsions as a potential delivery system for docetaxel. Int J Pharm 379:174–180
207. Puglisi G, Fresta M, Giammona G, Ventura CA (1995) Influence of the preparation conditions on poly(ethylcyanoacrylate) nanocapsule formation. Int J Pharm 125:283–287
208. Roney C, Kulkarni P, Arora V, Antich P, Bonte F, Wu A, Mallikarjuana NN, Manohar S, Liang H-F, Kulkarni AR, Sung H-W, Sairam M, Aminabhavi TM (2005) Targeted nanoparticles for drug delivery through the blood–brain barrier for Alzheimer's disease. J Control Release 108:193–214
209. Soppimath KS, Aminabhavi TM, Kulkarni AR, Rudzinski WE (2001) Biodegradable polymeric nanoparticles as drug delivery devices. J Control Release 70:1–20

210. Mainardes RM, Khalil NM, Gremião MPD (2010) Intranasal delivery of zidovudine by PLA and PLA-PEG blend nanoparticles. Int J Pharm 395:266–271
211. Kumari A, Yadav SK, Pakade YB, Kumar V, Singh B, Chaudhary A, Yadav SC (2011) Nanoencapsulation and characterization of Albizia chinensis isolated antioxidant quercitrin on PLA nanoparticles. Colloids Surf B Biointerfaces 82:224–232
212. Li R, Li X, Xie L, Ding D, Hu Y, Qian X, Yu L, Ding Y, Jiang X, Liu B (2009) Preparation and evaluation of PEG-PCL nanoparticles for local tetradrine delivery. Int J Pharm 379:158–166
213. Fessi H, Puisieux F, Devissaguet JP, Ammoury N, Benita S (1989) Nanocapsule formation by interfacial polymer deposition following solvent displacement. Int J Pharm 55:R1–R4
214. Khoee S, Yaghoobian M (2009) An investigation into the role of surfactants in controlling particle size of polymeric nanocapsules containing penicillin-G in double emulsion. Eur J Med Chem 44:2392–2399
215. Cho YW, Park JH, Park JS, Park K, Gad SC (2006) Pegylation: camouflage of proteins, cells, and nanoparticles against recognition by the body's defense mechanism. In: Gad SC (ed) Pharmaceutical sciences encyclopedia. Wiley, Inc., Hoboken
216. Govender T, Stolnik S, Garnett MC, Illum L, Davis SS (1999) PLGA nanoparticles prepared by nanoprecipitation: drug loading and release studies of a water soluble drug. J Control Release 57:171–185
217. Barichello JM, Morishita M, Takayama K, Nagai T (1999) Encapsulation of hydrophilic and lipophilic drugs in PLGA nanoparticles by the nanoprecipitation method. Drug Dev Ind Pharm 25:471–476
218. Byrappa K, Ohara S, Adschiri T (2008) Nanoparticles synthesis using supercritical fluid technology – towards biomedical applications. Adv Drug Del Rev 60:299–327
219. Sharma A, Sharma US (1997) Liposomes in drug delivery: progress and limitations. Int J Pharm 154:123–140
220. Jesorka A, Orwar O (2008) Liposomes: technologies and analytical applications. Ann Rev Anal Chem 1:801–832
221. Mozafari MR (2005) Liposomes: an overview of manufacturing techniques. Cell Mol Biol Lett 10:711–719
222. Paasonen L, Romberg B, Storm G, Yliperttula M, Urtti A, Hennink WE (2007) Temperature-sensitive poly(N-(2-hydroxypropyl)methacrylamide mono/dilactate)-coated liposomes for triggered contents release. Bioconjugate Chem 18:2131–2136
223. Jones MN (1995) The surface-properties of phospholipid liposome systems and their characterization. Adv Colloid Interface Sci 54:93–128
224. Sulkowski WW, Pentak D, Nowak K, Sulkowska A (2005) The influence of temperature, cholesterol content and pH on liposome stability. J Mol Struct 744:737–747
225. Takeuchi H, Kojima H, Yamamoto H, Kawashima Y (2000) Polymer coating of liposomes with a modified polyvinyl alcohol and their systemic circulation and RES uptake in rats. J Control Release 68:195–205
226. Ulrich AS (2002) Biophysical aspects of using liposomes as delivery vehicles. Biosci Rep 22:129–150
227. Semple SC, Chonn A, Cullis PR (1998) Interactions of liposomes and lipid-based carrier systems with blood proteins: relation to clearance behaviour in vivo. Adv Drug Del Rev 32:3–17
228. Bondurant B, Mueller A, O'Brien DF (2001) Photoinitiated destabilization of sterically stabilized liposomes. Biochim Biophys Acta-Biomembranes 1511:113–122
229. Coderch L, Fonollosa J, De Pera M, Estelrich J, De La Maza A, Parra JL (2000) Influence of cholesterol on liposome fluidity by EPR – relationship with percutaneous absorption. J Control Release 68:85–95
230. Hayashi H, Kono K, Takagishi T (1999) Temperature sensitization of liposomes using copolymers of N-ispopropylacrylamide. Bioconjugate Chem 10:412–418

231. Cho EC, Lim HJ, Kim HJ, Son ED, Choi HJ, Park JH, Kim JW, Kim J (2009) Role of pH-sensitive polymer-liposome complex in enhancing cellular uptake of biologically active drugs. Mat Sci Eng C-Biomimetic Supramol Syst 29:774–778
232. Kono K, Henmi A, Yamashita H, Hayashi H, Takagishi T (1999) Improvement of temperature-sensitivity of poly(N-isopropylacrylamide)-modified liposomes. J Control Release 59:63–75
233. Kono K (2001) Thermosensitive polymer-modified liposomes. Adv Drug Del Rev 53:307–319
234. Hayashi H, Kono K, Takagishi T (1996) Temperature-controlled release property of phospholipid vesicles bearing a thermo-sensitive polymer. Biochim Biophys Acta-Biomembranes 1280:127–134
235. Han HD, Shin BC, Choi HS (2006) Doxorubicin-encapsulated thermosensitive liposomes modified with poly(N-isopropylacrylamide-co-acrylamide): drug release behavior and stability in the presence of serum. Eur J Pharm Biopharm 62:110–116
236. Seki K, Tirrell DA (1984) Interactions of synthetic-polymers with cell-membrane and model membrane systems. 5. pH-dependent complexation of poly(acrylic acid) derivatives with phospholipid vesicle membranes. Macromolecules 17:1692–1698
237. Drummond DC, Zignani M, Leroux JC (2000) Current status of pH-sensitive liposomes in drug delivery. Prog Lipid Res 39:409–460
238. Kono K, Igawa T, Takagishi T (1997) Cytoplasmic delivery of calcein mediated by liposomes modified with a pH-sensitive poly(ethylene glycol) derivative. Biochim Biophys Acta-Biomembranes 1325:143–154
239. Sakaguchi N, Kojima C, Harada A, Kono K (2008) Preparation of pH-sensitive poly(glycidol) derivatives with varying hydrophobicities: their ability to sensitize stable liposomes to pH. Bioconjugate Chem 19:1040–1048
240. Leroux JC, Roux E, Le Garrec D, Hong KL, Drummond DC (2001) N-isopropylacrylamide copolymers for the preparation of pH-sensitive liposomes and polymeric micelles. J Control Release 72:71–84

Author Index

M.S. Mozaffari (ed.), *New Strategies to Advance Pre/Diabetes Care: Integrative Approach by PPPM*, Advances in Predictive, Preventive and Personalised Medicine 3, DOI 10.1007/978-94-007-5971-8, © Springer Science+Business Media Dordrecht 2013

D

E

H

L

R

S

T

Y

Subject Index

M.S. Mozaffari (ed.), *New Strategies to Advance Pre/Diabetes Care: Integrative Approach by PPPM*, Advances in Predictive, Preventive and Personalised Medicine 3,
DOI 10.1007/978-94-007-5971-8, © Springer Science+Business Media Dordrecht 2013

D

E

MIX
Papier aus verantwortungsvollen Quellen
Paper from responsible sources
FSC® C105338
FSC
www.fsc.org

If you have any concerns about our products,
you can contact us on
ProductSafety@springernature.com

In case Publisher is established outside the EU,
the EU authorized representative is:
Springer Nature Customer Service Center GmbH
Europaplatz 3, 69115 Heidelberg, Germany

Printed by Libri Plureos GmbH
in Hamburg, Germany